Proceedings of the Conference on

Pipelines In Adverse Environments II

Sponsored by the Pipeline Division
of the
American Society of Civil Engineers

Co-sponsors
Los Angeles Section, ASCE
San Diego Section, ASCE

San Diego, California
November 14-16, 1983

Mark B. Pickell, Editor

Published by the
American Society of Civil Engineers
345 East 47th Street
New York, New York 10017-2398

The Society is not responsible for any statements made or opinions expressed in its publications.

Library of Congress Catalog Card No.: 83-82079
ISBN 0-87262-385-8
Manufactured in the United States of America.

FOREWORD

The conference committee is pleased to be able to present these proceedings on behalf of the Pipeline Division and the Los Angeles and San Diego sections of the American Society of Civil Engineers for this 1983 Specialty Conference: Pipelines in Adverse Environments II.

We take great pride in this, the second in a series of successful conferences assessing the current state-of-the-art and future trends in pipeline design, construction, and maintenance in the less than text book environs, i.e., arctic, oceans, rivers, and harbors, deserts, and urban areas. The international participation at this conference compliments the industry, the authors, and the innumerous people who worked for over two years to bring this conference together. We wish we could individually name them all.

Each of the papers presented were carefully screened by at least two positive peer reviews from the applicants who submitted for presentation. All the papers are eligible for discussion in the *ASCE Transportation Journal* and are also eligible for the ASCE Awards.

On behalf of the committee:

Mark B. Pickell, P. E.
General Conference Chairman

Conference Committee:

Mark B. Pickell, General Chairman
Jey K. Jeyapalan, Deserts Division Chairman
Rodney Browning, Arctic Division Chairman
D. S. (Steve) Graham, Oceans, Rivers, Harbors Division Chairman
Peter K. Willerup, Urban Division Chairman
M. J. (Shelly) Shelton, West Coast Liaison
Francis J. Stastny, Publicity

CONTENTS

ADVERSE TERRAIN CONDITIONS
SESSION I
Michael Collins, Chairman

RIVER CROSSINGS I (Northern)
SESSION II
D. S. Graham, Chairman

MARINE PIPELINES
SESSION III
R. Hudspeth, Chairman

URBAN I
SESSION IV
Peter K. Willerup, Chairman

GEOTECHNICAL CONSIDERATIONS UNDER ADVERSE CONDITIONS
SESSION V
Wayne Dunlap, Chairman

ARCTIC I
SESSION VI
Rodney Browning, Chairman

RIVER CROSSINGS II
SESSION VII
D. S. Graham, Chairman

URBAN II
SESSION VIII
E. Earl Peebles, Chairman

SPECIAL PIPELINES
SESSION IX
Amster Howard, Chairman

ARCTIC II
SESSION X
Rodney Browning

OUTFALLS I
SESSION XI
D. S. Graham, Chairman

URBAN III
SESSION XII
Peter K. Willerup, Chairman

OUTFALLS II-GEOTECHNICAL ASPECTS
SESSION XIII
J. Michael Duncan, Chairman

ARCTIC III
SESSION XIV
Rodney Browning, Chairman

FORCES ON OCEAN PIPELINES
SESSION XV
R. Wiegel, Chairman

Case Histories:
Construction on Canyon Wall Terrain

By
Frank W. Pita, P.E., P.G., AM ASCE[1]
and
B. Louise Woodward, C.E.T.[2]

ABSTRACT

Sections of two gravity flow irrigation flumes built on steep canyon walls in Washington State were replaced by pipe. Design and construction methods had to be tailored for hand tools and material deliverey by helicopter only, because of the extremely difficult rocky terrain. The lifting capability of the helicopters controlled the size and weight of materials and limited the prefabrication units (span and number of supports). One replacement project was necessary because of imminent failure of an old, high-maintenance wood flume. The other project was the emergency replacement of a 350 cfs, fully operating, concrete flume. Construction of the exposed replacement pipelines involved similar problems and solutions.

INTRODUCTION

Many gravity-flow irrigation flumes were built near the turn of the century in Washington State, placed along steep canyon walls with hand labor and tools. Sections of two of these irrigation flumes, both located in the Eastern foothills of the Cascade mountains, were replaced with pipe in 1980. This paper presents case histories of the replacement projects and a discussion of the construction methods used successfully on these two difficult projects.

---

[1]Mr. Pita is Vice President of Rittenhouse-Zeman & Associates, Inc., 13837 N.E. 8th Street, Bellevue, WA, 98005. (Formerly with CH_2M-Hill as Civil/Geotechnical Engineering Department Manager.)

[2]Ms. Woodward is a technician in the Civil/Geotechnical Fngineering Department at CH_2M-Hill, Inc., 1500-114th Ave., S.E., Bellevue, WA, 98004.

Some construction problems encountered at these canyon wall sites were:

- Limited access
- Small work area, if any
- Dangerous conditions for workers
- Atypical geotechnical investigation
- Potential of surveying error due to difficult terrain
- Susceptability of work to inclement weather

The major difference between the two projects was the time available for design and construction. The Icicle Irrigation District (I.I.D.) project included the standard preparation of contract documents, bidding by the contractor, and construction during the irrigation off-season with sufficient time for completion. The repair work done by the Yakima-Tieton Irrigation District (Y.T.I.D.) was the emergency replacement of a washed-out section of the main feeder canal, at the height of the irrigation season. Because service had to be restored within three weeks, the district hired an engineer and contractor to begin work immediately. Despite this difference many construction methods and constraints were common to both projects. This paper addresses the similarities in design and construction of the two projects, as well as the differences related to the time constraint on the Y.T.I.D. emergency repair.

One of the authors, F. W. Pita, was one of the two lead design engineers and construction supervisors for the Y.T.I.D. project and geotechnical engineer/designer for the I.I.D. Project. The co-author, B. L. Woodward, provided technical support for the designs.

CASE HISTORY I: ICICLE IRRIGATION DISTRICT FLUME REPLACEMENT

The project location is on the northern valley wall above the Wenatchee River near the City of Dryden, Washington. The site is approximately 400 feet above the valley floor on an exposed sandstone rock slope dipping at a 35° angle. Figure 1 is a cross-section showing the wooden flume that was replaced by 1550 feet of elevated steel pipe. The flume had required continual maintenance and was in danger of imminent failure. The design of the replacement pipe was tailored for small helicopter installation and working with hand tools. Because of previous small rock falls and winter avalanches the pipe is elevated above the rock face so that slides can pass beneath the pipe without damage. Figure 2 shows a typical fitting and column support for the 30-inch diameter spiral-weld steel pipe. The supports range in height from 2-feet to 15-feet and are positioned at 30-foot intervals. Custom-designed concrete footings with rock anchors allow for irregular rock formations and varying grades of steepness on the face of the cliff. The steel supports were bolted down to ensure a solid foundation for the pipeline.

The survey crew had to be extremely precise in laying out footings so that all of the prefabricated material would fit together properly. Components of the new pipeline were designed to allow field adjustment at the base of each footing and at each cradle during final alignment.

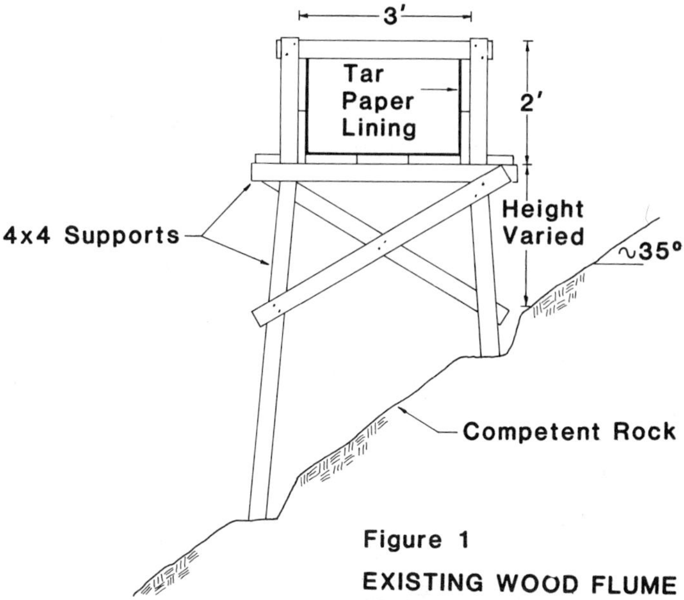

Figure 1

EXISTING WOOD FLUME
ICICLE IRRIGATION DISTRICT

The base plates of the steel supports were adjustable, and mortar was hand placed beneath the plate after final alignment of the entire pipeline. Significant savings were made by limiting the weight of the sections to 1800 pounds so that a small, locally based helicopter could be used for lifting and placing them. All of the prefabricated metal pieces were moved and set in place by helicopter as were three-quarter yard concrete buckets used for placing the concrete footings. The pipe sections were connected by using compression collars which were hand-placed and tightened. No field welding was performed on the pipe. Exact fabrication and field surveying were key elements in the success of the project.

The contract required that when the delivery system was taken out of service it must be during off-season time. In the State of Washington, this is the fall, winter and early spring. To avoid inclement weather it was imperative that the contractor be allowed to perform material fabrication during the irrigation season. At the time of shutdown the contractor was then able to mobilize quickly, remove the existing flume, and begin construction of the new facility. This plan enabled the contractor to use the good fall weather for the major portion of his construction.

The construction went according to plan and the contractor completed the work well before the deadline. The cost of the construction was approximately $400,000.

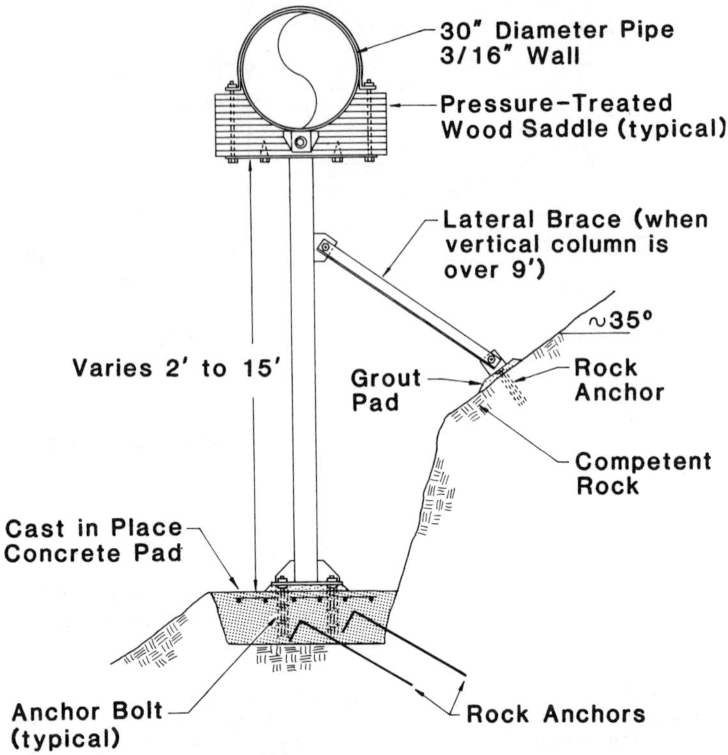

Note:
Support spacing at 30' interval.

Figure 2

TYPICAL PIPE SUPPORT
ICICLE IRRIGATION DISTRICT

CASE HISTORY II: YAKIMA-TIETON IRRIGATION DISTRICT'S EMERGENCY REPAIRS

Site and Failure Description

The main feeder canal of the Y.T.I.D. begins on the Tieton River at a diversion dam approximately five miles downstream from Rimrock Lake, and flows southeasterly along a moderate to steep mountain slope above the Tieton River and U.S. Route 12. The canal was constructed in 1906 by the Bureau of Reclamation. Throughout most of its 13-mile length, the canal is a horseshoe-shaped pre-cast reinforced concrete flume about 8-feet wide and 7-feet deep (Figure 3). The flume wall thickness is approximately 4 inches. The canal carries approximately 350 cubic feet per second of irrigation water to the 26,000 acres of orchards and farmland which make up the district.

The rock formations in which the Tieton Valley has been eroded range from andesite, basalt, conglomerate, mud flows, to tuff in the eastern portions to predominantly Yakima Basalt in the western portion. Loose talus is the primary surface material in the area of the canal break, forming a uniform slope of 1.3H:1V. Most of the canal is accessible only by foot (or by air) because of the steep slopes and loose talus.

Volcanic ash from the May 18, 1980, eruption of Mt. St. Helens covered the drainage basins upslope from the canal and was probably a major contributing factor to the canal break. The ash cover reduced the porosity of the soil, effectively preventing the ground-absorption of rainfall. Apparently, rainfall on June 16th caused a much greater volume of runoff than would normally be expected from a storm of that magnitude.

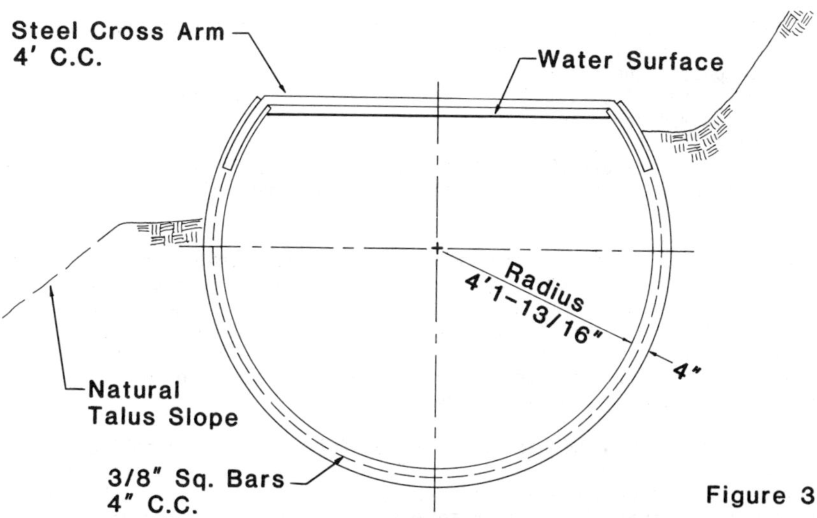

Figure 3

EXISTING CONCRETE CANAL
YAKIMA-TIETON IRRIGATION DISTRICT

The increased runoff caused erosion-debris-slides which filled the canal with rock, ash, trees, brush, and other upslope debris in many locations. Because the canal was in operation at the time, the blockage caused immediate overtopping of the canal.

The most severe damage occurred between canal-mile 9.13 and 9.15, where the overtopping eroded the loose talus slope on the downhill side of the canal, undermining a portion of the flume. A section approximately 100-feet long collapsed and the continued flow of water from upstream caused further erosion, resulting in a 300-foot wide by 100-foot deep plunge pool basin carved immediately below the flume alignment (Figure 4). At the canal level, the loose talus had not been eroded under the remaining exposed ends of the broken flume, although a 40-foot wide area where the water exited the upstream end of the old flume had been eroded down to bedrock. The bottom of the plunge pool area (approximately 100 feet below canal level) was relatively flat and the talus side walls were vertical and even undercut in some areas. Nearer to the valley floor (river level) the basin narrowed into a gully clogged with eroded talus rock. The broken remains of the collapsed 100 feet of flume were scattered and intermixed with trees and other debris down the slide area.

The complete interruption of the District's water supply endangered $100,000,000 worth of crops. Remedial efforts began immediately. Because time was so critical to the design and construction processes, a chronology table has been provided showing the sequence of events that followed the failure (Appendix A).

Design Approach in June 1980 for Emergency Repair

Within four days following the failure, the District and it's engineer, CH_2M-Hill, had accomplished the following:

- Inspected the site
- Met with Y.T.I.D. board to discuss problems and solutions
- Began mapping site by photogrametry
- Hired contractor, Turzillo, and helicopter service
- Began site cleanup - by District and volunteer crew
- Ordered 102" (9 feet) diameter CMP replacement pipe

The District had determined that the farmland it served would sustain severe crop damage if the canal was not repaired within three weeks. Because a repair engineered and constructed so quickly might not be sufficiently permanent, the design scope included both short-term repair and a more permanent solution. Many engineering risks were involved, such as lack of geotechnical data. The design could not be finalized until site conditions were better revealed by the completion of cleanup and scaling (loose rock removal) operations.

The replacement pipe had been ordered following the consolidation of several alternative solutions. The only economically feasible solution, and the only one possible within the time limit, was to bridge over the eroded area with a concrete-and-steel supported large diameter pipe. The 9-foot-diamter CMP was ordered before design of the supporting structure was final, to allow time for fabrication and delivery. The supporting structure was designed at the site, while hand-scaling was

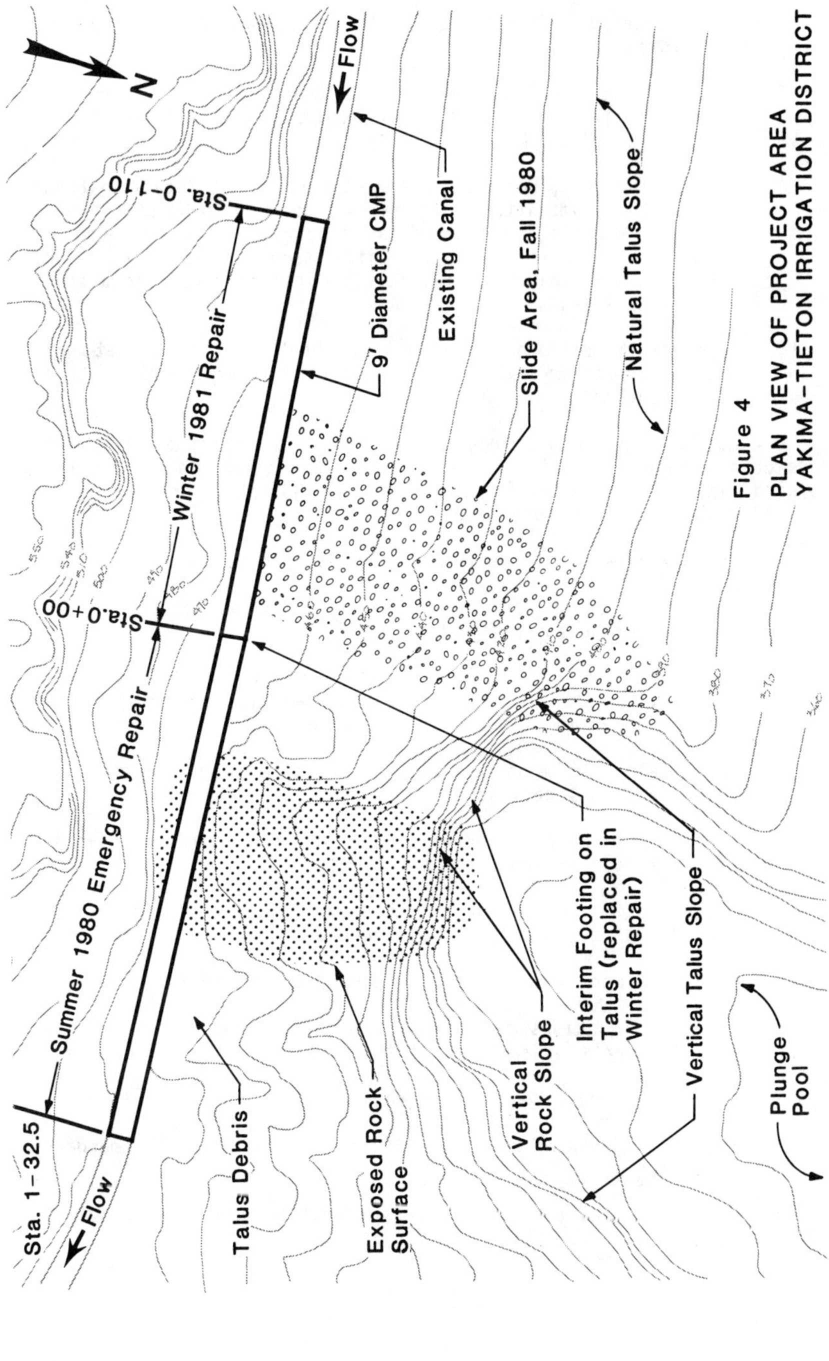

Figure 4
PLAN VIEW OF PROJECT AREA
YAKIMA-TIETON IRRIGATION DISTRICT

in progress. General design bases for the emergency pipe foundations are listed below:

- From Station 0+25 to 1+32.5 (the downstream end) permanent footings could be founded on the exposed bedrock, if reinforced with rock bolts.

- At station 0+00, the talus slope supporting the intact upstream end of the flume after the break was assumed to be stable on a short-term basis, even though the supporting slopes were not in a long-term stable condition. This assumption was a known risk, undertaken in order to complete the initial repair in a short period of time.

- Bedrock formations downstream of Station 1+32.5 that were supporting undamaged portions of the flume were assumed to be sound and not in danger of failure.

Because the upstream bridge footing on talus (Station 0+00) was considered adequate for the 1980 irrigation season only, a modification to stabilize it was planned for the winter of 1980-1981, and the design proceeded based on temporary stability.

The design allowed for the following constraints:

- The few feet of broken and cracked flume at each end of the break must be cut off and "squared".

- All steel and concrete components and welds must be designed to withstand the static and dynamic loadings imposed by the CMP pipe and maximum flow of irrigation water.

- Materials used must be readily available (e.g., in stock at suppliers in local cities).

- Repair must be constructable under extremely difficult access and work conditions (foot paths only).

- Size of components would be limited by lifting capacity (10,000 pounds) of the available helicopter.

- All site work must be done with hand or air tools.

- Invert elevations of the new pipe should match the existing canal inverts at each end of the break.

- Allowances must be made for thermal movement.

- No walkway would be available during the phase 1 emergency repair.

- The emergency portion of project must be complete by July 5, 1980.

Two important tasks were to be completed by the District during the irrigation off-season. These were coating the steel for protection from rust and lining the CMP flume to reduce the hydraulic headloss due to the increase in hydraulic roughness.

Emergency Construction

Repair work commenced with cleanup and scaling operations while the design was in progress. Two CH_2M-Hill senior design engineers supervised two shifts of design and construction onsite while many other engineers and Y.T.I.D. representatives supported the work as needed. Appendix A describes the mobilization and construction process.

The initial plan was to place a CMP pipe directly on steel columns anchored in grout-bag-formed foundations designed to support three welded steel bridge spans. Steel saddles were installed on the spans to cradle the pipe. Figures 5 and 6 show cross-sections of the repair at two key locations.

Field fabrication, helicoptor delivery of materials, and concurrent design and construction were successful and the emergency repair was completed before the three-week time limit, and irrigation resumed. The repair had cost approximately $600,000.

Off-Season Final Repair

The construction division of the Y.T.I.D. acted as the general contractor for the off-season repair work and hired the necessary specialty contractors for pipe fabrication and welding. CH_2M-Hill was engineer and construction manager for the project.

The talus material at the original failure site continued to erode as the slope approached it's natural angle of repose. This dictated an upstream extension of the steel-supported CMP flume. Exploratory hand-excavation was done by the Y.T.I.D. staff to a depth of about 7 feet, adjacent to the canal and approximately 20 feet upstream of the inlet end of the summer-placed CMP (Figure 2). Bedrock could not be located by hand-digging so drilling was done under the observation of CH_2M-Hill. Rock was encountered at varying depths in five bore holes. The borings showed that the depth to solid rock beneath the alignment of the flume was similar to that depth found at Station 0+25. The flume was not actually resting on bedrock until a point 110 feet upstream from Station 0+00 was reached. Analysis showed that as the unstable vertical talus slopes eroded to become stable again they would undermine approximately 100 feet of canal upstream from Station 0+00. To achieve stability, all of the loose talus material 110 feet upstream of Station 0+00 had to be removed down to bedrock. A replacement section was designed similar to the emergency repair section from Station 0+00 to 1+31.

A small Caterpillar-D3 bulldozer was winched up the slope to the site to excavate the loose material down to bedrock. A small Bobcat loader with a small backhoe attachment was driven through the canal to do scaling along the rock face. The construction that followed is

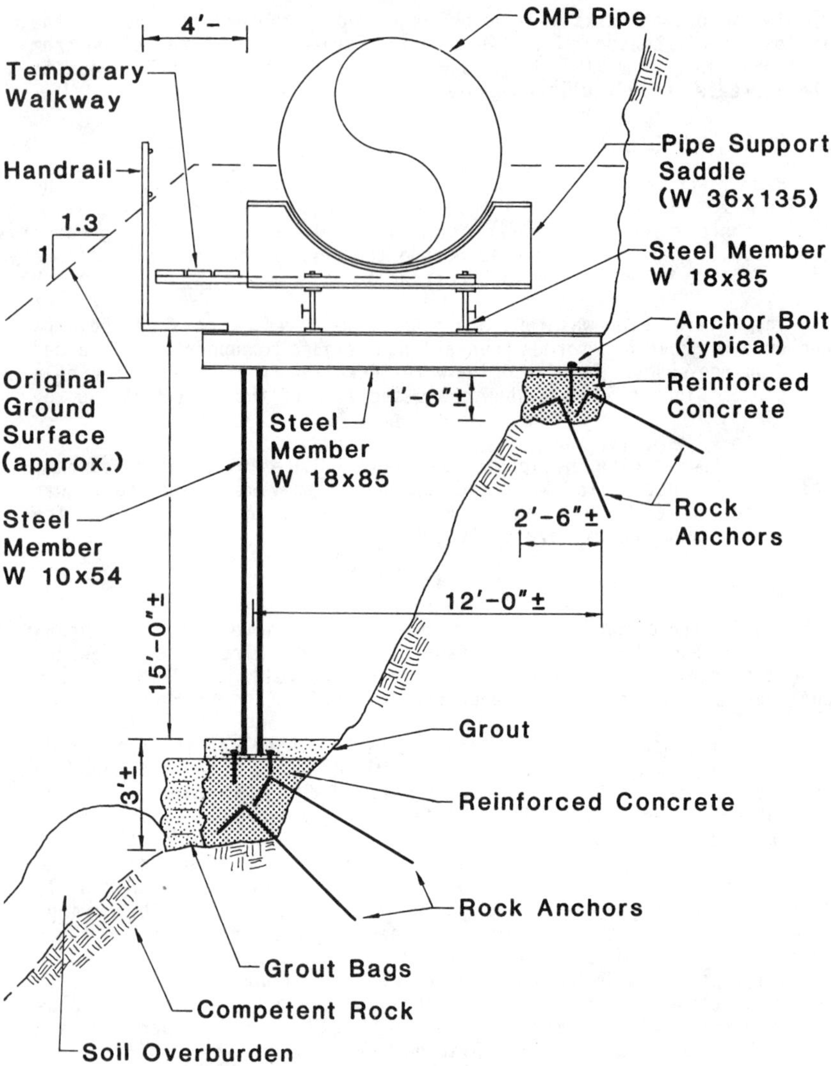

Figure 5

TYPICAL BRIDGE SUPPORT SECTION
YAKIMA-TIETON IRRIGATION DISTRICT

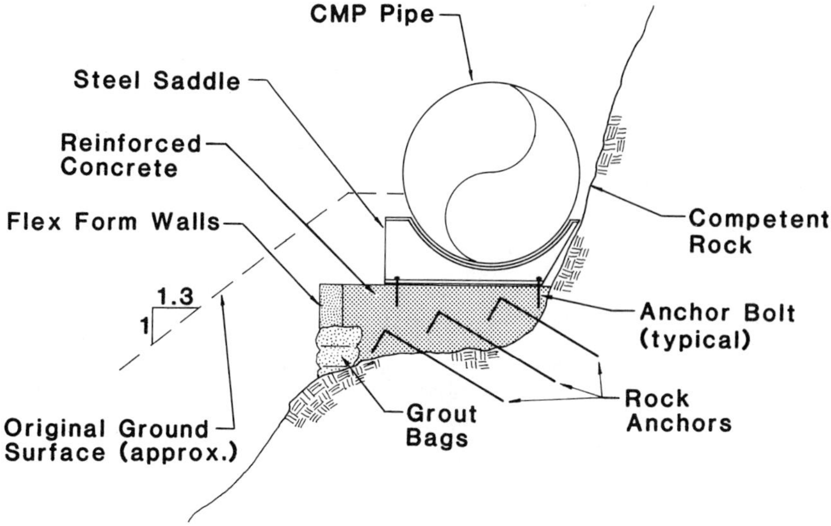

Figure 6

CONCRETE SUPPORT SECTION
YAKIMA-TIETON IRRIGATION DISTRICT

described in the appendix. Water was turned into the canal in April 1981, and the system functioned with no unforeseen problems.

During the following year's off-season the Y.T.I.D. staff has made additional repairs and tests to make the old and new repairs permanent. The additional pipe extention cost approximately $400,000.

SUMMATION

As these two case histories exemplify, for the construction of pipelines on canyon wall terrain, the designer must consider many factors that need not be considered when designing typical pipelines. When limited access is a problem the designer must consider the availability, cost and size of construction helicopters for lifting the materials in place, to achieve the most economical design for the owner. Since access to these sites may be difficult, even the best surveying techniques may not allow perfect alignment of all the system components. Therefore, the designer must incorporate field adjustment in the design so that precise alignment is possible on the site. When use of hand tools and a significant amount of hand labor are expected, the design components must be very simplistic since sophisticated equipment, such as for welding, may not be available without a significant premium cost. Furthermore, an experienced geotechnical engineer must be part of the onsite design and construction teams, to verify design assumptions or modify the design for variable subsurface conditions. Most importantly, the designer must be familiar with construction methods and sequence, and understand how a contractor would logically construct the pipeline.

APPENDIX A
CHRONOLOGY OF SIGNIFICANT EVENTS; Y.T.I.D. EMERGENCY REPAIR

DATE	EVENT
5-18-80	Mt. St. Helens volcano eruption leaves ash layer on hillsides surrounding canal.
6-16-80	Late afternoon thunderstorm and heavy rain triggers debris slides. Ninety-foot section of canal collapses down mountainside. Deep "plunge pool" and gully formed by erosion from broken canal water. Irrigation water delivery totally interrupted to all of Y.T.I.D. District manager, Warren Dickman, sees site in late p.m., calls Haapala, (CH_2M-Hill Irrigation Engineer), in p.m.
6-17-80	Haapala sees site a.m., calls Pita, (CH_2M-Hill Geotechnical Civil Engineer in Seattle office) in p.m.
6-18-80	Haapala, Pita and Y.T.I.D. board meet after site inspection, to discuss problem, solutions, risks, and funding. Concrete specialty contractor (Turzillo) and helicopter firm are contacted in p.m. CH_2M-Hill surveyors start setting air photo premarks for mapping (also 6-19-80).
6-19-80	Turzillo and helicopter representatives, (Columbia) arrive and plan lower staging area and repair. Pipe-bridge design selected by District and engineers.
6-20-80	102" dia CMP pipe (ARMCO) ordered. District and volunteer crews walk to site and work to clean plugged canal downstream of break.
6-21-80	Hand scaling of loose rock and debris begins. Turzillo moves in. Helicopter arrives and flies equipment in. Cole (CH_2M-Hill Senior Geotechnical Engineer), inspects site with Pita. High scalers expose bedrock suspended by ropes. ARMCO begins to roll pipe in Davis, California plant.
6-22-80	Explosives used to remove broken and undermined flume sections. Some sections lifted away by helicopter. Crews work to clear loose debris and talus. Grout bags placed and filled for lower footing at Station 0+50.
6-23-80	Crews continue scaling and cleaning slope. Two 1-yard concrete buckets delivered. Backup air compressor delivered.
6-24-80	Steel fabricator chosen, Erickson (CH_2M-Hill Senior Structural Engineer) designs steel with material which is readily available in Seattle supply yards and easily fabricated.

6-25-80 Contour site map completed (by photogrametry). DOE and USBR inspect site and review design concepts. Rock anchors drilled to anchor concrete foundation from Station 0+73 to Station 1+10.

6-26-80 Erickson arrives at site, designs footing rebar layout on site, and verifies steel design adequacy. Marker (CH_2M-Hill Senior Civil Engineer) inspects site and contributes to closure collar design.

6-27-80 Closure collar system design finalized on site, material ordered. Steel erection foreman arrives. Structural steel delivered to lower work site.

6-28-80 Steel erection crew prepares lower staging area work site; support bents fabricated. Pipe leaves Davis, California, by truck.

6-29-80 Flexform wall begins from Station 0+75 to Station 1+20. Pipe delivered to lower site. Bridge bents and bridge span sections placed by helicopter. All rock bolts and all rebar from Station 0+75 to 1+20 is placed.

6-30-80 Reinforcing steel placed for concrete downstream foundation (Station 1+20 to 1+32.5). Concrete (56 cubic yards) is placed in 2 hours from Station 0+75 to 1+32.51414. Three sections of pipe placed on bridge section. Upstream closure collar begun in p.m.

7-1-80 Pile support saddles placed on concrete (Station 0+75 to 1+32.5), aligned and bolted in place. Final pipe placed. Concrete form for upstream collar begun.

7-2-80 Upstream collar poured. Downstream collar formed and poured. Pipe joints sealed on inside pipe. Cleanup begun.

7-3-80 System inspected. Removal of equipment and cleanup completed.

7-4-80 Pipe tested. Flow increased gradually to 100 cfs then to 322 cfs by 50-cfs increments. Minor leaks, not considered severe enough to cause concern. System operating.

7-6-80 10" headloss measured through pipe.

7-7-80 Feasibility of adding freeboard upstream is studied and district decides to install it, so 350 cfs could pass through system without overtopping existing flume.

9-16-80 Site inspected by Pita, Haapala and Y.T.I.D. manager. Relatively minor leak in upstream closure collar.

10-1-80 Water turned off. End of season.

10-27 to 11-4-80	Y.T.I.D. excavates test hole by hand to locate bedrock upstream of Station 0+00.
11-4-80	Pita inspects test hole. Bedrock not found at 7 foot depth. Decision made by Pita and Y.T.I.D. manager to drill to locate rock and to establish foundation conditions.
11-22-80	Drilling begins in canal invert. Pita inspects.
11-24 : 26 & 12-1-80	Drilling continued. Shorey (CH_2M-Hill Engineering Geologist) inspects.
Mar 1981	Y.T.I.D. excavates all loose material to bedrock with D-3 bulldozer. Steel ordered and fabricated in Yakima shops.
3-19-81	Structural steel delivered to lower site.
3-23-81	Helicopter arrives. Pita and Beieler (CH_2M-Hill) act as onsite construction manager. Four columns and one bridge set in place. All remaining steel arrives.
3-24-81	All bridges and pipe in place. Final field welding continues. All large equipment removed by helicopter. Helicopter leaves.
3-28-81	Y.T.I.D. completes construction of structural steel and closure collars.
3-30-81	Structural inspection by CH_2M-Hill structural engineer.
4-7-81	Geotechnical engineering and geology inspection by Pita.
4-9-81	<u>Irrigation water turned on for 1981 season</u>.
1981-82	Y.T.I.D. staff mortar-lines pipe and coats steel for rust protection in next off-season.

HABIGANJ-ASHUGANJ PIPELINE PROJECT
- An Insignia of Appropriate Pipeline Technology for Bangladesh and the Near East.

Engr. Delawar Bakht* B.Sc.Engg.(Civil), FIE(B), MBIW(DAC), SPE(TEXAS).

This paper discusses the keys to the incredible success of Titas Gas in implementing its 12" dia 36 mile transmission line in extremely adverse terrain and climatic conditions at a good deal of savings through applying innovated techniques of semi-mechanized and manual means. Included are the evaluations of its prospects in such application, in similar conditions of other developing countries where human labour is cheap and unemployment figure in high.

Introduction

The success of implementation of the Habiganj-Ashuganj pipeline project provides a new solution to break through the prevailing line of thought that planning & execution of such project has to have a strong capital intensive logistics mostly drawing on hard currencies. This has been acting as a deterent factor for planning & materialization of massive hydrocarbon transportation projects in Bangladesh and the adjoining least developed countries of the near east.

Now with the trainings and experience gained by the Engineers of Titas Gas, the pioneer gas industry of Bangladesh in executing big inch pipeline in the South & far east, the middle east and in the west as well, they have been able to illiminate conventional use of heavy machineries & equipment for sophisticated pipeline construction technology, at the same time innovate the one which can make the most of local indegineous resources and gainful utilization of substantial number of the unemployed millions.

This paper is aimed to present a discussion of the different stages of execution of the project, review the special features of the construction technology applied and to evaluate its appropriateness. This must help launching such projects of greater magnitude in this perspective.

*Acting Manager, Regional Distribution Department, Titas Gas Transmission & Distribution Co.Ltd., 7, Motijheel C/A. Dhaka-2, Bangladesh, Cable: TITASGAS, Dhaka, Bangladesh Telephone : 401125(Office), 310551(Residence),

Project

Titas Gas T&D Co.Ltd. Dhaka have completed and commissioned its 36 mile 12" dia 1000 PSIG transmission line (Fig.1) to deliver 120 MMCFD Natural Gas from Habiganj Gas field to the Ist National Gas grid at Ashuganj on 6th April, 1981.

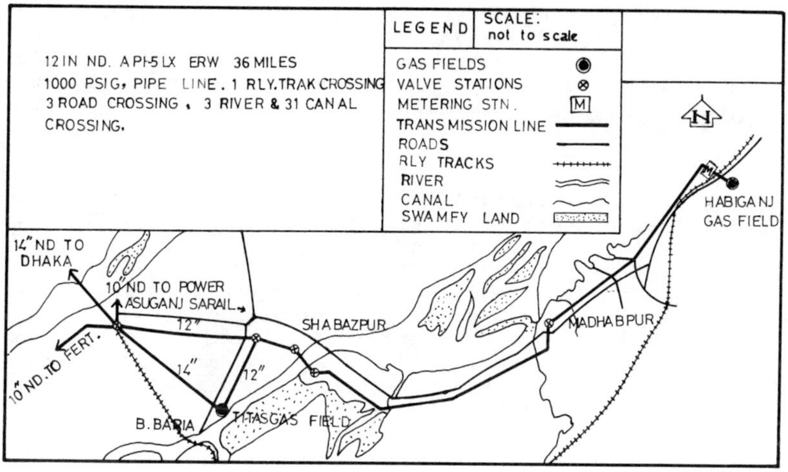

Figure 1: Asbuilt alignment of Habiganj-Ashuganj Pipeline Project.

The system comprised of installing 5 valve stations 1 metering station 1 railway track crossing 3 road crossings 3 river crossings and 31 canal crossings along the transmission pipeline laid mostly on swampy lands, rugged terrains and sometimes of extremely adverse climatic conditions.

Concept

Planning, engineering, materials procurement, construction management, logistic services, testing, inspection and commissioning and in fact, the total materialization of the project was done through mobilization & co-ordination of local resources and utilization of specialised know - how of the pipeline professionals of Titas Gas.

Costs involved in mechanized handling and laying of pipelines i.e. operating costs, depreciation, maintenance, spares, fuel, skilled operators wages and designed speed of progress vis-a-vis down time on movement and usage of such equipments in the extremely adverse terrain, climatic and ROW clearance conditions necessitated a second thought for limiting use of the mechanical means and encouraged devicing means for utilization of abundantly available low cost human labourers in thousands of numbers who could sustain enough strength proplonged for shifts together in all working conditions within reasonably acceptable wage range.

On instances, these labourers organized in groups, have strung hundres of lengths of line pipe along the ROW, dug miles of trenches and coated & wrapped 100's of lengths of pipes within couple of weeks. Under similar conditions this progress could hardly be possible through conventional capital intensive pipelaying technology.

CPM

In line with supply schedule of gas to the major consumer Zia Fertilizer & Chemical Complex and proposition of implementation of the project through optimum utilization of the prevailing logistic infrantructure, the CPM was spanned for an year for actions right from procuring right of way easement down to commissioning of the transmission line. The uniqueness of the CPM was characterised by the fact that no time of the year was considered a down time and one or the other activity at field was assigned even on the flooded right of way during rainy season.

Technology

Contract: For simultaneous construction of the pipeline during one working season, the total length was devided into 3 parts and 3 local contractors were awarded with with order for 1 part each inclusive of special crossings involved within these parts. One special condition was that each contractor had to have full strength for running 2 spreds at a time.

Delivery of materials: Arrangement was made by the Company to deliver all line pipes, valves, fittings and coat & wrap materials at points closest to the project site.

Hauling & stringing: At the time of launching the project at site, it was full rainy season and the advantage of the flooded ROW was taken by double jointing the pipes along the side of the highway and floating away these pipes in bunches right up to the ROW alignment by minimum usage of human labour. However, remaining pipes and materials were hauled at site by using small roller bearing locally made trailors and pushing & pulling by employing human labour.

Photograph 1 : Manual handling at the time of pipeline constructions.

Coating & Wrapping: Pipes were cleaned, coated & wrapped at locations near the right of way manually. Conventional primer-bitumen glass fiber inner wrap - bitumen - asbestos outer wrap materials were used. Each length was once Holiday inspected at wrapping yard and again on the trench.

Welding: As provided in the terms of contract each contract each contractor has been deploying 2 spreds of 3 welding machines & 3 welders each. They were working through ten hours shift and has been producing 25 welds a day per contractor in average during the welding works.

The reason for such apparently lower number of welds can be largely attributed to the interruption of works due to ROW problems and this is nothing unnatural for such projects where easements has to be procured from hundres of landowners along every mile of the pipeline routes. Since neither side boom nor similar other heavy handling equipments were used during welding, the cost of down time did never stood a significant one.

Before putting to line welds, each welder had to pass through high pressure line pipe welder's test as per API 1104 specifications and line weldings were 100% radiographed at the initial stage and minimum of 50% radiography was maintained through out with 100% radiography in above ground installations and special crossings.

For the purpose of NDT inspection, dark room facilities were installed at a central location of the project area and 3 radiography teams were deployed throughout the execution time.

Laying & backfilling: All trenching, laying and backfilling works were done manually just using shearlegs & chainpullies at a peak speed of 1 mile a day per spread providing a minimum cover of 4 ft.

Testing & commissioning: Pipeline was hydraulically tested at 1500 PSIG, pigged & purged after completion of each part and before commissioning of the project. However, individual sections at special crossing were hydraulically tested and pigged before laying.

Valve Station installations: Techniques used for installation of 12" ND line valves weighing about 1½ ton each at an average height of 8 to 10 ft. above vergin ground level without using any crane or side boom was another creative feature of project. In fact, temporary earth filling up to that high was done manually to form a platform for the valve to be installed. Thereafter, the valve and fittings, etc. were lifted, fitted and welded with the help of ordinary shearlegs & chain pullies.

Minor River Crossings: Bed of the 2 river of widths 400 ft. were trenched manually by putting earth filled cross dams & diverting the water flow through a temporary channel dug along one bank of each river.

The coated & wrapped line pipe was placed over one of the cross dams, welded, radiographed, hydraulically tested pigged and concrete coated. After adequate curing, the pumped out water from between the cross dams were refilled and the completed section of the line pipe was floated and aligned over the dug out trench, lowered and subsequently backfilled after putting specified sand cement bags.

The entire operation was done without assistance of any mechanized dredging, handling or lowering equipment. Just hunsewsa od human labourers did it upon guidance & instructions of the experienced pipeline crew and the project engineers.

Major River Crossing: The technology innovated for an applied to crossing the Titas River of 600 ft. width and at water depth ranging between 20 & 35 ft. during the crossing time, must stand significant towards programming such constructions in Bangladesh and the near east where human laboures are yet to be gainfully employed.

It is surprisingly true that apart from a dredger, when 5 to 6 side booms and at least 1 or 2 laybarge cranes are required inclusive of a lot of rails & trollys - for a conventional river crossing of this kind, the job was completed with just one dredger, one winch fitted motive unit, one speed boat and a few country boats.

Photograph 2: Fastening of floats prior to River Crossing

Though overall actions of welding, concrete coating, testing, ditching and dredging took about 4 weeks, the river crossing pipeline was floated and aligned in one day and lowered to the bottom of the trench at the bed of the river on the day following. Placing of sand cement bags and backfilling followed thereafter.

The different stages of the crossing has been outlined in fig. 2 and 3. After completion of the work the area has been backfilled & reinstated again manually to provide an acceptable environmental protection.

Equipment

This was a maiden project of its kind & magnitude in Bangladesh where no equipment or machinery was bought or hired from abroad. Foreign currency spent was just to bring the materials and accessories.

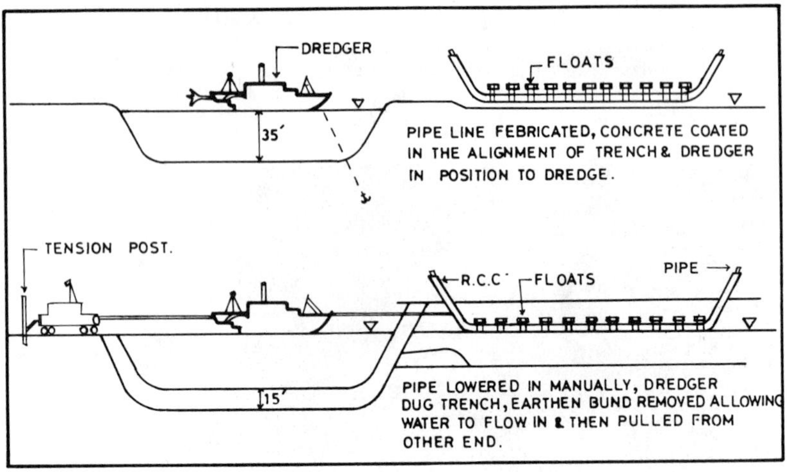

Figure. 2

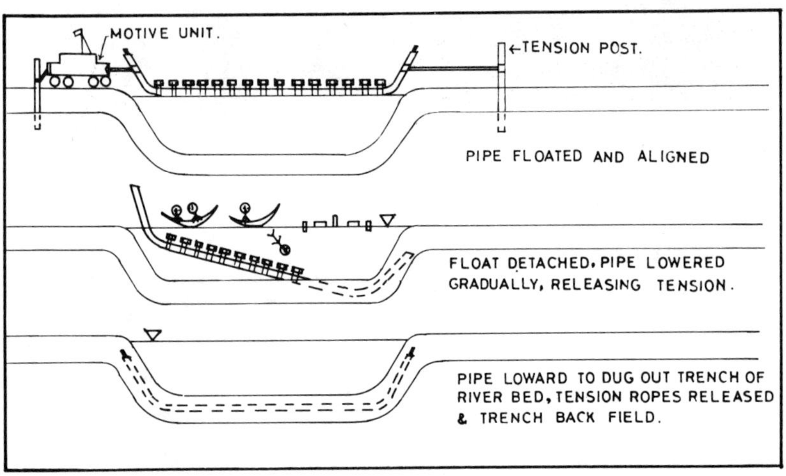

Figure: 3

Figure 2 & 3 : Steps of River crossing works at Titas River.

Photograph 3 : Lowering of pipeline at bed.

Titas Gas had in its possession, all heavy machineries and equipments like bull-dodgers, cranes, motive units, pipe trailers, side-boom, etc. for dodging right of way, handling line pipes, installing valve stations and laying and laying pipeline in the trench for executing project of such magnitude. But the most significant aspect of the technology innovated and applied in the spreads was that the project was taken right from begining as a labour intensive one and except occasional use of one side boom and a pipe trailer none of these heavy equipments and machineries were even mobilized to the sites.

Economics

Project was launched at an estimated cost of Tk. 142.8 Million (1 US $= Tk. 16/-) with foreign currency component of Tk.72.5 million and completed at Tk.120.2 million with foreign currency component of Tk.59.3 million inclusive of easement, materials, construction, civil works, logistics and interests on investment.

This let to a pipeline cost per mile at Tk.3.33 millions. It is unbelievable but true that the pipelaying and ancillary construction component inclusive of logistics and over head has been standing at Tk.1.10 millions only against current international costing of Tk.1.37 millions per mile of equivalent system for this region of the near east.

Conclusion

In course of planning & scheduling of projects of similar nature the key to the success of the technology applied to Habiganj-Ashuganj Pipeline can be taken into account for getting the best possible pipeline at the lowest cost within the prevailing logistic infrastructure and with due consideration to the socio - economic conditions of this region. In turn this must help growing local infrastructure at the same time encourage an easy view to launch the plan ned projects without any appreciable impact on the hard currency components.

Acknowledgement

The author wished to thank Mr. Mark B.Pickell, P.E., General Conference Chairman, for his invitation to present this paper in the November, 1983 ASCE Specialty Conference on pipelines in Adverse Environments, colleagues at Titas Gas for their contributions and help in its preparation and the General Manager, Titas Gas T&D Co.Ltd., Dhaka for his kind permission to present it.

Metropolitan Water District
San Diego Pipeline 5

Gary M. Snyder, Member ASCE[1]
William P. Lundy, Member ASCE[2]

Abstract

The Metropolitan Water District of Southern California (Metropolitan) is a public agency created by an act of the State Legislature in the late 1920s.

Metropolitan's San Diego Pipeline No. 5 (SD 5) is an 18.2 mile (29.3 km) long pipeline, 96 inches (2438 mm) in diameter extending from Lake Skinner in Riverside County, California, to a point in San Diego County about 5.71 miles (9.19 km) south of the Riverside-San Diego County line. It was constructed in two segments and completed in 1983. One segment is prestressed concrete cylinder pipe and the other is welded steel pipe. The pipeline traverses very rugged terrain with pipe slopes in some locations exceeding 50%. Much of the alinement is in nearly inaccessible areas while other areas traverse avocado groves where root-rot fungus (phytophthera cinnomoni) is present and measures had to be taken to prevent the spread of the fungus to uninfested groves. Much of the pipeline parallels Metropolitan's existing San Diego Pipeline No. 4 (SD 4) which was constructed in 1969 to 1971. That existing pipeline varies in diameter from 99" to 89" (2515 mm to 2260 mm, respectively). At some locations, the clearance between SD 5 and SD 4 is less than 40 feet (12.2 m). Much of the alinement is in rock where blasting was required which, due to the proximity of the existing pipeline, had to be done with special care.

The pipeline also includes 4 tunnels, one of which was at a slope of over 30%. This presented some extreme construction problems.

The design of the pipeline included a small hydro-electric power plant built in a remote location, 3 valve structures, 1 meter structure, 1 control structure and 115 manholes, air release, vacuum valve and blowoff structures.

1. Gary M. Snyder is a Principal Engineer in charge of Structural Design for the Metropolitan Water District of Southern California, P.O. Box 54153, Los Angeles, California 90054.
2. William P. Lundy is a Principal Engineer in Charge of Civil Design for the Metropolitan Water District of Southern California, P.O. Box 54153, Los Angeles, California 90054.

This paper will describe the planning, the design, and the construction of this major water distribution pipeline, constructed in extremely rugged terrain.

Introduction

The Metropolitan Water District of Southern California is a public agency created by an act of the California State Legislature in the late 1920s. Metropolitan is a wholesaler of water and serves water to 27 member agencies located in six Southern California counties. Those agencies cover an area of over 5,000 square miles (13,000 km^2) and serve a population of nearly 13 million people. Metropolitan's distribution system consists of 750 miles of pipeline, eight reservoirs, five filtration plants, numerous regulating structures, and other miscellaneous facilities. Metropolitan delivers about 1.5 million acre feet ($1.85 \times 10^9 m^3$) of water per year, 400,000 acre feet ($4.93 \times 10^8 m^3$) of which is delivered to San Diego County.

Metropolitan currently owns and operates five transmission pipelines which deliver both treated and untreated water to San Diego County, California. The first pipeline was built in the late 1940s. San Diego Pipeline No. 5 (SD 5), with which this paper is concerned, was completed in 1983. SD 5 traverses very rugged terrain. Along one reach of the pipeline, there is an elevation change of over 400 feet (122 m) in a distance of less than 1 mile (1.6 km) and pipe slopes of more than 50%. The problems caused by the rugged terrain were compounded by the fact that much of the alinement is in areas that are nearly inaccessible, while in other areas the presence of avocado groves infested with root-rot fungus (phytophthera cinnomoni) required special attention to prevent the spread of the fungus to uninfected trees.

SD 5 will deliver treated water from Metropolitan's Robert A. Skinner Filtration Plant in Riverside County to San Diego County. It is an 18.2 mile (29.3 km) long pipeline with an inside diameter of 96 inches (2438 mm). The pipeline was constructed under two separate contracts, both of which permitted the use of either prestressed concrete cylinder pipe or welded steel pipe. Contractors were permitted to bid either option and, as a result, the northern reach is prestressed concrete cylinder pipe and the southern reach is welded steel pipe.

This paper will be divided into three major sections: preliminary planning, final design, and construction.

Preliminary Planning

In December of 1978, Metropolitan's Planning Division completed its Report No. 933, titled "Facilities Required to Meet Forecasted Water Demands of San Diego County Water Authority." Prior to the issuance of that report, a series of studies had been conducted to determine the best means of meeting the future water demands within San Diego County. Those studies involved analysis of long-range water demands, the segregation of service into treated and untreated water delivery systems, matters of economy relating to new facilities, and the efficient use of existing and new resources, including pipelines, water treatment plants, and reservoirs.

San Diego County has experienced an unusually rapid population growth and it is anticipated that this trend will continue, although possibly at a lower rate than in recent years. Water demand projections were made and all showed a continued increase in overall demands, which resulted from the expected increase in population. The need to build a fifth pipeline to serve San Diego County became very apparent.

Three possible routes between Lake Skinner in Riverside County, California, and the San Diego County Water Authority's diversion structure in San Diego County, California, were identified. These were (1) the Red Mountain route, (2) the Rainbow Pass route, and (3) the Pala Pass route. Primarily for reasons of environmental sensitivity and high cost, the Pala Pass route was rejected.

After analyzing the various parameters with each alternative, the following design requirements for SD 5 were selected:

1. Capacity: 500 cubic feet per second (cfs) (14 m^3/s) maximum; and

2. hydraulic gradiant: sufficiently high to provide for the required flow in San Diego County Water Authority's system and to provide sufficient head for a small hydroelectric power facility on Metropolitan's portion of the system.

The final route selected was the Red Mountain route, since it was more accessible to those agencies most likely to utilize the pipeline, was less environmentally sensitive, and had the potential for construction of a small hydroelectric power plant. The hydroelectric power plant is currently being constructed under separate contracts and will not be addressed in this paper.

In January 1979, an environmental impact report for San Diego Pipeline No. 5 was prepared by Metropolitan, and in that same year Metropolitan's Board of Directors approved the project and final design commenced.

Final Design

Metropolitan's Civil Design Branch then began detailed alinement studies and other related work.

An aerial survey was made along the alinement to provide topographic mapping at a scale of 40 feet to the inch (0.48 m/mm) with two-foot (0.61 m) contour intervals. Horizontal and vertical control for the aerial survey was established by ground surveys. Because of the very rough terrain and difficulty of establishing pipe grades, a ground-profile survey was made along the centerline of the pipe alinement. Pipeline grades were established to fit the lateral and longitudinal slopes over the pipeline.

In addition to the right-of-way that was already owned by Metropolitan for its existing Pipeline No. 4, it was necessary to acquire only 14.1 acres (5.71 ha) of fee property and permanent easements and 52.5 acres (21.25 ha) of temporary construction easements. Although the required acreage was relatively small, there were 72 parcels of land that had to be obtained from approximately 51 owners with varied interests and concerns.

At crossings of two major river channels, the grades were established based on runoff and geological data. At the Murrieta River crossing, the subgrade was composed of a deep, porous, aluvial deposit subject to bed movement during heavy storm runoff. For protection, the pipeline was placed at depths of up to 30 feet (9.14 m) below the river channel invert and incased in reinforced concrete. The Santa Margarita River crosses the pipeline alinement in a deep gorge approximately 300 feet (91.4 m) below the natural terrain. The river flows on a bedrock formation. A pipeline trench with an invert approximately 18 feet (5.5 m) deep was blasted across the stream bed. The pipe was placed into the trench and incased in unreinforced concrete to a depth of 12 inches (305 mm) above the top of the pipe. The remainder of the trench to the river invert was backfilled with grouted rock riprap.

The pipeline design was compounded by many ecological problems. Of particular concern to many organizations was the crossing of the Santa Margarita River Gorge. The property in the area is owned by the State of California, and is under the jurisdiction of the San Diego State University. The university, as well as others, was very concerned that the sensitive ecological balance in this area be preserved. One of the reasons Metropolitan elected to traverse the sides of the canyon walls in the Margarita

River Gorge by means of tunnels was to insure that this ecological balance would be preserved.

As the pipeline construction was being completed in the fall of 1982, preparations were made to replant the work areas to prevent erosion and slides from the winter storms. Metropolitan's forces placed approximately 3 acres (1.2 ha) of jute mesh over areas of the pipeline alinement where erosion could damage adjacent properties, roads, and steep banks.

A contract was then awarded to hydromulch approximately 186 acres (75 ha) with a mixture of seed, fertilizer, organic stabilizer, mulch, and water. Two types of hydromulch, one composed of 1830 pounds per acre (2051 kg/ha) and the other of 2770 pounds per acre (3105 kg/ha) of mix, were used on the property. The heavier application was used on 86 acres (34.8 ha) composed of the steeper slopes and areas where soils were more subject to erosion. The lighter application was used on 100 acres (40.5 ha) of flatter slopes and in areas of more stable soil.

In each of the types of mixes, there were 10 varieties of seeds, totaling 80 pounds per acre (90 kg/ha). The seeds were selected to produce plants with one or more of the following characteristics: (1) good reseeding life cycle, (2) good soil-stabilizing root and/or canopy growth, (3) rapid germination, (4) environmental blending, and (5) colorful appearance.

The hydromulch was applied and germination began to take place immediately. Metropolitan's planning goals for hydromulching were successfully met, with little erosion occurring from the many heavy storms during the 1982/1983 rain season.

Upon completion of the preliminary civil design work, the Structural Design Branch began design of the pipeline and appurtenant structures. Designs were prepared for two types of pipe--welded steel pipe and prestressed concrete cylinder pipe. In both cases, the pipe was designed for internal pressure with no external load, external load with no internal pressure, and a combination of both. The most severe of the above conditions was then used.

For welded steel pipe, the allowable stress for internal pressure (ring tension) was 16,500 psi (113,800 kPa). External load design was based on the Spangler equation, with an allowable deflection of 3%. The prestressed concrete cylinder pipe was designed by classical methods, using Perris coefficients. In both cases, a soil density of 120 pounds per cubic foot (1920 kg/m^3) was assumed and a live load equivalent to AASHTO H20 loading with 50% impact was applied when the earth cover was 12 feet (3.7 m) or less.

The welded steel pipe was designed to have 1/2" (12.7 mm) internal mortar lining and was protected externally by a 3/4" (19 mm) mortar coating underlaid in some reaches by a layer of dielectric material.

The hydraulic thrust design for both the welded steel pipe and prestressed concrete cylinder pipe assumed no passive resistance from soil at bends in the pipeline. This requirement eliminated the use of thrust blocks. This design philosophy was employed as a safeguard against the possibility that the pipeline could be excavated at the bends, thereby rendering a thrust block useless and, also, during a seismic event it would be possible for a thrust block to separate from the surrounding soil mass as each moved in a different manner. This design requirement was handled by welding the pipe joints for a sufficient distance on each side of the bend to anchor the forces caused by the hydraulic thrust. In the case of prestressed concrete cylinder pipe, this necessitated the use of heavier embedded steel cylinders to withstand the longitudinal stresses. An allowable longitudinal stress of 10,000 psi (69,000 kPa) was used for both the prestressed concrete cylinder pipe and the welded steel pipe.

The design of the structures along the pipeline was based on classical methods of reinforced concrete design, assuming an equivalent fluid pressure of 60 pounds per cubic foot (960 kg/m^2). Seismic forces were also considered in the design.

Portions of the southern reach of the pipeline pass through avocado groves infested with root-rot fungus. Prior to the start of construction, extensive testing was undertaken to determine the extent of infestation. The fungus is easily spread from infected to uninfected trees, and special precautions had to be specified as part of the pipeline design to avoid this. A detailed decontamination procedure was established by Metropolitan to insure that the spread of the fungus would not occur. These requirements were strictly enforced by Metropolitan's resident engineer.

The pipeline specifications required that all areas to be excavated were to first be stripped. Stripping consisted of removing all vegetation and topsoil to a minimum depth of 6" (150 mm). The vegetation had to be removed from the work site, but the topsoil was stockpiled and spread over the surface after the backfilling had been completed. This requirement was included in the specifications in an attempt to return the area to its native state after completion of the construction work. The removed topsoil contained native seeds which, when replaced over the backfilled areas, sprouted and reestablished the area with native growth.

The geological conditions throughout a large portion of the pipeline route required excavation in rock; therefore, blasting was permitted under very strict control. The detailed blasting procedures were determined by the contractor; however, general restrictions were placed upon the contractor in the written specifications. When water was present in blast holes, a nonnitroglycerine explosive had to be used. This was done to prevent drillhole-to-drillhole propagation and assure that delay patterns were effective. The explosive loading per delay period was tied to a factor called "scaled distance." The scaled distance is defined as the distance in feet between a blast drillhole and the structure of interest, divided by the square root of the explosive loading per delay period in pounds. Where the SD 5 pipeline alinement was adjacent to Metropolitan's existing San Diego Pipeline No. 4 or to other pipeline structures, a scaled distance of 8.0 or more was required. In the areas where the SD 5 pipeline was in the vicinity of residences, a scaled distance of 40 or more was required. As an example, for a distance of 15 feet (4.6 m) from the blast drillhole to a structure along San Diego Pipeline No. 4, the maximum explosive per delay period in pounds would be 3½ pounds (1.6 Kg). If the distance of 15 feet were divided by the square root of 3.5, this would yield a scaled distance of 8.0. As an alternative to the scaled distance requirement, the contractor was given an option in lieu of conducting blasting operations at the scaled distance. He was allowed, at his expense, to perform instrumented testing to demonstrate that the blasting operations would not exceed the following limits with respect to the existing San Diego Pipeline No. 4:

1. Maximum ground motion particle velocity in pipe zone backfill, 6.0 inches per second (152 mm/sec);

2. maximum displacement of pipeline, 0.050 inch (1.3 mm); and

3. maximum tensile strain in mortar coating of pipeline, 0.000250 inch per inch (.000250 mm/mm).

In lieu of conducting blasting operations at a scaled distance of 40 or more with respect to residences, the contractor was allowed, at his option and expense, to perform instrumented seismographic testing to demonstrate that his blasting operations would not cause a ground motion particle velocity of more than 1.0 inch per second (25.4 mm/s) at the residences. All blasting submittals were reviewed prior to execution by Metropolitan. Detailed monitoring was conducted throughout all blasting operations.

The pipeline design included alternative routes, which allowed the construction contractor to choose among various combinations of tunnels, shafts, and open trenches.

Construction

The northern portion of the pipeline was constructed by Granite Construction Company, headquartered in Watsonville, California. The southern section was constructed by Artukovich-Hood, a joint venture, headquartered in South El Monte, California.

On the northern portion, the prestressed concrete cylinder pipe alternative was selected. The pipe was manufactured by United Concrete pipe at their plant in Etiwanda, California. The 96" (2438 mm) inside diameter pipe was fabricated in lengths of 24 feet (7.3 m) where possible, with wall thicknesses of 8½ and 12 inches (216 mm and 305 mm, respectively). The varying wall thickness was a result of the required pressure varying from about 45 psi (310 kPa) to 205 psi (1420 kPa). Each thick-wall piece weighed about 51 tons (46,000 kg). The cast concrete core contained an imbedded steel cylinder, with thickness varying from 16 gage (1.52 mm) to 5/16" (7.94 mm). Each core was then wrapped with high-tensile prestressing wire. The wire diameter varied from 8 gage (4.11 mm) to 5/16" (7.94 mm). There are about 3.5 miles (5.6 km) of prestressing wire in a typical 24-foot length. The design 28-day concrete strengths of the cores ranged from 5500 psi (37,900 kPa) to 7000 psi (48,300 kPa). After manufacturing, the pipe was transported about 50 miles (80 km) to the jobsite.

On the southern part of the pipeline, the welded steel pipe alternative was selected. The steel pipe was manufactured from ASTM 283 Grade D steel by Kaiser Steel at their plant in Fontana, California. It was fabricated in nominal lengths of 40 feet (12.2 m). The plate thickness varied from 1/2" (12.7 mm) to 1¼ inches (31.8 mm), with the weight of each piece varying from about 10 tons (9100 kg) to 26 tons (23,600 kg). The pipe was lined and coated as described under the final design section of this paper, and then transported about 50 miles (80 km) to the jobsite.

There are four major tunnels on the pipeline, ranging in length from about 400 feet (122 m) to 1140 feet (347 m). One of these tunnels has a slope of over 30%. The tunnel was driven from the top down because the lower end was nearly inaccessible for removing the muck. A muck car that could climb a 35% slope was used. The contractor drilled and blasted at the head of the tunnel. When the loaded muck car was traversing the tunnel, all personnel were required to be out of the tunnel for safety reasons. This process was done on a 24-hour basis five days a week.

Some portions of the tunnel did not require any support, while other portions were supported with 4-inch (102 mm) horseshoe-shaped steel wide-flanged members, at about 4-foot (1.2 m) centers. Upon completion of the tunnel, the pipe was lowered into the tunnel on rails, the joints were welded and the tunnel was grouted.

In the area where root-rot fungus was present, copper sulfate was used to prevent the spread of the fungus. Boxes of copper sulfate were provided to disinfect boots on people leaving the infested area, and all tools and equipment were disinfected prior to leaving. Additionally, the access road was treated with copper sulfate on a regular basis (up to three times a day) to avoid having the fungus spread by vehicles.

The contractors on both portions of the pipeline were very well organized, well staffed and well equipped. The construction was well planned and, as a result, both portions were completed ahead of schedule, without any serious injuries. It is expected that the first water will flow through the pipeline in 1984, carrying this vital resource to the factories, farms, and residences in San Diego County.

Sugar Pine Pipeline Design and Construction

David A. Aitken*, Associate Member ASCE

Abstract

Sugar Pine Pipeline presented unique design, construction, and safety problems which taxed the ingenuity of the contractor and Bureau of Reclamation design and construction personnel. Extremely adverse terrain and anticipated soft water corrosion problems dictated special designs, and the installation of this pipeline required imaginative and nontypical construction methods. The adverse terrain necessitated the design and construction of a short tunnel section in order to avoid a treacherous and costly bench cut and typical pipe installation.

Introduction

The severe drought of 1976 created critical water shortages throughout the State of California. The town of Foresthill, California, not immune from the drought, at the same time faced an additional dilemma - a critically inadequate water supply.

The Foresthill Public Utilities District is charged with the responsibility of providing water to Foresthill and the surrounding community. This became a formidable task because of the District's inability to obtain adequate storage capacity, compounded by a large influx of new families into the District in preceding years. In 1976, these multiple problems forced the District to impose stringent watering restrictions and defer the acceptance of any new customers. Lawn watering and the washing of cars were strictly prohibited during this crisis.

These measures were partially successful; for in June 1976, water consumption in the District was held to 4.6 million gallons as compared to the 8.6 million gallons consumed during June 1975. Commendable as these efforts were, the measures taken were insufficient for the District had virtually no water available for fire protection. Since the town of Foresthill lies in a heavily forested area, and a timber products plant is located within the District, the unavailability of fire protection water made an already poor situation into an extremely hazardous one. The potential for disaster was enormous. Fortunately, the District made it through this crisis.

---

* Civil Engineer, Bureau of Reclamation, Water Conveyance Branch, Division of Design, Engineering and Research Center, Denver, Colorado.

Foresight by the District in the early 1960's, in anticipation of the problems encountered in 1976, prompted the congressional enactment of Public Law 89-161 in 1965. Little public recognition was made of this seemingly minor occurrence which authorized the planning studies, design, and construction of Sugar Pine Dam and Main Conveyance System, of which Sugar Pine Pipeline is a part.

Features of this project, which include Sugar Pine Dam, Sugar Pine Pipeline, Foresthill Pipeline, and Foresthill Water Treatment Facilities, are part of the Bureau of Reclamation's immense and ongoing multimillion dollar Central Valley Project in California. The planning and feasibility studies of Sugar Pine Dam and the Main Conveyance System took many years to complete. Although the design and construction of the system were not sufficiently complete to avert the District's problems in 1976, the system will serve to alleviate similar crises in the future.

From the initial concept, Sugar Pine Pipeline was envisioned as a unique proposal and became one of the most challenging pipeline projects ever undertaken by the Bureau of Reclamation. Though not viewed as a long or large conveyance system by Reclamation standards, the pipeline alinement presented unusual design, construction, and safety problems which would tax the ingenuity of Reclamation design and construction personnel and the contractor.

General Description of Pipeline and Delivery System

Water impounded by Sugar Pine Dam and Reservoir flows by gravity through Sugar Pine Pipeline to a 40-acre-foot regulating reservoir, previously designed and constructed by the Bureau. On user demand, water is released from the regulating reservoir, flows through Foresthill Pipeline, is made potable at the Foresthill Water Treatment Facility and continues on to serve the town of Foresthill and surrounding area. Both Foresthill Pipeline and portions of the water treatment facility were designed and constructed under separate contracts by the Bureau of Reclamation.

Sugar Pine Pipeline consists of approximately 2.7 miles of 27-inch-diameter steel and 5.6 miles of 24-inch-diameter ductile-iron pipe with design pressure heads ranging from 50 to 800 feet. A delivery tee and sectionalizing valve are provided on the pipeline as well as manhole and blowoff structures, and air valve assemblies. Combination rate-of-flow controller, pressure-sustaining, remote-control valves; a pressure-reducing valve; and a pressure-relief valve are incorporated with steel manifold piping and placed in a concrete pressure-reducing structure located immediately upstream of the regulating reservoir.

Geologic Considerations

The pipeline traverses an area of the rugged Sierra Nevada in California, where temperatures range from 9 to 107 °F and annual precipitation (rain and snow) averages 54 inches. Over the past few years, major winter storms in the area have deposited 15 to 20 feet of snow in a matter of days.

Extremely adverse terrain was encountered in numerous areas along the pipeline alinement. Approximately 5.75 miles of the 8.3-mile-long pipeline is in areas of deep, steep canyons and along hillsides with severe side slope conditions. Five major canyon and creek crossings were contended with; the canyons varied in breadth from 700 to 1,900 feet. Slopes along the alinement and side slopes crossing the alinement ranged up to 145 percent. In addition, major rock outcroppings, heavily wooded areas, and combinations of all the previously identified problems encountered.

Despite the steep slopes, only one landslide was described in the geologic data. Vertically standing trees up to 50 feet tall suggested the landslide had been stable for 50 to 100 years. Vertical or nearly vertical cut slopes on roads near the pipeline alinement appeared essentially stable.

The geologic design data indicated the pipeline trench excavation would be essentially stable in alluvial materials except for occasional areas of slumping and rockfalls. For design purposes, it was assumed that the steepness of the excavation of the side slopes and deep canyons would be governed more by construction methods and safety regulations rather than by slope and pipe instability.

The geology underlying the Sierra Nevada was also significant to the final design. Reservoirs of the Sierra Nevada typically store water derived from snowmelt which percolates through areas underlain by essentially insoluble crystalline, igneous, and metamorphic rock. As a result, these waters are cold, very low in temporary hardness (i.e., soft water), and have moderate to high amounts of dissolved carbon dioxide and a pH of around 7.0. This combination of water quality parameters causes the Sierran waters to be very aggressive in its corrosive attack on portland cement, mortar, and steel.

Water for testing purposes was not yet impounded in Sugar Pine Reservoir when the design of the pipeline began. However, data from other reservoirs in similar geologic and geographic areas indicated that waters passing through Sugar Pine Pipeline would have similar aggressive traits.

Land Use

Land use of the area affected the design of Sugar Pine Pipeline to some degree. The pipeline traverses an area which is logged extensively. The alinement is crossed numerous times by what the local lumber company terms skid trails which are used when removing logs from the cutting areas. The company was concerned that skidding logs across the pipeline with a track-type tractor would gouge and consequently decrease the cover over the pipe, and hence, render the pipe more susceptible to damage, particularly during the spring when the ground is very wet. Therefore, the minimum depth of cover over the top of the pipe through these areas was set at 10 feet. The depth of cover at haul road crossings was based on standard logging operations and set at 5 feet minimum. Compacted backfill (95 percent standard dry density) to the normal ground surface and "B" wall pipe were specified beneath the haul road crossings.

The lumber company also owned and operated a loading and staging area through which the pipeline crossed. The pipe-soil system through this area had to be designed to withstand a 40,000-lb concentrated wheel load in the event the weight of the tractor equipment and its load were momentarily applied to one wheel. Final design requirements through this area were "B" wall pipe and a minimum of 6 feet of cover compacted to the normal ground surface. All pipe appurtenances were either offset or eliminated from the staging area by adjusting the pipe profile.

Soft Water and Pipe Restrictions Affects Design and Pipe Options

Concrete is subject to corrosion attack and consequently rapid deterioration in some cases when pH values are below 7.0, the attack becoming more severe as the pH decreases. Distilled water will leach concrete regardless of pH or carbon dioxide (CO_2) content. However, the CO_2 content of water controls water aggressivity, i.e., its ability to dissolve calcium carbonate ($CaCO_3$) and to a lesser degree, magnesium carbonate ($MgCO_3$) which are present in concrete. If water contains free CO_2, its aggressiveness is increased. Water with dissolved carbon dioxide acts as a weak carbonic acid (H_2CO_3) which reacts with relatively insoluble calcium carbonate and magnesium carbonate and changes the carbonates into readily soluble calcium (or magnesium) bicarbonate. This process causes variable amounts of corrosion of concrete surfaces, some liberation of concrete aggregates, and increases the hardness of the water.

To overcome these problems, two different design alternatives were considered. The first alternative was to construct an automated lime treatment facility at the beginning of Sugar Pine Pipeline near Sugar Pine Dam. This was a viable alternative even though it seemed superfluous to make soft water hard. The idea was eventually abandoned however because of the inaccessibility of the area through much of the winter. The other design alternative was to provide a type of pipe that would withstand the deleterious effects of the water.

The deleterious effect of the water on mortar and concrete is twofold: (a) loss of structural strength and (b) loss of corrosion-inhibiting capability. Corrosion of steel is likely since the soft water does not provide a protective scale. The potential loss of structural strength precluded any pipe which derives its structural integrity from the concrete or mortar from being considered as a pipe option for Sugar Pine Pipeline. Therefore, pipe such as AC (asbestos cement), RCP (reinforced concrete), prestressed concrete, and PT (pretensioned concrete cylinder) pipe were not allowed. In addition, at the time of design, PVC (polyvinyl chloride) and RTR (reinforced thermosetting resin) pipe were not yet approved by the Bureau in the sizes required on Sugar Pine Pipeline (the Bureau now approves PVC and RTR pipe in sizes up to 27 inches in diameter) and were not allowed as pipe options.

The Foresthill Public Utilities District also imposed its own restrictions on allowable pipe options by insisting that all specified pipe options be conducive to "hot tapping" for anticipated future hookups.

The combination of the adverse terrain, the agressive water, the high-head system, and the District's "hot tapping" criteria subsequently resulted in only three allowable pipe options. These are (1) mortar-lined and mortar-coated steel pipe, (2) mortar-lined and polyethylene wrapped ductile-iron pipe, and (3) RPM pipe.

Certain restrictions were placed on the use of RPM (reinforced plastic mortar) pipe. This type of pipe could only be used when the design head of the pipeline was 500 feet or less. Additionally, because RPM pipe was not considered conducive to hot tapping, special air valves incorporating blind flange tees would have to be installed along the pipeline; the number and location of which would be at the discretion of the District.

Concerns regarding the corrosive effects of the anticipated soft water and the need for a relatively long-life pipeline (50 years) prompted the Bureau to undertake design and laboratory studies on various types of pipe. These studies drew upon the Bureau's limited experience in this matter, pipe manufacturer's corrosion data, and available literature on the subject.

Though little information was received from the pipe manufacturers, the studies indicated that by doubling the standard thickness of mortar lining on both the steel and ductile-iron pipe, more than 50 years of pipe service, with normal maintenance, could be achieved.

Three access manholes were specified on the pipeline to permit inspection and collection of data regarding the corrosive effects of the soft water on the pipe and mortar lining. These data will be forwarded to the Bureau by the District and used in future design considerations.

Pipeline Design

The installation of the pipeline required special designs, imaginative and nontypical construction methods, and special safety measures. Anticipated substantial costs in constructing the pipeline in the deep canyons and steep hillsides as well as the aesthetic and environmental impact on this scenic area weighed heavily in the final route selection. Hydraulic criteria, topographic features, slope stability and erosion problems, land use, and geologic features all had significant impact on the final design.

The hydraulic design of the pipe system was state-of-the-art. Transient analyses were performed on the system using required fire hydrant flows and fire nozzle closure rates of 2 seconds and less. Design gradients were established based on criteria which produced the maximum and minimum surges anticipated on the system. Combination air vacuum and air-release valves were placed at high points along the pipeline to provide nominal surge protection and increase hydraulic efficiency. A pressure-relief valve located in the pressure-reducing structure affords the primary source of surge protection for the system. This valve is specified to be quick opening (3 seconds or less) and slow closing (10 seconds or more) to permit flow to pass through the pressure-relief line, thus preventing damage to the control valves located within the pressure-reducing structure.

Automatic control of the system is accomplished using a float assembly located in a floatwell structure on the regulating reservoir and six pair of telemetering cable extending from the regulating reservoir back to the outlet works of Sugar Pine Dam. The telemetering cable is installed within the pipe trench.

In anticipating the future needs of the community, the Foresthill Public Utilities District requested that the Bureau provide the means to supplement Sugar Pine Pipeline flows from a creek which crosses the pipe alinement. This was accomplished by designing a modified blowoff structure. Under normal operations, the blowoffs, which are placed at the low points in the profile, are used to drain the pipeline. Since the creek is located at a low point in the profile, the modified blowoff piping is daylighted close to the creek and can be used as a draining feature. Then, when additional water is needed, a blind flange can be removed from the modified blowoff and a pump installed to pump water from the creek directly into the pipeline.

The contractor elected to provide 27-inch-diameter steel and 24-inch-diameter ductile-iron pipe for Sugar Pine Pipeline. Mortar-lined and mortar-coated 27-inch-diameter steel pipe is considered flexible pipe. In order to develop passive soil resistance and prevent excessive vertical deflections under full trench loads, this pipe requires a high degree of compaction on its trench sidewalls. The minimum height of the compacted select material is specified to be 0.7 times the outside diameter of the pipe. In addition, a 4-inch bedding of uncompacted select material to cushion the pipe is required.

Ductile-iron pipe has been an acceptable pipe option in Bureau specifications for many years, but its use has been quite limited and confined primarily to smaller applications. Sugar Pine Pipeline is the Bureau's first major pipeline project which incorporates this type of pipe.

Ductile-iron pipe is regarded as semiflexible pipe in the 24-inch-diameter size and requires a somewhat smaller height of compacted select material on its trench sidewalls in resisting the vertical earth loads. This height is specified as 0.5 times the outside diameter of the pipe. Exceptions to this occur when the compacted material is located on the outside of horizontal curves. In these cases, the compacted material is used to resist lateral displacement and is placed to a height of 0.7 times the outside diameter of the pipe. The 4-inch bedding of uncompacted select material was not required for the ductile-iron pipe.

Both pipe types require a 6-inch-minimum overexcavation in all areas where low density soil, rock, or other unsuitable materials are encountered. This overexcavation is replaced with compacted backfill. Both pipe options also require the select (cohesionless) material to be compacted by means of saturation and internal vibration. Where pipeline grades exceed 0.3, cohesive in lieu of cohesionless material and mechanical compaction in lieu of saturation and internal vibration were permitted. Gradation of the select material was specified such that no more than 5 percent (by weight) could pass a No. 200 sieve or be larger than 0.75 inch.

Along the hillsides where severe side slope conditions would be encountered, a bench method of construction was anticipated. Material excavated from the bench cut was to be side cast down the hill, and the pipe installation was to hug the benched hillside as much as possble. V-shaped earth cross drains were to be excavated at 90° to the pipeline to allow for drainage across the pipe and to prevent erosion.

Pipe to be installed in the steep canyons was specified to be laid uphill with each preceding installed section supporting the next section of pipe being laid. Pipe installed in this manner averted the need for designing special collars for stability. Cross drains were also specified in the steep canyons to prevent channel erosion and piping along the pipeline. For the most part, the design measures taken were simple, practical, and economical.

Construction

The contractor was faced with many challenges prior to and during construction which stimulated his imagination in developing techniques and methods unique to pipeline construction and safety. He was required to deviate considerably from normal installation patterns and equipment use. Safety was of paramount importance and a close liaison between his proposed construction plan and actual site conditions was essential.

Equipment to be used in the deep canyons and along hillsides with severe side slopes was required to be equipped with roll bars and operator-safety cages. Where pipeline grades exceeded 60 percent, the specifications required the contractor to submit to the Bureau a "Hazard Analysis for Equipment Operation on Steep Slopes" for approval.

After his evaluation of the many different aspects and methods of trench excavation, placing of pipe, backfilling of the trench, final cleanup, and water barring of slopes, the contractor elected to use a logging-type, crawler-mounted, five-drum mobile Highlead/Skyline yarder with a 48-foot tower to accomplish the work (see fig. 1). This type of equipment is used extensively in the States of Oregon and Washington by the timber industry where clear cutting operations are predominant.

The machine chosen was a Madill model 071 equipped with four working drums with lines and one spare. These lines were used as the main, haulback, tag, and skyline. Three additional drums were obtained and used for powered guylines.

After right-of-way clearing operations were completed, the contractor chose to use a No. 215 Caterpillar track excavator (backhoe) to excavate the pipe trench. A restraining device was designed and fabricated using a 1-1/8-inch steel cable and a 1/4-inch box channel iron. This device formed a bridal arrangement which was attached to the backhoe, then in turn, connected to the yarder mainline 1-1/8-inch

Fig. 1. - Madill logging yarder set up at landing on downstream side of Temperance Creek crossing, used to transport materials by cableway which is anchored upstream side of crossing.

cable. With the restraining device in place and the yarder located at the high point of the slope, the backhoe was suspended and lowered into the canyons as the excavation progressed (see figs. 2 and 3).

Certain equipment modifications to the backhoe were necessary with this scheme of operation. Due to the steepness of the slopes, oil in the backhoe's engine was settling to the backside of the oil pan. To overcome this problem, additional baffles were placed in the oil pan and an additional amount of oil was added to provide an adequate source of lubrication at all times. This method proved satisfactory.

Once the excavation was completed to the bottom of a ravine, the backhoe was retrieved by pulling it back up the slope with the yarder, the hoe traveling adjacent to the excavated trench. An identical procedure was used in excavating the opposite side of the canyon.

Following excavation, the yarder was set up with a skyline. The skyline passed through the yarder drum and was anchored on both sides of the canyon. Three 1-1/8-inch guylines extending from the

Fig. 2. - View looking at Cat 215 restraint harness. Note protective screen on operator's cage.

Fig. 3. - View looking at Cat 215 backhoe and restraint system.

yarder tower also were anchored. The anchorages for the skyline and guylines consisted of 24-inch-diameter, 18-foot-long pine trees buried in 12-foot-deep trenches. Standing timber was not used as anchorages because of the need to maintain the safety and integrity of the system. The 7/8-inch mainline and 3/4-inch haulback cables of the yarder served as load lines.

Once the assembly was complete, all construction materials and equipment were delivered to the required location by a travel carriage along the skyline. This carriage, mounted on closed 12-inch rollers, had a block arrangement with the mainline running through it, thus, enabling the attached loads to be lowered into position. Trench preparation for pipe laying was very time consuming using this method. Whenever possible, native material was incorporated in the trench preparation. However, approximately 60 percent of the canyons were in rock material which required predrilling and blasting. Thus, very little native material could be used in these areas. When native soil could not be used, the earth material required was delivered to the point of use by an open-ended skip or bucket, approximately 1-1/2 cubic yards in size.

Pipe laying operations were extremely hazardous. Considerable care was required while bringing the pipe into position and alinement and in making joint closures. Mechanical come-along devices were used to make the final joint closures. Access to the pipe laying operation was difficult in areas where slopes exceeded about 60 percent. Workmen standing on these slopes had extreme difficulty in moving about and performing their respective job functions. In some cases, the use of safety lines was required.

Select backfill was delivered to the yarder location and transported to the trench using the open-ended bucket (spoon shaped) previously mentioned. Two lines were used for this operation, one for loading and lowering the bucket and the other, a tag line, used for dumping. Although this method of operation was very time consuming, it proved acceptable from a safety standpoint. Hand compaction of the soil in the pipe haunch area was accomplished using "J" bars. Once the soil in the haunch area was consolidated, gasoline-powered plate compactors were used to compact the remaining material to 85 percent standard dry density. Normal Bureau standards would have required the material to be compacted to 95 percent standard dry density. The contractor, however, was unable to achieve densities higher than 85 percent on the steep slopes without considerable effort. Design calculations indicated that the earth material being used could be compacted to 85 percent standard dry density and still provide adequate side support for the pipe. Because of this and the adverse working conditions, the normal Bureau standards were relaxed. On the steep slopes, consolidation by means of saturation and internal vibration was virtually impossible due to the migration of the materials down the slope.

The two most severe canyon crossings required the use of low slump concrete as the backfill material. The concrete was placed around the pipe and extended to 1 foot above the pipe, with cutoff collars installed at every other joint. Concrete cutoff collars were not considered in the original design but were added later, by change order, when the excavations revealed that rock in these areas broke slick and slabby with all surfaces facing down the slope into the canyons. The cutoff collars provide pipe stability and help prevent erosion of the concrete backfill by channeling water off the impervious layer. The concrete backfill was delivered and placed using the yarder operation previously described.

In a majority of the most difficult areas, ductile-iron pipe was used. Although the handling of the pipe, due to its weight, was somewhat cumbersome in the deep canyon operations, its durability during rough handling procedures was excellent. Backfill earth materials would not stay in place on steep slopes greater than 80 percent. This problem was compounded by the polyethylene sleeve required around the ductile-iron pipe. Due to a lack of contact friction between the sleeve and the surrounding backfill soil, earth materials near the sleeve would not stay in place on slopes exceeding 45 percent. This problem was resolved by constructing dikes, extra effort by contractor personnel, the relaxation of density requirements, and eventually the use of low slump concrete where the other remedies would not suffice.

The contractor also elected to use some 27-inch-diameter steel pipe in portions of the steep areas, using the same construction procedures as were used in installing the ductile-iron pipe. For corrosion protection, mortaring of the interior joints of the steel pipe was required after the pipe was in place and backfilled. When pipe is laid on relatively flat slopes, the mortaring operation is a fairly common procedure. However, on very steep slopes, this process is very slow, expensive, and hazardous. Due to the steepness of the terrain, the contractor used two men working together to accomplish this task. Small carts with wheels were used to transport these men, inspectors, materials, and equipment into the pipe. On the steep slopes, small power winches were used to lower and retrieve the men and equipment into and out of the pipe.

Construction along hillsides with severe side slope conditions was arduous and proved to be as equally challenging to the contractor as the placement of pipe in the canyon areas.

Bench excavation was required for approximately 12,000 linear feet, with about 5,500 linear feet of the total traversing through rock and rock outcroppings that required blasting (see fig. 4).

Most of the backslopes were drilled on 1/4:1 to 1/2:1, depending on the soundness of the rock, resulting in some slopes as much as 35 feet high. After excavation of the bench, numerous rockslides occurred which greatly increased the difficulty of construction and resulted in a higher cost.

Through specifications and preconstruction conferences, the contractor was informed that all spoil from the benched areas would be side cast down the hillsides. Environmental and ecological concerns which surfaced during construction resulted in a change order to the specifications requiring 3,400 linear feet of rock excavation to be

Fig. 4. - View showing bench cut safety scaling operation (laborer on rope) and bench excavation with Cat 977 trackloader.

hauled from the jobsite and wasted. The remaining portions of the 12,000 feet of bench spoil could still be side cast.

Of the 3,400 feet, 3,100 feet needed minor adjustments of both the horizontal and vertical alinement of the pipe to accommodate the change order requirements. Field cutting and repair of the pipe were required at approximately 22 locations. These changes increased the cost of the project appreciably since the contractor had already acquired nearly all the preengineered 27-inch-diameter cement mortar-lined and -coated steel pipe for installation through this area.

The remaining 300 feet presented even greater obstacles, primarily because the pipeline alinement traversed an area composed of a massive vertical rock outcropping. Under the scope of the change order, this area was to have been benched with the spoil material hauled and wasted. However, slow construction progress, the instability of the rock outcrop, the side casting restrictions and expense of rock disposal, and especially the safety of personnel working in the area prompted the Bureau to seek alternative methods of construction through this area. Tunneling, a bin wall, and various pipe-jacking methods were all considered. Based on the contractor's experience and input, the economics, and the time frame for project completion, the tunneling alternative was selected.

The tunnel scheme jointly agreed to by the Government, contractor, and the Utilities District is about 300 feet long with a 68- by 84-inch modified horseshoe-shaped cross section. It is excavated entirely in amphibolite (a fine-grained, schistose, metavolcanic rock of Mesozoic age) by conventional drill and blast methods. The rock, commonly referred to as blue granite, is hard, dense, and moderately fractured.

Split set, 60-inch rock bolts on 4-foot centers located in the crown at the 10:00, 12:00, and 2:00 o'clock positions were specified. Bearing plates were 6 by 6 by 1/4 inch. The ribs and crown were lined with 2 inches of pnuematically-applied concrete (shotcrete). No other support was required.

Following completion of the tunnel excavation and shotcrete application, an 8-inch layer of compacted sand was placed in the invert to serve as a cushion for the 24-inch-diameter ductile iron pipe the contractor chose to use (see fig. 5).

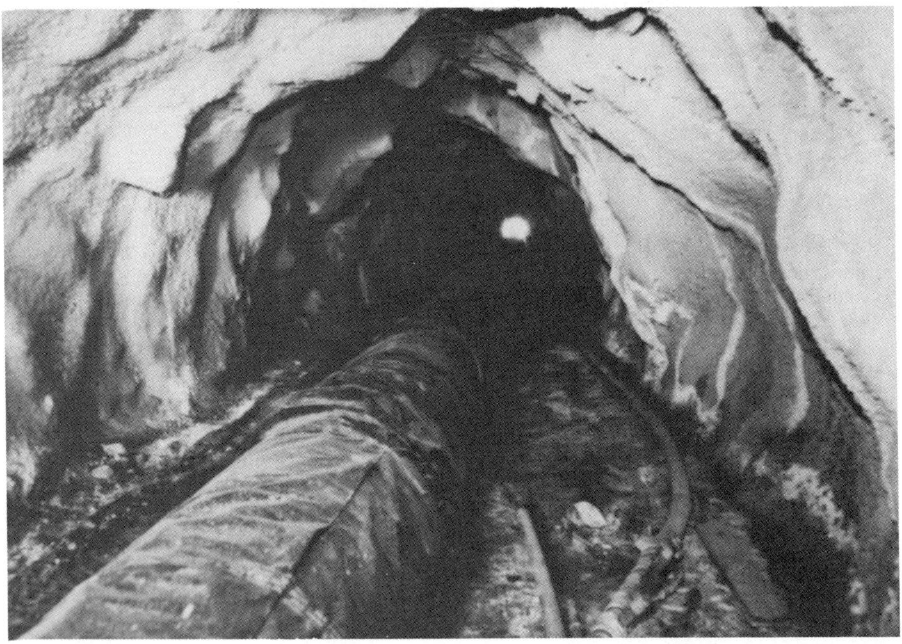

Fig. 5. - Looking upstream at downstream tunnel portal - 24-inch ductile pipe in vinyl wrap.

In essence, the tunnel serves as a casing for the carrier pipe. Once the pipe was installed, two 4-foot-long No. 10 reinforcing steel bars, placed on each side of each bell section, were drilled into the rock of the tunnel invert and grouted. The bars serve to prevent lateral displacement of the pipe. Select backfill was then placed and compacted to a height of 0.5 times the outside pipe diameter, and 2-inch-thick redwood bulkheads with drain holes were installed to seal the tunnel portals.

Revegetation Using Helicopters

Completion of the contract required right-of-way cleanup, revegetation, and erosion stabilization along the pipeline alinement. In the relatively flat areas, few problems were encountered in performing the work. However, in critical areas where slopes exceeded 45 percent and where the spoil was side cast, these normally simple tasks became a real challenge.

In the critical areas, water bars were erected across the disturbed rights-of-way, in conjunction with the final cleanup operations, before any reseeding could be performed.

Helicopters were used to deliver straw bales, jute netting, fertilizer, and seed to strategic locations, thus eliminating many physical and safety hazards to contractor personnel doing the same work manually. After hand broadcast seeding and fertilizing was accomplished, manual placement of straw and netting was required for 100 feet up slopes from creek beds in the bottom of the canyons.

Helicopters were used to broadcast seed and fertilizer over both the excessively steep slope areas and those portions of the alinement requiring restoration as a result of the side-casting operations.

Conclusion

Sugar Pine pipeline was filled, hydrostatically tested, and placed into operation in June 1982, and is performing to its designed standards and requirements. Potential problems during filling and operation, such as severe surges and joint leakage, have not occurred.

The Bureau, the Utilities District, and the contractor are satisfied that a safe and functional pipe system is installed and will remain operational for its designed 50-year life. Open effective communication and cooperation between the contractor and the Bureau were essential in completing the project. The contractor's positive attitude to change orders and stringent Government inspections is greatly appreciated by the Bureau. His unwillingness to accept less than what his equipment and crew could achieve in such an adverse environment is commendable.

Recent studies have indicated that many of the municipal and industrial and irrigation water systems in the United States, especially those in the eastern and northeastern States are, in excess of 100 years old, obsolete, dilapidated, and in dire need of replacement. Economically, the studies indicate that rehabilitation and replacement costs would run in the billions of dollars. Existing land use, environmental and ecological concerns, public safety, and the geography of the eastern/northeastern United States indicate that water conveyance systems perhaps would have to be constructed in similar terrain and fashion as the Sugar Pine Pipeline closed conduit system.

Obviously, the greatest beneficiaries of the project are the citizens represented by the Foresthill Public Utilities District, but the experience gained by the Bureau's engineering design and construction staff and the contractor is another important, tangible benefit. Hopefully, the pipeline construction industry in its involvement in future M&I systems and irrigation systems can benefit from the experience gained in the design and construction of Sugar Pine Pipeline as well.

Influence of Northern Pipelines on River Crossing Design

by E.K. Yaremko,[1] M.ASCE, and R.H. Cooper,[2] M.ASCE

1. INTRODUCTION

Over the past decade there has been an unprecedented level of work in the fields of river engineering and hydrology related to northern pipeline projects. This work has been undertaken by consultants and engineers on behalf of the various projects, by government and university research organizations, and by government agencies responsible for review and regulatory approval.

For the most part, the work has been directed at developing safe and environmentally acceptable designs for river and floodplain crossings. Reasons behind this strong interest in river engineering and hydrology include:

(i) Recognition that processes unique to northern rivers must be considered in establishing criteria for designing river and floodplain crossings.

(ii) Recognition that river and floodplain crossings have historically been areas of high risk with respect to pipeline exposure and failure.

Recognition by the industry of the need for specialist hydrologic and river engineering input to the design of safe river crossings has, in itself, been a significant advance in river crossing design technology.

This paper is a discussion of current technology for the design of buried river and floodplain crossings as it has been developed and applied on recent northern pipeline projects. Much of this technology has been adapted from river engineering design methodology developed originally for bridge crossing design, and for the design and stabilization of highway and railway embankments. Examples include prediction of scour, assessment of bank stability and the use of river training structures.

---

[1]Principal, Northwest Hydraulic Consultants Ltd., Edmonton, Canada.

[2]President, Northwest Hydraulic Consultants Ltd., Edmonton, Canada.

In the discussion that follows, a river crossing refers to a direct crossing of a river and its adjacent floodplains; a floodplain crossing refers to a segment of pipeline that is located within active river floodplain areas and aligned roughly parallel to the river.

For the most part, the river engineering and hydrologic design technology that has been used on recent northern pipeline projects is applicable not only to northern rivers, but to similar types of rivers located elsewhere. There are, however, a number of important design implications that have been identified as being unique to river and floodplain crossings constructed in arctic and sub-arctic climates. These include:

(i) Extreme flood events resulting from the outburst of glacier dammed lakes.

(ii) The necessity to route parallel to rivers within alluvial floodplains to avoid soils that are thaw unstable (hot-oil-pipeline) or frost-heave susceptable (chilled-gas-pipeline).

(iii) River ice processes, including river bed icings, and their impact on flood levels, channel shifting, river bed scour and the performance of river training structures.

(v) Impact of a chilled-gas-pipeline on sub-surface flows; impact of a hot-oil-pipeline on the ability of frozen stream banks to resist erosion.

2. DESIGN OBJECTIVES

The objective in designing a buried crossing of a river or floodplain is to bury the pipe such that it does not become directly exposed to the streamflow as a result of erosional processes likely to be experienced throughout the design life of the pipeline. The design task is then to predict the depth to which the river bed can scour and the lateral extent to which the river channel can migrate over the twenty to fifty year period when the pipeline will be in service. The pipe is then buried below the design scour elevation to a point outside the lateral limits of bank migration to effect a secure crossing design.

Until recently, little attention was given to predicting these limits of river erosion when designing buried river and floodplain crossings. Designs were often based on arbitrarily established depths of cover beneath the natural river bed as it existed at the time the crossing was constructed. Sagbends were set at nominal distances into the natural channel banks. The result is that pipe exposures in the banks and beds of rivers have been a common occurrence. When this happened, remedial measures would be implemented to restore and stabilize bank lines and bed levels. In many instances it has become necessary to replace entire crossings where less costly remedial measures were not feasible.

Today, river and floodplain crossings are usually engineered with the benefit of input from specialists in river engineering. It is the responsibility of these specialists to advise on selection of suitable crossing locations; to estimate river bed scour, bank erosion and channel migration; and to design river training structures and erosion control measures.

3. CROSSING LOCATION AND RIGHT-OF-WAY ALIGNMENT WITHIN RIVER VALLEYS

In locating river crossings, and pipeline right-of-way (ROW) that must be routed alongside stream channels, consideration should be given to the following factors:

(i) river engineering and geotechnical implications
(ii) environmental and sociological concerns
(iii) right-of-way acquistion
(iv) constructability
(v) cost

On large rivers where severe river engineering or geotechnical problems exist, locating a suitable crossing can be a key element to overall route selection. Fortunately, suitable crossing sites can be found on most streams by making only minor deviations to the mile-by-mile alignment.

The following are some of the factors that, from a river engineering viewpoint, should be considered in locating pipeline ROW within river valleys:

(i) Locating a river crossing on a straight reach versus a bend. Generally it is preferrable to locate a buried crossing on a straight reach of river where the rate of bank migration is relatively low in comparison to a location on a bend. In situations where the channel is laterally stable it may, however, be more economical to locate on a bend where the crossing is shorter and the thalweg (deepest segment of the channel) position is fixed. This will minimize required excavation volumes because the natural stream bed profile will tend to be near the predicted scour profile.

(ii) Direct crossings. River crossings should be aligned approximately at right angles to the river, to minimize crossing length, and to prevent channelization along the ROW.

Whenever possible, alignments paralleling the river and located within active channel or floodplain areas should be avoided. On the Trans Alaska Pipeline (TAPS) project, overriding concerns about soil-pipe interaction in thaw-unstable soils led to a radical departure from this practice. To avoid thaw unstable materials lengthy segments of the ROW were located in active alluvial floodplains and aligned parallel to the river. This raised a concern that major stream channels would migrate onto the ROW and expose shallow buried pipe or cause erosion at the foundations of vertical support members. The solution was to construct an extensive system of river training structures designed to prevent channels from migrating onto the ROW.

(iii) Alluvial fans. Because of their dynamic nature - rapid changes in bed levels and channel location - crossings of alluvial fans can be problematic for highways, railways and pipelines. Yet these alluvial features are frequently encountered where it is necessary to route within river valleys in mountainous terrain. On both the TAPS project and the proposed Alaska Highway Gas project there are numerous crossings of highly active alluvial fans. If it is necessary to cross an alluvial fan, the crossing should be located as far upstream on the fan as possible.

(iv) Braided rivers. Many of the larger rivers in Northern Canada and Alaska are braided or multi-channel, gravel-bed-streams. Crossings of these rivers can be difficult because deep burial is required across the entire active channel area which can be an order of magnitude wider than an incised channel conveying equivalent flows. The problem is further compounded because the trench must be excavated down, from bankfull level, over much of the crossing.

(v) Proximity to existing works. Care should be exercised when locating a pipeline crossing near existing works such as bridges, river training structures or material dredging sites. These works may have impacts on scour and erosion, at the crossing site, that are difficult to predict. Experience on the TAPS project has shown, that when selecting river material sites, close attention must be given to the effect that such sites can have on scour processes at nearby river crossings.

4. PIPE BURIAL DEPTH

The requirements for pipe burial should be determined on the basis of a river engineering investigation aimed at assessing the potential for river bed scour. The investigation should begin with an examination of river bed subsurface materials from the viewpoint of assessing their susceptability to erosion.

It is not uncommon in northern rivers to encounter bedrock or dense till at a shallow depth below the surface of the river bed. As these strata represent the historical limit of scour, shallow burial of the pipe within the bedrock or till would be a safe design practice.

Other northern rivers have a stable bed paving layer consisting of cobbles and boulders that are not transported even at extreme flood flows. Generally this layer was formed as the river down-cut to its present level through glacial till or outwash deposits. Where a stable paving layer is encountered, burial of the pipe beneath the layer, or at a nominal depth of say five feet below its surface, will result in a safe crossing.

At river crossings where significant depths of erodible alluvium are encountered, it is necessary to estimate the minimum level to which the bed can scour. Determination of scour is an art-in-itself requiring considerable skill and experience on the part of the engineer.

In estimating the total potential for scour, consideration should be given to (i) local bed scour, which refers to the fluctuation of local bed levels below mean bed level; and (ii) river bed degradation, which refers to a general lowering of the mean bed elevation over a significant reach of the river (see Fig. 1).

Estimation of Local Scour

Local bed scour can be estimated from the empirical formula:

$$d_s = zd_m \quad \cdots\cdots\cdots (1)$$

where: d_m = mean depth measured from a selected water level referenced to bankfull or design flood conditions

d_s = depth to maximum scour

z = a multiplying factor depending on channel geometry and the selected reference water level

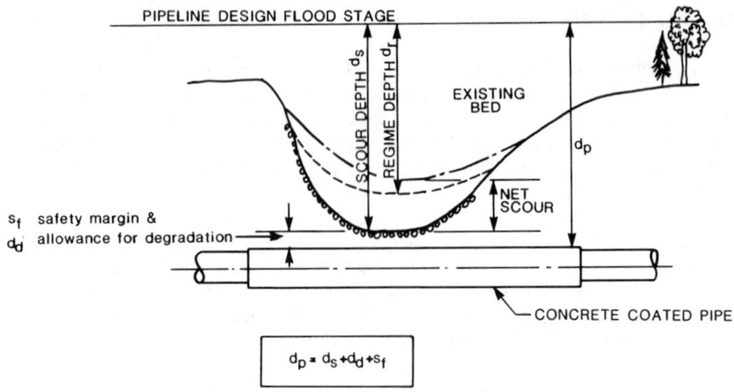

SCOUR DEFINITIONS

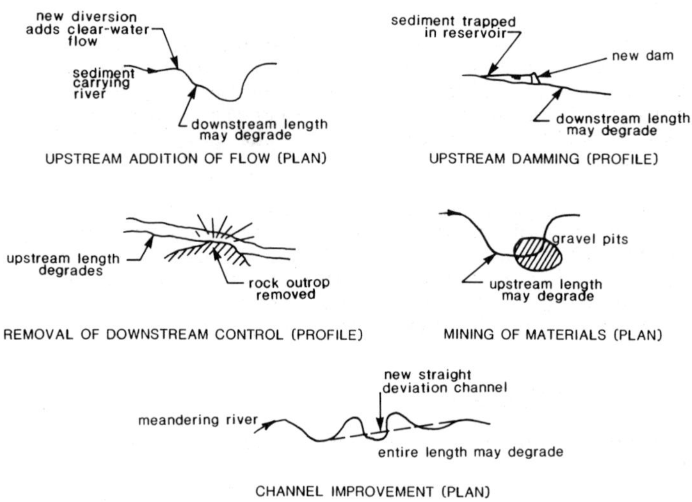

DEGRADATION

CAUSES

GENERAL SCOURING OF THE BED
OVER AN APPRECIABLE LENGTH OF CHANNEL.
POSSIBLE CAUSES LISTED ABOVE

FIGURE 1

The following steps are involved in applying Eq. 1:

(i) Determine the reference water levels corresponding to bankfull and/or design flood conditions. For spilling rivers, either reference level is appropriate. For incised channels, the design flood level should be used. This involves establishing a stage-discharge relationship either by direct measurement, slope-area computations, or backwater analysis. The resulting design levels should be compared to field observations of bankfull levels and historic high water marks.

(ii) Determine the mean depth of flow, d_m, at the selected reference stage by direct measurement from channel cross-section profiles, and alternatively, by applying empirical regime equations (5). Results should be compared and significant differences resolved.

(iii) Lastly, an appropriate multiplying factor, z, is chosen for the selected reference stage, and depth of scour is computed and subtracted from the reference stage to obtain the design scour level.

Table 1 lists values of z recommended in a recent Indian Bridge Code (3). In applying Eq. 1 using these values, the reference water levels and mean depth, d_m, are for the design discharge.

TABLE 1.
Empirical multiplying factors for maximum scoured depth

Nature of Location	Factor
Straight reach of channel	1.25
Moderate bend	1.5
Severe bend	1.75
Right-angled abrupt turn	2.0
Noses of piers	2.0
Alongside cliffs and walls	2.25
Noses of guide banks	2.75

Charbonneau (1), in an investigation undertaken for the TAPS project, utilized hydraulic model and field data on gravel rivers to develop recommended ranges of z. These ranges, listed in Table 2, can be referenced either to bankfull or design flood conditions.

TABLE 2.
Z-values for estimation of scour depth

Condition	Design Flood	Bankfull
Rigid bend	1.4 - 2.5	1.4 - 4.0
Free bend	1.4 - 1.75	1.4 - 2.5
Confluence	1.5 - 2.0	2.0 - 3.0
Obstruction	1.75 - 2.5	2.5 - 4.0
Impinging flow	1.75 - 2.5	2.5 - 4.0

Whenever possible, field observations of maximum local scour over a reach of the same river should be obtained. A comparison between computed and observed scour depths, at similar locations on the river, should be made, and major differences resolved, particularly if the field soundings indicate deeper scour.

On recent northern pipeline projects, a great deal of attention has been focused on the importance of accurately estimating an appropriate design discharge for use in determining depth of scour. Yet, it is clear from the above that the design procedure for estimating scour is much more sensitive to the application of engineering judgement, in selecting an appropriate multiplying factor for use in Eq. 1, than it is to the accurate determination of a design flood discharge.

Channel Degradation

Channel degradation refers to a general lowering of the average stream bed over an appreciable length of river. In Fig. 1, a number of natural and artificial causes of degradation are illustrated. For crossings of streams where any of these situations is identified, an additional allowance for degradation should be made in determining the pipe burial profile.

5. SAGBEND LOCATION

In the authors' experience, the predominant cause of pipe exposure at river crossings has been the result of lateral migration of channel banks. Good design practice requires that the sagbends be set a sufficient distance back from the channel banks to allow for lateral migration of the channel resulting from bank erosion.

There are no reliable quantitative methods to predict the rate at which bank erosion will occur. A given length of stream bank can appear to be relatively stable and able to withstand the erosive attack of floods for many years without experiencing noticeable erosion. Yet, in a single severe flood event, several hundred feet of lateral migration can take place.

Estimating the requirement for set-back of the sagbends places a heavy demand on the experience, skill and judgement of the engineer. Aerial photographs, old maps, and surveys can be valuable for providing a recent (20-30 year) history of bank movement. This should be supplemented with an assessment of the following factors:

(i) Flow intensity - the magnitude of erosive attack on a bank is closely related to flow intensity.

(ii) Channel configuration - erosive attack is naturally greater at locations where flow becomes concentrated such as the outside of a bend.

(iii) Bank geometry - steep banks are more susceptible to erosion than gently sloping banks. High banks tend to erode less rapidly than low banks because the quantity of material that has to be removed is greater.

(iv) Bank materials - cohesive materials can be highly resistant to erosion whereas sands and gravels offer little resistance.

(v) Vegetation on and adjacent to the banks often stabilize an easily erodible bank material. Removal of this vegetative cover often leads to rapid bank erosion.

(vi) Ice jams and log jams, river-bed-icings - on multi-channel rivers, grounded ice jams and log jams in a major flow channel can cause diversion of flow into secondary channels resulting in enlargement of these channels with associated bank erosion. Abrupt changes in channel pattern can occur when spring freshettes flow across, and erode through, extensive river-bed-icings. These changes in channel configuration can alter the intensity of erosive attack on channel banks.

(vii) Thermal regime - in permafrost areas, annual bank erosion can be limited to a shallow active layer even in erosion susceptible materials. Degradation of permafrost due to burial of a hot oil pipeline can decrease resistance to erosion in the immediate vicinity of the pipe. A solution is to riprap that portion of the river bank that will be subject to thaw.

6. RIVER TRAINING STRUCTURES

Traditionally, pipeline crossings have been designed without reliance on river training structures to ensure crossing integrity. This is in contrast to railway and highway crossings where extensive use has been made of river training and erosion control structures to (i) confine the river and thereby minimize length of the crossing and (ii) protect embankments aligned parallel to stream channels.

Design practices implemented on the TAPS project represented a radical departure to this earlier pipeline design philosophy. It was necessary on the TAPS project to route lengthy segments of the ROW along stream valleys, and within active river floodplain areas into which flow channels could be expected to migrate during the operating life of the pipeline. Deep burial of these segments, below design-scour-levels, would have been prohibitively expensive and disruptive to the river environment. The solution was to construct the pipeline, in a shallow burial or elevated mode, behind a line of spur dikes designed to prevent active river channels from migrating onto the pipeline ROW. This concept is illustrated in Figure 2. Another application of river training that was employed on the TAPS project was the use of guide banks and riprap-bank-revetment to control channel migration and allow for minimal sagbend setback at river crossings.

With an increased awareness by the pipeline industry of a need for river engineering input to the design of pipeline crossings, more frequent use of river training measures to reduce costs and improve designs can be expected.

7. HYDROLOGY

On recent northern pipeline projects much of the effort in river and floodplain crossing design has been focused on the determination of design floods.

For the TAPS project, design floods were based on the Standard Project Flood as defined by the U.S. Army Corps of Engineers. The HEC-1 watershed model was used to compute floods based on 50-percent of the maximum probable precipitation.

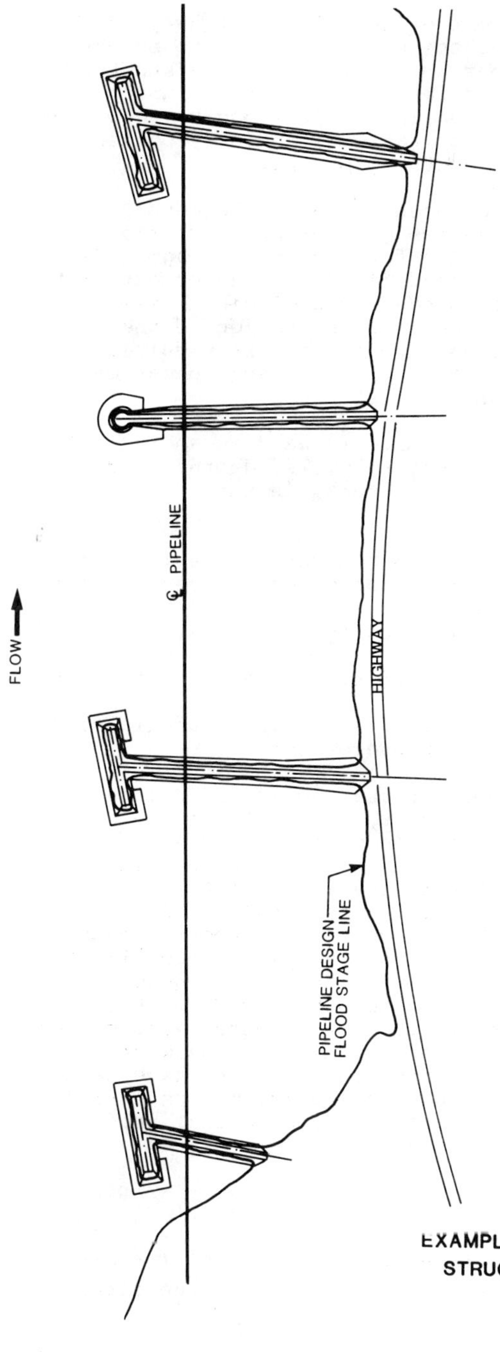

FIGURE 2

EXAMPLE OF USE OF RIVER TRAINING STRUCTURES FOR PIPELINE DESIGN

On the Canadian segment of the Alaska Highway Gas Pipeline project, river crossing designs were to be based on 100-year flood estimates in British Columbia and the Yukon, and 50-year flood estimates in Alberta. The rational for this distinction in design flood criteria was that the hydrologic data base was substantially better along the Alberta segment of the pipeline.

Criteria developed for both of these projects recognized the need to consider glacier outburst floods as extreme flow events that could far exceed the magnitude of 1:100 year, or Standard Project floods. Glacier outburst floods occur when water impounded in a glacier dammed lake is rapidly released due to melting or erosion of the glacier ice dam. Input from a specialist is required to determine the possibility of occurrence, and to estimate the magnitude of a glacier outburst flood.

The design discharge estimate is used as an input for the determination of design water levels, depths, and velocities of flow. These parameters, in turn, are required for

(i) delineation of floodplain boundaries for locating valves and establishing pipe weighting requirements.

(ii) estimating local, river bed scour levels.

(iii) establishing the height of river training structures.

(iv) assessing the stability of bed paving materials, and designing stable riprap protection.

It is the authors' opinion that design hydrology has been over emphasized on some of the recent northern pipeline projects. Determination of the key design parameters - bed scour levels and bank migration limits - is usually quite insensitive to the level of the design flood that is selected (2). Yet on several of these projects there has been heated controversy about acceptable design flood criteria (ie. Standard Project Flood, 1:100 year, 1:50 year), and about the methodology used to estimate design floods. The authors believe that there should be a shift in emphasis to obtain better information and improve methodology in those critical areas where design is based primarily on the judgement and experience of the engineer. These include estimating the potential for river bed scour and degradation, and lateral shifting of the channel banks.

8. RIVER ICE PROCESSES

Some of the concerns that have been identified with respect to crossing northern rivers relate to the effect of river ice processes on bed and bank scour.

During breakup, moving ice floes can cause scraping and gouging of the river bed and banks, and push-up of gravel, cobble and boulder ridges. These ice-shove ridges are a common occurrence along the banks of the larger rivers in the Canadian Arctic.

While direct action of ice on bed and bank materials is dramatic to observe, it is a relatively slow process in comparison to hydraulic erosion by flood flows. The authors know of no circumstance where removal of bed or bank materials by direct ice action would necessitate more conservative pipe burial than would normally be provided for by considering hydraulic scour and erosion.

River bed scour, caused by constricted flow beneath an ice jam or a frazil ice accumulation, can be an important consideration in sand bed rivers. This process occurs where a jam or frazil accumulation gradually thickens and reduces the flow area beneath the jam. This, in turn, causes upstream water levels to rise, and velocities in the constricted area to increase. Observations have been made of exceptionally deep scour holes at locations where frazil ice accumulations persistently form (ie. below a reservoir or at the downstream end of rapids that remain open through the winter).

For the Canadian Arctic Gas project the question of scour caused by a break-up jam on the Lower Mackenzie River was investigated theoretically (4) and a method was derived for predicting maximum scour. Unfortunately, there is a lack of field or experimental data to verify these predictions. From a practical standpoint, the designer need be concerned with ice-jam-induced scour only on those rivers where the following conditions prevail:

(i) River bed comprised of fine grained sediments.

(ii) Evidence of severe ice jamming during break-up.

(iii) Evidence of recurring frazil ice accumulations.

On northern rivers the most severe high water levels can occur as a result of ice jamming during break-up when the corresponding discharge may be substantially lower than an extreme flood flow. This means that design water levels (for locating valves, pump stations, etc.) should be based on field information about historic high water levels and not just on the levels corresponding to open water floods.

9. RIVER ICINGS

River icings are accumulations of surface ice that progressively grow and thicken as a result of overflooding due to seepage of water to the surface of the icing. They are common occurances in the shallow, braided-gravel rivers of the arctic, where the normal channel ice cover freezes into the bed thereby restricting flow to the underlying aquifer. The source of water is flow in the sub-bed aquifer that breaks through to the surface, or seepage from a spring located in the channel banks. River icings can grow to thicknesses of ten feet or more, and can extend across the entire width of the active channel area and into adjacent floodplains.

At locations where icings persistently form. dramatic changes in channel configuration can occur each spring as new flow channels erode down through the icing or where flow is diverted into secondary channels on the floodplain. This should be taken into account when assessing the susceptability of river banks to erosive attack, and when determining whether secondary flow channels are likely to develop and enlarge.

Experience on the TAPS project showed that icings can affect the performance of river training structures, particularly spur dikes, by blocking drainage, and causing overtopping and possible failure of the dike embankment. A solution to this problem is to base design elevations of the training structures on maximum ice levels rather than on design open-water levels.

A separate concern that has been identified for a chilled-gas-pipeline is the possibility that the freeze-bulb will induce an icing in a buried crossing of a shallow alluvial aquifer. This, in turn, could prevent sub-bed flows from supplying fish overwintering pools located downstream. Additional work is required to determine both the significance of this concern at specific locations and the practicality of proposed solutions such as heat tracing.

10. CONCLUSIONS

Today, both the pipeline industry and associated regulatory agencies recognize that there is a need for specialist river engineering and hydrologic input to the design of buried river crossings and pipelines located within river floodplain areas. To a large extent, this recognition has been brought about by the emphasis that has been focused on these disciplines on recent northern pipeline projects.

Most of the current, state-of-the-art technology for

the design of river crossings is applicable to rivers located anywhere, and not just in cold climates. A number of design implications are, however, unique to arctic and sub-arctic rivers. These include:

(i) The occurance of glacier outburst floods.

(ii) River ice processes, and their effect on flood levels and river bed scour.

(iii) Impact of a hot-oil-pipeline on permafrost banks and their ability to resist erosion.

(iv) The natural occurance of river bed icings, their impact on channel shifting and on the performance of river training structures.

(v) Impact of a chilled gas pipeline on winter flow that, on certain northern rivers, is conveyed entirely through the sub-bed aquifer.

Current river engineering technology for the design of buried pipeline crossings places heavy reliance on the experience, judgement and skill of the specialist. There is clearly a need to improve this technology particularly in critical areas such as estimating river bed scour and, predicting the potential for channel shifting and bank erosion.

APPENDIX I - REFERENCES

1. Alyeska Summary Reports, "River and Floodplain Design Criteria for the Trans Alaska Pipeline," 1974.

2. Foothills Pipelines (South Yukon) Ltd., "Assessment of South Yukon Flood Hydrology," Report prepared by C. R. Neill, Northwest Hydraulic Consultants Ltd., August, 1978.

3. Indian Roads Congress, "Standard Specifications and Code of Practice for Bridges," 1966.

4. Mercer, A. G., and Cooper R. H., "River Bed Scour related to growth of a Major Ice Jam," Proc. 3rd National Hydrotechnical Conference, Can. Soc. for Civil Eng., Quebec, 1977.

5. Neill, C. R. (Ed.), "Guide to Bridge Hydraulics," Roads and Transportation Association of Canada, University of Toronto Press, Toronto and Buffalo, 1973.

MATHEMATICAL MODELING OF SCOUR IN STREAMS

David R. Dawdy,*A.M., ASCE

Abstract

Whenever a pipeline crosses a stream, and a decision is to be made concerning burial and burial depth, an estimate of possible depth of scour must be made. Historically, such estimates have been based on static equations, such as a regime scour depth or a scour formula. Today there is considerable interest in hydrodynamic scour models based on the equations for open channel flow (1,3,4,11,19). A recent report of a committee of the National Academy of Sciences-National Research Council (17) reviewed the state-of-the-art of hydrodynamic scour models in the U.S.A. That report is summarized, shortcomings and needed improvements of such models are discussed, and suggestions for future needs in the development of such models are presented.

Classification of Hydrodynamic Scour Models

The various hydrodynamic computer models for estimating scour combine an equation of flow of water with a sediment discharge equation and a sediment continuity equation. The resulting sets of equations may be categorized according to several criteria. The first criterion is whether the flow model used is based upon steady state or unsteady state conditions. A second criterion is based upon the choice of the sediment transport equation. A third criterion is based upon the manner in which armoring of the stream bed is modeled, if at all. In addition, the method used for the handling of bank erosion or bank migration is of interest, although those processes are seldom included in the models available.

Steady Flow Models

Steady flow models combine flow continuity with flow energy to obtain a step-backwater model for the stage of thewater discharge. A sediment discharge equation is then applied at each cross-section in a downstream direction, and some means is used for estimating total scour and allocating that scour across the cross-section. HEC-6 is the standard program of this type (11), but several others exist, such as HEC2SR(17), KUWASER(19), and FLUVIAL-3 (3,4).

*Consulting Hydrologist, 3055 23rd Avenue, San Francisco, Calif. 94132

Each of these four models is similar in their method of flow routing, but they differ in other criteria. They differ in their sediment transport functions. HEC-6 gives a choice of Laursen (13), Toffaleti (20,21), or an empirical relation. HEC2SR uses a combination of Meyer-Peter-Mueller (15) and Einstein (7). KUWASER uses an empirical function based on the work of Colby (5,6). FLUVIAL-3 uses duBoys (1) for bed load and Graf (10) for bed material load (suspended load). Thus each of the steady state models has a different sediment transport function from each of the others. There are no criteria presented by the builders of any of the models as to how the particular functions were chosen. HEC-6 and HEC2SR include bed-armoring algorithms. KUWASER and FLUVIAL-3 do not, and therefore probably do not model very well gravel bed streams with a strong tendency for armoring. KUWASER is usable only for sand-bed streams. FLUVIAL-3 has a bed-material sorting algorithm which can induce a change of the bed material somewhat similar to armoring.

Bed Armoring and Scour Distribution for Steady-state Models

HEC-6 and HEC2SR differ in their armoring mechanism. HEC-6 computes an active layer for sediment transport. When all possible material is removed from the active layer, an armored layer exists, and no further scour occurs. With changing discharge, the stability of the armor layer is tested using Gessler's stability criterion (8,9). HEC2SR uses Shields' criterion (19) to determine a non-moving size, and a layer two grain diameters thick of the smallest non-moving particle armors the bed. The results of these two algorithms are entirely different, and there is no basis at this time for choice between them.

The means for allocating total scour across the cross-section determines how the section properties change during an event. The steady-state models are different in their approach to this problem. HEC-6 distributes scour and fill uniformly across the channel (actually, across the active portion of the channel, set by the modeler.) HEC2SR and KUWASER distribute scour and fill proportionally to relative conveyance in the cross-section. FLUVIAL-3 scours as HEC2SR, but fills in horizontal layers, leading to the development of a more or less rectangular cross-section after several episodes of scour and fill.

Unsteady Flow Models

The unsteady flow models are unsteady in the water discharge. They substitute conservation of momentum for conservation of energy to derive the water flow equations. Typical models are UUWSR, FLUVIAL-11, and SED-4H, all of which were reported upon in the NAS-NRC report (17), and a model by Brownlie (2), which was not. The latter model is included for comparison because it is the only fully coupled model.

As was the case with the steady-state models, the flow routing equations are generally similar. However, they differ in how they treat the energy slope. Also as with the steady-state models, choice of a sediment transport equation differs among the models. UUWSR is the same as KUWASER, with an empirical formula based on the work of Colby, and is thus most appropriate for use on sand-bed channel streams. FLUVIAL-11 is the same as FLUVIAL-3. SED-4H derives a function close to Einstein's bed-load function (1). SED-4H also assumes that suspended sediment particles are convected at the same velocity as the water. Therefore, it is most appropriate for use on sand-bed channels. Brownlie states that his model is for sand-bed channels, and he uses an empirical sediment transport equation he derived using data from such streams.

Distribution of Scour for Unsteady Flow Models

The distribution of scour and fill across the cross-section varies for the unsteady as for the steady-state models. UUWSR distributes by relative conveyance as KUWASER does. FLUVIAL-11 follows FLUVIAL-3, and distributes the scour and fill differently for scouring and for filling conditions. SED-4H and Brownlie's model do not have a description of their scour distribution algorithms published that is known, and so they cannot be compared to the rest of the models.

Treatment of Side Scour

The FLUVIAL models are the only ones discussed so far which handle side or bank scour. However, the bank scour algorithm uses the value for local energy slope directly. Such computed values of local energy slope have been known to be considerably in error. All errors of approximation in the water routing procedure are included in the computed local energy slope. The FLUVIAL programs do average over adjacent values, so that part of the instability in computed local slope values is overcome, but any use of such values makes any otherwise valid algorithm suspect. Slope is used in the sediment transport formulae for several models in the form of shear stress. This can be determined indirectly, as in HEC2SR. That model computes shear stress from resistance and depth, so that the error is on the order of that of depth. If shear stress is computed from local energy slope and used in the computation of sediment transport, the estimates of sediment transport can oscillate widely.

Other approaches to bank scour exist which are not included in present versions of scour models. Parker(12,18) uses stream curvature to determine a cross-stream velocity component. That value is then used to simulate long term meander migration under steady-state, bankfull conditions. Whether that approach could be used in event modeling (say, of a design storm), or of long term, variable discharges is not known, nor is it known whether such a use would be consistent with the assumptions of the model. Nanson and Hickin (16) present data which seem to validate the approach of Parker. Leopold (14) presents field data concerning secondary currents and the mechanism of meander development.

Stability Criteria

Only two of the models discussed mention problems of stability, but such problems exist for all the models. HEC-6 discusses the problem of stability, and suggests methods for testing stability of computations. Following their analysis, one can conclude that detailed geometric data may be useable for design storm analysis. Long term simulation requires long reaches, and can yield only long term average response. Brownlie mentions that his coupled, unsteady model is even stricter on stability grounds. A 100 meter reach length with a two meter flow depth requires a time step of 10 seconds. That may be why most applications programs are not fully coupled. The problem dictates the model, and the problem plus the model determine (or strongly influence) the data needs. Also, optimal data sets for step-backwater may not be optimal (or even adequate) for the prediction of sediment transport and resultant scour and fill.

Conclusions Concerning Future Model Needs

The NAS-NRC study concluded that "None of the movable-bed models evaluated was found to yield wholly satisfactory results." In order to improve the performance of such models, several processes must be better modeled. First, sediment transport functions must be better understood, and better criteria for their selection should be presented by the modelers. Second, the bed armoring process must be better understood, and the theoretical approaches used should be verified, where possible. Perhaps more field and laboratory data are needed on the processes of bed armoring. Bank erosion is an important problem, and it should be included in models for event modeling as well as those for long term response, such as meander migration. Stability problems should be discussed and presented in a general framework. Finally, the boundary conditions (stage down-stream, sediment input upstream) have a strong influence on model results. Their determination is an important part of every model application, although they are not a part of the model. Criteria for choice of boundary conditions should be given more emphasis by the modelers in the descriptions of their models.

Appendix.--References

1. Ariathurai, R., and Smith, D.J., "Flood Routing with Movable Bed", Resource Management Associates, March 1982.

2. Brownlie, W.R., "Unsteady Sediment Transport Modeling", paper presented at ASCE Water Forum '81,San Francisco,Calif.

3. Chang, H.H.,"Mathematical Model for Erodible Channels", Jour. Hydr. Div. ASCE,Vol 108,No. HY5,May 1982.

4. Chang, H.H., and Hill, J.C.,"Computer Modeling of Erodible Flood Channels and Deltas",Jour. Hydr. Div., ASCE, Vol.102, No.HY10, October 1976.

5. Colby, B.R.,"Relationship of Unmeasured Sediment Discharge to Mean Velocity", Trans. Amer.Geophys.Union, Vol.38, No.5,October 1957.

6. Colby, B.R.,"Discharge of Sands and Mean-Velocity Relationships in Sand-Bed Streams",U. S. Geol. Survey Prof.Paper 462-A, 1964.

7. Einstein, H.A.,"The Bed Load Function for Sediment Transportation in Open Channel Flows", U.S. Dept. Agric., Tech.Bull.No.1026,1950.

8. Gessler, J.,"Beginning and Ceasing of Sediment Motion", River Mechanics,Vol.1, Colorado State Univ.,Ft.Collins,1971.

9. Gessler, J.,"Self-Stabilization Tendencies of Alluvial Channels",Jour. Waterways and Harbors Div., ASCE, Vol.96, No.WW2, 1979.

10. Graf, W.H.,"Hydraulics of Sediment Transport", McGraw-Hill Book Co., New York, N.Y., 1971.

11. Hydrologic Engineering Center,"HEC-6 Scour and Deposition in Rivers and Reservoirs, Users Manual",US Army Corps of Engineers,March 1977.

12. Ikeda, Syunsuke, Parker,Gary, and Sawai,Kenji,"Bend Theory of River Meanders. Part 1.Linear Development",Jour. Fluid Mechanics,Vol.112,pp.363-377,1981.

13. Laursen,E.M.,"The Total Sediment Load of Streams", Jour. Hydr. Div. ASCE, Vol.54, February 1958.

14. Leopold, Luna,"Water Surface Topography in River Channels and Implications for Meander Development", in Gravel-Bed Rivers, John Wiley, 1982.

15. Meyer-Peter, E., and Mueller, R.,"Formulas for Bed Load Transport",Proc. Third Meeting IAHR,1948.

16. Nanson, G.C., and Hickin, E.J.,"Channel Migration and Incision on the Beatton River", Jour. Hydr. Engr., Vol.109, No.3, pp.327-337, March 1983.

17. National Research Council, Advisory Board on the Built Environment,"An Evaluation of Flood-Level Prediction Using Alluvial-River Models", Committee on Hydrodynamic Computer Models for Flood Insurance Studies, 1983.

18. Parker,Gary, Sawai,Kenji, and Ikeda,Syunsuke, "Bend Theory of River Meanders.Part 2.Nonlinear Deformation of Finite Amplitude Bends", Jour. Fluid Mechanics, Vol.115, pp.303-314, 1982.

19. Simons,D.B., Li,R.M., and Brown,G.O., "Sedimentation Study of the Yazoo River Basin-Users Manual for Program KUWASER",Dept.Civil Engr., Colo.State Univ., Ft. Collins, Rept.No.CER79-80DBS-R,1979.

20. Toffaleti,F.B.,"A Procedure for Computation of the Total River Sand Discharge and Detailed Distribution, Bed to Surface", Tech. Rept. No.5, Comm. on Channel Stabilization, U.S.Army Corps of Engineers, November 1968.

21. Toffaleti, F.B., "Definitive Computation of Sand Discharge in Rivers",Jour. Hydr. Div., ASCE, Vol.95, No.HY1, Jan.1969.

Arctic Pipeline River Crossings - Design Trends and Lessons Learned

Wim M. Veldman, M. ASCE*

ABSTRACT

Recent major Arctic pipeline projects have resulted in extensive research into the hydrologic, hydraulic and river engineering behavior of streams. For the design of pipeline crossings, sophisticated mathematical models have been developed to predict scour and bank migration. Although applicable in some instances, they must be combined with sound Arctic knowledge and qualitative judgments by experienced engineers. Permafrost, aufeis and snow impose dramatically different design conditions than those encountered in nonArctic environments.

INTRODUCTION

Completed Arctic projects such as the 800 mile long trans-Alaska oil pipeline and the Prudhoe Bay gathering facilities have resulted in unique design solutions to meet design, environmental and regulatory constraints. Pile-supported elevated line, insulated buried segments, animal crossings - both buried and elevated - and long segments of shallow floodplain burial protected by river training structures were some of the state-of-the-art design concepts developed for the Alyeska project. Projects presently in the planning or final design stages such as the Northwest Alaska (NWA), Polar Gas and Dempster Lateral gas pipelines and the Norman Wells and Mackenzie Valley oil pipelines have all required extensive innovative designs and research to meet the unique Arctic conditions.

The work for these projects has significantly improved our understanding of arctic hydrology, river engineering and hydraulics - essential components in the design of river crossings. Whereas in some instances the Arctic design conditions are more severe than in non-Arctic areas, the opposite may also be true when comparing design conditions.

This paper will first briefly review the design considerations for Arctic pipeline river crossings and floodplains, describe alternative crossing modes, assess present-day design trends, and with this background, discuss the lessons learned relative to scour, bank stability and water levels and floods.

* President, Hydrocon Engineering, Calgary and Fairbanks.
220-4014 Macleod Trail, S.E., Calgary, Alberta, Canada. T2G 2R7

GENERAL DESIGN CONSIDERATIONS

Since the intent of this paper is not a detailed discussion of the design criteria, considerations and approaches that can be used for river crossings, they are discussed only briefly herein to provide the necessary background for subsequent sections.

From a river engineering viewpoint, the prime objective in the design of buried pipeline river crossings is to minimize the risk of exposure due to bed scour and bank migration. The river engineer must therefore assess the depth of burial and the location of the sagbends.

The depth of burial, assuming a non-exposure criteria is used, is set at the lowest level to which any portion of the riverbed may be scoured. This scouring may be due to local effects such as strong secondary currents at channel bends, turbulent eddies produced at the confluence of subchannels, progressive channel profile degradation or ice-induced scour. The severity of the scour at any point along a river is dependent on the flow conditions which produce the greatest hydraulic forces acting on the riverbed. For rivers having banks that are too high to be overtopped during extreme flood events, the maximum open-water scour generally occurs during the design flood. However, for rivers with low banks subject to overtopping, maximum scour may occur at a bankfull stage which is considerably less than design flood conditions.

Major factors affecting the pattern and extent of channel switching and erosion are flows, resistance of bank materials to erosion, and the amount of sediment transported by the river. Ice or log jams can also have a major impact. From the past behavior of a river - as charted from comparative aerial photographs - its most probable future behavior can be predicted within broad limits. Locating sagbends requires a great deal of judgment by experienced river engineers.

For elevated crossings, the major design considerations are design water level, abutment protection and bed scour at in-stream piers. Channel shifting and bank erosion are important in establishing bridge lengths and in the design of protective structures to align and confine the flow within the waterway.

ALTERNATIVE CONSTRUCTION MODES

The alternative crossing modes are:

- deep burial across the entire main channel and floodplain,
- deep burial in the main channel portion flanked by shallow burial in the floodplain areas protected by river training structures,
- elevated crossings of various types across all or part of the river channel and floodplain. River training structures will generally be required if the flow is to be confined through the waterway opening.

Where the pipeline is parallel to and in the floodplain or main channel (as opposed to a crossing), it can be either deep buried or in a shallow buried mode protected by river training structures. An elevated floodplain portion with or without river training structures is another option.

The choice of buried versus elevated crossings - the latter first used extensively on the Alyeska project - depends on technical, economic, environmental and safety considerations. In the case of a hot-oil pipeline, an elevated mode is required for thaw-unstable soils.

Excessive burial depths due to a highly mobile riverbed may favor an elevated mode over a buried one. Design uncertainties related to flood sizes (from a glacier-dammed lake as an example) or scour (the impact of a potential channel cutoff) may again favor an elevated crossing. Schedule restrictions due to environmental reasons may necessitate an elevated crossing which requires less in-stream activity than a buried one. At streams located in the bottom of narrow deep rock gorges, a bridge or pipe span would substantially reduce grading requirements compared to that for a buried crossing. From a safety viewpoint, a buried mode is the preferred method for a gas pipeline unless the other factors indicated above dictate an elevated mode. At smaller crossings where in fact an elevated mode may be more economic, a buried mode might be preferred to permit continuous pipe laying activities using a single spread.

Elevated crossing types are suspension or, girder or pile-supported sections. The pipeline and roadway can also be combined into a single structure such as the 2000 foot long Yukon River bridge in Alaska.

TRENDS IN DESIGN APPROACH

River crossings have been typically designed and constructed using various rules of thumb and previous experience. Although this approach may be successful if prudently applied in regions with numerous operating pipelines, an analysis of over 1500 reported pipeline failures in the United States between 1950 and 1974, (O'Donnell, 1977) concluded that although river crossings constituted less than 0.5% of the total onshore pipeline mileage, 3.4% of the failures occurred at crossings. A crossing is thus seven times more likely to fail than mile-by-mile construction. Sixty-three percent of the crossing failures were attributed to floods, scour and washouts. Whereas a mile-by-mile failure can usually be repaired quite readily, a large river crossing failure may be very difficult and time-consuming to repair particularly since it may occur during high velocity and depth conditions.

The environmental and security-of-supply concerns for recent major projects have resulted in design stipulations requiring non-exposure of the pipeline during extreme flood events for the life of the project. The federal and state review bodies established for projects such as TAPS and NWA have necessitated detailed hydrologic and river engineering studies, analysis and documentation - a first in the pipeline industry. This has resulted in the development of novel and extensive techniques, approaches and mathematical models to quantify the hydrologic, hydraulic and river engineering conditions.

The design pendulum for recent major Arctic projects appears to be swinging too far in the direction of trying to develop elegant mathematical models. Designers are expounding various models to predict scour and even bank migration and in their attempts to find elegant mathematical solutions - and thus "improve" on methodologies such as the Blench regime method - overlook the fact that many of the river engineering processes cannot be quantified. Furthermore the mathematical models often require the same type of "fudge factors" and assumptions as the oft-criticized empirical methods. This desire for quantitative and precise solutions may, to a degree, be attributed to governmental review requirements. There is no question that quantitative analyses are easier and quicker to review whereas qualitative ones, depending on the experience of the designer and reviewer, may at times lead to long debates, the end result of which is a compromise which satisfies neither party. This however should be accepted as a fact of life with river engineering. It is not a pure science but mainly an art.

This desire or need to produce quantitative techniques and results can also result in a disproportionate amount of the designer's and reviewer's time spent on some of the less important design aspects from an integrity viewpoint. For example, computing the precise magnitude of the design flood for a buried crossing of a braided river or its burial depth - two factors which can be quantified to a degree - is less important from an integrity viewpoint than assessing the potential for channel switching or subchannel development - factors which cannot be quantified. Also an inherent danger in mathematical models is to analyze only the immediate crossing area. With qualitative approaches, the behavior of the river over a much longer reach can be assessed. Nanson and Hickin (1983) have identified the shortcomings of predicting migration rates for specific areas from data at individual bends.

LESSONS LEARNED

Detailed field investigations and office analysis for major Arctic projects have revealed some interesting and surprising comparisons between the behavior of Arctic and temperate rivers and the impact of this on the design of river crossings. These comparisons as well as some comments on the methodologies used by various designers are presented in the following sections. This critique is by no means exhaustive - this has already been done in part by researchers such as Graham (1979, 1980, 1982) - it is merely intended to highlight the applicability of some of the techniques as applied to Arctic conditions.

Scour

The following has been learned from observations of Arctic gravel bed rivers:

- ° little or no bed movement occurs during spring breakup when the bed is frozen and/or covered with ground-fast ice (aufeis).

- ° Arctic Coastal Plain streams, whose dominant hydrologic process is spring breakup when the riverbed and banks are frozen, are characterized by very minor bed changes.

- ° streams originating in the Arctic Foothills or Brooks Range or southern climates, can experience major summer floods and can therefore exhibit rapid bed changes. This is especially true as a result of a late summer storm when the thalweg, sub-channels and floodplain area are thawed to their maximum depth.

- ° overbank flow during spring breakup when the ground is still frozen results in little or no development, movement or enlargement of channels. This is particularly true for streams whose peak flows occur during breakup.

- ° unusual bedforms such as shown on Photo 1 may exist on Arctic streams. Similar bedforms have been shown for the Babbage River in the Yukon Territory, Canada by Lewis and McDonald (1973). In the photo shown, the top of the bare gravel bedforms, up to 4.5 feet high, were only one foot below the elevation of the main channel bank. A comparison of airphotos (Figure 1) indicates very few changes over 30 years even in these bare gravel bar areas. This could be due to the shallow depth of flow over the bars or their frozen nature. Flow under and over aufeis ridges, could perhaps have produced unusual flow patterns and resultant bedforms. Lewis and McDonald do not postulate on the formative factors for these bars, however they do indicate that the depth of the depressions is governed by the frost table.

In view of the above, the difficulty of predicting scour in Arctic gravel-bed rivers is obvious and drastically different approaches by various researchers are understandable. Graham (1979) has presented a detailed comparison of the Blench regime method versus his tractive force (rational) theory. Other techniques are the U.S. Army Corps of Engineers HEC-6 and the Colorado State University water-sediment routing model (Shen, 1979). Some comments are:

- ° all the techniques, to be properly applied, require a great deal of judgment by experienced river engineers.

- ° the local scour computed from multiplication or Z-factors (the latter as used with the Blench regime method) is much more significant than general scour computed from regime equations

PHOTO 1 - Kadleroshilik River, Alaska. Bedforms, maximum height of 4.5 feet and top at 1.5 feet below main channel bank, during low flow summer conditions. See Figure 1 for aerial photo.

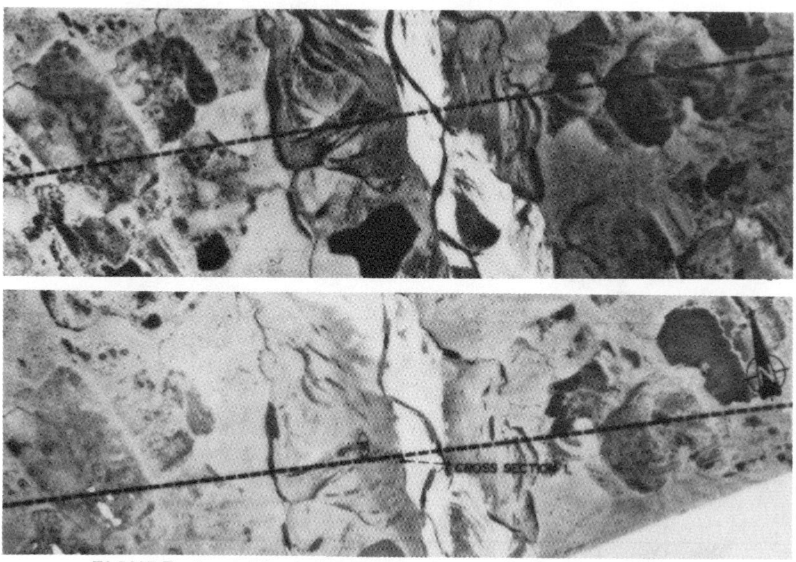

FIGURE 1 - Airphoto comparison of Kadleroshilik River on the Alaska north slope. Even subchannels shown in foreground of Photo 1 have changed little over 30 years.

or other techniques. These Z or shape factors, although often attributed only to the Blench method, have been documented in the field and laboratory by numerous researchers. The range of Z values typically presented for say confluences or bends can be refined somewhat by plotting the values versus angles and surface width/outer radius respectively. This has been done by various researchers (Nwachukwu, 1973) and is the subject of a research project by Hydrocon Engineering (1983). River surveys should include the delineation of local scour holes.

- results from mathematical models are very dependent on the selection of proper input data and assumptions. Much of this data such as incoming sediment loads, is lacking for Arctic conditions. Simply omitting a cross-section at a subchannel confluence - the critical scour section for a braided stream - could lead to erroneous and unsafe results. The models, particularly the CSU and HEC-6 are probably more suited to evaluating long term aggradation and degradation than local scour. Using HEC-6 for multi-channelled braided streams is questionable.

- if a mathematical model is used, its results should be compared to that obtained from Z-factors, with or without linking them to the Blench equations.

Bank Migration

The following comments can be made based on field observations of Arctic conditions:

- based on thirty years of photo comparisons for the Alaskan north slope, permafrost banks and overflow channels generally exhibit lower rates of migration and changes than similar non-permafrost conditions. This has also been documented by Cooper (1973) in his studies of Arctic rivers in Canada.

- this stability can be attributed to the lack of debris and frozen subchannel beds and is particularly true for streams where the major hydrologic process is in the spring when the bed and banks are frozen.

- groundfast (ice) aufeis can cause local severe bank erosion. If the aufeis in the cross-section is not horizontal - as in the case where the aufeis is generated by tributaries entering the main channel from one side - the total initial breakup flow is along one bank. Heat absorption by bare channel banks or river training structures can further result in initial flow concentration along one bank.

- permanent channel switching does not generally occur during the breakup season unless the high flows are of sufficient duration to thaw the overbank areas. Streams whose major floods occur after breakup (rainfall induced) will be less stable than Arctic coastal streams which have little or no flow in the open water period following breakup.

- aufeis formations which persists well into the summer (Photo 2) will reduce the macro channel changes and switches although it could promote increased local bank erosion.

PHOTO 2 - Canning River, Alaska.
Remnant aufeis in mid-July affecting channel formations and bank erosion.

- minor streams in the lake-strewn Arctic Coastal Plain may "capture" minor lakes resulting in a variable width stream having a beaded appearance.

Estimating bank migration rates and thus sagbend setbacks for buried pipelines requires qualitative judgments by experienced river engineers. Time lapsed aerial photography should be the basis of this judgment. However as indicated by Nanson and Hickin (1983) in their detailed analysis of the Beatton River in northern Canada, "..... prediction of migration rates for individual bends based on short term measurements (20 - 30 years) are highly suspect". Mathematical models to predict lateral migration rates during single flood events are questionable in their applicability to Arctic rivers at this time. Certainly they cannot be used to establish sagbend locations in isolation from a qualitative review of the behavior of the river over a considerable length and without the benefit of considerable Arctic river engineering experience.

Water Levels

For elevated crossings of Arctic streams, the critical design condition may be moderate flows over snow or aufeis rather than the design open water flood level. The latter, which is usually critical for buried crossings, is computed in a manner similar to non-arctic conditions.

The presence and extent of aufeis at a crossing location should be determined from several years of field observations. Computing a site specific "design" or maximum aufeis level is not possible at this time. Although the factors producing the optimum conditions for the maximum generation of aufeis (cold temperature and little snow amongst others) are well known, predicting site-specific aufeis levels is guesswork at best. Take the example of a tributary stream or impermeable section of the main stream riverbed - conditions which can generate aufeis - located one mile upstream of a proposed bridge site. Observations during an extreme winter indicate maximum aufeis levels one-half mile upstream of the bridge with little ice at the site itself. Observations the following year - a warm winter with heavy snowfall - show extreme aufeis levels at the bridge - an apparent violation of well-accepted principles. Why? The tributary or impermeable riverbed section caused the flow to surface in the main stream one mile upstream of the bridge. During the cold winter, the overflow froze almost immediately whereas in the warm winter, freezing did not occur until the overflow had reached the bridge site.

The bridge crossing, by locally reducing the snow cover on the river, can create a block point causing the flow to surface and thus increasing aufeis levels. This is particularly true for a low girder type bridge or where the road and pipeline bridges are proximate. Selecting the magnitude of flow over the aufeis before it downcuts or lifts is the second uncertainty facing the bridge designer. Experience on the Sagavanirktok River at Prudhoe Bay, Alaska shows that if flows increase dramatically over groundfast ice, the ice will generally lift in a deteriorated state in about 5 days. If flows are moderate, downcutting (Photo 2) rather than lifting occurs. The design criteria for setting the low steel elevation could be the maximum recorded or estimated breakup flow over the historic or estimated maximum aufeis level. Terraces or floodplains may of course limit the maximum water or ice level that can be achieved.

Various techniques have been tried in the Arctic to reduce ice levels or to promote the earlier disintegration of the ice at bridges. These include dusting and cutting of the ice. With the latter method, the aim is to propogate the early deterioration and lifting of the ice. If the ice lifts but does not move out of the waterway opening (Photo 3), - due to insufficient flows or grounding on the ice downstream - it could in fact result in increased water levels and local scour at bridge piers.

PHOTO 3 - West Channel Sagavanirktok River, Prudhoe Bay, Alaska. Ice was cut with a "Ditch Witch" before commencement of breakup. In deep part of channel where it was not grounded, it lifted after several days of overflow. Flows and/or elevations of downstream ice prevented movement of ice out of the waterway opening. This could result in increased local water levels thus defeating the purpose of the cutting program.

Aufeis, compared to thermal ice covers on large continuously flowing rivers, imposes significantly lower forces on bridges during breakup. The ice which lifts after 5 days of overflow is badly deteriorated with the larger pieces generally becoming grounded in the shallow depths of the braided streams. Ice jams on braided streams are uncommon as the blockage of a channel simply results in the flow bypassing the jam without a significant increase in head.

Wind-blown snow on the barren and flat Arctic north slope can block a waterway opening particularly if multiple bridges and a vertical roadway embankment are proximate. This snow, as shown in Photos 4 and 5 is surprisingly erosion resistant during the early stages of breakup and unless cleared at some sites, could produce very high - albeit short duration - water levels. The use of snow fences at sites where snow is a problem is worthy of study.

The above considerations complicate the application of commonly-used backwater programs such as the Corps of Engineers HEC-2 model. The model is a useful tool for design floods during open water conditions and to assess the impact of existing facilities on proposed

PHOTO 4 - West Channel Sagavanirktok River, Prudhoe Bay, Alaska. View of paths cleared in snow with D-9 cat.

PHOTO 5 - Close-up of the above. Even under high velocities (note headloss in path in foreground) the flow areas did not significantly widen.

bridge crossings. With regard to the location of a roadway bridge relative to a pipeline bridge, the former should generally be located upstream. Economic considerations will affect the length of a roadway bridge more significantly than a pile-supported or even girder type pipeline bridge. Floodplain causeway sections of the roadway can be designed to be overtopped thus reducing the design water level for the bridge. This overtopping, assuming it is limited and of short duration, would have little or no impact on a pile-supported elevated pipeline downstream. Upstream of the roadway beyond the main channel (during normal flows) or in the low-velocity backwater area (during high flows), a pile-supported structure can again be used for the pipeline.

River Training Structures

Arctic conditions affecting the design and construction of river training structures, - often required at crossings and for shallow-buried pipe - are as described by Veldman and Yaremko (1978). The lessons learned may be summarized as follows:

- river training structure may be an attractive alternative to deep burial for pipelines for reasons of economics, scheduling and construction.

- river training structures may be mandatory to prevent the development of subchannels along a cleared right-of-way or to prevent excavation into and/or grading of an ice-rich bank.

- break-up flows over aufeis are often the critical design elevation for the structures whereas open water design floods will govern armor sizes and depth of aprons.

- where structures block subchannels, the crest elevation should be a minimum of 5 feet above the floodplain elevation to prevent overtopping due to impoundment of the subchannels. This condition, rather than that computed for the main channel using the HEC-2 backwater or similar model (with or without aufeis) may be the critical design condition particularly for steep braided streams.

- the structure armoring must be flexible to allow for structure fill or foundation settlement. Structures constructed under winter conditions are particularly susceptible to settlement even with every attempt made to maintain compaction specifications and limiting the use of frozen fill. Summer compaction and topping off is required for structures built in winter.

- rock riprap is the easiest material to place and repair under Arctic conditions. Alternatives such as gabions or articulated concrete mats, although economic if riprap haul distances are excessive, are more difficult to place and repair under Arctic conditions. Articulated concrete mats (up to 8 feet by 60 feet) could be effectively used for emergency repairs under flowing water conditions. They can also be used for temporary structures and then re-used elsewhere.

CONCLUSIONS

- Arctic pipeline river crossings require unique design solutions.

- The present trend by designers and reviewers on major projects is to quantify, through the use of various mathematical models, the scour and bank migration processes.

- Mathematical models cannot be used to determine site specific bank migration rates and from these, sagbend setbacks. Qualitative assessments by experienced river engineers analyzing the overall behavior of the river are required.

- The "fudge factors" required for these models are supported by less hard Arctic data than the oft-criticized empirical methods such as the Blench regime equations.

- The magnitude of general degradation or scour is relatively minor compared to local scour at bends and confluences - usually the critical design conditions for buried pipelines. The scour multiplication or shape factor (known as the Z-factor in the Blench method) has been extensively studied in the field and laboratory. By relating this factor to the specific characteristics of the bends or confluences, the range of approximate values is decreased and thus the quality of the answers improved.

- Bank migration and channel switching is less for Arctic rivers particularly those for which the main hydrologic event is spring breakup when the bed and banks are frozen. Aufeis can lead to local severe bank attack but may also if prolonged, minimize channel changes even well into the open water period.

- Flow over aufeis or snow is often, rather than the open water design flood, the critical water level condition for elevated crossings.

- River training structures can be effectively and economically used as an alternative to deep burial or to shorten the length of elevated crossings. The armoring must be flexible and easy to repair under Arctic conditions.

REFERENCES

1. Cooper, R.H. and Hollingshead, A.B., May 1973. "River Bank Erosion in Regions of Permafrost". Fluvial Processes and Sedimentation, Proceedings of Hydrology Symposium, University of Alberta, Edmonton.

2. Graham, D.S., January 1979. "Hydrodynamic Criteria for Submarine Pipeline Burial, Regime versus Rational Theory". Pipelines in Adverse Environments, Proceedings of the ASCE Pipeline Division Specialty Conference, New Orleans.

3. Graham, D.S., March 1980. "Pipeline River Crossings: A Design Method". ASCE Transportation Engineering Journal.

4. Graham, D.S., May 1982. "Numerical River Scour Model for Pipeline Crossings". Proceedings of the Annual Conference, Canadian Society for Civil Engineering, Edmonton, Alberta.

5. Hydrocon Engineering (Continental,) Ltd., 1983. Unpublished report for Alberta Environmental Research Trust.

6. Lewis, C.P., and McDonald, B.C., May 1973. "Rivers of the Yukon North Slope". Fluvial Processes and Sedimentation, Proceedings of Hydrology Symposium, University of Alberta, Edmonton.

7. Nanson, G.C. and Hickin, E.J., March 1983. "Channel Migration and Incision on the Beatton River". ASCE Journal of Hydraulic Engineering.

8. Nwachukwu, B.A., May 1973. "Laboratory Study of Scour at Channel Bends". Proceedings of Canadian Hydraulics Conference, University of Alberta, Edmonton.

9. O'Donnell, H.W., October 1977. "Design, Construction and Inspection of Pipeline River and Stream Crossings". ASCE Fall Convention and Exhibit, San Francisco.

10. Shen, H.W., 1979. "Water and Sediment Routing in Rivers". Chapter 10 of "Modelling of Rivers", edited by H.W. Shen, John Wiley and Sons.

11. Veldman, W.M. and Yaremko, E.K., May 1978. "Design and Construction of River Training Structures". Proceedings of the ASCE Conference on Applied Techniques for Cold Environments, Anchorage, Alaska.

Facility Designs in the Arctic for the Kuparuk Pipeline System

Victor Manikian[1], Kenneth J. Nyman[2], U. J. Baskurt[3]

Introduction

Some 64 km (40 mi) west of Prudhoe Bay lies the Kuparuk River oil field, the newest development on the North Slope. The field came on production in December 1981. A 43 km (27 mi) long, aboveground 40.6 cm (16 in.) pipeline carries the Kuparuk crude oil production to the Trans Alaska Pipeline at Pump Station 1. The common carrier pipeline is presently operating at a flow rate of approximately 90,000 barrels per day. The system also includes aboveground pipe supports, six river crossings, workpad/access roads, a pump station, a power plant, and a connection facility at TAPS.

The pipeline is entirely aboveground because of arctic permafrost conditions. All tundra sections are supported 1.5 m (5 ft) above the surface to allow for caribou migration. Pipe supports consist of a carbon steel pipe pile driven in the tundra and a structural steel crossbeam on which the pipe rests. There is a total of 2,644 supports, spaced at 18.3 m (60 ft) intervals. The various piles are 27.3 cm (10.75 in.) diameter on the tundra; 45.7 cm (18 in.) diameter for two pile anchors; 61 cm (24 in.) diameter at secondary river crossings, and 76.2 cm (30 in.) diameter on the Kuparuk River floodplain.

In the traverse between the production facility and Pump Station 1 the pipeline crosses the Kuparuk River. Along the pipeline route a gravel access road is provided for the construction, operation and maintenance of the pipeline system. In addition, a major portion of this road system is designed to accommodate the transport of heavy prefabricated oil production modules, off loaded from barges in Prudhoe Bay to production locations in the Kuparuk field.

[1]Staff Engineer, Civil-Geotechnical, ARCO Alaska, Inc., Anchorage, Alaska 99510

[2]Senior Geotechnical Engineer, ARCO Pipeline Co., Alaskan Division, 515 South Flower Street, Los Angeles, California 90071

[3]Manager, Engineering, ARCO Pipeline Co., Alaskan Division, 515 South Flower Street, Los Angeles, California 90071

A new heavy duty 91 m (300 ft) long bridge, which also crosses the main channel of the Kuparuk River, was designed as part of this road system.

During the construction of this pipeline system unusual designs were developed for this arctic environment. This technical paper presents some of these designs and construction techniques.

Pile Driving and Load Tests in Permafrost for the Pipeline System

Placement of structural foundation support piles on Alaska's North Slope generally has been performed using drilled holes and slurry backfill techniques. The process of drilling, placing, aligning and slurrying piles is time consuming and requires considerable manpower and equipment. A method of accomplishing the same results more quickly, using fewer men and less equipment, and the early success of specially modified driven H-pile structural shapes in permafrost (Crory 1973, Crory 1975, Nottingham 1981, Rooney 1976), prompted ARCO Alaska, Inc. to initiate a series of driving and load tests starting in February 1980.

These field tests were designed to verify the practicality of driving piles in cold permafrost on a production basis using thermally-modified predrilled pilot holes. This technique employs a small drilled pilot hole, generally 15.2 to 21.6 cm (6 to 8.5 in.) in diameter depending on pile diameter, filled with water, heated to a minimum of 66°C (150°F). The pilot hole is filled with the hot water 30 min before driving in frozen granular soils and 60 min before driving in frozen fine-grained soils, in order to raise the surrounding soil temperature and facilitate the pile placement. Thermal modification and pile driving procedures were refined to suit the arctic pile driving. Ultimate adfreeze strengths of these piles were determined by load tests after isolating and eliminating the effects of end bearing and the active layer. In Prudhoe Bay, heaving forces from silty soils in the active layer have controlled the design of lightly-loaded piles (Crory 1965).

Driving tests were conducted in three phases, with a total of 45 piles being driven. In Phase 1, 27 pipe piles were driven to a depth of 4.6 m (15 ft) in frozen silty sand using a Delmag D15 diesel hammer and predrilled, thermally-modified pilot holes. In Phase 2, eight reinforced H-piles were driven to depths ranging from 6.7 to 7.6 m (22 to 25 ft) in frozen silts and sands, using thermally-modified pilot hole procedures similar to those used in Phase 1. Phase 2 also included the installation of two drilled and slurry pipe piles. In Phase 3, ten pipe piles with cast steel tips were driven into frozen river gravelly sands using the Phase 1 procedures.

Short-term pile load tests were performed on seven piles and long-term (creep) tests were performed on five piles. For comparative purposes the two slurried piles were subjected to short- and long-term (creep) load tests under conditions similar to those used for driven piles. All piles were loaded until an apparent failure occurred. In the long-term load tests, the piles were loaded until the load deformation plots indicated that irreversible progressive failure, greater than 0.5-in. (1.3 cm) downward movement, had occurred.

The following paragraphs describe each of the three testing phases and results obtained from them.

Phase 1 - Pile Driving and Load Tests for Pipe Piles in Frozen Ice-Rich Silty Sand Soils.

Pile Driving: The general approach was to establish a method using hot water-filled, drilled pilot holes that would provide the quickest pile driving without affecting pile integrity. Various pile wall thicknesses and tip reinforcing were used, along with variable thaw times for water in the pilot holes.

A decision was made to use a D15 Delmag, single-acting diesel hammer, with a maximum rated energy of 3,750 kg-m (27,100 ft-lb), which turned out to be large for the smaller 27.3 cm (10.75 in.) diameter piles. Driving proved extremely easy, when using the hot water filled pilot holes, if water remained in place for sufficient time. With continuous, uninterrupted driving and proper pilot hole thermal modification, an average driving time of about 5 min was required for the 4.6 m (15 ft) penetration.

Of the 27 piles driven in Phase 1, using the parameters indicated above, 10 piles were driven without deformation. From the results of various methods tried, Table 1 has been prepared to show the relationship of parameters affecting driving piles in permafrost. Based on these test results, installation of driven pipe piles in frozen silty sands requires pilot holes and, prior to driving, a thermal-modification time between 40 and 60 min to prevent tip damage.

Short-Term and Long-Term Vertical Load Tests: Since few adfreeze data exist on piles driven in permafrost and there are virtually no data on piles driven in Prudhoe Bay permafrost conditions, short-term tests were planned to establish starting points for better definition of long-term load and creep rate tests. The purpose of these long-term tests was to add information to the developing data base on design criteria. As experience is gained from long-term sustained pile loading in permafrost, it becomes apparent that in certain designs, creep-related settlements control rather than adfreeze strength. The design of pipeline supports, with large short-term live loads, has adfreeze strength criteria which will limit the creep to a design limit. The results of the load tests are given in Tables 2 and 3.

Phase 2 - Pile Driving and Load Tests for Slurry Pipe and Driven H-Piles in Frozen Ice-Rich Silty Sand Soils.

Little literature is available regarding the strength of driven piles or the comparative strength of driven versus slurry piles in cold permafrost. For this reason, in Phase 2 it was planned to drive two test H-piles 6.7 m (22 ft) below grade, and place two slurry test pipe piles 6.7 m (22 ft) below grade. Six H-piles were also driven 7.6 m (25 ft) below grade for anchor piles. H-piles were structurally designed to resist variable tip loading and localized buckling under driving conditions. Table 1 summarizes the installation records. The results of short-term and long-term load tests are given in Tables 2 and 3.

Phase 3 - Pile Driving and Load Tests for Pipe Piles in Frozen Gravel.

Ten 27.3 cm (10.75 in.) diameter by 1.27 cm (0.5 in.) wall pipe piles were driven using APF 0-14001 tips in frozen gravel and 15.2 cm (6 in.) diameter pilot holes, drilled 0.61 m (2 ft) deeper than the nominal tip embedment.

Concerns had been expressed regarding driving piles in frozen gravelly soils. As a result, these tests were planned primarily to drive pipe piles with cast steel tips to depths of 4.6 m (15 ft) in frozen gravelly materials. The test site was adjacent to the Sagavanirktok River, where typical riverbed soils included frozen sandy gravels. Following driving, load tests were performed to establish data concerning adfreeze strengths in these soils. Tests were initially conducted during the summer, when ground temperatures are generally warmest and adfreeze values are lowest.

A summary of driven piles for Phase 3 is shown in Table 1. The results of short-term and long-term load tests are given in Tables 2 and 3.

Conclusions and Recommendations

The results of these field tests to verify the practicality of diesel pile driving in cold permafrost, on a production basis using thermally-modified, predrilled pilot holes, are encouraging. Subsequent to these tests the use of

TABLE 1 Summary of Driven Piles

Pile No.	Size Dia. & Wall (cm)	Size Dia. & Wall (in.)	Tip Size (cm)	Approx. Pilot Hole Fill Time (min)	Approx. Pilot Hole Thaw Time (min)	Driving Time* (min)	Pile Condition After Driving 0 Sound x Failed	Initial Pilot Hole Water Temp. (°C)	Initial Pilot Hole Water Temp. (°F)	Thaw Remarks Just Prior to Driving
Phase 1										
1	27.3x1.27	10.75x0.500	None	10	60	15	0	54	130	Reheated to 21oC(70°F)
2	27.3x1.27	10.75x0.500	None	10	18	19	0	54	130	No reheat
3	27.3x1.11	10.75x).438	None	10	40	13	0	34	94	Cooled to 4°C(39°F)
4	27.3x0.93	10.75x0.365	1.9(0.75")	10	15	16	0	37	98	No reheat
5	27.3x0.93	10.75x0.365	None	0	0	55	x	-	-	Dry hole
6	27.3x0.93	10.75x0.365	None	10	20	15	x	43	110	No reheat
7	27.3x0.93	10.75x0.365	None	10	15	9	x	40	105	No reheat
8	32.4x1.27	12.75x0.500	None	10	42	9	0	32	90	Cooled to 7°C(45°F)
9	32.4x1.03	12.75x0.406	None	10	12	8	x	49	120	No reheat
10	32.4x1.27	12.75x0.500	None	10	18	24	x	14	58	No reheat
11	32.4x0.95	12.75x0.375	None	10	19	7	x	38	100	No reheat
12	32.4x1.27	12.75x0.500	None	10	17	9	x	27	80	No reheat
13	32.4x1.03	12.75x0.406	None	10	19	88**	x	43	110	No reheat
14	32.4x0.95	12.75x0.375	None	10	20	17	x	43	110	No reheat
15	32.4x1.27	12.75x0.500	None	10	20	12	x	44	112	No reheat
16	32.4x1.27	12.75x0.500	None	10	15	7	x	49	120	No reheat
17	45.7x1.27	18.00x0.500	None	20	78	14	0	43	110 Est.	Cooled to 0°C(32°F)
18	45.7x0.95	18.00x0.375	None	20	30	36	x	43	110 Est.	No reheat
19	32.4x1.27	12.75x0.500	2.54(1.0")	0	70	10	0	18	64	Reheated to 21°C(70°F)
20	32.4x1.03	12.75x0.406	None	0	41	9	0	31	88	Reheated to 17°C(62°F)
21	32.4x1.27	12.75x0.500	None	0	58	11	x	23	74	Reheated to 32°C(90°F)
22	32.4x0.95	12.75x0.375	None	0	53	7	x	23	74	No reheat
23	27.3x1.27	10.75x0.500	None	0	67	4	0	18	64	Reheated to 21°C(70°F)
24	27.3x1.27	10.75x0.500	None	0	43	6	x	14	58	Cooled to 3°C(38°F)
25	27.3x1.27	10.75x0.500	None	0	41	5	x	6	43	No reheat
26	-	-	-	-	-	-	-	-	-	-
27	27.3x0.93	10.75x0.365	1.9	0	19	5	x	-	-	No temp. measurement
28	27.3x1.11	10.75x0.438	1.9	0	20	4	0	-	-	No temp. measurement
Phase 2										
1	HP 10x57		APF 75750	10	35	14	0	82	180	No reheat
2	HP 10x57		APF 75750	10	26	8	0	66	150	No reheat
3	HP 10x57		APF 75750	10	61	12	0	71	160	No reheat
4	HP 10x57		APF 75750	10	60	15	0	93	200	No reheat
5	HP 10x57		APF 75750	10	64	12	0	72	162	No reheat
6	HP 10x57		APF 75750	10	81	25	0	93	200	No reheat
7	Slurry Pile		-	-	-	-	-	-	-	-
8	HP 10x57		APF 75750	10	59	19	0	93	200	No reheat
9	Slurry Pile		-	-	-	-	-	-	-	-
10	HP 10x57		APF 75750	10	50	13	0	88	190	No reheat
Phase 3										
1	27.3x1.27	10.75x0.500	1.9	10	-	13	0	-	-	-
2	27.3x1.27	10.75x0.500	1.9	10	20	11	0	71	160	No reheat
3	27.3x1.27	10.75x0.500	1.9	10	15	12	0	66	150	No reheat
4	27.3x1.27	10.75x0.500	1.9	10	29	12	0	80	175	No reheat
5	27.3x1.27	10.75x0.500	1.9	10	60	11	0	85	186	No reheat
6	27.3x1.27	10.75x0.500	1.9	10	52	15	0	66	150	No reheat
7	27.3x1.27	10.75x0.500	1.9	10	20	10	0	66	150	No reheat
8	27.3x1.27	10.75x0.500	1.9	10	20	11	0	66	150	No reheat
9	27.3x1.27	10.75x0.500	1,9	10	20	7	0	66	150	No reheat
10	27.3x1.27	10.75x0.500	1.9	10	20	8	0	66	150	No reheat

* Driving time includes intermittent hammer firing. Piles 23 through 28 drove without significant interruption.

** Values not reliable due to hammer malfunction.

TABLE 2 Summary of Short-Term Vertical Load Test Results: Phases 1, 2, and 3.

Test Phase	Pile No.	Size Dia. & Wall (cm)	Size Dia. & Wall (in.)	Soil Type	Placement Method	Monitored Soil Temp. at Failure (°C)	Monitored Soil Temp. at Failure (°F)	Maximum Adfreeze Strength (kg/cm²)	Maximum Adfreeze Strength (psi)	One-Day Reheal Test Adfreeze Strength (kg/cm²)	One-Day Reheal Test Adfreeze Strength (psi)	Test Duration (min)
1	2	27.3x1.27	10.75x0.500	Ice-Rich Silty Sand	Driven	-8	18	7.4	105	4.0	57	536+121[3]
1	17	45.7x1.27	18.00x0.500	Massive Ice and Ice-Rich Silty Sand	Driven	-6	21	5.1	72	4.8	68	429+162[3]
1	19	32.4x1.27	12.75x0.500	Ice-Rich Silty Sand	Driven	-8	18	8.2	116[1]		-	275
1	23	27.3x1.27	10.75x0.500	Ice-Rich Silty Sand	Driven	-8	18	9.1	129[1]		-	124
2	2	HP 10x57 w/Angles		Ice-Rich Silty Sand	Driven	-6	22	7.0	100[2]		-	150
2	9	45.7x1.27	18.00x0.500	Ice-Rich Silty Sand	Slurry	-6	21	6.3	89		-	197
2	4	HP 10x57 w/Angles		Ice-Rich Silty Sand	Driven	-5	23	7.9	112[1,2]		-	140
2	7	45.7x1.27	18.00x0.500	Ice-Rich Silty Sand	Slurry	-5	23	6.6	94[1]		-	106
3	4*	27.3x1.27	10.75x0.500	Gravelly Sand	Driven	-3	27	2.0	28		-	2
3	7	27.3x1.27	10.75x0.500	Gravelly Sand	Driven	-6	22	3.7	52		-	90

1 Short-term loading at completion of long-term test.
2 Encompassed perimeter area is considered rather than actual contact area, since creep will tend to occur on that basis.
3 Initial test plus reheal test duration.
* Load test attempt failed 2 min after load application. Pile was partially frozen (2.1 m, or 7 ft, from tip) and ground temperatures may have been higher than initially measured due to thermal modification.

Phase 1 load tests were conducted as downward tests.
Phase 2 & 3 load tests were conducted as uplift tests.

TABLE 3 Summary of Long-Term Vertical Load Test Results: Phases 1, 2, and 3.

Test Phase	Pile No.	Size Dia. & Wall (cm)	Size Dia. & Wall (in.)	Soil Type	Placement Method	Monitored Soil Temp. at Failure (°C)	Monitored Soil Temp. at Failure (°F)	Maximum Adfreeze Strength (kg/cm²)	Maximum Adfreeze Strength (psi)	Test Duration (days)
1	19*	32.4x1.27	12.75x0.500	Ice-Rich Silty Sand	Driven	-8	18	3.4	48[2]	103
1	23	27.3x1.27	10.75x0.500	Ice-Rich Silty Sand	Driven	-8	18	4.0	57[2]	103
2	4	HP 10x57		Ice-Rich Silty Sand	Driven	-5	23	2.1	31[1,2]	49
2	7	45.7x1.27	18.00x0.500	Ice-Rich Silty Sand	Slurry	-5	23	2.5	36[2]	62
3	2	27.3x1.27	10.75x0.500	Gravelly Sand	Driven	-11	12	3.4	49	28
3	9	27.3x1.27	10.75x0.500	Gravelly Sand	Driven	-9	15	2.2	32	20

1 Encompassed perimeter area is considered rather than actual contact area, since creep will tend to occur on that basis.
2 Failed during short-term testing subsequent to long-term tests (see Table 2).
* Failed during long-term testing. Results are documented graphically in Figure 1.

vibratory hammer for driving production piles has proven successful. The technique of employing a small drilled pilot hole filled with water heated to a minimum of 66°C (150°F) prior to placement is effective. Thaw time before driving was approximately 30 min for frozen granular soils and 60 min for frozen fine-grained soils in order to raise the surrounding soil temperature and improve the pile driving.

The results of the pile load tests conducted during this program provided useful information for development of pile design criteria. The pile load testing equipment, especially designed for arctic conditions worked well. The development of design criteria should recognize permafrost peculiarities such as loading conditions and creep. For purposes of comparison, Tables 2 and 3 are provided as a summary of short-term and long-term (as compared to short-term) load test data for all load tests.

Generally, differences between soil types were more important than installation techniques in the determination of adfreeze strengths. Driven pile placement methods produced comparable strengths to typical slurried piles in ice-rich silty sandy soils. However, driven piles in frozen gravelly soils indicate lower adfreeze values than ice-rich silty sands due to their location near the rivers and the warmer ground temperature effects.

Based on results of these tests, it was determined that pipe and H-piles could be driven in permafrost with acceptable accuracy, providing ARCO Alaska, Inc. with a technically sound alternate pile placement method. The thermally-modified pile driving method was ultimately selected in 1980 for the aboveground Kuparuk Pipeline system design.

Kuparuk River Pipeline Crossing

The Kuparuk River pipeline crossing is about 5,600 ft in length. It consists of 41 bents supported by 82 piles 1.07 m (42 in.) in diameter set in 1.37 m (54 in.) diameter grouted holes drilled 24 m (80 ft) into the river bed. The piles are 30.5 m (100 ft) long and the 6.4 m (21 ft) long crossbeam is capable of accommodating 5 pipelines.

An aboveground river crossing design was selected after analysis of soils data indicated that a buried system would not be feasible. This was due to the fact that massive ice and ice-rich frozen silty soils were encountered throughout this location. It was then determined that the most economical and constructable structure would be the use of 35 m (115 ft) pipeline spans, supported on piers. The structural design problem then became one of designing a pier to support the pipeline which would resist the ice forces expected in the Kuparuk River.

The installed river pier design represents the final result of several preliminary designs. The design schemes that were investigated prior to arriving at the final design are the following: single pile support cantilevered from the earth; two pile rigid frame; and two pile rigid frame with ice-breaking wedge.

Both the single and double piles cantilevered from the earth had very large moments below the soil line and required piles that were prohibitively large. The two pile rigid frame reduced the maximum moment in the structure by distributing moment to the rigid connection at the top of the pile. However, excessively large piles were still needed.

In an attempt to lower ice loadings, a design using an ice breaking wedge with a flat enough slope to ensure ice failure in bending was investigated. Without an extra pile, this design was vulnerable to lateral loads. To maintain the ice breaking wedge in a position to defend against ice, as the river scours over the years, would require embedding the inclined wedge below the existing river bottom. This would require either a battered pile or large amounts of excavation, neither of which was acceptable construction technique during the project period of January through April.

The river pier design is a two pile "X"-braced bent. This design satisfied the ice criteria of 1.5 m (5 ft) in thickness with a crushing strength of 28 kg/cm^2 (400 psi) and a main channel stream velocity of 3 m/sec (9.8 fps). The pile embedment is designed for an uplift force due to winter ice frozen to the piers being forced up by the rising water level during spring thaw. This force exceeded any which might be expected from frost jacking in the river channel. All pile computations are based on the assumption of a thawed soil with properties of sandy gravel.

The attached photos in Figure 1 of the Kuparuk River pipeline crossing were taken during the summer breakup conditions.

Kuparuk River Module Crossing Bridge

The Kuparuk River module crossing located between Prudhoe Bay and the Kuparuk oil field may be the strongest bridge of its type in Alaska. It is a steel box girder system with bolted precast concrete deck panels, designed to carry properly distributed loads up to 2,091,000 kg (4,600 kips). The structure is designed for ice floes 1.2 m (4 ft) in thickness under working stress condition and 1.7 m (5.5 ft) in thickness under ultimate stress condition, failing against the bridge supports with a crushing strength of 28 kg/cm^2 (400 psi).

FIGURE 1 - KUPARUK RIVER PIPELINE CROSSING

The entire foundation has been placed in an active floodplain underlain by permafrost. Construction of approach embankments now restricts the major annual river flow through a 91 m (300 ft) wide area. This is believed to increase the potential thaw depth around foundation piles. Mechanical refrigeration was used during construction for freezeback of thawed soil zones as well as to generally cool the soil to provide adequate early pile strengths for the initial planned 1,582,000 kg (3,480 kips) module crossing in August 1981. Passive refrigeration devices (heat pipes) were installed to help maintain ground temperatures on a long-term basis and provide an added margin of safety against pile settlement.

The bridge was constructed in less than three months. The box girder design was selected in order to provide redundancy of design and ease of construction without worries of bad weather stand-by times or waiting for concrete to set up.

The Kuparuk River bridge has three spans. The bridge overall length is 91 m (300 ft) and the deck is 9.9 m (32.5 ft) in width. To allow crossing of extra wide low loads the bolt-on tubular guardrail can be easily unbolted and removed.

Construction of the foundation piles started in March 1981. These were fourteen 1.37 m (54 in.) diameter holes drilled from the frozen river ice surface. In-stream eight pier piles were drilled 23.5 m (77 ft) deep and six abutment piles were drilled 17.4 m (57 ft) deep. Drilled hole alignment was critical along the longitudinal bridge axis since the steel pile caps were designed to fit perfectly and welded into the foundation piles.

Pile foundations for the bridge were 1.2 m (48 in.) diameter Grade X-65 and X-70 surplus Trans Alaska Pipeline sections. Piles were welded with end plates and installed to design embedment depths. The piles were slurried in place with a dense saturated sand slurry. Special modifications were made to the piles for the installation of the refrigeration and heat pipe system. Each pile was fitted with 16.8 cm (6.625 in.) diameter pipes installed the full length of the pile. The pipes were then used to house a mechanical refrigeration piping loop in which a cooled refrigerant was circulated and a heat pipe designed to maintain cooled soil conditions long-term.

The structural steel components (girders, cross bracing, diaphragms and steel caps) were purchased and fabricated in Japan. Concrete abutment backwalls and deck panels were precast at the North Slope in a building to protect them from the arctic environment.

Around the approach embankment slopes and the slopes of a 305 m (1,000 ft) long low water crossing area, east of the Kuparuk River bridge, concrete block flexible armoring was placed to reduce embankment erosion due to high water. The revetment system consist of concrete blocks tied together with galvanized wires running the full length of each mat, which were assembled in approximately 2.4 m by 12 m (8 ft by 40 ft) in size. A filter fabric was installed on the slopes prior to installation of mats to prevent soil loss from underneath. The bottom ends of revetment mats were embedded in the soil below expected scour depths.

As the attached photos in Figure 2 indicate the Kuparuk River bridge has provided safe transportation to heavy modules and materials in its short period of operation.

Drainage Culvert Design

The Kuparuk projects construction pads and access roads cross numerous small drainages. These drainages are divided into two distinct types. The first type of drainage is sheet flow occurring in undefined channels. This flow becomes significant only because the natural drainage is blocked by the construction pad or road. The second type of flow occurs in well defined stream channels. The magnitude of discharge is not effected by the gravel roads.

For both types of flow, culverts were hydraulically designed to accommodate the most severe condition resulting from either the 25 year rainstorm or snow melt. Because of the uncertain knowledge of the region's hydrology, and of the effects of ice and snow blockage, a 100% safety margin was adopted. The minimum diameter of any culvert, regardless of the estimated discharge the culvert must convey, was 61 cm (24 in.).

Sheet Drainage Culverts

Sheet flow culverts are spaced from 61 m to 366 m (200 ft to 1,200 ft) apart depending on local topography. In order to prevent erosion of the embankment the outfall is extended beyond the toe. In some cases both the inlet and the outlet of the culverts were equipped with standard end sections to further reduce erosion. Sheet flow culverts were typically underlain by two 3.8 cm (1.5 in.) thick sheets of board insulation. This insulation covered an area the length of the culverts, usually 12 m (40 ft), and 4.3 m (14 ft) in width. The insulation mat is extended to prevent thaw and degradation of the permafrost soils underlaying the culverts. The insulation mat was placed 7.6 cm (3 in.) below the original tundra so that the outflow would not severely erode the tundra. The minimum cover over the culverts were 91 cm (3 ft) in order to adequately distribute traffic wheel loads, as indicated on the attached Figure 3.

FIGURE 2 - KUPARUK RIVER MODULE CROSSING BRIDGE

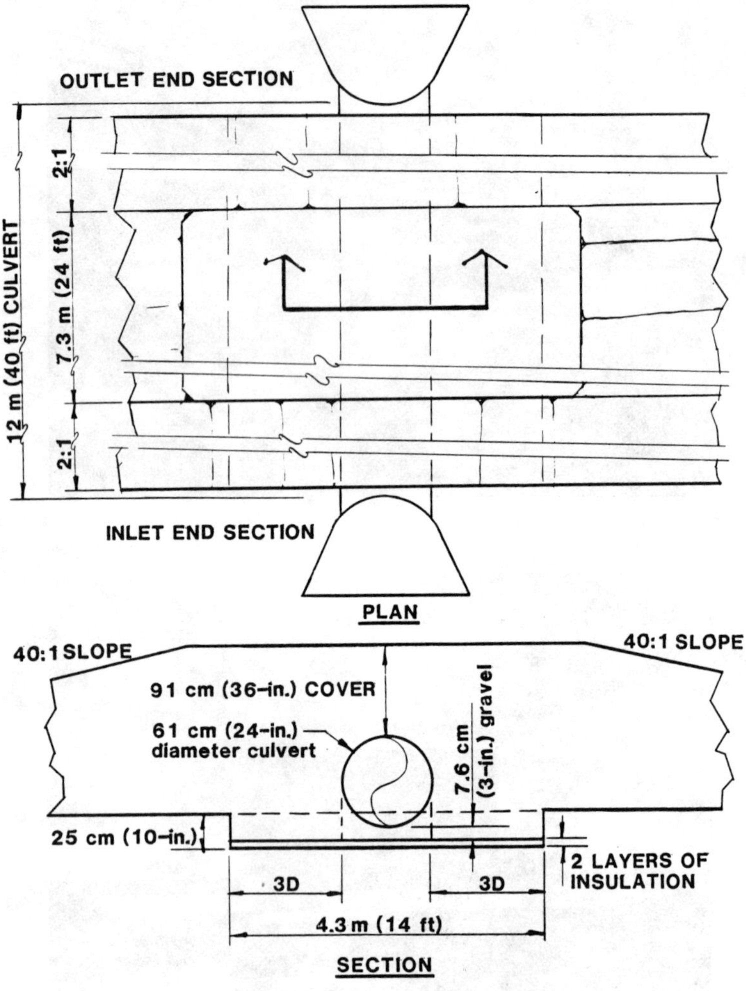

FIGURE 3 - SHEET FLOW CULVERT

Culverts in Defined Channels

There were 6 crossings of defined channels. These crossings consisted of closely spaced batteries of culverts sized to accommodate the design flow. Because these culverts were founded on relatively thaw stable gravel an insulation blanket was not used. The culverts were backfilled with gravel fill materials compacted to 95%.

Wind Induced Pipeline Vibration Analysis

The configuration of simply supported aboveground pipelines introduces exposure and vulnerability to local wind effects. Similar to vertical cylindrical chimneys or stacks, horizontal pipelines are affected by shedding of alternating wind vortices provided that the flow separates from the structure's surface and the wind flow regime is subsonic in nature. Vortices are formed when the fluid pressures acting on the structure are not sufficient to force the boundary layer to remain in contact with the surface of the structure. This shedding process causes the pressures on the structure to fluctuate and can induce an elastically supported member to vibrate perpendicular to the direction of wind flow.

The wind velocity which produces steady state vortex shedding along a cylinder is related to the Reynolds Number:

$$Re = \frac{UD}{V} \quad \text{(1)}$$

Where: Re = Reynolds Number

U = free stream velocity component normal to axis of cylinder

D = cylinder outside diameter

V = kinematic viscosity of fluid (air)

Based on empirical studies, researchers (Blevins 1977, Loiseau 1972) have isolated several specific regions of fluid flow as presented in Figure 4. Regimes 1 and 2 represent conditions in which fluid flow velocity is not sufficient to separate the boundary layer from the body or cause alternate shedding of vortices. Regimes 3 and 4 correspond to a range in Reynolds Numbers which produce steady state shedding of alternating vortices from the body. Regime 5 indicates that the boundary layer wake is completely turbulent with no definite vortex formation. The primary concept being that the steady state periodic vortex shedding is limited to a finite range in wind velocities for a given structure geometry. The potential for an aboveground pipeline to exhibit wind induced vibration occurs

Figure 4 — Regimes of Fluid Flow
(After Loiseau and Szechenyi, 1972)

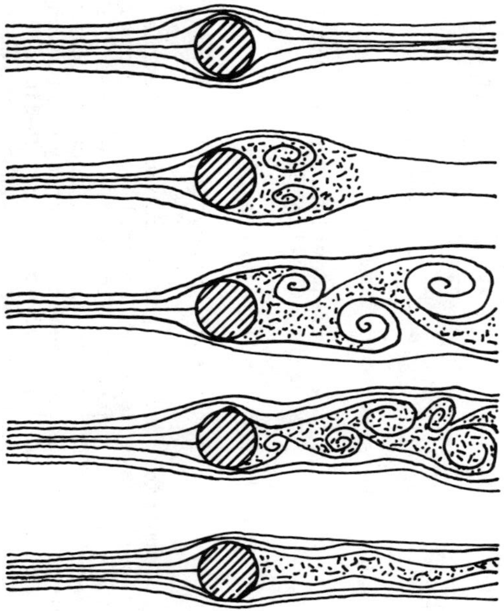

1)	$R_e < 1$	Laminar unseparated flow
2)	$1 < R_e < 40$	Laminar symmetric separation
3)	$40 < R_e < 3\times10^3$	Subscritical regime — laminar separation with alternating Von Karman vortices
4)	$3\times10^3 < R_e < 6\times10^6$	Transcritical regime — separation between laminar and turbulent with random vortex formation
5)	$R_e > 6\times10^6$	Supercritical regime — turbulent separation with some oscillation of wake

when the frequency of steady state vortex shedding approaches the natural or resonant frequency of the simply supported pipeline segments.

The frequency of vortex shedding of a circular cylinder is defined as

$$f_s = \frac{SU}{D} \quad \ldots \ldots \ldots \ldots \ldots \ldots \ldots \quad (2)$$

Where: f_s = vortex shedding frequency

S = Strohal Number

U = wind velocity

D = pipe O.D.

The Strohal Number is a function of Reynolds Number (or wind velocity) and ranges from approximately 0.19 to 0.25 for non-turbulent flow regimes (Loiseau 1972).

An approximation for the fundamental frequency of the pipeline segments between supports can be derived from the following equation.

$$f_p = \frac{n^2 \pi}{2} \sqrt{\frac{EI}{mL^4}} \quad \ldots \ldots \ldots \ldots \ldots \ldots \quad (3)$$

Where: f_p = natural frequency

n = mode number (1,2,3...)

E = modulus of elasticity

I = moment of inertia

m = mass per unit length

L = length of span

The assumption is made that each segment acts as a simply supported beam isolated from effects of adjacent segments. Alternately, more sophisticated analytical techniques can be used to better assess natural frequency of a system of continuous supports subjected to dynamic excitation.

The wind velocity required to produce steady state vortex shedding compatible with the resonant frequency of the pipeline system ($f_s = f_p$) is found by substituting the

pipe natural frequency obtained by Equation 3 into

Equation 2 and solving for wind velocity:

$$U = \frac{f_p D}{S} \quad (4)$$

Where: f_p = natural frequency of the pipe span (Eq. 3)

D = pipe O.D.

S = Strohal Number

If the wind velocity found from Eq. 4 is greater than the design wind for the geographic area, then the potential for wind induced vibrations is low. If the wind velocity determined from Eq. 4 is less than the design wind, then there is a potential for the pipeline system to experience repeated bending stress reversal due to wind induced vibration.

The magnitude of cyclic bending stress is proportional to the amplitude of steady state vibration. Two semi-empirical methods are available for determining the response amplitude of a cylinder to vortex shedding, the wake oscillator model and the correlation model. A rigorous derivation and presentation of these methods is not within the scope of this paper. This information can be obtained from publications (Blevins 1977, Blevins 1976, Bishop 1964, Iwan 1975). The wake oscillator model assumes the self-excited vortex shedding to be due to a simple nonlinear oscillator coupled to the structure. This model was first introduced in 1964 (Bishop 1964) and has since been expanded by others. As vortices are shed from a body there is some component of fluid velocity in the wake which will be perpendicular to the free stream fluid flow. The method of solution uses the forces due to this change in fluid momentum as the driving force in the equation of motion of an elastically mounted, viscously damped cylinder.

Blevins and Burton introduced the correlation model based on a random vibration approach to predicting vortex induced vibration that accounted for correlation effects and amplitude dependence of vortex forces (Blevins 1976). These effects are incorporated into the equation of motion for a cylinder under a uniform force, through use of an empirical coefficient.

Both methods require an assessment of system damping against vertical pipe vibration. Analyses indicate that the vibration amplitude (and resulting cyclic bending stresses) predicted by these methods is extremely sensitive to the selection of damping ratio used to represent the pipeline system:

$$\beta = \frac{C}{Cc} \quad (5)$$

Where: β = damping ratio
C = actual system damping
Cc = critical system damping

The damping ratio, $\frac{C}{Cc}$, is a measure of the dissipated energy per cycle compared to the total energy of the vibrating system.

Unfortunately, information regarding damping ratios for pipelines or circular cylinders is sparse. Newmark and Hall recommend values ranging from 0.01 to 0.02 for dynamic analysis of bare pipelines (Newmark 1982). Insulated pipelines that are foam encased with a protective jacket may have somewhat of an increased damping characteristic associated with the composite cross-section.

Alternately, the damping ratios can be calculated from direct measurement of a vibrating systems and is proportional to the natural logarithm of the ratio of amplitudes of two successive cycles of the lightly damped structure in free decay.

The maximum pipe vibration amplitudes are used to calculate maximum bending stresses and the potential for cumulative fatigue damage. The effects of fatigue can be assessed in accordance with standard codes of practice which account for stress intensities for welds.

The selection of various approaches to mitigate against adverse fatigue damage will be dependent upon cost and the estimated time to reach fatigue.

The critical fatigue time is computed from the natural frequency of the system (rate of cyclic vibration) and the allowable number of cycles obtained from the design fatigue curve for the pipe steel. If, for example, the fatigue time is estimated to occur within one or two equivalent weeks of prevailing design wind for a given support spacing and pipe cross sectional area, it may be prudent to redesign the support spacing or pipe geometry to prevent adverse vibration. On the other hand, if analyses indicate that pipe fatigue would occur only after one equivalent year of prevailing design wind, the the support spacing and pipe cross section used in the analyses may be adequate if alternate mitigative methods are employed when adverse vibration is detected during operations monitoring. A flow diagram is suggested in Figure 5 for the design of pipelines against wind induced vibration.

Figure 5 — Preliminary Design Procedure for Vortex Shedding Analysis

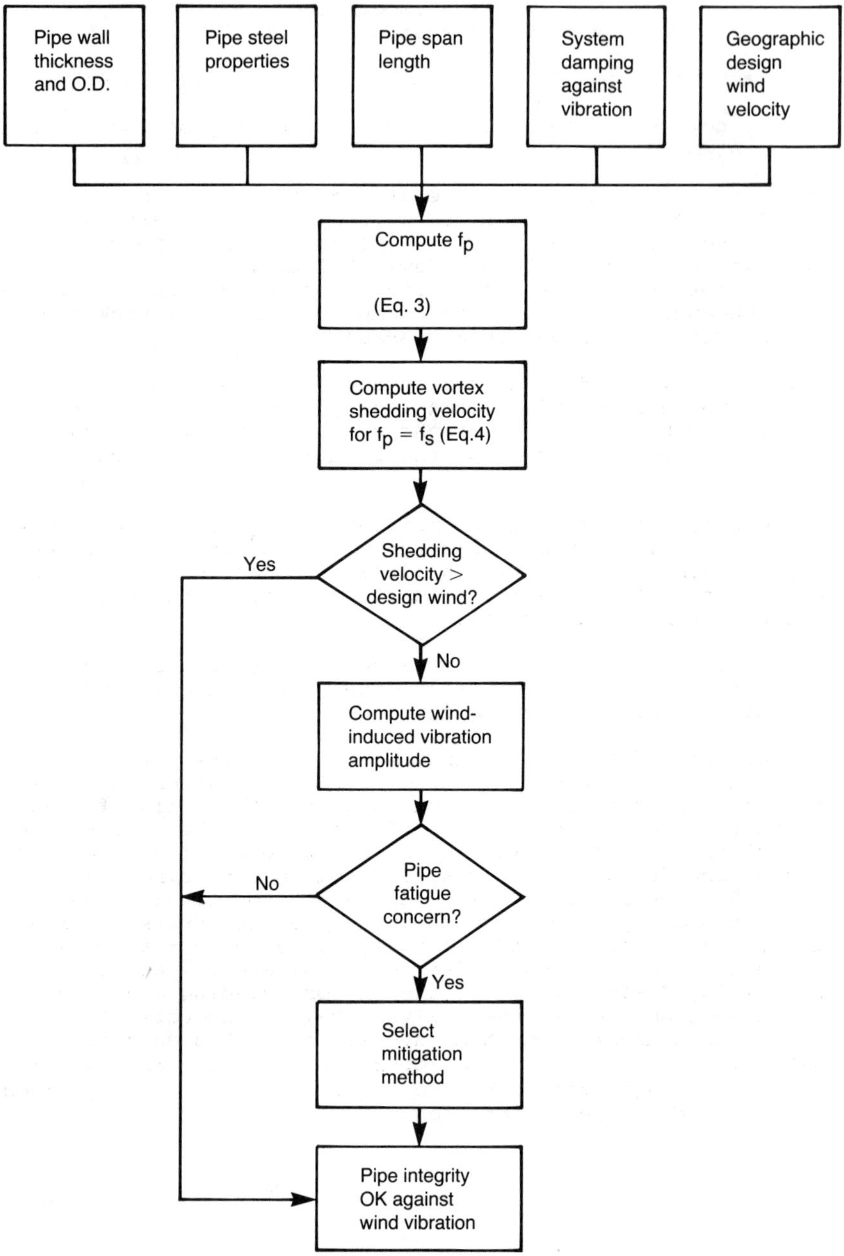

Design alternatives to prevent adverse wind induced pipeline vibration are classified as direct or indirect methods. Direct methods address the vortex shedding phenomenon itself, whereas indirect methods are associated with optimizing the pipeline properties such as increasing damping or pipe stiffness to decrease the effects of vortex shedding on the pipe. Direct methods to mitigate adverse wind induced vibration are application of a strakes or wind spoilers to the structure. These devices perturb the formation of alternating vortices by modifying the nature of the cylinder surface boundary. The most common wind spoiler configuration is the helical strake which has been successfully used to prevent wind induced vibration of vertical chimneys or stacks. The helical strake was incorporated along the Kuparuk pipeline system at the Kuparuk River crossing and along other non-river crossing segments where pipe vibration was detected (Figure 6). Other potential wind spoiler designs include the sawtooth vortex splitter and ribbons as presented in Figure 7. It is believed that the proximity of two or more parallel pipelines on a pipe rack or support system may have inherent self-mitigating capability against wind induced vibrations. If the space required to develop alternating downstream vortices is greater than the separation space between the pipelines, then the proximity of the two pipelines prevent formation of vortices all together as illustrated in Figure 8.

Indirect methods to mitigate effects of wind induced vibration include increasing the pipe span stiffness through increased pipe wall thickness or in combination with decreasing the span length between supports. Both alternatives increase the natural frequency of the system (Eq. 3), thus requiring a higher wind velocity (Eq. 4) to produce a compatible vortex shedding frequency. Theoretically, the span stiffness can be increased to a value such that the wind velocity required to produce steady state vibration is greater than the design wind associated with the geographic location, or has a flow regime which is turbulent.

Acknowledgements

The permission of the owners of the Kuparuk Pipeline System to publish this paper is gratefully acknowledged.

The design of the Kuparuk Pipeline facilities as described in this paper were accomplished through the efforts of Williams Brothers Engineering Company, Tulsa, Ok.; Santa Fe/Braun and Co., Orange, Ca.; Structural Mechanics Associates, Houston, Tex.; and Peratrovitch, Nottingham and Drage, Anchorage, Ak.

FIGURE 6 - PIPELINE WIND SPOILERS FOR THE KUPARUK RIVER CROSSING

Figure 7 — Alternate Wind Spoiler Configurations

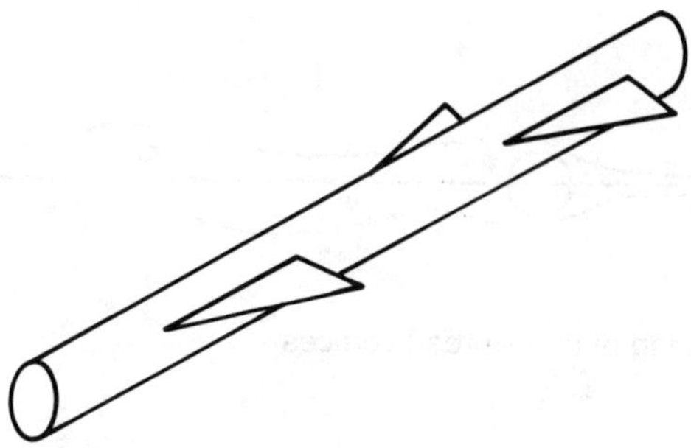

Pipe fitted with vortex splitters

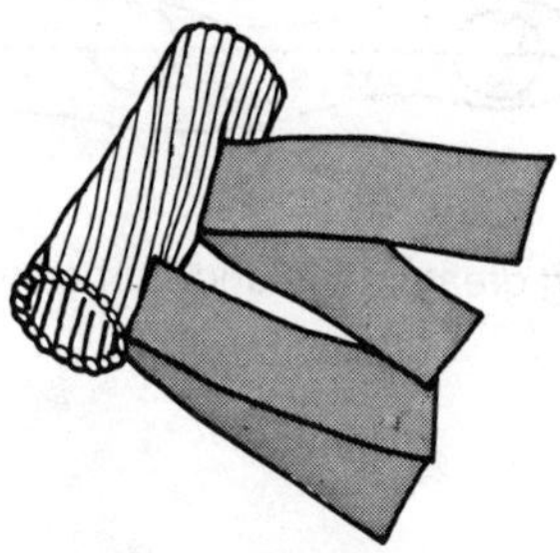

Cable equipped with ribbons

Figure 8 — Effects of Adjacent Pipeline on Vortex Shedding

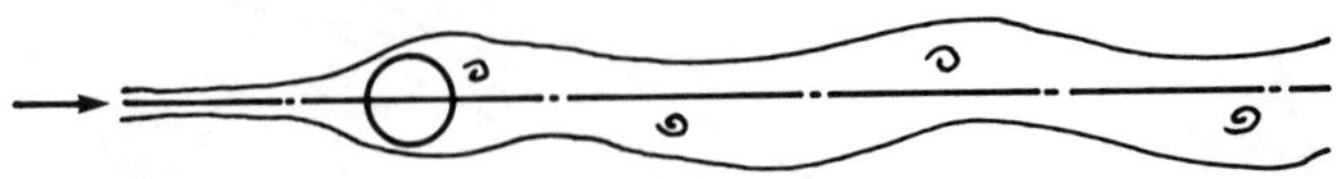

Spacing of undisturbed vortices

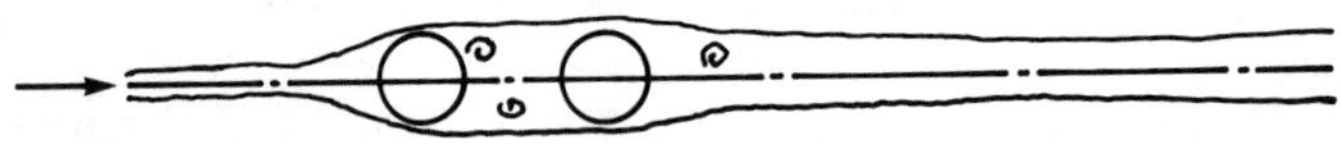

Pipe spacing less than undisturbed vortex spacing

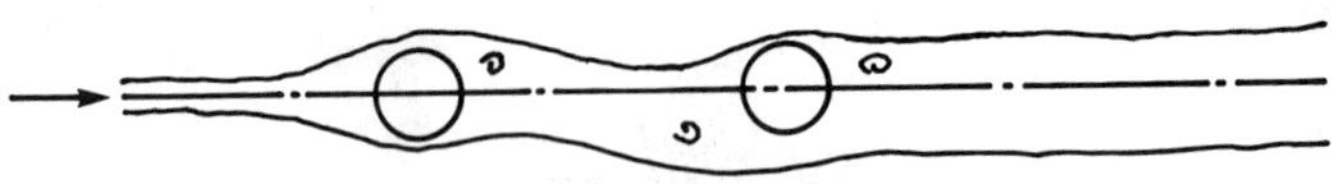

Pipe spacing greater than undisturbed vortex spacing

References

1) Blevins, R., 1977, Flow-Induced Vibrations, Von Nostrad Reinhold, NY.

2) Blevins, R. D., and Burton, T. E., 1976, "Fluid Forces Induced by Vortex Shedding", J. Fluid Engineering, ASME, 95.

3) Bishop, R. E. D., and Hassan, A. Y., 1964, "The Lift and Drag Forces on a Circular Cylinder in a Flowing Field", Proc. Roy. Soc. (London) Ser. A, 277, pp. 51-75.

4) Crory, F. E., 1973, "Installation of Driven Test Piles in Permafrost at Bethel Air Force Station, Alaska", U. S. Army Cold Regions Research Engineering Laboratory, Technical Report 139.

5) Crory, F. E., 1975, "Bridge Foundations in Permafrost Areas", U. S. Army Cold Regions Research Engineering Laboratory, Technical Report 266.

6) Crory, F. E., and Reed, R. E., 1965, "Measurement of Frost Heaving Forces on Piles", U. S. Army Cold Regions Research Engineering Laboratory, Technical Report 145.

7) Iwan, W. D., 1975, "The Vortex-Induced Oscillation of Elastic Structural Elements", J. Eng. Industry, ASME, 97.

8) Loiseau, H., and Szechenyi, E., 1972, "Analyse Experimentale des Portance sur un Cylindre Immobile Soumis a un Ecoulement Perpendiculaire a son axe des Nombres Reynolds Eleves", La Recherce Aerospatiale, Bulletin 13, L'Office National d'Etudes et de Recherches Aerospatials, France, No. 150.

9) Newmark, N. M., and Hall, W. J., 1982, Earthquake Spectra and Design, Earthquake Engineering Institute.

10) Nottingham, D., 1981, "Method and H-Pile Tip for Driving Piles in Permafrost", U. S. Patent 4,297,056, October 27.

11) Rooney, J. W., Nottingham, D., and Davison, B. E., 1976, "Driven H-Pile Foundations in Frozen Sands and Gravels", Proceedings Second International Symposium on Cold Region Engineering, Fairbanks, Alaska, 00 169-188.

12) Simiu, E., and Scanlan, R. H., 1978, Wind Effects on Structures: An Introduction to Wind Engineering, John Wiley and Sons, NY.

Scouring Induced Vibrations
on Sub-Sea Pipelines

Amit B. Ghosh*

Louis M. Laushey**
F. A.S.C.E.

Abstract

Theory and experimental results are presented to suggest that some pipeline failures might have been caused by vibrations, either because of excessive deflections or fatigue. Laboratory experiments showed that vibrations of large magnitude occurred in circular pipes when the clearance between the pipe and a fixed boundary was less than one pipe diameter. At resonance, Strouhal numbers were in the range of 0.19 to 0.28, near the accepted 0.20 for cylinders in an infinite fluid. Asymmetrical vortex shedding was deemed to be a major cause of the vibrations at resonant Strouhal numbers. Design parameters to prevent vibrations are suggested, and include preventing gaps smaller than one pipe diameter and avoidance of Strouhal numbers between 0.2 and 0.3. Experiments showed that in these critical ranges, large vibrations and resonance occurred. Scaling relationships are presented to suggest methods of analysis of submarine pipelines that might destruct due to vibrations.

Introduction

The most economical method of bringing offshore oil onshore is through pipelines laid on the seabed. Submarine pipelines are also used for disposal of industrial wastes in lakes and oceans.

These pipelines are exposed to bottom currents, seismic loads, wave actions during storms, liquefaction and slumping of the bearing soil and various other

*Graduate Student, Department of Civil and Environmental Engineering, College of Engineering, University of Cincinnati, 735 Rhodes Hall, Cincinnati, Ohio 45221.

**Executive Officer and Professor, College of Engineering, University of Cincinnati, 651 Baldwin Hall, Cincinnati, Ohio 45221.

environmental effects. The seafloor is often irregular. Modifying the laying route to more regular profiles is not practical. Therefore pipelines are buried in trenches wherever possible. Otherwise, sections are either laid on the seabed (Figure 1) or supported on trestles to keep the bending stresses below the design limits.

Various pipeline failures are being investigated for probable causes. This work considers vibration of the pipelines, and the resulting fatigue, as an additional probable cause for some pipeline failures.

Experimental studies have identified conditions under which a suspended pipeline can vibrate because of bottom currents and erosion of soil from underneath the pipe. The modes of vibration under certain conditions are periodic and intense, of such a magnitude that at resonance they could probably cause a direct failure, or an ultimate fatigue failure.

Recommended practices, rules and standards regarding scouring around submarine pipelines have been proposed by the American Petroleum Institute, American National Standards Institute, Det Norske Veritas and a few others.

Most of such publications however, merely state in very general terms that whenever scouring is a possibility, the effect of currents, waves, seismic activity and the like, should be included in the design.

Such guidelines therefore are insufficient and can provide little aid in selecting an adequate design or installation procedure.

This project has been aimed at developing safe installation criteria for a situation where soil is being eroded from underneath the pipe.

The intent has been to predict a range of critical clearances at which the pipe might vibrate excessively and fail in fatigue. Further, to suggest remedial measures that can be taken, or insure in the specifications that these clearances are avoided. Experiments have been conducted in a flume containing circular cylinders at or near the bottom of planar and irregular profiles.

The variables chosen were: depth of water, flow velocity, pipe diameter, skin roughness of the pipe, bottom profile and clearances of the pipe above the floor.

Under defined relationships between the variables, vibrations have been observed. These vibrations are large and periodic, at or near the resonance frequency of the pipeline system. Research has been successful in relating the variables.

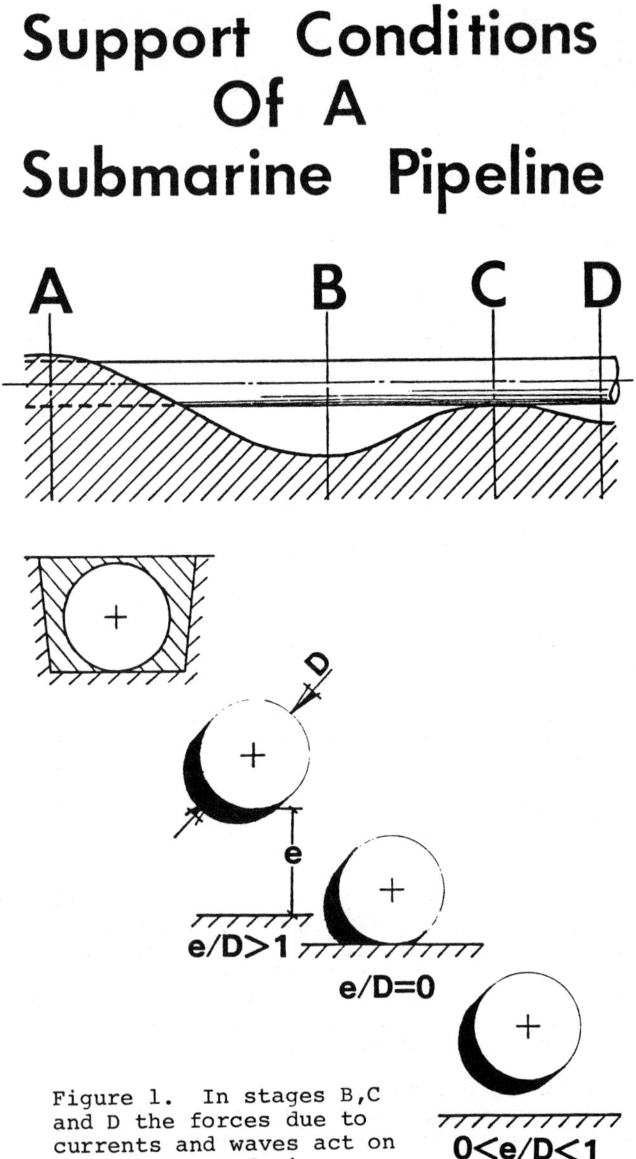

Figure 1. In stages B,C and D the forces due to currents and waves act on the unprotected pipe.

Literature Review

The dynamics of fluid flow around cylinders has been studied by many authors for tubular strength members in offshore structures.

The action of forces has been modelled for various boundary conditions, for example, at the air-sea interface for wave slamming effects.

Studies in areas which in some ways complement this work and so are briefly reviewed are:

(a) Wave forces on cylinders near a plane boundary.

(b) Wave-induced scour around offshore pipelines.

(c) Forces of currents and waves on submarine pipelines.

(a) Essentially two-dimensional approaches by Chakraborti[2], Yamamotto and Nath and Slotta[5], Beattle and Brown and Webb[1] and others, derive lift, drag and added mass coefficients from hydrodynamic force equations which are functions of velocity and temporal acceleration. Despite numerous available experimental data, the assumption of "rigid" cylinders and plane boundaries are inadequate.

(b) An extensive laboratory study by Herbich[4] evaluating scour patterns around the pipe with the pipeline buried at various depths below the ocean bed.

Analysis indicates that depth of scour around pipelines under wave conditions is greater than anticipated. The pipe itself causes additional scour.

Wave action is significant mainly in the surf zone and in shallower waters even in the strongest of gales. The study is thus limited to buried pipelines in shallow waters.

(c) Different load situations have been analyzed by Hafner[3] for pipes lying in contact with the seabed and at various heights above it.

Force coefficients have been obtained from experiments for various skin and bottom roughnesses and compared with the theoretical values predicted by Yamamotto[5]. Comparisons show good agreement between theory and experiments. However, this work also lacked reduction of the elasticity according to the geometrical scale (see "Modelling for hydoelastic similitude" later), like the studies listed in (a) and hence ignored the possibility of vibration.

Experimental Procedure

Apparatus: Experiments have been conducted in a 1.5 ft. (45.7 cm.) deep, 1.29 ft. (38.1 cm.) wide and 24 ft (731.5 cm.) long glass-walled flume, with the pipe suspended transverse to the direction of flow.

The reduced rigidity of the pipeline system according to the geometrical scale, was realized by carefully chosen sets of springs. The suspension system was capable of holding pipes of different diameters at desired heights above planar or contoured bottoms (Figure 2).

Measurements: For smooth or roughened pipes of different diameters, readings were taken of heights of pipes above bottom, rates of flow, depths of flow, velocities of flow, intensities and frequencies of vibration.

The heights above the bottom were measured at the center of the span and also at the ends for planar bottom profiles. For contoured bottoms they were measured at the center of the span.

The height of pipe above the bottom was increased in stages. At every stage the rate and depth of flow were gradually increased and the pipe was observed for vibrations. Rate and depth of flow were increased beyond the state of maximum vibration to observe the damping effects that were induced. The decay in amplitudes and frequencies of oscillations was very consistent in most cases.

The rates of flow were calculated from the readings taken on an orifice meter, while the depth was measured at the side of the flume and at the centerline section of the pipe. The calculated velocity was the average velocity of flow approaching the pipe.

The states of vibration were divided into five categories: at rest to low, low to moderate, moderate to high, high to peak and peak.

The frequencies of vibration were measured using a steel wire connected to the axis of the pipe such that it could open or close a micro-switch with the oscillation of the pipe.

A digital counter in the circuit then displayed the number of times the switch opened or closed and hence the number of oscillations of the pipe.

At every stage the time required for fifty oscillations was recorded by an electric stopwatch to obtain the frequency in cycles/second. The micro-switch

APPARATUS

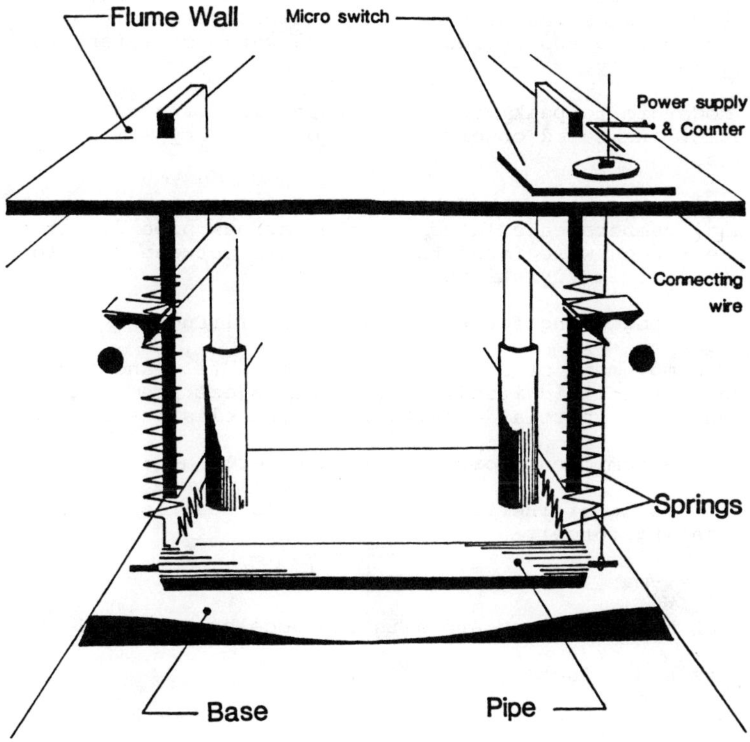

Figure 2 .

had a "play" of one-eighth of an inch (3.2 mm.) between its open and closed positions so any vibration with amplitude higher than one-sixteenth of an inch (1.6 mm.) could be recorded. In most cases the amplitudes were much larger.

In a few cases, the amplitudes of oscillation were lower than required to activate the switch. At the same time the frequencies of such oscillations were only of the order of one to two cycles/second and were convenient to record manually.

Moderate to peak vibrations were always large and periodic. Repeated observations produced very similar results.

As a measure of the phenomena of vortex shedding, the Strouhal numbers calculated for the various pipe diameters and clearances were found to have values between 0.19 to 0.28 at maximum vibrations.

Modelling for Hydroelastic Similitude

The movement of an elastic body in a flow and the motion of the fluid itself around the object are influenced by inertia, gravitational and elastic forces.

These component forces derived briefly are:

1. Force of inertia (D' Alembert's reaction to any accelerating force).

$$F_i = -ma$$

in which m is the mass and a is the acceleration.

Its scale factor:

$$c_i = \frac{-m_p . a_p}{-m_m . a_m} = \frac{\rho_p . v_p . a_p}{\rho_m . v_m . a_m}$$

$$= c_\rho . c^3 . c . c_t^{-2}$$

$$= c_\rho . c^4 . c_t^{-2}$$

$$= c_\rho . c^2 . c_v^{-2}$$

The subscript 'p' refers to the prototype, the subscript 'm' refers to the model, c is the geometrical scale ratio, c_ρ, c_t and c_v are the scale ratios of density, time and velocity respectively.

2. Gravitational force: $F_g = mg$.

Where g is equal to the acceleration due to gravity.

Therefore,

$$c_g = \frac{m_p . g_p}{m_m . g_m} = \frac{\rho_p . v_p . g_p}{\rho_m . v_m . g_m} = c\rho . c^3$$

(Scale ratio of gravitational acceleration can be assumed to be equal to one).

3. Force due to elasticity,

in solid bodies: $F_e = EA \frac{\Delta L}{L}$

in which E is the Young's modulus of the material, A is the cross-sectional area and ΔL is the elongation. The scale ratio for the elastic force:

$$c_e = c_E . c^2 .$$

For dynamic similarity of the model to the prototype the following relationship has to be satisfied:

$$c_i = c_g = c_e$$

from which we can write the independent equations

$$c_i = c_g$$

or $c_\rho . c^2 . c_v{}^2 = c_\rho . c^3$

or $c_v = \sqrt{c . c_g}$ (an expression of Froude's laws)

or $c_v = c^{\frac{1}{2}}$ ...(A)

That is, to satisfy Froude's law the scale ratio of velocities must equal the square root of the scale ratio of dimensions.

For example: for a geometrical scale ratio of 36, a current of 5 knots has to be reduced to

$\frac{5x1.685}{6}$ (1 knot = 1.685 ft/sec = 0.514 m/sec)

= 1.40 ft/sec (0.427 m/sec)

for the model.

Satisfying $c_i = c_e$

we get $cp . c^2 . c_v^2 = c_E . c^2$

After simplifying and combining with (A)

$$c_E = c \quad \ldots \quad (B)$$

In other words, the elasticity of the participating materials should be reduced according to the dimensional ratios.

Therefore, for a geometrical scale ratio of 36, the stress acting on the model of same magnitude as would be applied to the prototype, should produce 36 times the strain or deformation in the model.

However, considerations of elasticity would also imply that the elasticity of the fluid should also be reduced for model a investigation.

The elasticity of water is evidently beyond control, however in fluids,

$$\text{Felast} = EA \frac{\Delta V}{V} .$$

Also the change in the volume of water is negligible in comparison with the deformation (bending, torsion) of the structure.

Consequently, it is sufficient to deal with correct modelling of structural elasticity.

Effects of Currents and Waves

The proximity of a boundary causes asymmetry in the flow around the pipe. For an unsupported span where there is a gap between the pipe and the boundary, the water particle velocity near the lower part of the pipe becomes higher than that near the upper part. Consequently, a lower pressure is created near the lower part and the pipe is subjected to a downward force acting perpendicular to the boundary.

The vortex in the wake behind the pipe creates circulation which also generates a force perpendicular to the boundary of magnitude equal to the product of the mass density of the fluid, the circulation and the steady velocity of the flow.

The sum of the two forces acting perpendicular to the boundary and their combination with dynamic pressure (Froude-Krylov) force, where both waves and currents are present, can provide the initial excitation force to the suspended pipeline system.

This excitation force when vortices are being shed alternately from the top and bottom of the pipe, at or near the resonance frequency of pipeline system, can lead to extremely large and periodic oscillations.

The forces and the vibrations they induce are considerably reduced for larger gaps, particularly for gaps larger than one pipe diameter (see Figures 3A through 8).

Effects of Scouring

The presence or starting of scour around the pipe can considerably aggrevate the situation described in the previous section. The pipe itself causes the additional instability and scour around it.

Where the pipe touches the bottom, a net force exists away from the bottom. However, even if a very small gap exists between the pipe the bottom, a large net force exists towards the bottom. The solution is obtained by imposing a distribution of doublets on a cylinder radius between the centre and the edge below [5].

Such an upward acting force can create a suction underneath the pipe, where the pipe is touching the bottom, can lower the compaction of the soil, to develop scouring.

Moreover, once scour develops and a gap is created or in a case where a gap already exists between the pipe and the soil, the increased water particle velocity under the pipe and the repetitive reversal of forces when the pipe starts to vibrate, can also enhance erosion. It has been confirmed by the experiments that for clearances larger than one pipe diameter, the pressure difference between the upper and lower parts of the pipe and thus the vibrations are considerably diminished.

It would seem therefore that beyond such clearances the pipe-induced sediment movements will be greatly reduced.

Analysis of Results

The results obtained have been plotted in graphs of Frequency vs. Velocity (see Figures 3 through 8) and Strouhal Numbers vs. Velocity (see Figures 9 through 11).

The effect of clearance 'e' of the pipe above bottom, on the modes of vibration has been shown distinctly, for each plot represents a certain value of e/D.

When clearances are greater than one pipe diameter, there are marked reduction in frequencies and amplitudes of oscillations. Most plots are of shape characteristic

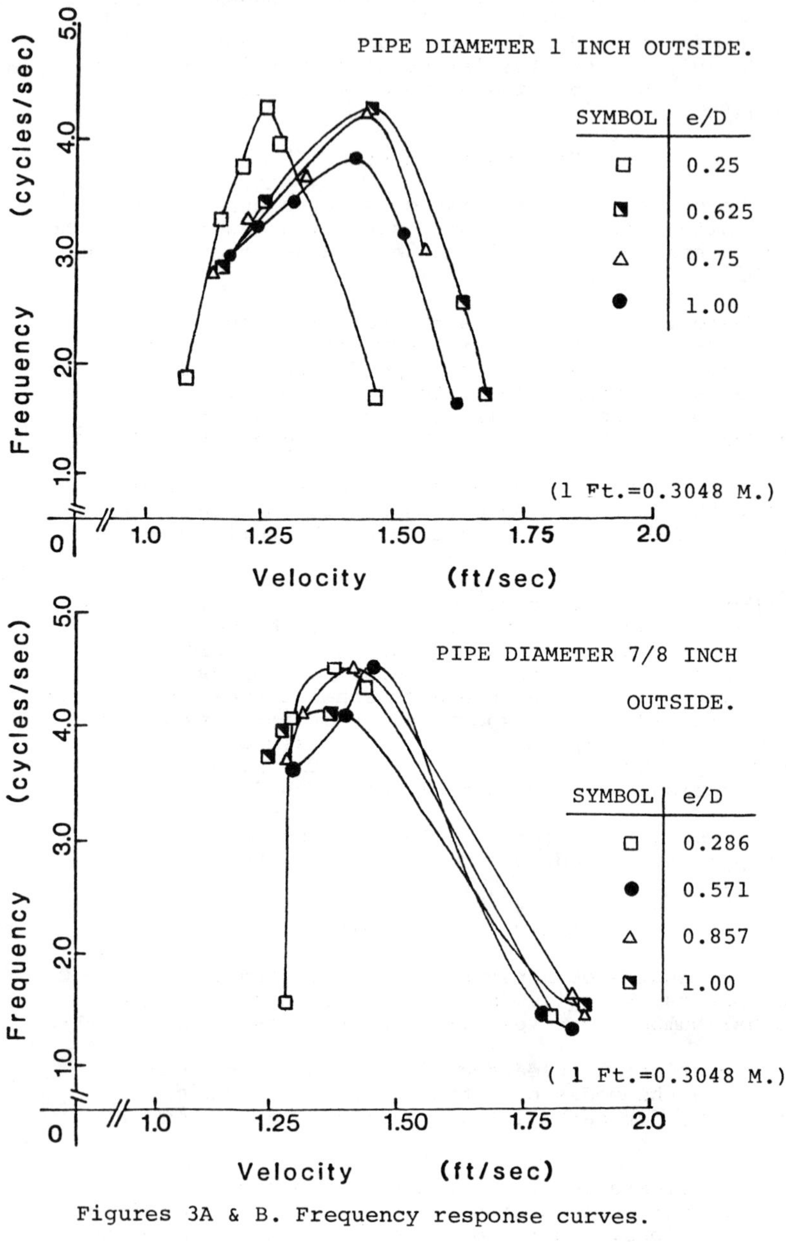

Figures 3A & B. Frequency response curves.

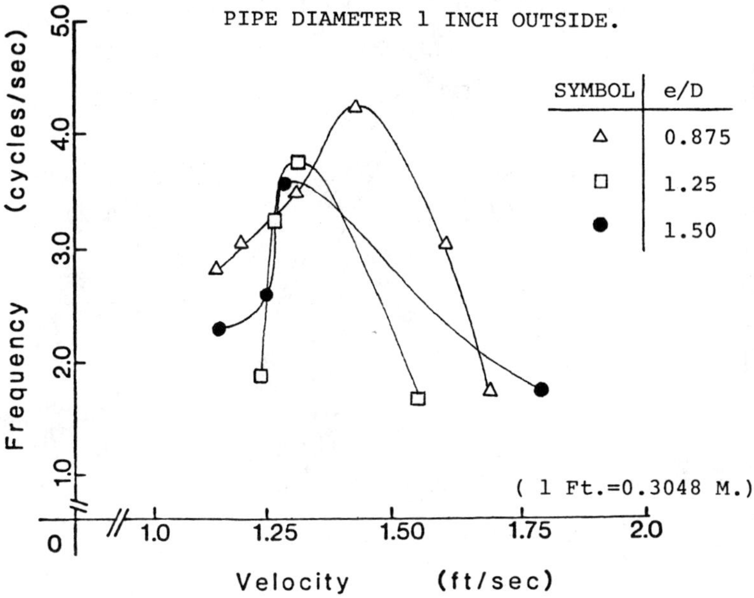

Figure 4. Frequency response curves for smooth pipe. See figures 3A and B also. Note the reduction in the maximum frequencies of oscilla--tions as the Clearance 'e' increases.

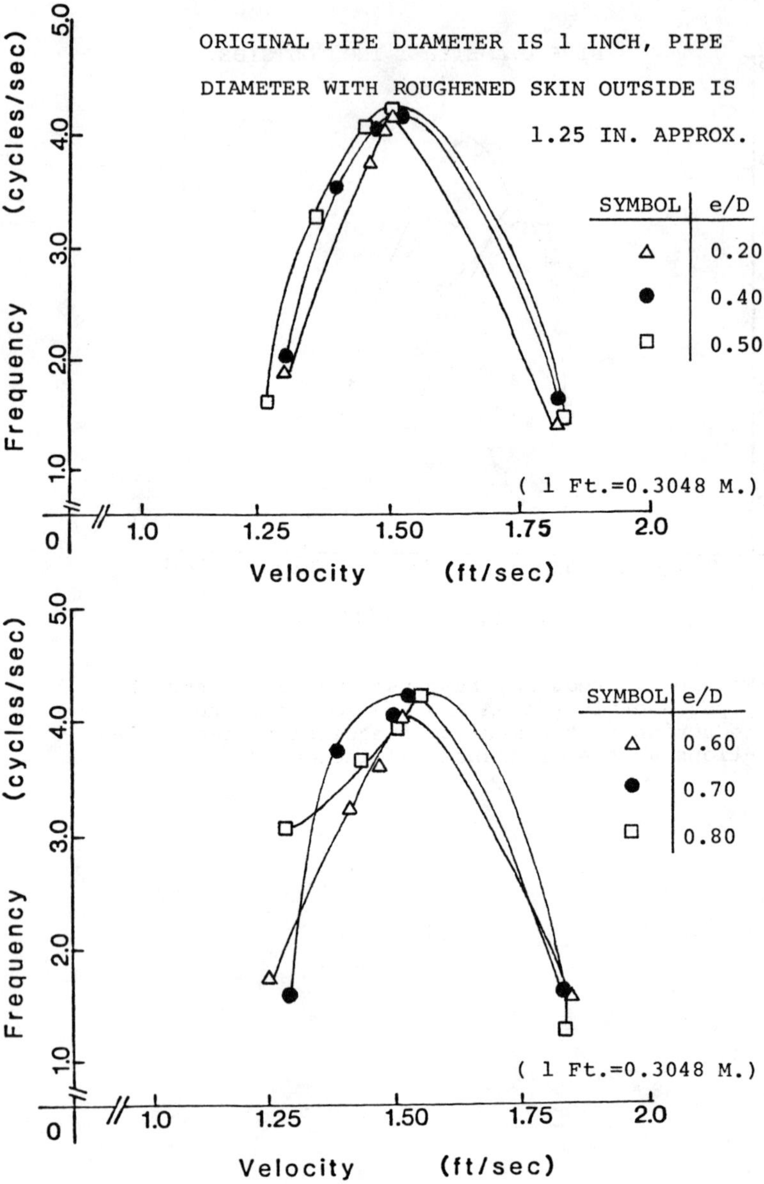

Figures 5A & B. Response curves for roughened pipe.

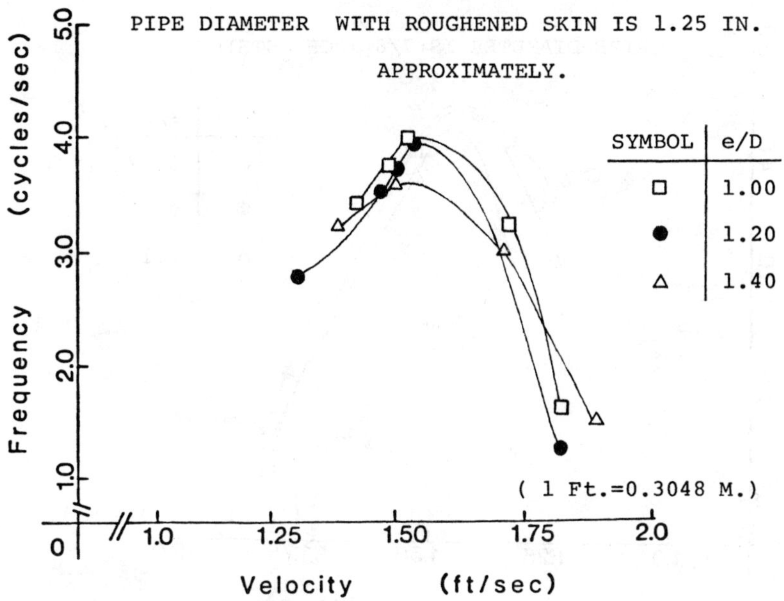

Figure 6. Response curves for a roughened pipe. See figures 5A and B also. Note the reduction in the maximum frequencies of oscillations as the Clearance 'e' increases.

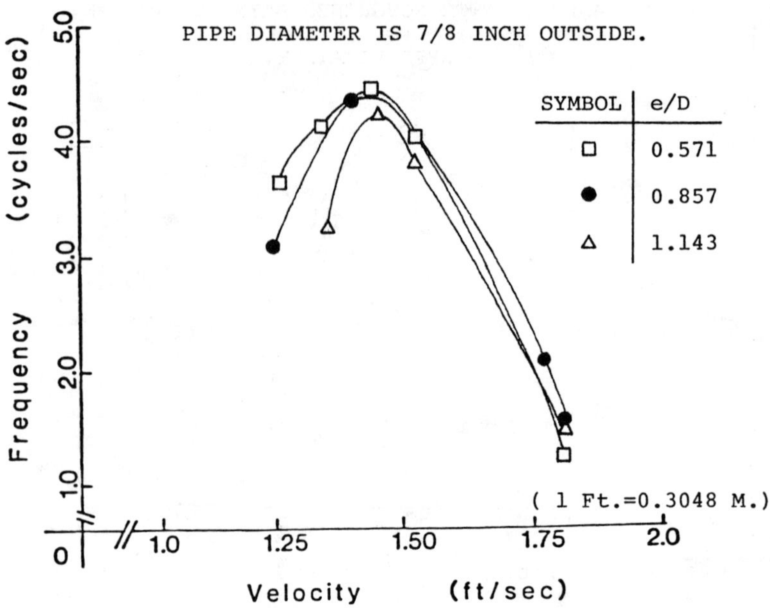

Figure 7. Response curves for a smooth pipe in the flume with contoured (flared U) bottom. Clearances are measured at the centre of the span.

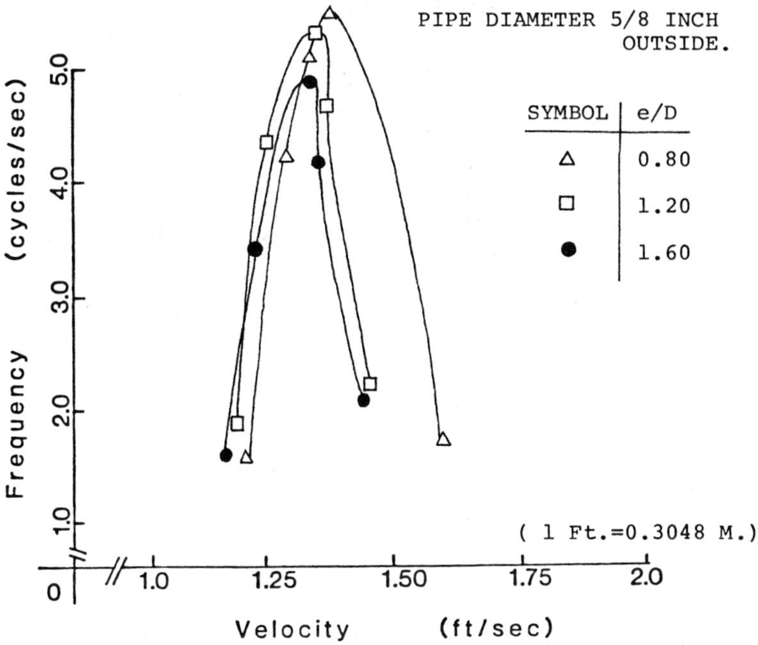

Figure 8. Response curves for a smooth pipe in the flume with contoured (flared U) bottom. Clearances are measured at the center of the span.

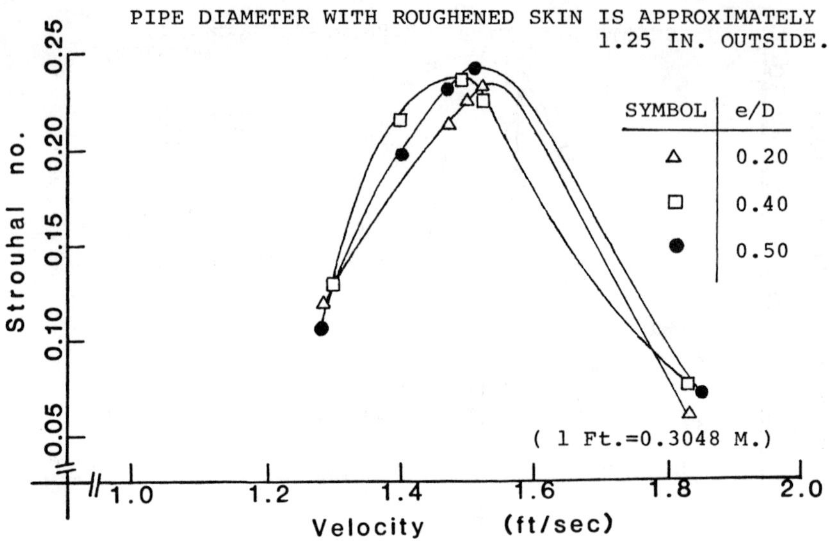

PIPE DIAMETER WITH ROUGHENED SKIN IS APPROXIMATELY 1.25 IN. OUTSIDE.
SYMBOL
e/D
0.20
0.40
0.50
(1 Ft.=0.3048 M.)
Strouhal no.
0.25
0.20
0.15
0.10
0.05
1.0
1.2
1.4
1.6
1.8
2.0
Velocity (ft/sec)

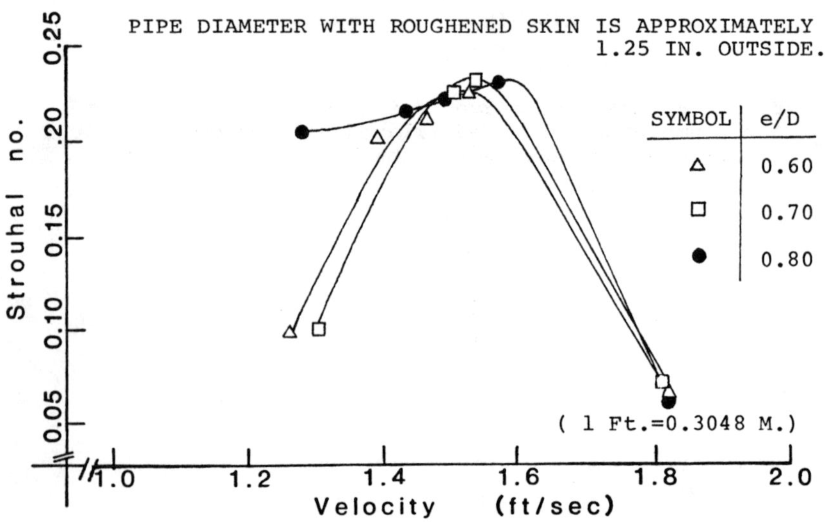

PIPE DIAMETER WITH ROUGHENED SKIN IS APPROXIMATELY 1.25 IN. OUTSIDE.
SYMBOL
e/D
0.60
0.70
0.80
(1 Ft.=0.3048 M.)
Strouhal no.
0.25
.20
0.15
0.10
0.05
1.0
1.2
1.4
1.6
1.8
2.0
Velocity (ft/sec)

Figures 9A & B.

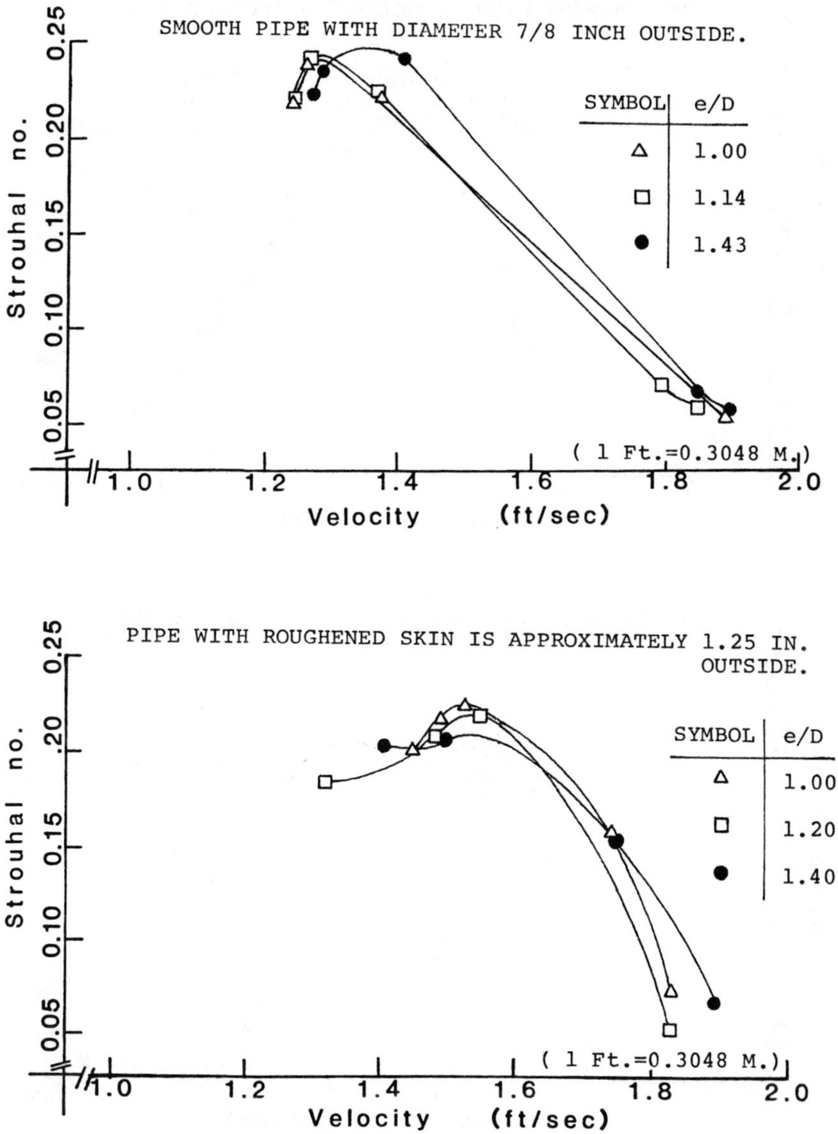

Figures 10A & B.

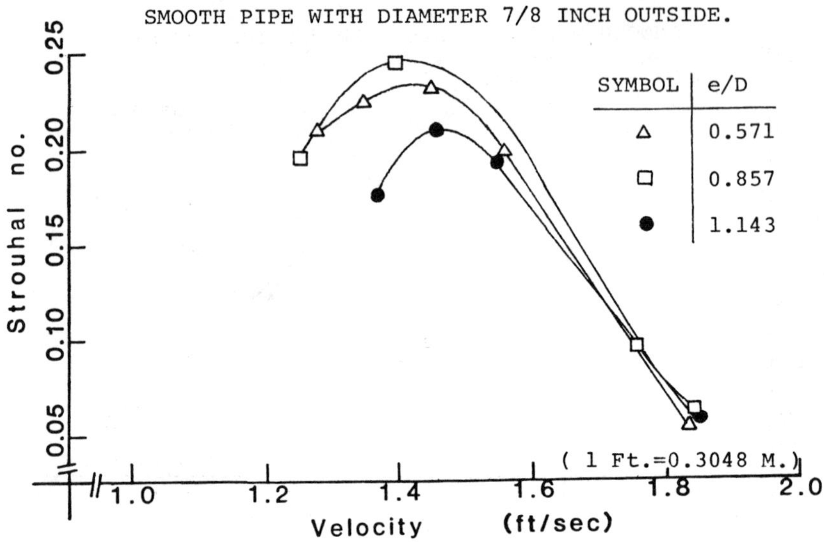

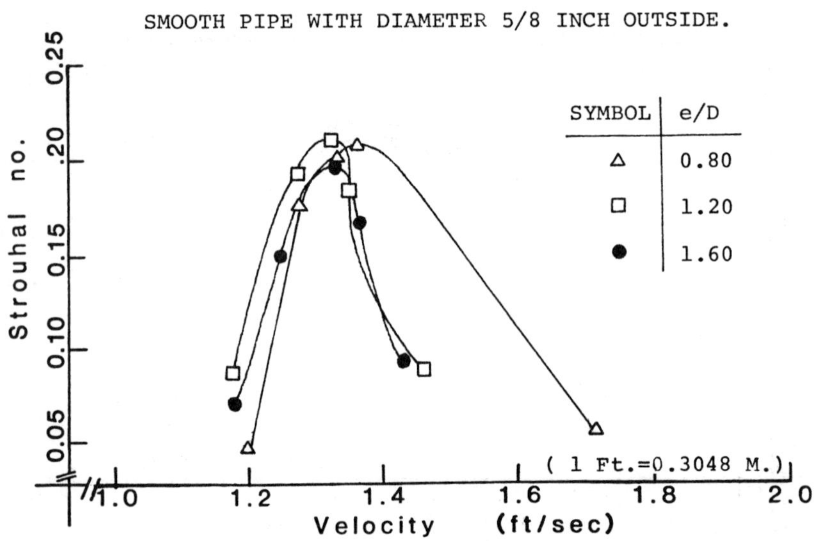

Figures 11A & B. Pipes in flume with countoured bottom.

of forced oscillations with damping. The points of high frequency values also correspond to states of moderate to peak vibrations and large amplitudes.

Plots of lift coefficient vs. bottom distance (e/D) by Hafner[3] show that the lift coefficient gradually decreases to very close to zero at e/D equal to one, starting with maximum positive values at e/D equal to zero.

This implies that as the clearance increases to one pipe diameter the net force or its time-average reduces to a very small magnitude.

This therefore, is an agreement with the findings of this study, that for clearances smaller than one pipe diameter the forces arising due to the proximity of a boundary and thus the likelihood of vibrations are significant.

The Strouhal numbers for maximum vibrations in all plots have values as listed below. They would indicate, as compared to Strouhal number of 0.20 for a cylinder in an infitite fluid, that at or near the resonance frequency the phenomena of vortex shedding is critical.

Summary: Strouhal Numbers at Resonance

7/8"	(22.2 mm.)	O.D.	Smooth pipe,	0.239 to 0.247.
1"	(25.4 mm.)	O.D.	Smooth pipe,	0.236 to 0.284.
1 1/4"	(31.8 mm.)	O.D.	Roughened pipe,	0.219 to 0.241.
5/8"	(15.9 mm.)	O.D.	Smooth pipe on contoured bottom	0.194 to 0.209.
7/8"	(22.2 mm.)	O.D.	Smooth pipe on contoured bottom	0.21 to 0.238.

Concluding Remarks

Based on the study the following conclusions have been made:

1. Pipeline systems should be designed to avoid natural frequencies that may lead to Strouhal numbers between 0.2 and 0.3.

2. When the pipeline rests uncovered on the bottom or when there are clearances in the range of one pipe diameter or less, means to cover the gap and stabilize the bottom should be adopted.

3. Further investigation is suggested of Strouhal numbers for cylinders, smooth and rough, in the proximity of boundaries.

Appendix

References

1. Beattle, J.F., Brown, L.P., Webb, B., "Lift and Drag forces on a submerged circular cylinder", Offshore Technology Conference, Paper OTC 1358 (1971), pages 319/28.

2. Chakraborti, S.K., "Wave forces on cylinders near plane boundary", Discussions, Proceedings-American Society of Civil Engineers, WW4 (1975), pages 303/05.

3. Hafner, E., "Forces of currents and waves on submarine pipelines", Mecrestechn. Nr. 2, April (1980), pages 51/58.

4. Herbich, J.B., "Wave induced scour around offshore pipelines", Offshore Technology Conference, Paper OTC 2968 (1977), pages 79/90.

5. Yamamotto, T., Nath, J.H., Slotta, L.S., "Wave forces on cylinders near plane boundary", Proceedings-American Society of Civil Engineers, WW4 (1974), pages 345/59.

6. American Society of Civil Engineers, "Pipelines in the ocean", Final report of the task committee, A.S.C.E. Pipeline Division.

WAVE-INDUCED PRESSURES ON BURIED PIPELINES

Susan H. Davidson[1], William G. McDougal[2], A.M. ASCE,
and Charles K. Sollitt[3], M. ASCE

ABSTRACT

Ocean waves induce a dynamic pressure response in permeable seabeds which induce an unsteady load on buried pipelines. In the present study the pressure is modeled by potential theory and an analytical solution is obtained employing conformal mapping techniques. Theoretical results for pressure on the pipeline surface are compared with small and large scale experimental measurements.

INTRODUCTION

Wave-induced failures of buried marine pipelines have been well documented (e.g., Brown, 1957; Herbich, 1977). Failures may occur as a result of wave-induced erosion, liquefaction or slope instability. This paper addresses the dynamic pore pressure development associated with the onset of a liquefaction or flotation type failure. Pipeline flotation has long been recognized as a problem for pipelines buried in saturated sediments (ASCE, 1966). Methods of stabilization include armor stone and concrete caps (Layton and Scott, 1979), flexible mattresses (Crowhurst, 1983) and geotextiles (McDougal and Sollitt, 1983). To predict conditions which are susceptible to failure and select optimum stabilization methods requires an understanding of wave-seabed-pipeline interactions.

BACKGROUND

The dynamic response of marine soils has received considerable attention. Early theories were based on the assumptions of an incompressible pore fluid, a rigid soil skeleton and the application of Darcy's Law. Conservation of pore fluid resulted in a pore pressure model governed by LaPlace's equation. Solutions of this type have been developed by Putnam (1949), Reid and Kajiura (1957), Hunt (1959),

[1] Graduate Student, Ocean Engineering Program, Department of Civil Engineering, Oregon State University, Corvallis, OR 97331.
[2] Assistant Professor, Ocean Engineering Program, Department of Civil Engineering, Oregon State University, Corvallis, OR 97331.
[3] Associate Professor, Ocean Engineering Program, Department of Civil Engineering, Oregon State University, Corvallis, OR 97331.

and others. A second analytical approach has been to assume that the soil behaves as a viscous liquid, as an elastic solid or as a visco-elastic medium. Solutions of this type have been developed by Gade (1958), Mallard and Dalrymple (1977), MacPherson (1980), and others. These models may be appropriate for very soft sediments such as those near the Mississippi River delta. A third approach has been to model the soil as a linearly elastic porous medium using Biot consolidation theory (Biot, 1941). This approach has been employed by a number of investigators in the marine environment (eg. Yamamoto, 1977; Yamamoto, et al, 1978; Madsen, 1978; McDougal and Sollitt, 1982). Solutions have been developed to address a variety of marine applications including wave damping, layered soils, geotextiles, and wave-soil-caisson interaction. This system of equations, for most applications, appears to adequately represent the small amplitude dynamic response of the soil for a variety of wave-soil interaction phenomena. However, it has been observed for silts and sands that the full Biot equations are not required and the potential pressure model is adequate (Madsen, 1978; Sleath, 1970). For these highly permeable and rather stiff materials, the soil deformations are small and do not significantly influence the pore pressure. For this reason, and due to the complex geometry, most work on the buried marine pipeline has been based on the potential pressure model. One notable exception to this is the work by Matsui (1982) in which the full Biot equations are solved by a hybrid numerical-analytical solution technique which minimized errors in the boundary conditions.

The potential pressure model has been addressed numerically employing finite difference, finite element and boundary integral methods (Lai, et al, 1974; Liu and O'Donnell, 1979; Lennon, 1983). Solutions have also been developed analytically by MacPherson (1980) using bipolar coordinates and Monkmeyer, et al, (1983) using the method of images. When compared with available data, the potential pressure model appears to provide reasonable estimates of pore pressure development (Nyhus, 1977; Monkmeyer, et al, 1983).

In the present paper, the potential pressure model is addressed analytically employing conformal mapping techniques. In the transformed space, the geometry is separable and an exact solution is obtained. Theoretical results are then compared with small and large scale wave tank measurements.

FIELD EQUATION AND BOUNDARY CONDITIONS

The objective of this paper is to develop an analytical model for predicting wave-induced pressures on buried marine pipeline in seabeds of sand and silt. Because these geologic materials are relatively stiff, it is appropriate to decouple the fluid flow and soil displacements. For this case, the fluid flow may be modeled by Darcy's equation

$$\vec{q} = - K \nabla\left(\frac{p}{\gamma} + y\right) \tag{1}$$

in which $\vec{q}$ is the vector discharge velocity, K is the permeability, ∇ is the gradient operator, p is the dynamic pore water pressure, γ is the weight density of sea water and y is the elevation. Invoking conservation of mass for an incompressible fluid yields LaPlace's equation for the pore water pressure

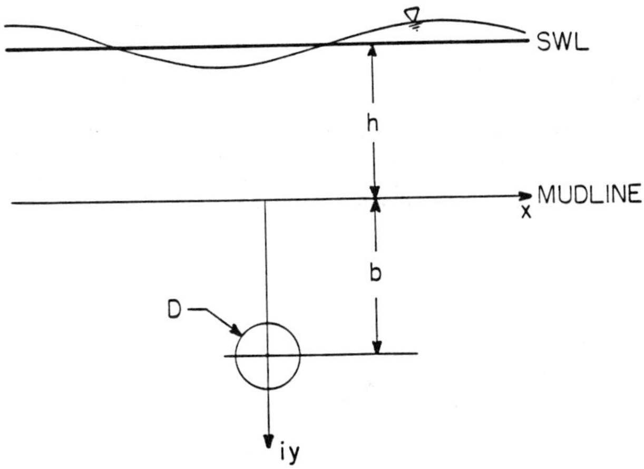

Figure 1. Definition sketch

$$\nabla^2 p = 0 \qquad (2)$$

in which ∇^2 is the two-dimensional LaPlacian operator.

The wave-induced dynamic pressure along the mudline drives the system. This pressure is taken to be

$$p = p_o f(x)\, e^{-i\omega t} \qquad ; \; y = 0 \qquad (3)$$

in which p_o is the amplitude of the bottom pressure fluctuation, $i = \sqrt{-1}$, ω is the radian wave frequency, t is time and f(x) is a dimensionless arbitrary shape function for the mudline pressure distribution. The coordinate axes are defined in Figure 1. The pipeline is assumed to be impermeable which yields a no-flow condition normal to the cylindrical pipe surface,

$$\frac{\partial p}{\partial n} = 0 \qquad ; \; x^2 + (y-b)^2 = \frac{D^2}{4} \qquad (4)$$

in which n is the normal to the pipeline, b is the depth of burial to the center line of the pipe and D is the diameter of the pipe. The seabed is assumed to be of infinite thickness and, therefore, a bounded solution is required at great depth. The response of the soil is assumed to be strictly periodic in time. Therefore, only the solutions for the spatial dependency of the pore pressure will be sought.

SOLUTION TECHNIQUE

The boundary value problem defined by Eqs. (2) (3) and (4) is difficult to analyze in rectangular coordinates. However, this difficulty may be overcome by transforming to a space in which the variables become separable. One such transformation is given by (Kober,

1952)

$$w = \frac{z-i\sigma}{z+i\sigma} \qquad (5)$$

This transformation maps the problem in the z plane into an annulus centered at the origin in the w plane (see Figure 2). This transformation is analytic and conformal throughout the problem domain. The parameters σ and ρ are determined from the problem geometry.

$$\sigma = \left[b^2 - D^2/4\right]^{1/2} \qquad (6)$$

$$\rho = \frac{D}{2(b+\sigma)} \qquad (7)$$

The inverse transformation is given by

$$z = -\sigma i \left(\frac{w+1}{w-1}\right) \qquad (8)$$

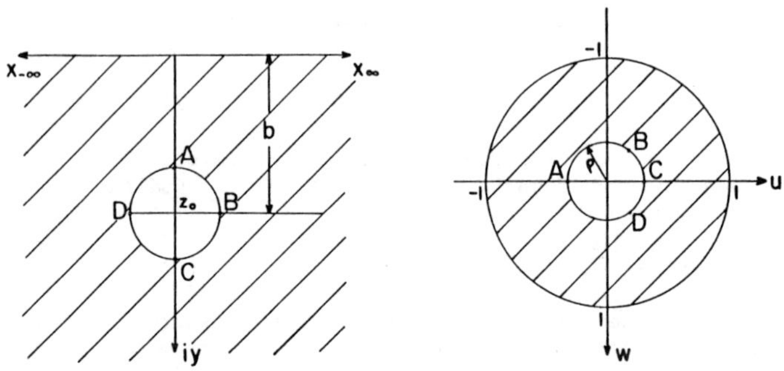

Figure 2. Transformation from physical plane to w plane

Since a harmonic function in the w plane has a corresponding harmonic function in the z plane when the mapping function is analytic, the governing equation in the w plane is

$$\nabla^2 p(u,v) = 0 \qquad (9)$$

in which u and v are the rectangular coordinates in the w plane as defined in Fig. 2. The no-flow boundary condition on the pipe surface is transformed to the condition

$$\frac{\partial p}{\partial n} = 0 \qquad ; u^2 + v^2 = \rho^2 \qquad (10)$$

The mudline boundary condition is slightly more difficult. Rewriting the transformation given by Eq. (8)

$$x = \frac{-2\sigma v}{(u-1)^2 + v^2} \tag{11}$$

$$y = \frac{-\sigma(u^2+v^2-1)}{(u-1)^2+v^2} \tag{12}$$

leads to the condition

$$p = p_o\ f\ \left[\frac{-2\sigma v}{(u-1)^2+v^2}\right] \qquad ;\ u^2+v^2 = 1 \tag{13}$$

It is convenient to change to polar coordinates with

$$u = r\ \cos\theta \tag{14}$$

$$v = r\ \sin\theta \tag{15}$$

The boundary value problem may now be written as

$$\nabla^2 p = \frac{\partial^2 p}{\partial r^2} + \frac{1}{r}\frac{\partial p}{\partial r} + \frac{1}{r^2}\frac{\partial^2 p}{\partial \theta^2} = 0 \tag{16}$$

$$\frac{\partial p}{\partial r} = 0 \qquad ;\ r = \rho \tag{17}$$

$$p = p_o\ f(\theta) \qquad ;\ r = 1 \tag{18}$$

Solving by separation of variables and nondimensionalizing the pressure by the amplitude of the mudline pressure response yields

$$\frac{p}{p_o} = a_o + \sum_{n=1}^{\infty} \left[r^n + \left(\frac{\rho^2}{r}\right)^n\right]\left[a_n\cos(n\theta) + b_n\sin(n\theta)\right] \tag{19}$$

in which the coefficients are given by

$$a_n = \frac{1}{\pi(1+\rho^{2n})} \int_{-\pi}^{\pi} f(\theta)\cos(n\theta)d\theta \ ;\ n \geqslant 0 \tag{20}$$

$$b_n = \frac{1}{\pi(1+\rho^{2n})} \int_{-\pi}^{\pi} f(\theta)\sin(n\Theta)d\Theta \ ;\ n \geqslant 1 \tag{21}$$

This solution in the w plane is transformed back to the z plane by

$$r = \frac{\left[(x^2 + y^2 - \sigma^2)^2 + 4\sigma^2\ x^2\right]^{1/2}}{x^2 + (y + \sigma)^2} \tag{22}$$

$$\theta = \tan^{-1}\left(\frac{-2\,\sigma x}{x^2 + y^2 - \sigma^2}\right) \tag{23}$$

SOLUTION BEHAVIOR

The simplest form of the mudline boundary condition, Eq. (3), is that obtained from linear wave theory

$$p = p_o\left[\cos(kx) + i\,\sin(kx)\right] \tag{24}$$

in which k is the wave number determined in accordance with the linear wave theory dispersion equation and p_o, the amplitude of the mudline pressure response, is given by

$$p_o = \gamma\,\frac{H}{2}\,\frac{1}{ch(kh)} \tag{25}$$

where H is the wave height and h is the still water depth. For this boundary condition, the solution for pressure due to the real part of Eq. (24) may be written as

$$\frac{p}{p_o} = \frac{2}{\pi} I_o + \frac{4}{\pi}\sum_{n=1}^{\infty}\left[r^n + \left(\frac{\rho^2}{r}\right)^n\right]\frac{I_n}{(1+\rho^{2n})}\cos(n\theta) \tag{26}$$

in which

$$I_n = \int_0^{\pi/2} \cos\left(\frac{k\sigma}{\tan\alpha}\right)\cos(2n\alpha)\,d\alpha \tag{27}$$

The integrals I_n depend only on the value of n and the wave-geometry factor $k\sigma$. I_n is shown in Figure 3 for two representative values of $k\sigma$. The series solution converges quickly and typically less than 5 terms are required. A similar solution is obtained for the imaginary part of Eq. (24) except the cosines are replaced by sines.

EXPERIMENTAL APPARATUS

Experiments were conducted at the Oregon State University Wave Research Laboratory. The large wave channel at the laboratory is 104 m (342 ft) long, 3.66 m (12 ft) wide and 4.57 m (15 ft) deep. The generator is of the hinged flap variety, generating waves up to 1.5 m (5 ft) high in a water depth of 3.35 m (11 ft). The useful period range is 2 to 5 seconds.

A test section for the wave soil pipeline experiments was constructed in the middle third of the wave channel as shown in Figure 4. Waves generated at the waveboard propagate past the test section and were absorbed on a 1:3 slope rubble beach. Reflected waves were reduced to approximately 20% of the incident energy.

The test section was fabricated from 1.9 cm (0.75 in) exterior plywood, reinforced with 5x10 cm (2x4 in) vertical and horizontal bracing. The test section was 11 m (36 ft) long, 1.22 m (4 ft) wide and 1.19 m (3.9 ft) deep. A typical cross section, without the buried pipeline, is shown in Figure 5. The pipeline was buried at various depths in the fine sand. The upper gravel layer (d_{50} = 10.5 mm) was

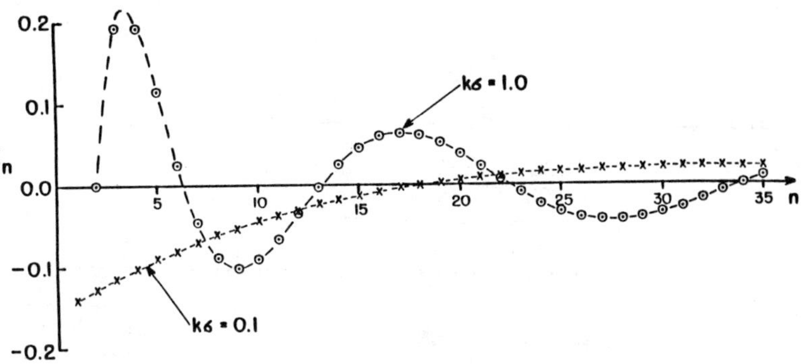

Figure 3. Typical values of the integrals given by (27)

necessary to provide a stable cover for the large waves tested. This gravel was very permeable and provided satisfactory pressure transmission to the fine sand layer. The lower sand (d_{50} = 0.3 mm) was very uniform and selected because of its potential for dynamic liquefaction (Seed and Idriss, 1967).

An aluminum pipeline test section 0.22 m (0.72 ft) in diameter and 1.07 m (3.50 ft) long was constructed. The pipeline was instrumented with four pressure transducers and four total stress cells mounted flush with the cylinder surface. These instruments were mounted in pairs at the quarter points around the cylinder. The pressure transducers (Druck model PDC410) were pre- and post-calibrated for each test configuration and showed very little drift. The pore pressure at the mudline was also measured and used as a reference pressure. The total stress cells were fabricated by technicians at OSU. The stress cell results were somewhat more scattered than the pressure measurements, and will not be discussed in this paper. The cylinder was supported by two vertical cables. Tension in the cables was measured with strain bars mounted inside the cylinder. This vertical tension was varied to simulate different degrees of buoyancy. The free surface elevation was measured with a sonic profiler.

TEST PROCEDURES

The theory assumes uniform properties throughout the sand layer. Consequently, careful preparation of the lower layer of fine sand was an important aspect of the experiments. This was accomplished by fluidizing and reconsolidating the sand layer in a repeatable manner. The cylinder was placed before the fluidizing process.

Wave tests were run at a water depth of 2.44 m (8 ft). Wave heights and periods were selected to span the useful range of intermediate wave numbers and steepnesses. A sufficient number of waves

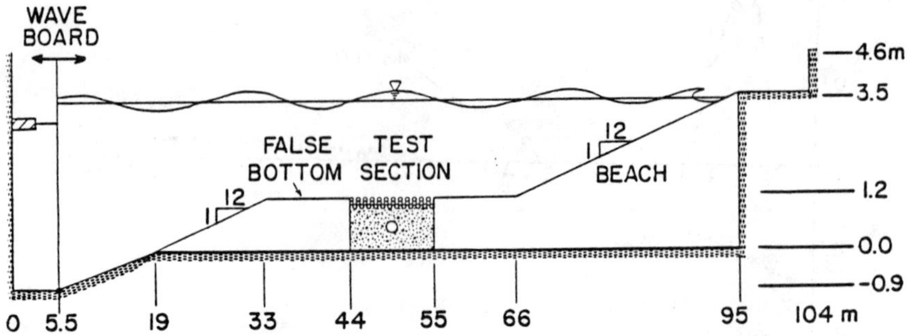

Figure 4. Test section configuration

were run, typically thirty, for the mean pressure to attain a constant value. The free surface, pressure, total stress and lifting force signals were passed through a bank of matched filters, an analog-to-digital converter and stored on magnetic tape. These files were later analyzed to determine the amplitude and phase of each signal relative to the mudline pressure.

RESULTS

Typical measurements of the dynamic pressure response of the cylinder are given in Figure 6. Each point plotted is the average of 7, 14 or 21 runs at different wave steepnesses and simulated pipeline bouyancies. The wave steepnesses were selected to correspond to 25%, 50% and 75% of the breaking wave height in accordance with Dean's stream function wave theory (Dean, 1974). In general, the wave steepness had little influence on the pressure when scaled by the corresponding mudline value. However, the errors in measurement tended to increase for the shorter wave lengths because the pressures induced by the smaller waves were near the threshold resolution of the pressure transducers. For fairly short waves (h/L_o = 0.5), the 90% confidence intervals were approximately ±39% of the mean. These errors decreased with increasing wave length to ±1.5% for h/L_o = 0.02. The level of simulated bouyancy appeared to have little influence on the dynamic pore pressures as long as this bouyancy was below the breakout force. In this particular set of runs a woven monofilament geotextile was placed between the gravel armor and the fine sand to provide separation. The curves in Figure 6 reveal several anticipated trends; the amplitude of the dynamic pore pressure

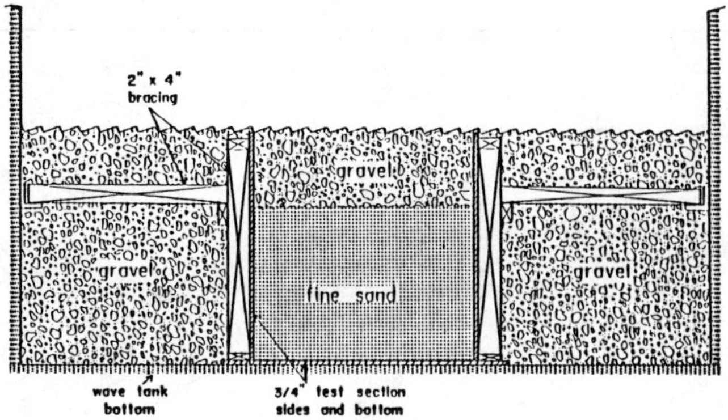

Figure 5. Cross-section of test section

is largest on the top of the cylinder and smallest on the bottom and the shorter waves are more highly attenuated with depth.

Theoretical results are compared with measurements made by Nyhus (1977) in Figure 7. The comparisons are reasonable for both waves with an average error of 8%. The theory is compared with a set of large scale experimental observations made in the present investigation in Figure 8. These measurements are for runs in which the gravel layer was 0.305 m (1.0 ft.) thick, no geotextile was used and the cylinder was buried 0.69 m (2.27 ft.) below the mudline. The measured pressure at the gravel-sand interface was used as the boundary condition for the analytical model. The theory generally underpredicts the measured pressures but the same dependence on wave number is noted for pressure transducers 1, 2 and 3. Transducer number 4 might be suspected of being in error in this sequence of runs because the amplitudes of the signals from the P4 experimental observations tend to exceed the theory. However, there are several factors which complicate the interpretation of the measured response. The theory is for an infinitely deep soil, but in the experiments, in all cases the soil was shallow with respect to the wave length. Infinitely deep soils are known to cause greater dynamic pressure attenuation from finite depth soils, consequently, the theory for infinitely deep soils should fall below the experimental observations. The finite depth potential model, by Monkmeyer and Mantovani (1983) predicts that P2 and P4 are not the same. Experimental measurements by Vincent (1979) suggest that the pressure at the quarter points may even be less than the pressure at the bottom. A second complication is the displacement of the cylinder. The fine sand was prepared by fluidizing and then reconsolidating back to the desired level. For most tests the relative density of the find sand was around 60%. In this condition the sand is fairly weak and the large test waves may induce significant displacements of the test cylinder. Although the signal from transducer 4 is suspect, at this point in the study, it may not be disregarded.

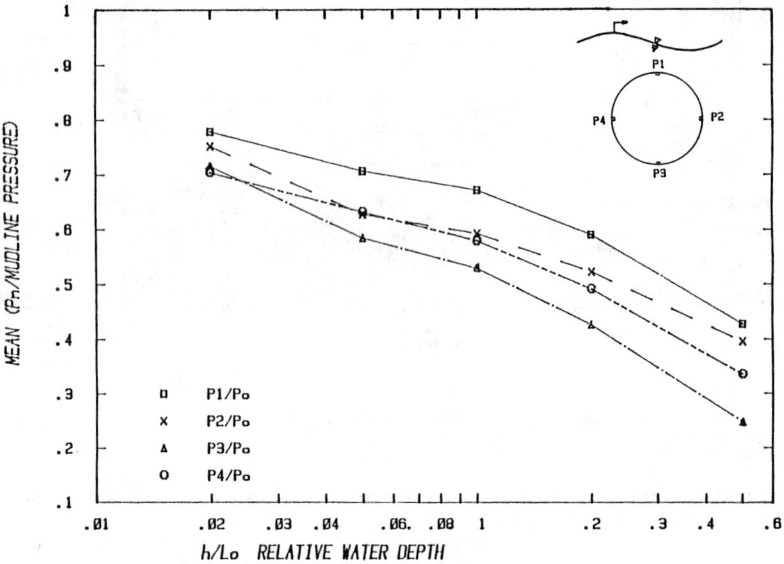

Figure 6. Typical pressure measurements

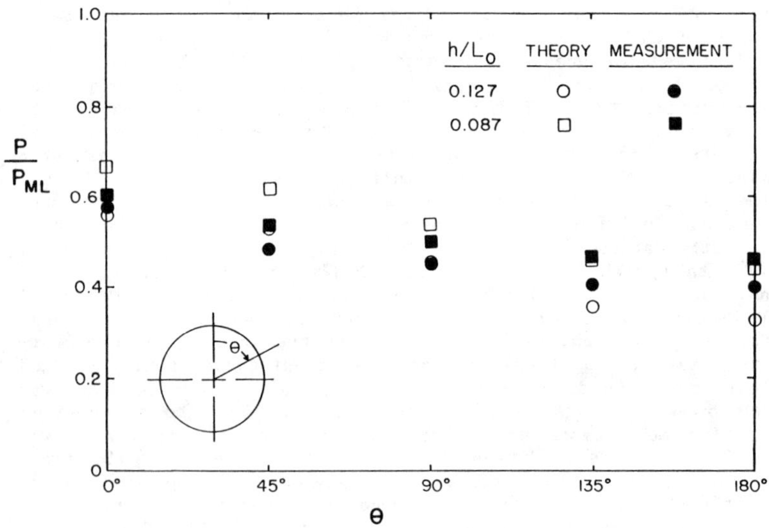

Figure 7. Comparison of theory and measurements by Nyhus (1977)

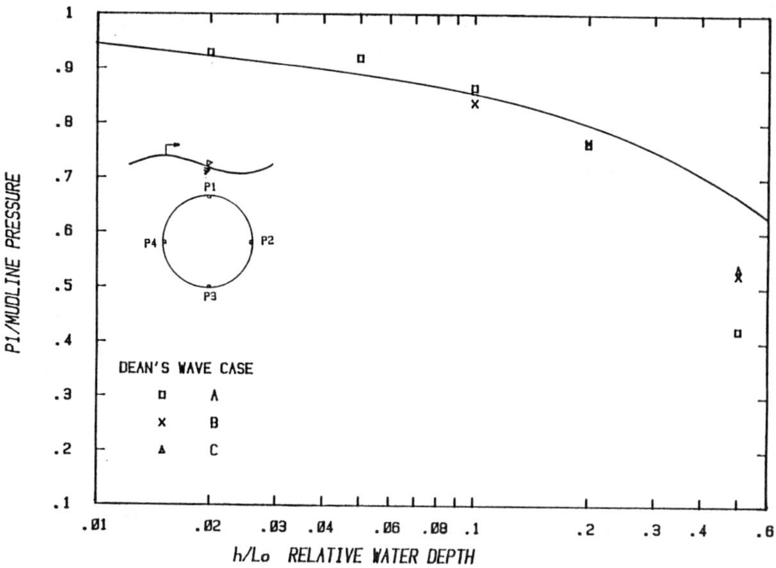
P1/MUDLINE PRESSURE
h/Lo RELATIVE WATER DEPTH
DEAN'S WAVE CASE
A
B
C
P1
P2
P3
P4

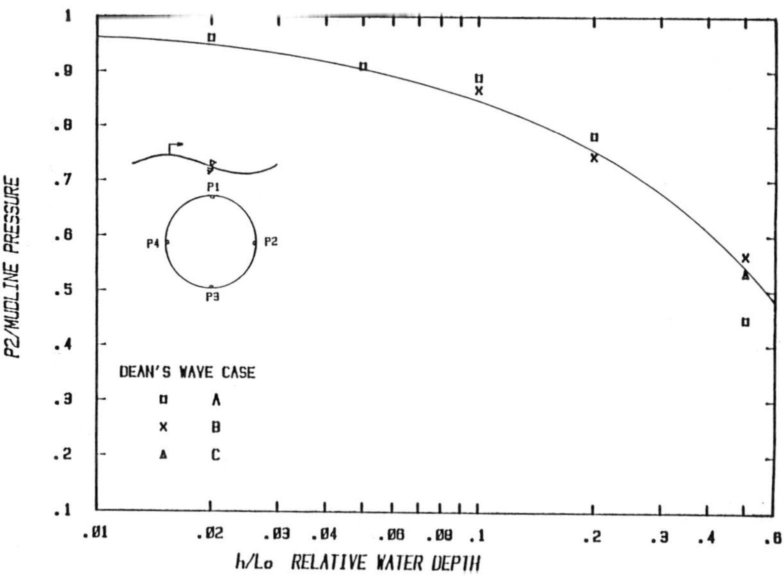
P2/MUDLINE PRESSURE
h/Lo RELATIVE WATER DEPTH
DEAN'S WAVE CASE
A
B
C
P1
P2
P3
P4

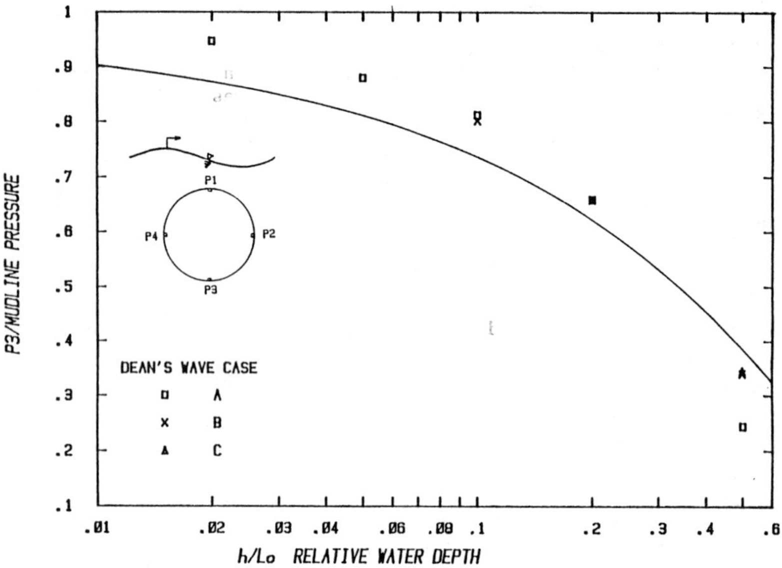

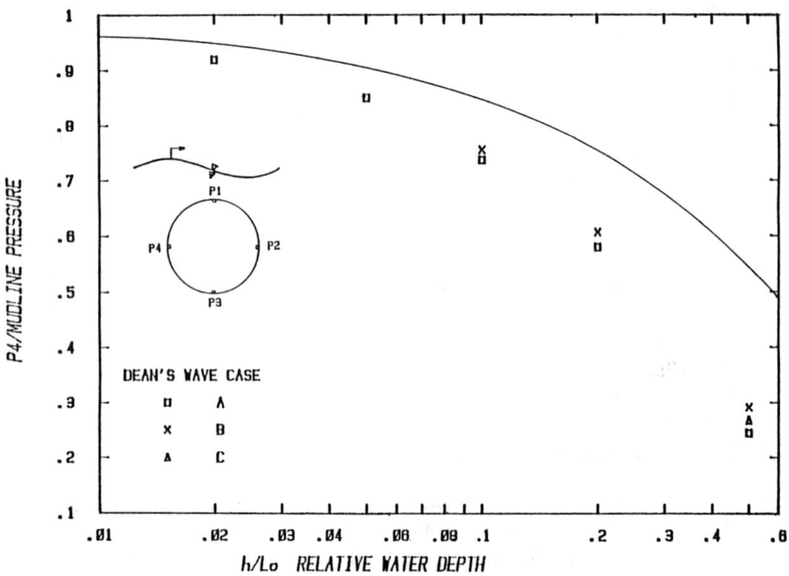

Figure 8. Comparison of theory and large scale tests

DISCUSSION

A potential theory model has been developed to predict wave-induced pressures on buried marine pipelines. An analytical solution is obtained employing conformal mapping techniques. The model is compared with both small and large scale experiments. Model predictions and small scale test measurements are in reasonable agreement. However, there is some disparity between the theory and the large scale tests. This difference may have been the result of instrumentation malfunctions or it may be due to finite depth conditions in the experiments. In tests with small waves using a dense coarse sand, the potential pressure model appears to be adequate. However, in large scale tests (or in the prototype) in which large waves and a finer, lower density backfill material are used, the dynamic pressure response may be more complicated than predicted by the potential pressure model.

ACKNOWLEDGEMENTS

This research was supported by the Oregon State University Sea Grant College Program, National Oceanic and Atmospheric Administration Office of Sea Grant, Department of Commerce, under Grant No. NA79AA-D-00106.

REFERENCES

ASCE (1966) ASCE preliminary research on pipeline research, J. of the Pipeline Div., ASCE, 92:27-71.

Biot, M.A. (1941) General theory of three-dimensional consolidation, J. Appl. Phys., 12:155-165.

Brown, R.J. (1957) Soil mechanics important in marine pipeline construction, Oil and Gas J., Sept. 16, 151-155.

Crowhurst, A.D. (1983) Marine pipeline protection with flexible mattress, Proc. Coastal Structures '83, 370-378.

Dalrymple, R.A. and P. L-F Liu (1982) Gravity waves over a poroelastic seabed, Proc. OSDS '82, 181-195.

Dean, R.G. (1974) Evaluation and Development of Water Wave Theories for Engineering Application, Special Report No1, U.S. Army Corps of Engineers, Coastal Engineering Research Center, 534 pp.

Gade, H.G. (1958) Effects of a nonrigid, impermeable bottom on plane surface waves in shallow water, J. of Marine Res., 16:61-82.

Herbich, J.B. (1977) Wave-induced scour around offshore pipelines, Proc. Offshore Tech. Conf., 79-90.

Hunt, J.N. (1959) On the damping of gravity waves propagated over a permeable surface, J. Marine Res., 16:61-82.

Kober, H. (1952) Dictionary of Conformal Representations, Dover Publications, Inc., 205 pp.

Layton, J.A. and J.L. Scott (1979) stabilization requirements for submarine pipelines subjected to ocean forces, Proc. Civil Engr. in the Oceans, IV, 61-76.

Lai, N.W., R.F. Dominguez and W.A. Dunlap (1974) Numerical Solutions for Determining Wave-Induced Pressure Distributions Around Buried Pipelines, Texas A & M University, Sea Grant Pub. No. TAMU-SG-75-205, 93 pp.

Lennon, G.P. (1983) Wave-induced forces on buried pipelines, Proc. Coastal Structures '83, 505-518.

Liu, P.L.F. and T.P. O'Donnell (1979) Wave-induced forces on buried pipelines in permeable seabeds, Proc. Civil Engr. in the Oceans IV Conf., 111-121.

MacPherson, H. (1980) The attenuation of waves over a nonrigid bed, J. Fluid Mech., 97:721-742.

Madsen, O.S. (1978) Wave-induced pore pressures and effective stresses in a porous bed, Geotechnique, 28:377-393.

Mallard, W.W. and R.A. Dalrymple (1977) Water waves propagating over a deformable bottom, Proc. Offshore Tech. Conf., 141-146.

Matsui, K. (1982) Pressure and Stress Distribution Around a Pipeline Buried in a Poro-Elastic Seabed, Dissentation in Civil Engineering, University of Houston, 112 pp.

McDougal, W.G. and C.K. Sollitt (1982) Geotextile stabilization of seabeds: theory, Proc. OSDS '82, 211-227.

McDougal, W.G. and C.K. Sollitt (1983) Geotextile applications in buried marine pipelines, J. of the Trans. Div., ASCE, in review.

Monkmeyer, P.L., P. Mantovoni and H. Vincent (1983) Wave-induced seepage effects on a buried pipeline, Proc. Coastal Structures '83, 519-531.

Nyhus, K.A. (1977) Hydrodynamiske Krefter Pa Nedgravd Ror, River and Harbour Lab., Norwegian Inst. of Tech., 73 pp.

Putnam, J.A. (1949) Loss of wave energy due to percolation in a permeable sea bottom, Trans. Am. Geophys. Un., 30:349-356.

Reid, R.D. and K. Kajiura (1957) On the damping of gravity waves over a permeable seabed, Trans. Am. Geophys. Un., 38:662-666.

Seed, H.B. and I.M. Idriss (1967) Analysis of soil liquefaction: Nigata earthquake, J. of Soil Mech. and Found. Div., ASCE, 93:83-108.

Sleath, F.A. (1970) Wave-induced pressures in beds of sand, J. Hyd. Div., ASCE, 96:367-378.

Vincent, H.B. (1979) Wave-Induced Seepage Effects on a Horizontal Pipe Buried Beneath the Sea, unpublished Independant Study Report, Dept. of Civil and Environ. Engr., Univ. of Wis.-Madison.

Yamamoto, T. (1977) Wave-induced instability in seabeds, Proc. ASCE Spec. Conf., Coastal Sediments, 898-913.

Yamamoto, T., H. Koning, H. Sellmeijer and E. Van Hijum (1978) On the response of a poroelastic bed to water waves, J. Fluid Mech., 78:193-206.

HAZARDS TO MARINE PIPELINES IN THE BREAKING WAVE ZONE

J.T. Powers*
Member
ASCE

ABSTRACT

Hazards to a marine pipeline passing through the surf zone are outlined. Methods of protecting the pipelines are discussed. The practical application of currently used design procedures for determining the length of pipeline to bury and the depth of burial through the surf zone is examined in detail. Cautions on the use of these currently used procedures are presented.

INTRODUCTION

One of the most adverse environments that a marine pipeline must traverse is the nearshore wave breaking zone and water-land interface. Stability and mechanical protection for the pipeline are the key concerns. This paper discusses the hazards which must be designed for, the design decisions required, and examines proposed solutions to these design problems, including the vertical stability design problem.

Most shore approaches include a combination of buried and unburied pipeline unless the entire line is to be buried. Choosing the optimum length and depth of burial is the critical design decision which this paper addresses.

HAZARD AND DESIGN CONSIDERATIONS

Chief hazards to a pipeline in the shore approach zone are damage by floating objects and damage from natural forces. Floating objects include ships, debris, and ice floes. Hazards from natural forces include:

- -shoreline erosion (leaving the pipeline exposed)
- -seabed sediment transport (removing protective overburden)
- -pipeline movement during storms due to high water particle velocity (for unburied lines)
- -liquefaction of seabed in breaking wave zone

*PI Tech, Inc./Post Office Box 740369/Houston, Texas 77274

Mechanical protection from floating objects is usually provided by burial, at least through the land-sea interface, although rip-rap and other means have been used with varying degrees of success. In practice, it is usually a specified condition that the pipeline be buried as it crosses the shoreline. Reasons include aesthetics, maintenance of grade (for instance, an outfall), or mechanical protection. In most areas of the United States burial is a regulatory requirement. For example, oil and gas pipelines in the Gulf of Mexico must be buried out to the 200 foot water depth contour. Most of the countries bordering the North Sea and the country of Australia also require burial. Where burial outside the surf zone is not required designers must choose the optimum point where the cost of burial exceeds its benefits.

Failure to design for natural hazards usually results in a loss of on-bottom stability, either vertically or laterally. Lateral instability for unburied pipelines means the pipeline weight is insufficient to withstand the water current forces tending to move it laterally. For the nearshore buried or trenched portion the hazard becomes vertical instability, either downward due to soil bearing failure or upward due to soil liquefaction caused by breaking waves or earthquakes.

A review of past problems illustrates needed areas of concern. At least 2 major natural gas pipelines were laid through the shore approach/surf zone in 2 to 3 meter deep dredged trenches with specific gravities between 1.2 and 1.3. After winter storms the pipelines had worked their way to the seabed surface and were exposed on the seabed. Millions of dollars were spent to rectify the situation, in one case with little positive results. When faced with laying another pipeline through the same area the pipeline owner asked his design staff to reconsider their shore approach design criteria to ensure that the line was heavy enough to stay buried. When this subsequent design was presented to installation contractors they were reluctant to bid the job because the pipeline was too heavy. After numerous delays, charges of poor design, and countercharges a compromise was reached. This is obviously a situation most of us want to avoid.

Pipelines in Arctic areas present more difficult decisions due to the ice pack or iceberg grounding problem which gouges deep holes in the shallow water seabed. Pipeline designs for one of the Canadian Arctic gas pipelines included a tunnel through the surf zone. Statpipe's proposed natural gas system offshore Norway also includes tunnels through the shore approach in its design. This illustrates the extent to which designers must go for safety in extreme environments.

LENGTH OF BURIAL

If offshore burial is not specified designers must choose the point where burial is not essential. Dredging/trenching costs increase substantially with water depth. Since the objective is to provide a safe but economic design, designers should choose the minimum length of burial consistent with this objective.

Length of burial is usually based on a balance of concrete weight coating required for unburied stability further offshore versus the amount of dredging required. On-bottom stability of unburied pipelines refers to the amount of in-water pipeline weight required to withstand the water currents trying to move it. The basic stability equation is:

$$\frac{W}{D_c} = \left(C_L + \frac{C_D}{F}\right) \frac{w\,V^2}{2g}$$

where

- W = submerged weight (#/ft) of pipeline
- D_c = outside diameter of weight coating (ft)
- C_L = lift coefficient
- C_D = drag coefficient
- F = friction coefficient
- w = water density (#/ft³)
- V = current velocity (feet/per second)
- g = 32.2 ft²/sec

If a sensitivity analysis is performed on the above equation it becomes obvious that the most critical variable is current velocity "V". To ensure that one remembers that "V" is not only steady state current but also that oscillating current induced by waves passing overhead, this paper will refer to "V" as water particle velocity rather than current velocity.

Many technical papers have been written on the lift and drag coefficients to be used in the above stability equation. Reference (3) presents the most used values. Designers should avoid using coefficients from handbooks since these are normally not based on pipelines lying on the seabed but on suspended or vertical pipes with water flowing around each side. Lift and drag coefficients of $C_L = 0.6$ and $C_d = 0.8$ are accepted values for Reynolds numbers above 3×10^5.

Friction coefficients quoted in literature vary from 0.2 to over 1.0. Reference (4) makes a strong case for using 0.6 to 1.0 in most cases where lateral stability is being analyzed. This value represents an effective resistance to lateral movement rather than a true friction coefficient as would appear on a soils lab report. If the seabed is soft the pipeline sinks into the soil, requiring a lateral shear failure of the soil before movement can occur, thereby giving a higher effective resistance to movement. A stability analysis on an empty pipeline (referred to later as an installation case) which has not been filled with hydrostatic test water should consider the fact that the pipeline may not have "settled" into the mud enough to develop the full effective lateral resistance so a smaller coefficient may be indicated. If the bottom is sand the coefficient of friction normally used is 0.65 to 1.0.

Water particle velocity ("V") is both the most sensitive parameter and the most difficult to accurately assess. This is particularly true for shore approaches. As the waves enter shallow water two changes begin to occur which are of interest to marine pipeline designers. First the wave encounters obstructions such as points of land, reefs, etc. and the wave direction changes. A shoaling analysis is used to determine the new wave directions. Reference (2) discusses shoaling analysis and other techniques used to determine what wave parameters are applicable as the wave approaches the shoreline.

The second changes is the increase in bottom water particle velocity "V" as the wave enters shallow water. This increase is due to the increased effect of the wave-induced component. Figure 1 illustrates the magnitude of this increase. As "V" become larger the weight coating of unburied lines must be increased to prevent movement. Note the sharp increase in water particle velocity as the water depth decreases to about 10 meters. This increase is characteristic of all locations where significant waves are present.

On Figure 1, "Operation Case", refers to maximum design conditions which are expected over the pipeline's design life. A one percent chance of recurrence storm (usually called a 100-year storm) is commonly selected. The pipeline is assumed to be full of whatever liquid or gas it is transporting. "Installation Case" refers to the installation phase when the pipeline is being pulled or laid into position. It is normally empty during installation and therefore lighter. A 5 to 15 percent chance of recurrence storm (5 year to 15 year storm) is selected for this condition, depending on site conditions and installation method.

After performing a shoaling analysis to obtain wave directions at some water depth increments (3 to 5 meters usually) and a wave theory calculation to determine the near-seabed water particle velocity at the same water depth increments, designers must now make a decision on whether to use water particle travel directions resulting from the shoaling analysis or take the conservative position and

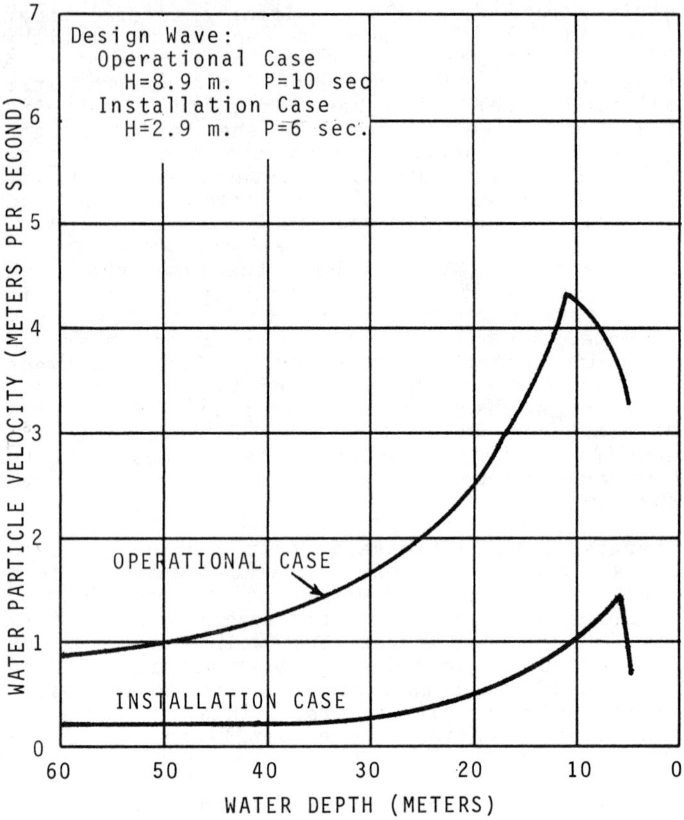

FIGURE 1
DESIGN WATER PARTICLE VELOCITIES

assume the full velocity acting at 90° to the pipeline route. In view of the need to have continuous pipeline service many facility owners prefer the conservative position while installation contractors and/or lump sum bidders prefer the design solution which results in the smallest pipeline weight. This less conservative position means taking the wave travel direction from the shoaling analysis and using only the component of the velocity which acts at 90° to the pipeline route for design.

Substitution of the selected design velocities "V" for the Installation (empty pipeline) Case and the Operational (full pipeline) Case into the stability equation for each water depth increment results in a required pipeline weight for each water depth increment. From this the thickness and density of concrete weight coating can be calculated. Figure 2 illustrates how the results may be presented for maximum clarity. At some water depth the required concrete coating usually becomes impractical for large design waves. It is at or before this point that burial should begin and proceed through the shoreline. The exact point is selected on the basis of a proper balance between weight coating thickness and dredging quantities. Using Figure 2 as an example the empty pipeline case requires a minimum of 4.375 inches of concrete for stability outside the surf zone. The Operational Case requirement exceeds the Empty Pipeline Case weight coating thickness at about the 20 meter water depth and continues to sharply increase, indicating the we gain very little by increasing the coating thickness. In this case the decision was to apply 4.5 inches of concrete and begin trenching at 20 meters.

A note of caution should be observed on selection of weight coatings. Experience has proved that continuous weight coatings are far superior to intermittent, set-on block or bracelet type weights. Intermittent type weights often cause scour and the end result is a long unsupported length of pipeline with a concentrated load at midspan.

Densities of concrete weight coatings vary from about 140 to 190 pcf. In the higher densities, aggregate usually must be imported from other areas so the concrete cost will be greater for the higher densities.

A final decision on length of burial must be made on the shore end. Designers must consult records and talk with longtime residents to determine if the shoreline is receding or growing seaward. If it is receding, some estimate should be made of rate and the pipeline buried below the expected future beach elevation. The length of extra burial on shore would be coincident with the rate of shoreline erosion expected over the pipeline's design life.

To summarize, design steps required to select an optimum length of burial are :

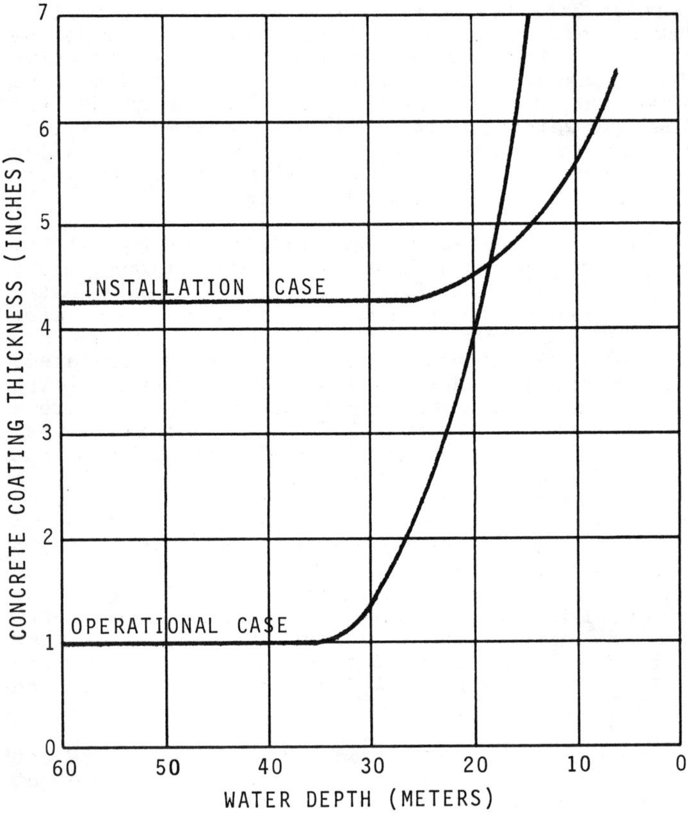

FIGURE 2

MINIMUM THICKNESS OF 140 PCF CONCRETE COATING TO PROVIDE STABLE PIPELINE FOR VELOCITIES IN FIGURE 1

a. Select design wave and current characteristics for the installation or empty pipeline condition (usually a 5 to 15 year storm return period although 25 year periods are sometimes used for extreme conditions) and the operational or full pipeline conditions (usually a 100-year storm return period).

b. Select a pipeline route and pipe and corrosion coating weights, excluding weight coating.

c. Perform a shoaling analysis to determine wave directions as the wave approaches the shoreline. Directions should be selected for about 5 meter water depth intervals.

d. Using applicable wave theory (Stokes, Cnoidal, Solitary, etc.) translate surface wave data into water particle velocities near the seabed, again for no greater than 5 meter water depth intervals.

e. Select wave-induced water particle directions to use at the 5 meter water depth increment points (conservative designers use the full velocity acting at 90° to the pipeline, others use the results from the shoaling analysis).

f. Combine steady current (tidal, etc.) components with wave-induced water particle velocities and select design water particle velocity "V" to insert in stability equation.

g. Calculate required weight coating thickness for each water depth increment and each pipeline condition (full or empty) and each weight coating density being considered.

h. Plot the calculated valves on a graph similar to Figure 2 and select the most economical coating density and water depth at which burial should begin.

DEPTH OF BURIAL

A decision on how deep to bury a pipeline in the shore approach depends on:

- -extent of exposure to floating hazards
- -soil type and characteristics
- -presence of rock or coral

As mentioned earlier, if maximum protection from floating hazards is necessary, a protected tunnel may be used. Directional drilling through the surf zone also is an established technique. Although expensive, these solutions will provide maximum protection and designers eliminate the need for selecting a pipeline burial depth.

If shore approaches are solid rock or coral a trench can be blasted through the wave breaking zone to provide lateral stability and protection from floating objects. This trench should be deep enough to provide at least 0.3 meters between existing seabed surface and top of pipeline coating. If the pipeline is in an area subject to the jurisdiction of the U.S. Corps of Engineers deeper burial may be required.

A majority of pipelines crossing the shore do not fall in either of the above categories. They cross shorelines composed of sand, clay, silt or mixed soils compositions and designers must choose a depth of burial which will provide protection from hazards such as seasonal depth variations due to sediment transport and soil liquefaction due to earthquake of breaking waves. An acceptable solution to the depth variation problem is to make depth measurements at varying times of the year to define the trend and magnitudes of change, then base designs on the most critical condition. To minimize construction costs the installation period should obviously coincide with the period when the sediment has moved out.

Protection from soil liquefaction is a more complicated problem. One of the first problems is obtaining adequate soil data. Soil data gathering in shallow water presents problems in that large data gathering vessels will not enter shallow water so smaller boats must be used. Small boats do not have space or facilities to accommodate the larger soil sampling devices. The unfortunate result of this is that designers must often make decisions on this most critical pipeline zone based on insufficient data.

Marine pipeline designers are concerned about liquefaction because:

a. The pipeline may "float" out of a liquefied sediment if the specific gravity of the sediment is greater than that of the pipeline.

b. In granular soils the pipeline may rise slightly due to a lessening of pressure under wave troughs, allowing sand particles to "run" under the pipe, eventually "jacking" the pipeline out of its trench.

c. In very soft, under-consolidated sediments liquefaction may be so deep that a heavy pipeline will sink so far into the sediments that failure will occur due to differential settlement.

d. In mudslide areas the entire seabed, down to 10 or 20 meters below the seabed surface, may move downslope, taking the pipeline with it.

Design in mudslide areas is beyond the scope of this paper, so it will not be discussed further. Numerous OTC papers discuss the problems, principles, and design solutions.

Nataraja and Gill (7) and Chang and Ford (2) present simplified procedures to evaluating the potential depth of liquefaction due to passing surface waves. Although the values and details suggested by the authors vary slightly, both essentially (a) calculate the pressure induced on and in the seabed due to waves passing overhead, (b) calculate the soil's resistance to liquefaction, and (c) select the distance beneath the seabed where these two values are equal. Rate of increase of seabed pressure due to passing waves is shown on Figure 3. This increasing pressure increases the pore water pressure and therefore the stress on the seabed sediments. Obviously the pore water pressure increase due to the overhead waves decreases with depth below the seabed and the soils strength increases with overburden so they will cross at some depth, unless the strength at the seabed is greater than the wave-induced pressure. When plotted the values take the form shown in Figure 4.

Some unanswered questions from these analyses are:

a. Should the top or bottom of the pipeline be placed at the trench depth where soil resistance equals wave-induced stress? For large diameter lines this is a significant difference in dredging.

b. Since this analysis is based on in-situ strengths, what is the effect of the disturbance caused by trenching? It appears logical to assume that the soil used to replace the excavated soils (or that naturally replacing them) would have different liquefaction potential.

c. If the bottom of the line is to be placed at the trench depth where soil stress equals soil resistance or above, what weight should the pipeline have to resist "floating" out of the heavy liquefied sediment?

When reviewing the available references another problem faces the designer. Different equations are presented by different authors for both the wave-induced pressure at the seabed and the pressure at some point below the water/soil interface.

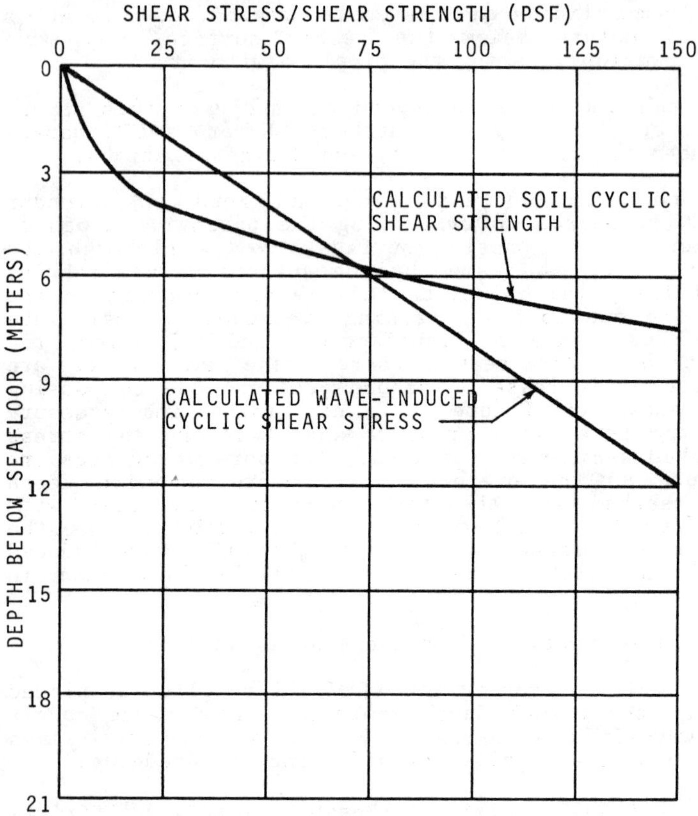

FIGURE 4

TYPICAL PLOT
SOIL SHEAR STRENGTH AND WAVE-INDUCED SHEAR STRESS
VS. DEPTH BELOW SEAFLOOR

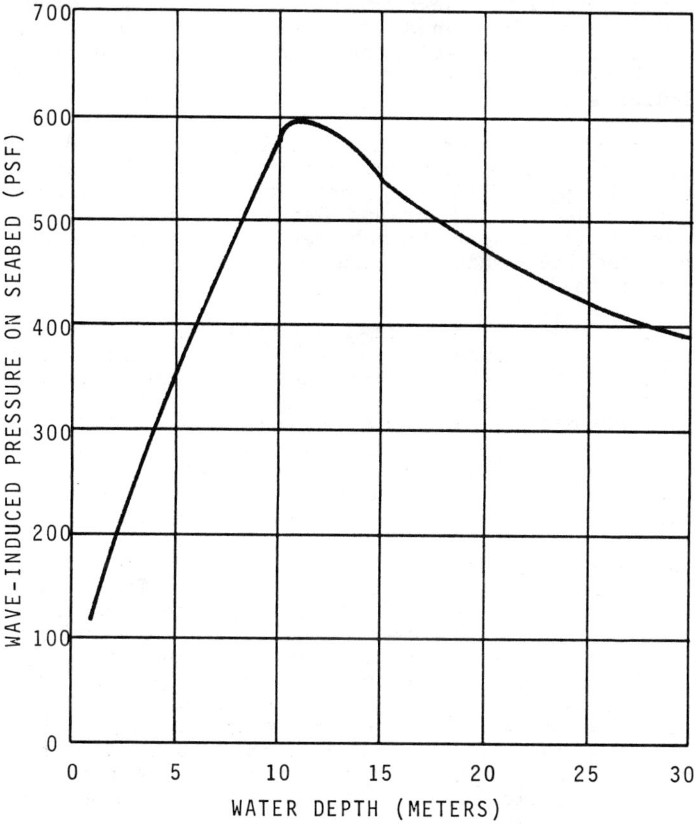

FIGURE 3
TYPICAL EXAMPLE
WAVE INDUCED VERTICAL PRESSURE
ON SEABED

Some equations are more appropriate to porous sediments while others were developed to predict pressure in non-porous seabeds. After soils data gathering is complete designers are better able to select the formulae best suited for a particular site.

After selecting a set of formulae and calculating the depth of potential liquefaction, designers must decide if burial to that depth is practical. If not, what is the proper balance between pipeline weight and depth of burial? Costs become excessive if a deep trench depth is specified and very heavy pipeline weights are difficult to nearly impossible to install.

When viewed in terms of potential hazards, it is highly probable that the maximum storm will occur during the pipeline's operational life as opposed to the installation phase. Therefore, for liquid filled pipelines the linefill plus pipe weight usually results in sufficient weight to overcome the buoyancy of a liquefied soil. For liquid lines, it logically follows that there is little justification for placing the bottom of the pipeline below the depth where soil resistance equals soils stress.

Gas lines present a different problem because the filled weight is near the empty pipeline condition. It is gas lines which have "floated" out of sandy surf zone crossings. If one examines the equations used to calculate wave-induced pressure at the seabed it is preceded by a ±, plus under the wave crest and minus under the trough. In sandy soil, the pressure differential both plus and minus reaches deep into the sediment as shown by Tsui (9) and Moshagen (6). When the minus pressure is added to the pipeline's buoyancy this can be enough to cause a slight cyclic rise in the pipeline at about the wave period. As each rise occurs, some sand slides under the pipeline and eventually it "jacks" its way to the surface.

This understanding of an upward pressure extending for a considerable distance into porous seabeds is borne out by the model test results by Tsui and Helfrich (9). These authors compared their model results with some of the more popular formulae and found that pressures measured in porous seabeds (models) were higher than predicted theory. Suhayda et al (8) also reported that field measurements in soft sediments indicated that wave-induced pressures on and below the seabed were up to 35% higher than predicted by linear theory. Since the backfill and a newly laid pipeline will probably be both soft and porous it seems reasonable to conclude from the model tests and field measurements that the plus and minus pressures may be even higher than predicted by currently used design equations. Deeper burial does not appear a practical solution due to the excessive dredging required.

In the cases of gas lines in soft or sandy seabeds designers must then decide on a proper balance between burial depth and pipeline weight. A reasonable approach, if we accept the procedures used in references (2) and (7) to calculate the increase in positive or downward pressure with depth below the seabed, is to use the same procedure to estimate the negative or upward pressure at given depths below the seabed. If one then adds the buoyancy of the pipeline to this negative upward pressure this gives the minimum required downward pressure or pipeline weight to prevent vertical movement. If the trench is backfilled, the weight of this backfill may be subtracted from the required downward pressure in establishing pipeline weight. If the trench is not to be backfilled the value of wave-induced pressure at the ditch depth should be used.

In areas where the trench backfill (either natural trench filling or machine applied) will have a high liquefied specific gravity a further check should be made to ensure that gas lines will not "float" in the event of liquefaction. A procedure for estimating the liquefied specific gravity of a soil was presented by the ASCE Committee on Pipeline Installations (1961). It recommends extending the liquid limit curve, which is a plot of number of blows vs. water content, until it crosses the 0.01 blows value. Supposedly, at this water content, the soil/water mixture acts as a fluid with little or no shear strength. With the water content corresponding to 0.01 blows one enters the water content vs. mass specific gravity graph to obtain the estimated maximum liquefied specific gravity. This value can be used to check the pipeline weight calculated using the procedure outlined in the previous paragraph. Liquified specific gravities as high as 2.0 have been reported but 1.2 to 1.3 are more common.

SUMMARY AND CONCLUSIONS

The methods and procedures discussed herein for protecting marine pipelines as they cross the surf zone and water/land interface will provide protection from the natural and man-made hazards.

Design procedures are available for calculating the length of burial and depth of burial through the shore approach or surf zone. However, designers must recognize that many of these procedures were developed for analyzing the seabed as a foundation, and therefore analyze its in-situ condition. Pipeline installation and trenching modifies the seabed and the analysis calculations must be performed on the backfilled material, not the in-situ material.

A consolidation of the various formulae for calculating seabed and below seabed pressures into a comprehensive procedure for different seabed compositions is the next logical step. This has progressed substantially for mudslide areas with highly under-consolidated material but there is not substantial agreement in the literature for porous and soft natural backfill type materials which are the most commonly encountered.

APPENDIX I - REFERENCES

1. ASCE Committee on Pipeline Installations, "Rational Design for Pipelines Across Inundated Areas", February 1961, Proceedings of the ASCE Pipeline Division.

2. Chang, Henry H. and Ford, J. (1982) "Analysis and Design for Pipeline Shore Approach", ASCE 1982 Annual Meeting, New Orleans, LA.

3. Jones, Warren T. (1976) On-Bottom Pipeline Stability in Steady Water Currents", Offshore Technology Conference, 1976, OTC Paper No. 2598, p. 763-777

4. Liu, (1973) "Damping of Water Waves Over Porous Bed", Journal of the Hydraulics Division, ASCE, No. HY12, December 1973, p. 2263-2271.

5. Lyons, C.G. (1973) "Soil Resistance to Lateral Sliding of Marine Pipelines" Offshore Technology Conference, 1973, OTC Paper No. 1876, Vol II. p. 479-484

6. Moshagen, H. and Torum, A. (1975) "Wave Induced Pressures in Permeable Seabeds", Journal of the Waterways, Harbors and Coastal Engineering Division, February 1975, p. 49-57.

7. Nataraja, M. S. and Gill, H.S. (1980) "Ocean Wave-Induced Liquefaction Analysis, A Review of Available Procedures" ASCE Convention Exposition, Florida, October 27-31, 1980, Preprint 80-571

8. Suhayda, J.H.; Whelan, T. III; Coleman, J.M.; Booth J.S.; Garrison, L.E. (1976) "Marine Sediment Instability: Interaction of Hydrodynamic Forces and Bottom Sediments", Offshore Technology Conference, 1976 OTC Paper No. 2426, p. 30-33.

9. Tsui, Yuet and Helfrich, Steven C. (1980) "Wave-Induced Pore Pressures in a Submerged Sand Layer" ASCE Convention Exposition, Florida, Oct. 22-31, 1980, Preprint 80-570.

COLLAPSE OF THICK WALL PIPE IN ULTRA DEEP WATER

By Edward A. Verner,[1] M. ASCE, Carl G. Langner,[2]

and Michael D. Reifel[3]

Introduction

The feasibility of laying offshore pipelines in "deep waters", out to 3,000 ft (914 m) depth, has been studied and established as feasible by the offshore industry. The industry is now investigating the feasibility of laying pipe in "ultra deep waters", defined as depths beyond 3,000 ft (914 m), motivated primarily by recent U.S. East Coast OCS lease sales. Current interest in the Baltimore Canyon area is primarily in the range of depths from 3000 to 8000 ft. (914 to 2,438 m.). A thorough understanding of the buckling and collapse behavior of pipe will be necessary before undertaking to lay pipe in these great depths.

This study is concerned with the situation of a pipeline on or near the seafloor, subjected to a large external hydrostatic pressure. It is concerned primarily with the construction phase in which a long pipe span is suspended from a surface pipelay vessel. Pure external pressure, pressure plus tension, and pressure plus bending are the most probable loadings and are discussed in that order in the following three sections of this paper. The combination of bending, tension, and pressure is not considered in this study since large tension stresses and large bending stresses rarely occur simultaneously in deep water pipelaying operations.

The type of failure of concern in this study is the buckling or yield type of failure, rather than fracture. Buckling of pipes may occur in the form of wrinkles due to bending and/or axial compression, or it may occur in the form of flattening of the cross-section due to external pressure and/or bending. This latter form of buckling is usually referred to as collapse.

While the above considerations establish the types of loading to be considered, the behavior of the pipe will depend on the diameter to thickness ratio and the yield strength. In order to further bound the problem, the diameter to thickness range of 10 to 50 has been selected, and the yield range will depend upon the materials available.

---

1. Vice President, TERA, Inc., P. O. Box 740038, Houston, Texas 77274
2. Staff Research Engineer, Transportation Research and Engineering Department, Shell Development Company, P. O. Box 1380, Houston, Texas 77001
3. President, TERA, Inc., P. O. Box 740038, Houston, Texas 77274

Materials being considered for this study include oil well casing as well as conventional high strength line pipe. While the practicality of casing as pipeline material in ultra deep water has not been established, it is considered in this study because (1) it is available in higher strengths and relatively thicker wall than line pipe (although in limited sizes), (2) screwed connection technology is attractive from the installation viewpoint, and (3) a wealth of data is available with respect to collapse testing and design.

The purpose of this study is to determine what knowledge, data, design guidelines, theories, etc. are available in the literature, and to compare briefly some of the data with the theories. The literature review was completed in 1981 and thus includes no literature beyond that date.

Collapse Due to Pressure

Elastic Collapse. - Collapse of a long tube by external pressure is physically similar to collapse of a ring by an external line load normal to the surface of the ring. This problem was apparently (see Southwell (1914), p. 209) first solved by Bresse in 1866. It is also often attributed to Levy (1884). For this problem, the elastic bifurcation pressure P_e is given by:

$$P_e = 3EI/r_m^3 \tag{1}$$

where E is Young's modulus, I is the area moment of inertia of the ring cross-section, and r_m is the radius to the centroid of the circular ring. Southwell gives a rigorous solution for the case of lateral pressure on a long thin tube, and attributes the first solution for the long tube to Bryan in 1888. The bifurcation pressure for thin wall tubes is given by:

$$P_e = 2E/\{(1-\nu^2)(\xi-1)^3\} \tag{2}$$

where,

$$\xi = D/t \tag{3}$$

is the diameter to thickness ratio, ν is Poisson's ratio, D is the outside diameter, and t is the wall thickness.

For elastic collapse of long thick wall tubes, there seems to be some question still remaining as to the correct formula. The current formula used for casing, (American Petroleum Institute, 1980), is based on the formulation by Clinedinst (1939):

$$P_e = 2E/\{(1-\nu^2)\xi(\xi-1)^2\} \tag{4}$$

Holmquist published a discussion (see Clinedinst (1939)) saying that there was an error in the derivation and that Eq. 2 above was the correct result. The error in Clinedinst's result can be readily appreciated by carrying out his derivation with internal pressure P_i as well as external pressure, to obtain

$$(P-P_i) + 2P_i/\xi = 2E/\{(1-\nu^2)\xi(\xi-1)^2\} \qquad (5)$$

For the case of equal internal and external pressure, this shows that the tube will collapse when

$$P_i = P = E/\{(1-\nu^2)(\xi-1)^2\} \qquad (6)$$

Additional evidence that Eq. 2 is accurate even for thick walls was published by Boresi (1955), where he showed that there is little difference between Eq. 2 and the exact solution based on elasticity rather than curved beam theory. It is also noted that the solution for very thick wall hyperelastic materials, Hill (1975), tends to vary with $(\xi-1)^3$ much more closely than with $(\xi-1)^2\xi$, in the diameter to thickness range of 22 to 4 for which numerical results are given.

As a practical matter, there is usually little consequence attached to the choice between Eq. 2 and Eq. 4, since the elastic bifurcation pressure only controls for large diameter to thickness ratios where both equations give nearly equal results. The only exceptions are the very high strength materials such as some casing material, and the applications to inelastic collapse through the tangent modulus approach.

The solution for stress in terms of pressure was given by Lame' in 1852:

$$\sigma_r = P(1 - r_i^2/r^2)/(1 - r_i^2/r_o^2) \qquad (7)$$

and

$$\sigma_h = P(1 + r_i^2/r^2)/(1 - r_i^2/r_o^2) \qquad (8)$$

where r is the radius to a point in the tube wall, r_o is the outside radius, r_i is the inside radius, σ_r is the radial stress, and σ_h is the hoop stress. At the collapse pressure (Eq. 2) the average hoop stress is:

$$\overline{\sigma}_{he} = E/\{(1-\nu^2)(\xi-1)^3/\xi\} \qquad (9)$$

For the diameter to thickness range of interest (10 to 50) it is noted that the average elastic collapse hoop stress is in the range of 14 to 452 ksi (97 to 3 116 MPa) assuming Young's modulus is 30,000 ksi (206 843 MPa) and Poisson's ratio is 0.3. Thus the full range of elastic to plastic collapse is of interest.

Plastic Collapse. - An excellent survey of the general topic of plastic buckling is given by Hutchinson (1974). The survey by Sewell (1972) also contains an extensive bibliography. The reader is referred to these articles as an inroad into the complex field of plastic instability. This section is concerned only with the problem in hand which has simple geometry and can be treated by a tangent modulus approach.

The tangent modulus approach is applicable to compressed rings as well as to columns, and its use has been recommended by many authors (see, e.g. Timoshenko (1956), (1961), and Holmquist (1939)), for both rings and tubes. Using this approach the inelastic lateral collapse pressure (no end pressure) is given by multiplying the elastic result by a reduction factor defined by:

$$\eta \equiv P_c/P_e = E_t/E \tag{10}$$

where P_c is the collapse pressure, and the tangent modulus E_t is defined at the average hoop stress at bifurcation.

Tubes and plates are, however, slightly different from bars in that Poisson's ratio deformations are restrained during wall or plate bending, and this creates a two dimensional stress field resulting in some bending stiffness remaining even at the yield stress (Bijlaard, 1949). Bijlaard's results are given by

$$\eta = (1-\nu^2)\{(1-2\zeta)^2 E + 3E_t E/E_s\} \div \{\{(5-4\nu_t)(1+\zeta^2)$$

$$+ 2(5\nu_t-4)\zeta - 3(1 - E/E_s)\gamma^2\}E \tag{11}$$

$$- (2\nu_t-1)\{(2\nu_t-1)\gamma^2 + 3\zeta E/E_s\}E_t\}$$

where E_s is the secant modulus at the bifurcation stress, ν_t is a transition value of Poisson's ratio,

$$\zeta = \sigma_a / \bar{\sigma}_h \tag{12}$$

is the ratio of axial stress σ_a to average hoop stress, and

$$\gamma = \hat{\sigma} / \bar{\sigma}_h = \sqrt{1 - \zeta + \zeta^2} \tag{13}$$

is the ratio of effective stress $\hat{\sigma}$ to average hoop stress by von Mises' yield theory. Here the effective stress is the stress at which the tangent and secant moduli are evaluated. A simplification is to substitute 1/2 for ν_t and then multiply the whole expression by $(3/4)/(1-\nu_t^2)$ as a transition effect for Poisson's ratio. This results in the expression:

$$\eta = \{(\gamma - 0.75)E_s + 0.75E_t\}(1-\nu^2)/\{(1-\nu_t^2)E\ \gamma^2\} \tag{14}$$

which, for lateral pressure ($\zeta = 0$) was obtained by Gerard (1957). Allen (1980) suggests

$$\nu_t = \nu_P - (\nu_P - \nu)E_s / E \tag{15}$$

for the transition value of Poisson's ratio where ν_P is the plastic value for Poisson's ratio.

Huang (1980) gives plastic collapse results which also account for inelastic plate bending in the wall. His results are given both for deformation theory and for flow theory, and also include the effect of axial tension. For the preferred deformation theory, the reduction factor is:

$$\eta = (1-\nu^2)\alpha_{22}/(\alpha_{11}\alpha_{22} - \alpha_{12}{}^2) \tag{16}$$

where α_{11}, α_{12}, and α_{22} are nondimensional stress-strain coefficients:

$$\alpha_{11} = E/E_s + \tfrac{1}{4}(E/E_t - E/E_s)(2-\zeta)^2/\ \gamma^2 \tag{17}$$

$$\alpha_{12} = -\nu_t + \tfrac{1}{2}(1 - E/E_s) + \tfrac{1}{4}(E/E_t - E/E_s)(2-\zeta)(2\zeta-1)/\ \gamma^2 \tag{18}$$

$$\alpha_{22} = E/E_s + \tfrac{1}{4}(E/E_t - E/E_s)(2\zeta-1)^2/\ \gamma^2 \tag{19}$$

It can be shown that this is the same as Bijlaard's result (Eq. 11).

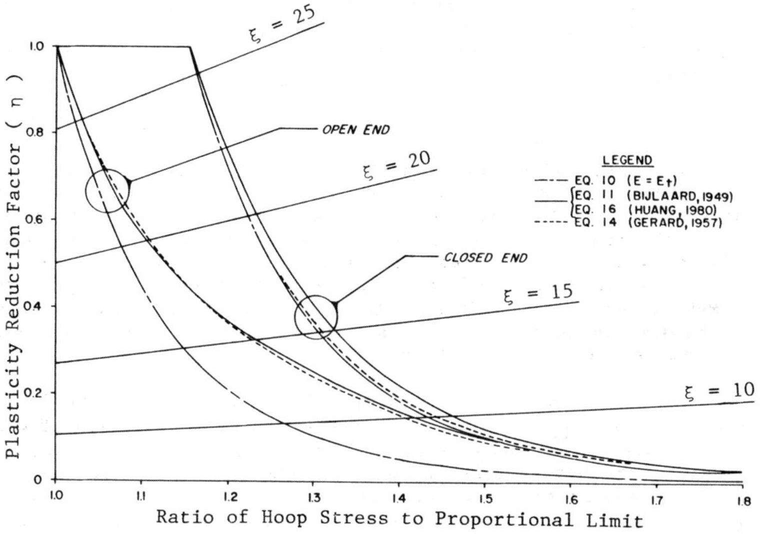

Fig. 1. – Plasticity Reduction Factors

In order to demonstrate the effect of the higher order terms in these equations, the plasticity reduction factors for the various theories are shown in Fig. 1 for an example corresponding to X65 steel. The stress-strain curve for X65 steel above the proportional limit (as proposed by Tvergaard (1976)) can be approximated by:

$$\varepsilon = \{\{(\sigma/\sigma_p)^n - 1\}/n + 1\}\sigma_p/E \tag{20}$$

where σ_p is the proportional limit stress, σ is uniaxial stress, ε is uniaxial strain, and n is an empirical exponent. The figure reflects the data used by Fabian (1981) with 29,700 ksi (204 774 MPa) for Young's modulus, a proportional limit stress corresponding to 0.0016 strain, and an exponent of 9.62.

Reduction factors are plotted in Fig. 1 for both lateral (open-end) pressure and hydrostatic (closed-end) pressure. For lateral pressures, the simplest form (Eq. 10) gives results slightly below the other equations, especially for low diameter to thickness ratios. For example, at a diameter to thickness ratio of 10, the difference in reduction factors is 11%. For hydrostatic pressure, the differences between the various cases are quite small. These observations apply only to the particular stress-strain curve chosen and may not be generally true over the range of steels of interest.

Residual stresses have long been recognized as an important factor in plastic collapse of columns because of early yielding and reduction of the bending stiffness. The potential exists, of course, for the same phenomenon in collapse of long tubes. Very little has been published about this aspect of tubulars. Residual stresses which do not vary circumferentially can be measured by the "slit ring" method (see Mehdizadeh (1976)). In this procedure, a 2-inch (51 mm) ring is cut from the pipe and slit axially. The movement across the slit gives a measure of the residual stress:

$$\sigma_R = (8/9)UE/(\xi-1)^2 \tag{21}$$

where U is the ratio of movement to thickness. Opening of the ring indicates compressive stress on the inside and is called positive. Mehdizadeh found some residual stresses induced in casing by quenching and rotary straightening, with rotary straightening being far more important. Casing with a few ksi residual stress may have 20 to 40 ksi (138 to 276 MPa) residual stress after rotary straightening according to the data of Mehdizadeh (1976). Other methods may cause much less residual stress. Mehdizadeh reported much less residual stress and higher collapse resistance with gag straightening.

The effect of a residual stress that is compressive on the inside wall is to cause early yielding on the inside and thus reduction of wall stiffness. This is accentuated by the nonlinear stress distribution. Eq. 8 predicts the hoop stress to be 22% higher on the inside than the outside at a diameter to thickness ratio of 10, which is at the low end of the present range of interest. Reductions in collapse resistance on the order of 10 to 30 percent have been reported in the literature. Nara (1981) reported tests on the effect of rotary straightening and compared them with a theory based on residual stresses. He also showed some recovery for specimens aged after rotary straightening.

Presumably, uniform residual stresses could be treated by the tangent modulus approach if the wall bending properties in the presence of residual stresses can be defined in terms of the measure σ_R of residual stress. Nonuniform residual stresses are probably more suitably treated in terms of an equivalent ovality.

Imperfection Sensitivity. - Apparently the first solution for the elastic postbuckling of rings was presented by Carrier (1947). This solution gives curvature in closed form at specific load levels up to 80% in excess of the bifurcation load. The curvature ϕ is given by:

$$r_m \phi = 1 + a_1\{a_2 + a_3 \mathrm{cn}(\alpha\theta, k)\}/\{1 + a_4 \mathrm{cn}(\alpha\theta, k)\} \tag{22}$$

where α is a parameter defined by

$$\alpha = 4K(k) / \pi \tag{23}$$

and θ is the Lagrangian angle or coordinate, cn is the elliptic cosine, K is the complete elliptic integral of the second kind, and a_1, a_2, a_3, a_4, and k are parameters which depend on the load level. Parameters given by Carrier in terms of the load coefficient c, are shown in Table 1, and the load is defined in terms of the load coefficient by

$$P = cEI/r_m^3 \tag{24}$$

TABLE 1. - Coefficients for Postbuckling Ring Solution Presented by Carrier (1947)

c (1)	a_1 (2)	a_2 (3)	a_3 (4)	a_4 (5)	k (6)	$a_1(a_2-a_3)/(1-a_4)$ (7)
3.000	0	0	1.0	0	0	0
3.060	.475	-.0280	.971	-.0577	0.1	-.45
3.227	.959	-.0540	.939	-.115	0.2	-.85
3.593	1.53	-.0830	.900	-.184	0.3	-1.27
4.076	2.13	-.115	.854	-.261	0.4	-1.64
5.408	2.98	-.160	.773	-.387	0.5	-2.00

Carrier concluded that the elastic ring is insensitive to imperfections in the shape of the buckling mode. It is noted that since $a_2 + a_3$ and $1 + a_4$ are equal, then a_1 is the relative increase in curvature at the point of maximum curvature and increases to 300% at a load which is 80% more than the bifurcation load. At the point of minimum curvature, the relative change is $a_1(a_2-a_3)/(1-a_4)$ and this changes sign and increases to 200%, indicating the typical "dogbone" shape. Fabian (1977) reports a numerical result that the sides of the tube overlap in the dogbone shape at about 83% above the bifurcation pressure. Kyriakides (1981) found a value of 75% above bifurcation buckling and reported a solution by Tadjbakhsh and Odeh which showed overlap at 68% above bifurcation buckling.

The primary imperfection in tubes is measured by ovality which is defined here as

$$\delta = (D_{max} - D_{min}) / (D_{max} + D_{min}) \tag{25}$$

where D_{max} and D_{min} are maximum and minimum outside diameters. Kyriakides showed solutions for initially oval tubes subject to external pressure that indicated considerable sensitivity of further ovalization to initial ovality. For example, an initial ovality of 0.8% induced additional ovality of 5% at 80% of the bifurcation pressure, increasing to 25% additional ovality at 100% of the bifurcation pressure.

When there is yielding in a structure at the time of bifurcation or at relatively small deflections after bifurcation, the load capacity can drop off rapidly with increasing deflection. This will produce an imperfection sensitivity in the plastic range even if the structure is insensitive in the elastic range. Hutchinson (1974) discussed the possibility of using the general Koiter (1963) theory for nonlinear elastic systems as an approximation for cases where yielding starts before bifurcation. However, it is well known that imperfections can cause significant load reductions even in structures that bifurcate well below nominal yield. Thus, it seems that the surest theoretical approach to imperfection sensitivity near the yield range requires large deflection numerical analyses with material nonlinearities.

Kyriakides (1981) presented such analyses for elastic-plastic rings with strain hardening. Although mean hoop stress and inelastic unloading appear to be neglected in the moment curvature relations, some of the conclusions should still be valid. For the example that he chose (a diameter to thickness ratio of 35.72, and nearly perfectly plastic), considerable imperfection sensitivity to initial ovality was shown even though the yield stress was 4.47 times the elastic bifurcation stress. The imperfection reduction factor was $\eta = 0.80$ for an ovality of 0.8% and $\eta = 0.60$ for an ovality of 2.4%.

Fabian (1981) presented numerical results which also showed unstable postbuckling behavior for tubes with a diameter to thickness ratio of 40 and X52 steel. He showed a drop in load at a deformed ovality of about 1% for an initially round tube. He also showed that initial ovalization drastically reduced the bending moment or curvature capacity in the presence of high pressure. This implies sensitivity of collapse pressure to initial ovality when it is noted that the mechanism of reduction in moment capacity is the increase in ovality due to the applied curvature, followed by collapse into the "dog bone" shape.

Various authors have developed approximate expressions for the effect of ovality. Timoshenko (1956, 1961) obtains the maximum stress for oval tubes and sets it to the yield stress σ_Y to obtain the collapse reduction factor for ovality:

$$\eta = P_c / P_e = 0.5(a+b) - \sqrt{0.25(a+b)^2 - a} \tag{26}$$

where

$$a = P_P / P_e \tag{27}$$

$$b = 1 + 3\delta\xi \tag{28}$$

and

$$P_P = 2\,\sigma_Y / \xi \tag{29}$$

Here P_P is the external pressure which causes yield over the entire wall thickness, a is a plastic-elastic collapse pressure ratio, and b is an ovality parameter. In the plastic range it is sometimes customary to express reduction factors as a fraction of the yield pressure. This factor β is simply

$$\beta \equiv P_c/P_P = \eta / a \tag{30}$$

For the example investigated by Kyriakides (1981) with $\xi = 35.72$ and $a = 4.47$, Eq. 26 predicts a reduction factor of $= 0.81$ for 0.8% ovality and $= 0.60$ for 2.4% ovality. These values agree very well with those obtained by Kyriakides.

Nishioka (1976) suggested a reduction factor in the plastic zone:

$$\beta = 1 / b = 1 / (1 + 3\delta\xi) \tag{31}$$

This would not apply in the elastic zone.

Lohmeier (1973) used limit analysis rather than first yield to estimate the effect of ovality. He used a bilinear interaction equation between hoop stress and hoop moment:

$$M/M_P + (1/2)P/P_P = 1 \qquad \text{for} \quad (P/P_P < 0.5) \tag{32}$$

$$(2/3)M/M_P + P/P_P = 1 \qquad \text{for} \quad (P/P_P > 0.5) \tag{33}$$

where M_P is the plastic moment capacity of the tube wall given by

$$M_P = \sigma_Y\, t^2 / 4 \tag{34}$$

His expression for the upper bound ovality reduction factor in the plastic range is

$$\beta = 1 / \{1 + \delta + (4/3)\delta(\xi-1)\} \qquad \text{for} \quad (\delta < 3/(4\xi-1)) \tag{35}$$

with different expressions for higher ovality limits. The initially oval tube was modeled as two semicircles separated by two flat spots on opposite sides of the tube. Small (1977) reworked this problem, and obtained an exact result:

$$\beta = 1 / \{1 + \delta + (4/3)\delta(\xi-1)(1+\delta)\} \tag{36}$$

Livsey (1976) also used the theory of Lohmeier, but substituted a total ovality of

$$\delta = \delta' / (1-\eta) \tag{37}$$

where η is based on first yield (see Eq. 26), and δ' is the initial ovality.

For later comparisons, and simplicity, the terms which are small compared to ξ in Eq. 35 or 36 are dropped and Eq. 37 is incorporated to obtain a simplified form of the Lohmeier-Livsey-Small result:

$$\beta = 1 / \{1 + (4/3)\delta\xi/(1-\eta)\} \tag{38}$$

where, again, η is obtained from Eq. 26. Going one step further and taking δ as the total elastic ovality at the collapse load instead of first yield (combine Eq. 30 and Eq. 38), the result is

$$\eta = 0.5(a+b) - \sqrt{0.25(a+b)^2 - a} \tag{39}$$

where

$$b = 1 + (4/3)\delta\xi \tag{40}$$

This is referred to here as the modified Lohmeier-Livsey-Small result and is similar to Timoshenko's result for first yield, but $\delta\xi$ is less prominent.

Haagsma (1981) and de Winter (1981) modified the Timoshenko solution slightly by requiring a full plastic hinge instead of looking at first yield. They give the criteria for a full plastic hinge in the presence of hoop stress as

$$(P/P_P)^2 + M/M_P = 1 \tag{41}$$

This results in the following cubic equation for the reduction factor:

$$\eta^3 - \eta^2 - \eta a(a + 2\delta\xi) + a^2 = 0 \tag{42}$$

For the Kyriakides examples, this gives reduction factors of 0.88 and 0.72 which are 10 to 20% higher than Kyriakides or Timoshenko. Higher values would be expected since Kyriakides neglected hoop stress and Timoshenko's criteria was first yield.

An empirical equation is suggested in API RP 2A (American Petroleum Institute, 1981) in the commentary:

$$\eta = \beta = 1 - 40\delta \tag{43}$$

This formula gives reduction factors of 0.68 and 0.04 for the examples, the latter indicating it is not applicable to larger ovalities.

A comparison of the equations discussed above is shown in Fig. 2. the reduction factors are plotted against the slenderness ratio λ :

$$\lambda = \sqrt{a} = \sqrt{P_P/P_e} = \sqrt{(1-\nu^2)(\sigma_Y/E)(\xi-1)/\xi}\ (\xi-1) \tag{44}$$

The reduction factors are shown for values of relative ovality Δ of 0.01, 0.1, and 1.0, where

$$\Delta = \delta\xi = (D_{max} - D_{min}) / (2t) \tag{45}$$

It is estimated that a value of 0.01 is probably uncommonly small even for thick wall casing, whereas 0.1 would not be unreasonable for stock line pipe. Of course, pipe can be purchased to specification, so this is not to imply one is better than the other, only to give meaning to the curves. The limit analyses of Haagsma (1981) and the modified version of Lohmeier (1973), Livsey (1976), and Small (1977) agree fairly well, except at large relative ovality. The numerical analyses of Kyriakides (1981) however, agrees with the more conservative first yield result of Timoshenko (1961).

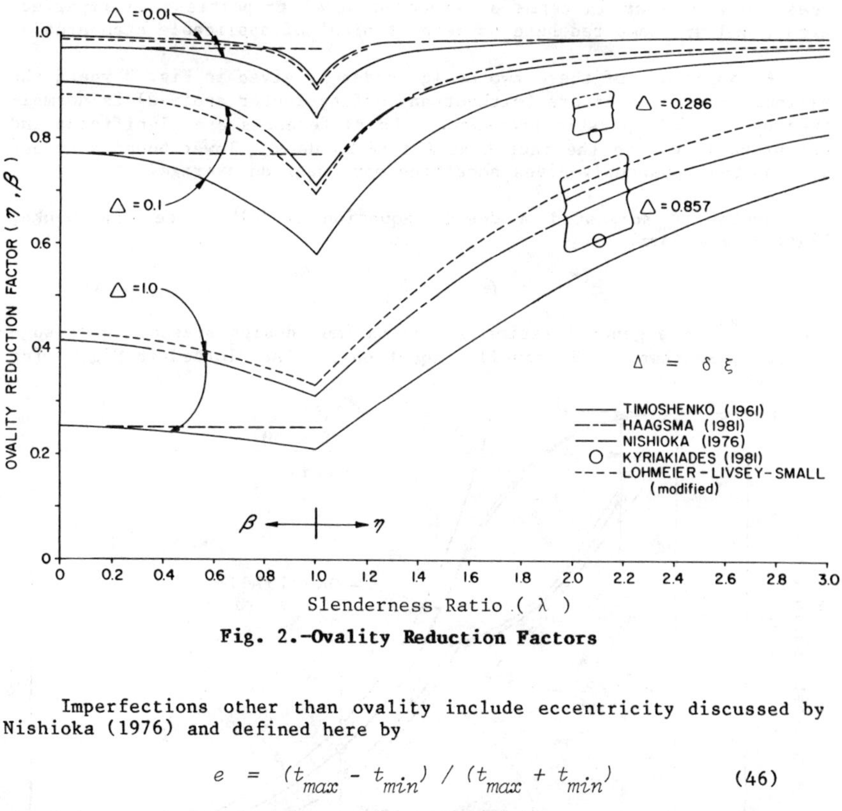

Fig. 2.-Ovality Reduction Factors

Imperfections other than ovality include eccentricity discussed by Nishioka (1976) and defined here by

$$e = (t_{max} - t_{min}) / (t_{max} + t_{min}) \qquad (46)$$

where t_{max} and t_{min} are the maximum and minimum wall thicknesses at a section. Eccentricity in seamless pipe can be as high as 12.5% and still be within specifications. Also noted in the literature are the effects of circumferential variation in wall thickness such as "flat spots" reported by Nordgren (1968), the effects of slack quenching reported by Mehdizabeh (1976), and the effects of wall thickness erosion reported by Livsey (1976). The extent and significance of such imperfections is apparently not well known at present.

Design Equations. - There are numerous equations adopted for design purposes. Design equations are usually aimed at a specific product such as casing, line pipe, or structural pipe, for which typical imperfections and steel properties are fairly well controlled or for which reasonable limits are known. For simplicity, the effect of these imperfections or properties are usually included implicitly in the design equations, rather than explicitly. Sometimes such effects will be included by taking a lower bound and sometimes the central or most probable value will be used. This makes design equations somewhat difficult to compare. The values of DNV (Det Norske Veritas, 1976), and API RP 2A are reasonably similar in terms of expected steel properties and imperfections, and are compared here as being typical of applicable standards.

A comparison of these two design codes is given in Fig. 3 where the maximum external pressure (without any safety factor applied) is normalized by the full plastic pressure. The differences are significant and are probably due to the fact that API RP 2A uses a lower bound approach whereas DNV apparently gives something more near an average.

Southwell suggested a design equation in 1915 (see Timoshenko, 1961) of the form:

$$P^{(n)} = (P_P^{-n} + P_e^{-n})^{-1/n} \tag{47}$$

where $P^{(n)}$ is a generalization of the maximum design pressure $P^{(1)}$ suggested by Southwell. Southwell's equation is also plotted in Fig. 3 for

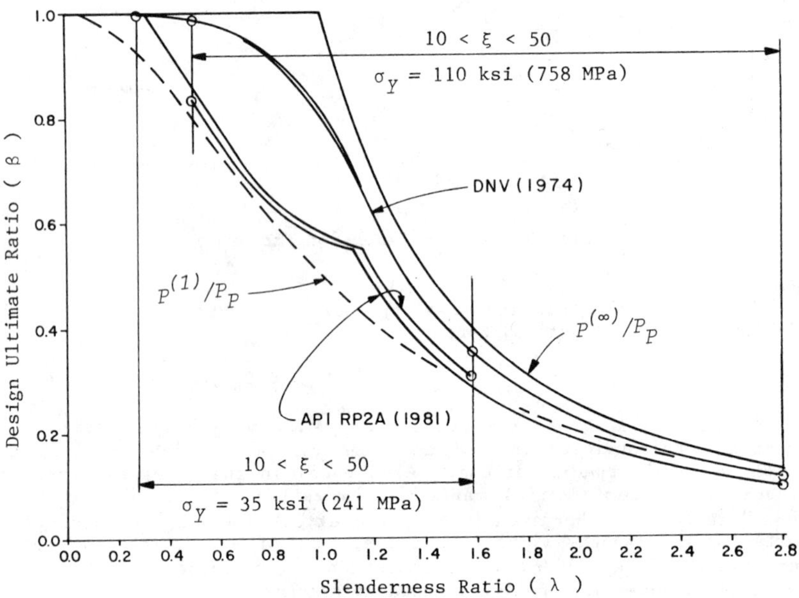

Fig. 3. - Design Criteria for Lateral Pressure

comparison. These equations are noteworthy because there is only one equation for the full range of slenderness ratio, and by selecting larger values of η the curve can be moved closer to the nominal upper bound indicated by $P^{(\infty)}$.

Test Data. - Most of the test data found in the literature on long thick wall cylinders subject to lateral or hydrostatic pressure comes from the well casing industry. The problem also has received some attention from those concerned with boiler tubes, especially in relation to nuclear power plants. Only recently has there been some activity directed toward deep submarine pipelines.

Holmquist (1939) reports tests on brass tubes which he used to evaluate a reduced modulus theory. There were four specimens made from commercial brass tubing which had extremely low ovality, and after annealing (if required) there were essentially no residual stresses. The data appears to be very complete.

An extensive set of data on casing was compiled by the American Petroleum Institute (Clinedinst, 1963) for use in setting minimum collapse properties for casing in the plastic region. This data consists of 2,700 tests (not all collapsed) reported by manufacturers. The data contains actual yield strengths and dimensions for each specimen. However, quantities that correlate with a reduction in collapse pressure, such as ovality, shape of stress-strain curve, or residual stresses, are not included. This data has been grouped by diameter, weight, grade, and type of test. The test type was either long ($L/D > 7$) or short ($L/D \simeq 2$), where L is the length of the test specimen. The long tests were sometimes open ended, and sometimes closed ended tests. The short tests were always open ended with a special flexible seal to allow the short test to behave like a wide ring. There is some evidence that the short-open test gives higher results than the long-open tests and for that reason, they have been omitted here along with the ones that did not collapse.

Sumitomo Metal Industries supplied some test results to Shell on their "SM95T" High Collapse Casing (Marlow, 1978). This data contains ovality, yield strength, hardness, and collapse pressure. Here, three different wall thicknesses were tested at length to diameter ratios of 5.5, 8, and 10, with three specimens for each combination. The data for a ratio of 5.5 is not applicable, and the thick wall specimens did not collapse. Thus, of 27 possible tests, 12 are applicable.

Kyogoku (1981a) presents collapse test results on casing with and without tension. Detailed test results were supplied to Shell (Kyogoku, 1981b) by Sumitomo. Results included ovality, eccentricity, and residual stress. In all, 29 tests are applicable to the lateral pressure case.

Nippon Steel Corporation has also recently provided casing collapse data to the American Petroleum Institute (Nippon Steel Corp., 1981). This data included 136 tests, with 22 at zero tension. The data also includes ovality, eccentricity, and residual stress.

The data available from those concerned with boiler tubes has mostly been intended to evaluate ovality effects. The material is Inconel, with Young's modulus of 31,000 ksi (213 737 MPa), and Poisson's ratio of 0.3. The data presented in Lohmeier (1973), Livsey (1976), and Small (1977) covers a range of ovality from 0 to 5%. Some specimens have residual stress, and others have been annealed. Small indicates considerable variability in yield strength of as-manufactured tubes (unannealed) which accounts for some of the scatter in the data since individual specimen yield strengths were not reported.

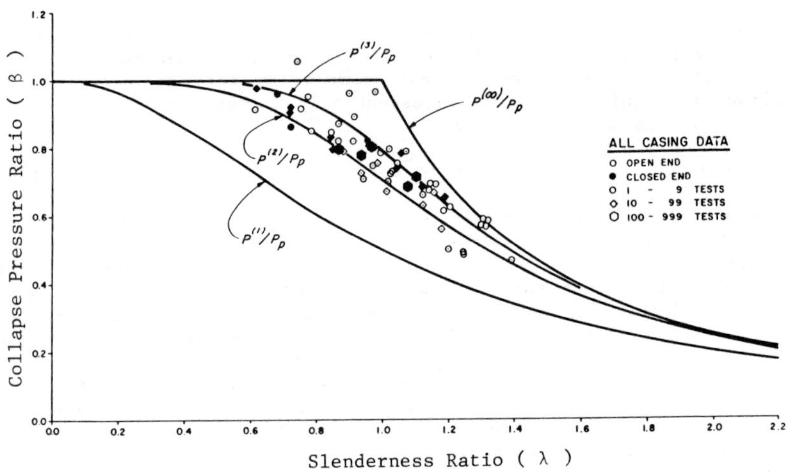

Fig. 4. - Collapse Data for Casing

The only other suitable data obtained from the literature reviewed are the three tests on ASTM A106 pipe by Haagsma (1981).

All available data as described above was tabulated for review and evaluation. Casing data was grouped by size, weight, grade, and test type. Average and coefficient of variation were tabulated for grouped data. Casing data as a class, is plotted in Fig. 4. The ratio of pressure at collapse to full yield capacity is plotted against the slenderness ratio which is proportional to the mean diameter to thickness ratio, but allows materials of different yield strength or modulus to be compared. As background reference, the Southwell type equations for $P^{(1)}$, $P^{(2)}$, $P^{(3)}$, and $P^{(\infty)}$ have been plotted. In this case, most of the data seems to fall in the $P^{(2)}$ to $P^{(3)}$ range.

Casing data with low ovality (0.2% or less) are shown in Fig. 5. These are all open end tests, so there may be a bias. There is a slight shift in the tendency of the data to fall more centrally around the $P^{(3)}$ line. However, there is little noticeable difference between the points with lower ovalities and the sum of all low ovality points. One might conjecture from this plot that at ovalities of around 0.2% and below, considerations such as the shape of the stress-strain curve, residual stresses, or even test procedures become more important than ovality in determining reported collapse pressure.

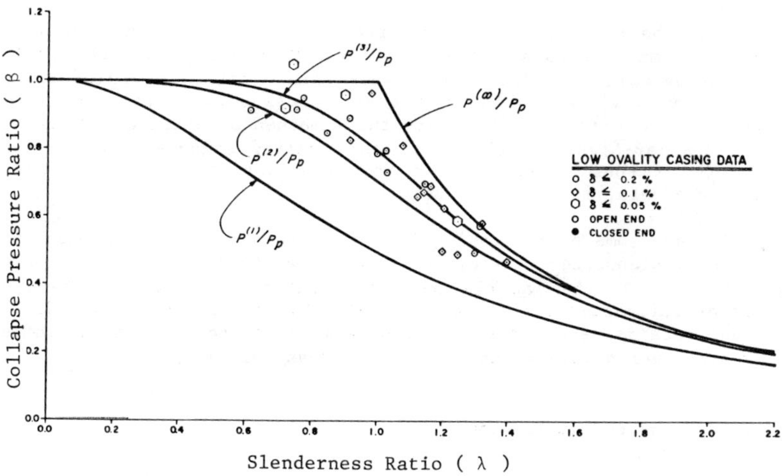

Fig. 5. - Collapse Data for Low Ovality Casing

The rest of the data is predominantly boiler tube data for a wide range of ovalities. The tubes are all either 5/8 inch (16 mm), 3/4 inch (19 mm), or 7/8 inch (2 mm) diameter with a diameter to thickness ratio

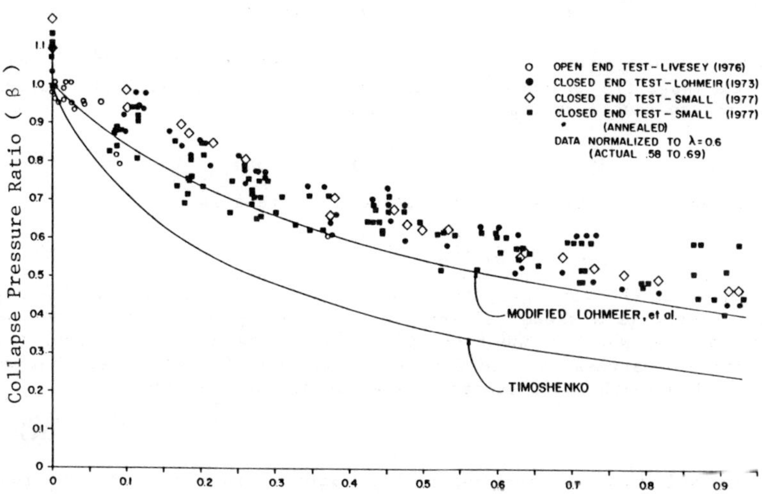

Fig. 6. - Ovality Data for Thick Wall Inconel Tubes

in the range of 17.36 to 17.50. The material is Inconel with a reported yield stress in the range of 45 to 63 ksi (310 to 434 MPa). Thus the slenderness ratios for all this data only vary between 0.58 and 0.68, which places it well into the yield zone.

Since the range of slenderness ratios was so small, the boiler tube data was normalized to a slenderness ratio 0.60 using the modified Lohmeier-Livsey-Small result. The results are plotted in Fig. 6 against relative ovality, which in this case is essentially proportional to ovality. Also shown in Fig. 6 are the Timoshenko equation and the modified Lohmeier-Livsey-Small result for ovality effects. It is clear that both of these equations are too conservative, as applied to the boiler tube data, in the yield zone.

Although some of the casing data could be used to span a much larger range of slenderness ratio, it happens that those tests with reported ovalities are in the low ovality range mentioned previously where any number of influences may obscure the effects of ovality. Obviously additional data is needed for line pipe and casing to evaluate the effect of ovality over a larger range of slenderness ratios.

Combined Pressure And Tension

The middle part of the suspended pipe in ultra deep water pipe laying would be subject to appreciable combinations of both pressure and tension only if it is required to sustain significant tension at the seafloor. For zero effective tension at the seafloor, the steel stress due to tension is

$$\sigma_a = \gamma_s h - 0.25\ \gamma_w H\xi^2/(\xi-1) \tag{48}$$

where h is the height off bottom, H is the water depth, and γ_s and γ_w are the specific weights of steel and water. Similarly, the average hoop stress due to hydrostatic pressure is

$$\overline{\sigma}_h = -0.5\ \gamma_w (H-h)\xi \tag{49}$$

For a von Mises yield theory the effective stress is

$$\hat{\sigma} = \sqrt{\sigma_a^2 - \sigma_a\sigma_h + \sigma_h^2} \tag{50}$$

The shape of the von Mises yield surface for constant values of equivalent stress, is a skewed ellipse when σ_a and σ_h are taken as the axes. This is true whether the highest stress at the inside surface or the average hoop stress is used for σ_h. Lubinski (1975) showed that when σ_a' is taken as the hoop stress at the inside surface, and

$$\sigma_a' = \sigma_a - 0.5\ \sigma_h \tag{51}$$

is the well known effective tension stress, then the ellipse is no longer skewed, and is defined by

$$\hat{\sigma}^2 = (\sigma_a')^2 + 0.75\, \sigma_h^2 \qquad (52)$$

Lubinski refers to $-\frac{1}{2}\sigma_h$ as the neutral axial stress, or the value of axial tension stress at which the effective tension in the pipe is zero. Since the effective tension stress is often the parameter of interest in analyzing pipeline behavior, this suggestion has some practical application.

Thus for zero effective tension on bottom, the following facts can be observed. For a diameter to thickness ratio of 10, the maximum effective stress is $.64\,\gamma_s H$ at the top, drops off to about $0.41\gamma_s H$ at $h/H = 0.4$, and is $0.57\gamma_s H$ at the bottom. The axial stress does not become tensile until about .36 d off the bottom. For a diameter to thickness ratio of 26, the effective stress is maximum at the bottom ($1.47\gamma_s H$) and drops off rapidly. The axial stress does not become tensile until about $.88H$ off the bottom, and at mid depth, the effective stress is only 50% of the effective stress at the bottom.

Thus, it is unlikely that tension will have a large influence on pipe collapse during laying unless effective tension is considerably higher than zero at the seafloor. If it does have a large influence, it is most likely to occur some distance off bottom where bending is small.

Theory. - Holmquist (1939) suggested the use of von Mises' yield theory for casing collapse predictions in the yield region. He also noted that tension would make little difference in the elastic region. Edwards (1939) compared numerous test data to the maximum strain energy theory of yielding given by

$$\hat{\sigma}^2 = \sigma_a^2 - 2\nu\sigma_a\sigma_h + \sigma_h^2 \qquad (53)$$

and found that von Mises' theory ($\nu = 0.5$) fit the data better. This comparison was based on the cases where failure was by large plastic deformation rather than collapse. He noted also that tension did not affect elastic collapse very much.

Bijlaard (1949) presented plasticity reduction factors in the presence of tension, based on a deformation theory with von Mises yield (see Eq. 11). Huang (1980) also derived the same result (Eq. 16) and showed tension to be of relatively little influence until the effective stress approaches yield. He compares his theoretical results with the extensive data by Edwards (1939) with excellent results. Pattillo (1980) also compares these results with full scale casing data and obtains reasonable agreement. Heise (1970) also proposes the use of the tangent modulus approach in the presence of tension, as does Nara (1981).

The tangent modulus approach is difficult to verify because of the difficulty of determining stress-strain curves and then adjusting to match collapse data influenced by ovality and residual stress as well. Pattillo (1980) suggests using the form of equation indicated by the tangent modulus approach together with the extensive existing collapse data on casing to determine empirical stress-strain curves which match the data.

De Winter (1981) presented a simplified model which provided the following expression

$$\eta^3 - \eta^2 - \eta a\{a + 2\delta\xi + f(1 - af)\} + a^2(1-f^2) = 0 \qquad (54)$$

where f is the ratio of axial stress to yield. This includes the effect of both ovality and axial load.

Design. - Casing design for the effects of tension is usually determined according to API Bulletin 5C3 (American Petroleum Institute, 1980). That formula is of the form

$$\hat{\eta}^2 + \hat{\eta} f + f^2 = 1 \qquad (55)$$

where $\hat{\eta}$ is a reduction factor on the allowable collapse pressure in the yield, plastic, or transition zone. Elastic collapse may also control. Pattillo (1980) pointed out the conservativeness of this equation.

Clinedinst (1981) proposed a modification in which von Mises' yield theory is applied to determine a new effective hoop yield stress:

$$\sigma_Y^* = \sigma_Y \{\sqrt{1 - 0.75f^2} - 0.5f\} \qquad (56)$$

for use in the collapse equations for hoop stress alone. This approach showed better agreement with test results in at least one example case and also agreed with a reduced modulus approach.

The design formula for the effect of tension on structural members as given in API RP 2A is similar to the form (Eq. 55) used for casing when factors of safety are ignored. The only real difference is that the maximum strain energy formula (Eq. 53) is used instead of the von Mises yield formula:

$$\hat{\eta}^2 + 2\nu\hat{\eta} f + f^2 = 1 \qquad (57)$$

Both the formulas for casing and structures are overly conservative for values of axial stress that are low enough to allow elastic collapse. Until axial stress approaches a value to cause yield, the collapse load should not decrease very much. Both of these formulas cause reductions in the allowable as soon as axial stress increases above zero.

Combined Pressure and Bending

Bending And Ovality. - Brazier (1927) published a theoretical result on pure bending of elastic circular tubes that indicated a progressive flattening of the cross-section as curvature increased. He showed that there exists a point at which the cross-section becomes so flattened that the bending moment drops off with increasing curvature. The compressive longitudinal stress on the side toward the center of curvature, and the tensile stress on the other side both have a component toward the center of the tube that is proportional to the curvature. It is these forces that tend to flatten the tube.

For thin wall tubes, Brazier obtained ovality of

$$\delta = (2/3)x^2 \tag{58}$$

where

$$x = \sqrt{(3/32)(1-\nu^2)}\ \ \xi D/R \tag{59}$$

is the nondimensionalized curvature parameter and R is the radius of curvature in bending. In nondimensional form, the moment-curvature relation was found to be

$$m = x(1-x^2) \tag{60}$$

where

$$m = M_x \sqrt{(3/32)(1-\nu^2)}\ \ \xi D \;/\; (EI_x) \tag{61}$$

is the nondimensionalized moment corresponding to the moment M_x in the tube cross-section, and I_x is the area moment of inertia of the tube cross-section. The moment is maximum at a dimensionless curvature parameter of

$$x_\ell = \sqrt{1/3} \simeq 0.577 \tag{62}$$

The bending strain is

$$\varepsilon_b = 0.5\ (1-\delta)D/R \tag{63}$$

For most line pipe and casing, the material will yield well before this limit moment is reached. For the bending strain to remain elastic at the limit moment for 110 ksi (758 MPa) material, the diameter to thickness ratio would have to be at least 210.

The factor significant for ovality reduction is

$$\Delta = \delta\xi = 0.25\ \varepsilon_b{}^2\xi^3\,(1-\nu^2) \;/\; (1-\delta)^2 \tag{64}$$

where the right hand quantity is obtained by combining Eq. 58, Eq. 59, and Eq. 63. For example with a diameter to thickness ratio of 26 and 45 ksi (310 MPa) bending stress and neglecting the term in brackets, one obtains a value of 0.01, suggesting a potential reduction in collapse resistance of 10% (see Fig. 2).

Reissner (1959) added pressure to the problem, and also obtained a second order result, where as Brazier's result was first order. Reissner's expression for moment in nondimensional form was

$$m = x\ \{1 - x^2 - (2/3)x^4\ \} \tag{65}$$

where,

$$x = \sqrt{(3/32)(1-\nu^2)/(1 - P/P_e)}\ \xi D/R \tag{66}$$

and

$$m = M_x \sqrt{(3/32)(1-\nu^2)/(1 - P/P_e)}\ \xi D / (EI_x) \tag{67}$$

The nondimensional moment is largest for

$$x_\ell = 0.5\ (\sqrt{201} - 9)^{1/2} / \sqrt{5} \simeq 0.509 \tag{68}$$

In this case the limit curvature and limit moment become arbitrarily small as the pressure approaches the bifurcation collapse pressure. Also, the limit curvature for zero pressure is 15% less than Brazier's while the limit moment is 10% less.

Reissner's expression for ovality is

$$\delta = x^2(2/3) + x^4(71 - 12P/P_e)/(135 - 27P/P_e) \tag{69}$$

which has a value at the limit curvature of 0.208 without pressure and 0.209 in the presence of the elastic collapse pressure. Thus, the ovality due to bending at which the limit moment is reached is practically independent of pressure. This result seems intuitively reasonable, but does not agree with the numerical results of Fabian (1977).

For this analysis, a tube with a pressure of half of the elastic collapse pressure (about as high as expected in practice), the previous example would require a diameter to thickness ratio of 133 to remain elastic at the limit moment (neglecting biaxial stress effects).

Kyriakides (1982) notes also that Kogakusi obtained a result for bending and pressure in 1940. Kyriakides' numerical results for the linear case agree more closely with Brazier than Reissner. Gellin (1980) also obtained numerical results for the linear case. His results also were closer to Brazier's than Reissner's. The comparison for nondimensional moment, curvature, and ovality are listed in Table 2

TABLE 2. - Comparison of Results for Brazier Effect

Author (1)	m (2)	m/m^* (3)	x (4)	x/x^* (5)	δ (6)	δ/δ^* (7)
Brazier (1927)*	.3849	1.00	.5774	1.00	.2222	1.00
Reissner (1959)	.3544	.92	.5088	.88	.2078	.94
Gellin (1980)	.3738	.97	.5841	1.01	.2574	1.16
Kyriakides (1982)	.3657	.95	.5525	.96		
Fabian (1981)	.378	.98	.583	1.01		
de Winter (1981)			.496	.86		

* Value obtained by Brazier (1927)

including the results of Fabian (1981), and the approximate results of de Winter (1981).

De Winter (1981) presented a result based on a simple model including pressure, in which the curvature relationship was of the same form as Reissner (1959), and the critical nondimensional curvature parameter was .496 which agrees fairly closely with Reissner, but with very few other results (see above).

Ades (1957) developed a numerical approach for calculating the Brazier effect for inelastic behavior. Jirsa (1972) presented test results that compared favorably with these predictions of ovality.

Reddy (1979) found experimentally, that tubes bent into the plastic range, retained the same rate of ovalization with curvature, although his experimental results on steel tubes were about 50% lower than the Brazier result. The two steel tubes for which ovalities were reported were relatively thin with diameter to thickness ratios of 52 and 68. In the early loading history, there are only a few points, but these seem to agree with Brazier's result. At about the yield region, the ovality drops off, then continues increasing with curvature to the second power. Shell has data which they report to indicate that the ovality in the plastic range is similar in magnitude and rate to Reddy's results. This data consists of 27 tests of 4-inch (10 mm) and 6-inch (152 mm) steel pipe in the diameter to thickness range of 18 to 58. Reddy's specimens buckled locally, because of the relative thinness and did not reach a limit moment of the Brazier type.

Gellin (1980) investigated ovalization and moment-curvature relations in the inelastic range with a numerical approach somewhat less restricted than that of Ades. In order to get better bifurcation results, and simplify the solution technique, he used deformation theory instead of flow theory. As a measure of the importance of plastic effects at the limit point, he identified the limit moment parameter.

$$\psi = 0.866 \sqrt{1-\nu^2} \;\; \xi \sigma_Y / E \qquad (70)$$

which is proportional to the ratio of yield strain to strain at the limit moment. If Brazier's limit moment is used, a coefficient of .825 would be appropriate instead of .866. A value for the limit moment parameter of 1.0 would indicate the elastic limit moment should be reached if buckling does not occur first. For values of ψ as low as 0.02 (a diameter to thickness ratio of 12 for 60 ksi (414 MPa) material) he found that the inelastic limit curvature only dropped to about 67% of the Brazier limit curvature.

Fabian (1981) adds pressure to the problem, and presents numerical results in the inelastic range. He notes that the imperfection sensitivity of collapse to ovality in the inelastic range contributes to the sensitivity to curvature for the bending-pressure problem. In this inelastic case, he points out that ovality is the operative parameter. This conclusion comes from a comparison of two cases that collapsed at the same ovality, but in one case, half of that ovality was initially present, and the other half was caused by curvature (less than half the curvature required without initial ovality). His results are based on considerable effective tension (zero material tension).

The presence of pressure in the bending problem amplifies any ovality (initial or caused by bending). The combination of pressure and inelastic bending is not surprisingly a more nonlinear problem. Fabian's results indicate a highly nonlinear increase in ovality with curvature, as opposed to the parabolic behavior noted by Reddy (1979) for pure bending in the inelastic range. This is to be expected, of course, because of the imperfection sensitivity to ovality in the inelastic range.

Kyriakides (1982) also reports numerical analysis procedures for inelastic bending plus pressure at zero material tension. He compares his results with experimental results on aluminum tubes at zero effective tension and found reasonable agreement, with a little over-prediction of limit moment, and to a lesser extent, limit curvature. He also compares his theory to the results of Johns (1975) (made at zero effective tension) with a fairly good fit, except at high diameter to thickness ratio where buckling may have been more important than ovalization.

Haagsma (1981) also presented some test results, but at zero material tension.

As a practical matter, numerical analyses of the type mentioned above are more useful when related to more simple equations or models. De Winter (1981) attempted this, but noted the sensitivity of results to assumptions because of the previously mentioned assumptions.

The majority of analyses dealt with the zero material tension case; the bending of pipe under pressure at zero effective tension did not receive much attention in the literature reviewed.

Axial Buckling. - Compressive bending forces in a round tube can cause wrinkle type buckling. Seide (1961) showed that this occurs elastically for tubes that remain round, at about the same stress in bending as would cause elastic buckling under axial compression alone. The elastic axial buckling stress for a cylinder is

$$\sigma_{ae} = 2E / \{\xi\sqrt{3(1-\nu^2)}\} \quad (71)$$

The cross-section shape results of Reissner (1959) give a radius of curvature on the flat side of

$$r = 0.5\,D / (1 - 2x^2) \quad (72)$$

Assuming that the axial result applies approximately to stresses resulting from bending, and if the flattened radius is used in lieu of the actual radius, then the expression for the buckling stress due to bending is

$$\sigma_{be} = \sigma_{ae}(1 - 2x^2) \quad (73)$$

However, the bending stress is a function of the curvature also

(Eq. 63). Combining Eqs. 63 and 73 with the results of Brazier's theory for ovality (Eq. 58) and Reissner's expression for the curvature parameter (Eq. 66), results in the expression

$$2\sqrt{2(1 - P/P_e)}\, x^3 - 6x^2 - 3\sqrt{2(1 - P/P_e)}\, x + 3 = 0 \qquad (74)$$

The solution for zero pressure gives a curvature parameter of 0.467 which corresponds to a 44% reduction from the pure axial stress at buckling.

Comparing the curvature parameter at buckling (0.467) with the value at the limit moment (0.496 to 0.5774) as shown in Table 2, indicates that buckling would occur slightly before the limit moment was reached.

Aksel'rad reportedly (Almroth, 1975) showed in a similar manner in 1965 that the ovaling of the cross-section in long tubes results in about 50% of the stress at pure axial buckling. Almroth (1975) confirms this elastic behavior numerically, and so does Fabian (1977, 1981) and Gellin (1980). Gellin, however, shows buckling at somewhat less curvature than does Fabian.

When external pressure is added to the problem, the results of Reissner (1959) would indicate a reduction of 43% for buckling due to bending stress as opposed to pure axial stress. The limit bending stress would be reduced to 50% on the same basis at 3/4 of bifurcation pressure. This suggests that even under considerable pressure, elastic axial buckling occurs at about the same bending stress as the limit moment does. This behavior is also confirmed by Fabian (1977, 1981) with numerical solutions to the elastic problem.

Fabian (1977) also shows that the buckling due to bending is unstable and thus imperfection sensitive, even in the elastic range. This is not unexpected in view of the above discussion and the fact that axial compressive buckling of cylinders is highly imperfection sensitive.

The significance of buckling becomes less pronounced at lower diameter to thickness ratios where plastic effects are more important. For zero pressure, Gellin (1980) shows numerically that the limit moment is very nearly attained, and the buckling curvature is only about 20 to 30% below the limit curvature for the range of parameters he considered.

With pressure added to the problem, Fabian (1981) shows numerically that for a diameter to thickness ratio of 40 and X52 steel there is little change in behavior even for 2/3 of the classic collapse pressure. He points out that there is much more sensitivity to ovality than axial buckling. For thin wall tubes, he points out that the reverse is true.

Johns (1975) presented a bending-pressure theory based on the assumptions that axial buckling was the failure mode and that the bifurcation strain could be approximated by the bifurcation strain for axially loaded tubes. Both of these assumptions are not unreasonable in view of the previously mentioned numerical analyses which considered ovality. However, such an approach cannot predict the effects of initial ovality.

Conclusions

The theory of elastic stability of a tube or ring under pressure should be based on the mean diameter. The difference between mean and outside diameter is only significant for thick wall pipe which doesn't ordinarily fail elastically. However, for some high strength casing materials, and for the application of tangent modulus theories, the use of thick wall elastic collapse equations may be required.

Plasticity effects, especially prevalent in thick wall pipe, induce collapse imperfection sensitivity with respect to initial ovality. While ovality is probably the most important parameter, nonlinear stress-strain behavior, residual stresses, and even eccentricity in seamless pipe can be significant in thick wall pipe. There is little data on any of these effects, especially for line pipe. The most notable data in this regard is the ovality data for Inconel boiler tubes. Ovality is so important, especially when combined with bending, that its effect should be explicit in any design criteria.

There is very little data on collapse of line pipe. At low ovalities, such as might be specified for ultra deep water pipelines, the effects of the stress-strain behavior and residual stresses become relatively more prominent. Most of the data that is available is for casing that has a significantly different stress-strain behavior and residual stress mechanism than line pipe.

Tension in combination with pressure seems to be reasonably amenable to a tangent modulus approach (Bijlaard 1949, or Huang, 1980) or to an equivalent hoop yield stress approach (Clinedinst, 1981). Considerable recent testing on casing adds weight to this argument. Some testing of the chosen approach should be done with line pipe.

Bending in combination with external pressure is the least well documented, and potentially the most critical problem. Plasticity effects make thick wall pipe imperfection sensitive to ovality to begin with. Bending adds significant additional ovality. Considerable work has been done computationally in modeling these effects (notably Fabian 1981) but little data is available to evaluate the results. Most work has been done at zero material tension instead of zero effective tension which is the case in point for ultra deep water pipelines.

More experimental work on the effects of bending is needed. Analyses such as those of Fabian (1981) may be used to fill in between test data points and at the same time promote the understanding neccesary to arrive at an optimum design criteria.

Acknowledgements

This review and summary effort was funded by Shell Development Company, and TERA, Inc.

Appendix I. - References

Ades, C. S. (1957), "Bending Strength of Tubing in the Plastic Range," Journal of the Aeronautical Sciences, Vol. 24, pp. 605-610.

Allen, H. G. (1980), and Bulson, P. S., "7.5 Cylindrical Shell Under External Pressure," Background to Buckling, McGraw-Hill Book Company (UK) Limited, London, pp. 526-563.

Almroth, B. O. (1975), and Starnes, J. H., Jr., "The Computer in Shell Stability Analysis", Journal of the Engineering Mechanics Division, ASCE, Vol. 101, No. EM 6, Dec., pp. 873-888.

American Petroleum Institute (1980), API Bulletin on Formulas and Calculations for Casing, Tubing, Drill Pipe, and Line Pipe Properties, (API BUL 5C3), Production Department, Dallas.

American Petroleum Institute (1981), API Recommended Practice for Planning, Designing, and Constructing Fixed Offshore Platforms, (API RP 2A), 12th ed., Production Department, Dallas.

Bijlaard, P. P. (1949), "Theory and Tests on the Plastic Stability of Plates and Shells," Journal of the Aeronautical Science, Vol. 16, No. 9, Sept., pp. 529-541.

Boresi, A. P. (1955), "A Refinement of the Theory of Buckling of Rings Under Uniform Pressure," Journal of Applied Mechanics, ASME, Vol. 22, No. 1, March, pp. 95-103.

Brazier, L. G. (1927), "On the Flexure of Thin Cylindrical Shells and Other Thin Sections," Proceedings of the Royal Society, London, Series A, Vol. 116, pp. 104-114.

Carrier, C. F. (1947), "On the Buckling of Elastic Rings," Journal of Mathematics and Physics, Vol. 26, pp. 94-103.

Clinedinst, W. O. (1939), "A Rational Expression for the Critical Collapsing Pressure of Pipe Under External Pressure," Drilling and Production Practice, API, p. 383.

Clinedinst, W. O. (1963), "Development of New Collapse Pressure Formulas," API, Production Department, Dallas.

Clinedinst, W. O. (1981), report of January 5, 1981 meeting, Task Group on Performance Properties of Casing, Tubing, and Drill Pipe, API, held at Bal Harbour, Florida.

Det Norske Veritas (1976), Rules for the Design, Constuction and Inspection of Submarine Pipelines, and Pipeline Risers, 1976, Oslo, Norway.

de Winter, P. E. (1981), "Deformation Capacity of Steel Tubes in Deep Water," Proceedings, 13th Offshore Technology Conference, Vol. 2, pp. 257-266.

Edwards, S. H. (1939), and Miller, C. P., "Discussion on the Effect of Combined Longitudinal Loading and External Pressure on the Strength of Oil-Well Casing," Drilling and Production Practice, API, pp. 483-502.

Fabian, O. (1977), "Collapse of Cylindrical, Elastic Tubes Under Combined Bending, Pressure and Axial Loads," Int. J. Solids and Structures, Vol. 13, pp. 1257-1270.

Fabian, O. (1981), "Elastic-Plastic Collapse of Long Tubes Under Combined Bending and Pressure Load," Ocean Engineering, Vol. 8, No. 3, pp. 295-330.

Gellin, S. (1980), "The Plastic Buckling of Long Cylindrical Shells Under Pure Bending," Int. J. Solids and Structures, Vol. 16, pp. 397-407.

Gerard, G. (1957), "Plastic Stability Theory of Thin shells," Journal of the Aeronautical Sciences, Vol. 24, No. 4. April, pp. 269-274.

Haagsma, S. C. (1981), and Schaap, D., "Collapse Resistance of Submarine Lines Studied," Oil and Gas Journal, Feb. 2, p. 86.

Heise, O. (1970), and Esztergar, E. P., "Elastoplastic Collapse of Tubes Under External Pressure," Journal of Engineering for Industry, ASME, Series B, Vol. 92, No. 4.

Hill, J. M. (1975), "Buckling of Long Thick-Walled Circular Cylindrical Shells of Isotropic Incompressible Hyperelastic Materials Under Uniform External Pressure," J. Mech. Phys. Solids, Vol. 23, pp. 99-112.

Holmquist, J. L. (1939), and Nadai, A., "A Theoretical and Experimental Approach to the Problem of Collapse of Deep-Well Casing," Drilling and Production Practice, API.

Huang, N. C. (1980), and Pattillo, P. D., "Collapse of Oil Well Casing," presented at the ASME August 12-15, 1980, Century 2 Pressure Vessels & Piping Conference, held at San Francisco.

Hutchinson, J. W. (1970), and Koiter, W.T., "Postbuckling Theory," Applied Mechanics Reviews, Vol. 23, pp. 1353-1366.

Hutchinson, J. W. (1974), "Plastic Buckling," Advances in Applied Mechanics, Vol. 14, pp. 67-144.

Jirsa, J. O. (1972), Lee, F. H., Wilhoit, J. C., and Mervin, J. E., "Ovaling of Pipes Under Pure Bending," Proceedings, 4th Offshore Technology Conference, Vol. 1, pp. 573-578.

Johns, T. G. (1975), Mesloh, R. E., Winegardner, R., and Sorensen, J. E., "Inelastic Buckling of Pipelines Under Combined Loads," Proceedings, 7th Offshore Technology Conference, Vol. 1, pp. 635-646.

Koiter, W. T. (1963), "Elastic Stability and Post-Buckling Behavior," Nonlinear Problems (R.E. Langer, ed.), The University of Wisconsin Press, Madison, pp. 256-275.

Kyogoku, T. (1981a), Tokimasa, K., Nakanishi, H., Okazawa, T., "Experimental Study of the Effect of Axial Tension Load on the Collapse Strength of Oil Well Casing," Proceedings, 13th Offshore Technology Conference, Vol. 3, pp. 387-395.

Kyogoku, T. (1981b), Tokimasa, K., Nakanishi, H., and Okazawa, T., "Summary of Collapse Test Result Under Axial Load," Sumitomo Metal Industries, Ltd, May 6.

Kyriakides, S. (1981), and Babcock, C. D., "Large Deflection Collapse Analysis of an Inelastic Inextensional Ring Under External Pressure," Int. J. Solids and Structures, Vol. 17, No. 10, pp. 981-993.

Kyriakides, S. (1982), and Shaw, P. K., "Response and Stability of Elastoplastic Circular Pipes Under Combined Bending and External Pressure," Int. J. Solids and Structures, Vol. 18, No. 11, pp. 957-973.

Levy, M. (1884), "Memoire sur un Nouveau cas Integrable du Probleme de l'elastique et l'une de ses Applications, Journal de Mathematiques Pures et Appliquees, Series 3, Vol. 10, pp. 5-42.

Livsey, W. O. (1976), and Juenjo, A., "The Collapse of Heat Exchanger Tubes with Ovality and Simulated Defects," Int. J. Press. Vessel Piping, Vol. 4, pp. 47-61.

Lohmeier, A. (1973), Small, N. C., Borin, H. J., and Gonnet, B., "Collapse of Ductile Heat Exchanger Tubes with Ovality Under External Pressure," Proceedings, Second International Conference on Structural Mechanics in Reactor Technology, Berlin, Germany, Sep. 10-14, Vol. 2, Div. F, Sess. F6, Pap 4.

Lubinski, A. (1975), "Influence of Neutral Axial Stress on Yield and Collapse of Pipe," Journal of Engineering for Industry, ASME, Vol. 97, Series B, No. 2, May, pp. 400-407.

Marlow, R. S. (1978), "Collapse Testing of Sumitomo's Seamless High Collapse Casing," Southwest Research Institute Report No. 03-5248-001 to Sumitomo Metal Industries, Ltd.

Mehdizadeh, P. (1976), "Casing Collapse Performance," Journal of Engineering for Industry, ASME, Vol. 98, Series B, No. 3, August, pp. 1112-1119.

Nara, Y. (1981), Matsuki, N., Furugen, M., Ohyabu, K., "Theoretical Study on Casing Collapse," Sumitomo Metal Industries, Ltd., Osaka, Japan.

Nippon Steel Corporation (1981), "Summary of Collapse Test Results Under Axial Stress," Products Research & Development Laboratories.

Nishioka, K. (1976), Nara, Y., Kyogoku, T., Hirakawa, K., Nishikawa, K., and Tokimass, K., "An Experimental Study on the Critical Collapsing Pressure of a Seamless Steel Tube for Well Casing Under External Pressure," The Sumitomo Search, No. 15, Sumitomo Metal Industries, Ltd, Osaka, Japan.

Nordgren, R. P. (1968), and Murphey, C. E., "Elastic Stability of Circular Tubes of Circumferentially Variable Thickness Under External Pressure," Journal of Engineering for Industry, ASME, Paper No. 68-Pet-29.

Pattillo, P. D. (1980), and Huang, N. C., "The Effect of Axial Load on Casing Collapse," presented at the Sept. 21-24, 1980, 55th Annual Fall Technical Conference and Exhibition of the Society of Petroleum Engineers of AIME, held at Dallas, Texas, Paper 9327.

Reddy, B. D. (1979), "An Experimental Study of the Plastic Buckling of Circular Cylinders in Pure Bending," Int. J. Solids and Structures, Vol. 15, pp. 669-683.

Reissner, E. (1959), "On Finite Bending of Pressurized Tubes," Journal of Applied Mechanics, ASME, Sept., pp. 386-392.

Seide P. (1961), and Weingarten, V. I., "On the Buckling of Circular Cylindrical Shells Under Pure Bending," Journal of Applied Mechanics, ASME, Vol. 28, pp. 112-226.

Sewell, M. J. (1972), "Chapter 5, A Survey of Plastic Buckling," Stability, (H. H. E. Leipholz, ed.), University of Waterloo Press, Ontario.

Small, N. C. (1977), "Plastic Collapse of Oval Straight Tubes Under External Pressure," presented at the Sept. 18-22, 1977, Energy Technology Conference and Exhibit, held at Houston, Texas.

Southwell, R.V. (1914), "On the Theory of Elastic Stability," Philosophical Transactions, Royal Society of London, Vol. 213, pp. 187-244.

Timoshenko, S. (1956), Strength of Materials, Part II Advanced Theory and Problems, D. Van Nostrand Company, Inc, Prinston, New Jersey, pp. 178-193.

Timoshenko, S. P. (1961), and Gere, J. M., Theory of Elastic Stability, McGraw-Hill Book Company, New York, pp. 175-206, pp. 287-297, pp. 474- 485.

Tvergaard, V. (1976), "Buckling of Elastic-Plastic Oval Cylindrical Shells Under Axial Compression," Int. J. Solids and Structures, Vol. 12, pp. 683-691.

Appendix II. - Notation

The following symbols are used in this paper:

a = plastic - elastic collapse pressure ratio;
a_i = parameters in Carrier (1947) solution;
b = ovality parameter;
c = load coefficient for Carrier (1947) solution;
D = outside diameter of tube;
e = eccentricity of tube;
E = Young's modulus;
E_s = secant modulus;
E_t = tangent modulus;
f = ratio of axial stress to yield stress;
h = height off bottom;
H = water depth;
I = area moment of inertia of ring cross-section;
I_x = area moment of inertia of tube cross-section;
k = parameter in Carrier (1947) solution;
L = length of collapse test specimen;
m = nondimensional moment parameter for tube cross-section;
M = moment in tube wall;
M_x = moment in tube cross-section;
n = exponent;
P = pressure (external unless noted otherwise);
$P^{(n)}$ = generalized Southwell design collapse pressure;
r = radius to a point in tube wall or ring;
R = radius of curvature of tube axis;
t = wall thickness of tube;
U = movement to thickness ratio for slit ring test;
x = dimensionless curvature parameter of tube axis;
α = parameter in Carrier (1947) solution;
α_{ij} = nondimensional stress-strain coefficients;
β = reduction factor on full plastic pressure;
γ = ratio of effective stress to hoop stress;
γ_s = specific weight of steel;
γ_w = specific weight of water;
δ = ovality measure;
δ' = initial ovality;
Δ = relative ovality;
ε = strain;
ζ = ratio of axial to hoop stress;
η = reduction factor on elastic collapse pressure;
$\hat{\eta}$ = reduction factor on design collapse pressure;
θ = angular coordinate of point on ring;
λ = slenderness ratio for ring or tube wall;
ν = Poisson's ratio;
ν_t = Poisson's ratio (transition);
ξ = outside diameter to thickness ratio;
π = 3.14159 ....;
σ = stress;
$\hat{\sigma}$ = effective stress (stress intensity);
$\overline{\sigma}$ = average stress;

σ_Y^* = effective hoop yield stress in the presence of axial tension;
σ_a' = effective tension stress;
σ_R = residual stress measure;
ϕ = curvature of tube wall;
ψ = limit moment parameter;

with subscripts:

a = in axial direction due to axial force;
b = in axial direction due to bending;
c = at ultimate strength in collapse mode;
e = at elastic bifurcation;
h = in hoop direction;
i = at inside surface;
ℓ = at elastic limit moment;
m = at middle surface of tube or centroid of ring;
max = maximum at a section;
min = minimum at a section;
o = at outside surface;
P = at full plastic value;
p = at proportional limit;
r = in radial direction; and
Y = at yield.

GEOFABRIC STABILIZATION OF SOFT BACKFILL MATERIALS FOR PLASTIC SEWER PIPE INSTALLATION

by

Jey K. Jeyapalan,[1] M. ASCE

ABSTRACT

The economy and ease of construction with plastics have resulted in a tremendous increase in the use of plastic pipelines for transporting sewage. In urban environment, however, there are many instances where the backfill at site is soft that the use of flexible plastic pipes become impractical. The lateral support capacity of soft soils is insufficient to resist the lateral movement of the spring line of large diameter plastic pipes. Because of this, in many cases, the designers avoid use of plastic pipes and instead use steel or concrete pipes. The purpose of this paper is to indicate the potential of using synthetic geofabric layers along the pipe trench walls to increase the lateral bearing strength of the native soft soils. Thus, by using geofabrics to stabilize the soft soils at site, the designers can install flexible large diameter plastic pipes even under marginal site conditions.

Introduction

The economy and ease of construction with plastics have resulted in a tremendous increase in the use of plastic pipelines for transporting sewage. In urban environment, however, there are many instances where the backfill at site is soft that the use of flexible plastic pipes become impractical. The lateral support capacity of soft soils, is insufficient to resist the lateral movement of the spring line of plastic pipes. By using geofabric along the trench walls, the lateral support capacity can be increased. Although there have been a number of studies (1, 11, 12, 13, 15, 20, 21) on the behavior of plastic pipes, the effectiveness of geofabric has never been analyzed. Even the research on soil-culvert interaction (2, 3, 4, 5, 6, 7, 8, 9, 10, 14, 18, 19, 22) has never addressed this problem. The purpose of this paper is to present the results of finite element analyses of the pipe-soil-geofabric interaction problem using the techniques outlined in another paper by Jeyapalan, Osequeda and

---

[1]Assistant Professor of Civil Engineering, Texas A&M University, College Station, Texas, U.S.A., 77843.

Horn (17).

Geometry

A cross-section through the backfill and the pipe used in this study is shown by its finite element representation in Fig. 1. The plastic pipe is modeled by linearly elastic beam elements and the geofabric was included in the analyses by a series of linearly elastic bar elements with no compressive strength. The backfill was modeled by two-dimensional isoparametric elements. A no-slip condition was simulated at the interfaces between the pipe and the backfill and between the geofabric and the backfill.

Analysis Procedure

The interactions between the flexible pipe, the surrounding backfill, and the geofabric were studied using the finite element analysis procedures. The analyses described in this paper were performed using a hyperbolic stress-strain relationship for the backfill and native materials. This relationship, which is described in detail in (2, 5) model the nonlinear stress dependent stress-strain behavior of soils incrementally, by varying the values of Young's modulus and bulk modulus in each element according to the previously calculated stresses. The analyses herein were performed incrementally, simulating the field construction operations. The backfill was placed around and over the pipe one layer at a time after placing the geofabric along the walls of the trench. The analyses were performed with and without the placement of geofabric.

Material Properties

The properties of the pipe used in the study are summarized in Table 1. The properties of the native soil and the controlled trench backfill are given in Table 2. The properties of the geofabric used in the analyses are summarized in Table 3.

Axial Forces

The distributions of axial force in the pipe calculated without and with the geofabric are shown in Fig. 2. It is interesting to note that there is an increase in axial force due to the presence of the geofabric indicating that the loads are carried more in the form of ring compression than in bending.

Bending Moments

The distributions of bending moment in the pipe without and with geofabric are shown in Fig. 3. It appears that the crown and haunch bending moments are significantly higher in the installation without geofabric in comparison to that with geofabric.

Stresses

The distributions of stresses in the pipe wall induced by the

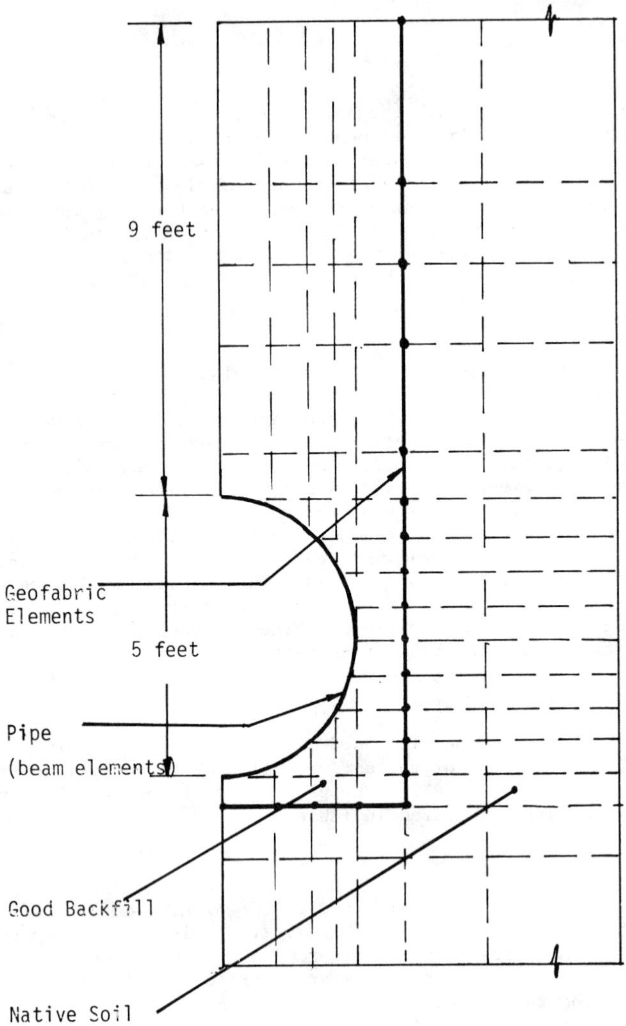

Fig.1 Pipe Embedment Problem

Table 1. - Properties of Plastic Pipe

Property	Value Used in Analyses
Mean Diameter (in.)	60
Area (in.^{2}/in.)	0.744
Moment of Inertia (in.^{4}/in.)	0.226
Young's Modulus (psi)	110,000
Yield Stress (psi)	3,000

Table 2. - Properties of Backfill Materials

Property	Values Used in Analyses	
	Native Soil	Trench Backfill
Unit Weight (pcf)	115	124
Cohesion (psf)	100	0
Angle of Friction (deg)	15	33
(deg)	0	5
k	50	350
n	0.25	0.25
k_b	30	280
m	0.2	0
R_f	0.7	0.7

Table 3. - Properties of Geofabric

Property	Values Used in Analyses
Cross-sectional Area (ft^2/ft)	0.01
Young's Modulus (ksf)	10,000 to 100,000

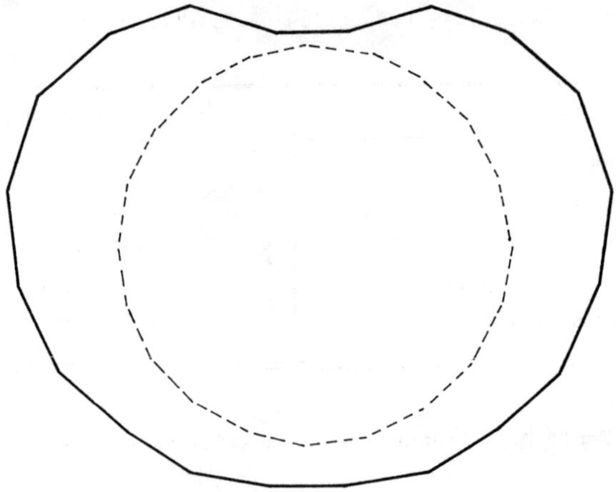

Pipe Installed Without Geofabric

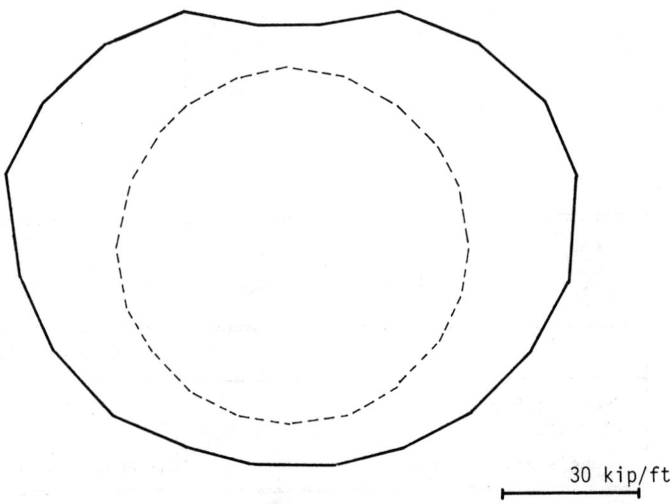

Pipe Installed With Geofabric

Fig.2 Distribution of Axial Forces

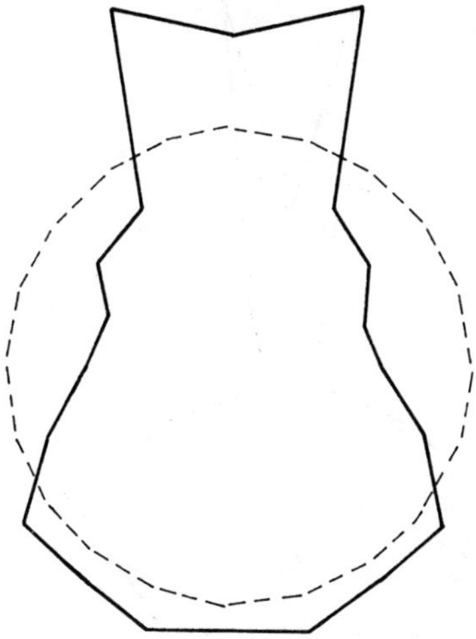

Pipe Installed Without Geofabric

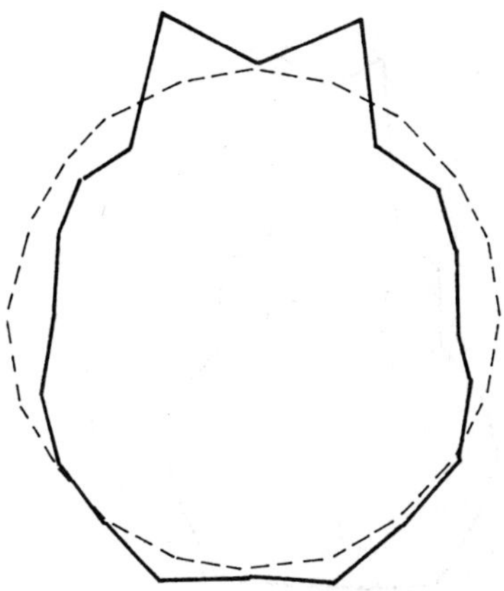

Pipe Installed With Geofabric

20 kip ft/ft

Fig.3 Distribution of Bending Moments

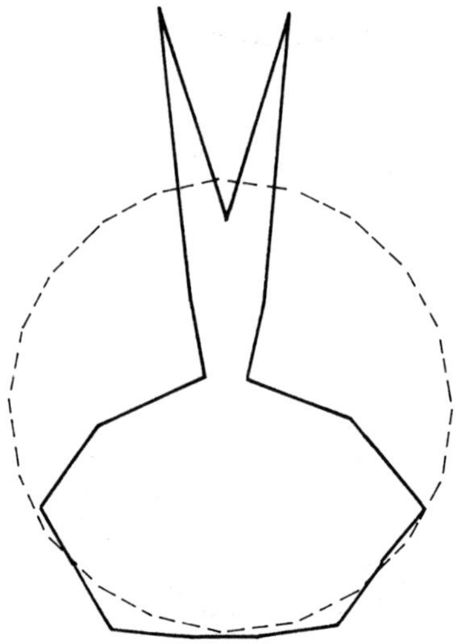

Pipe Installed Without Geofabric

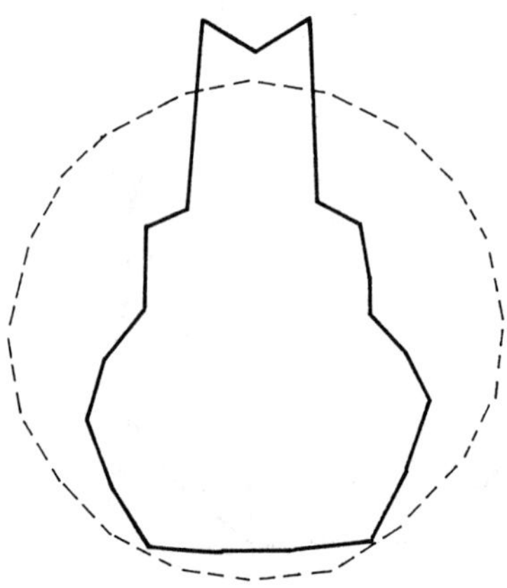

Pipe Installed With Geofabric

100 kip/ft^2

Fig. 4 Distribution of Stresses in Pipe Wall

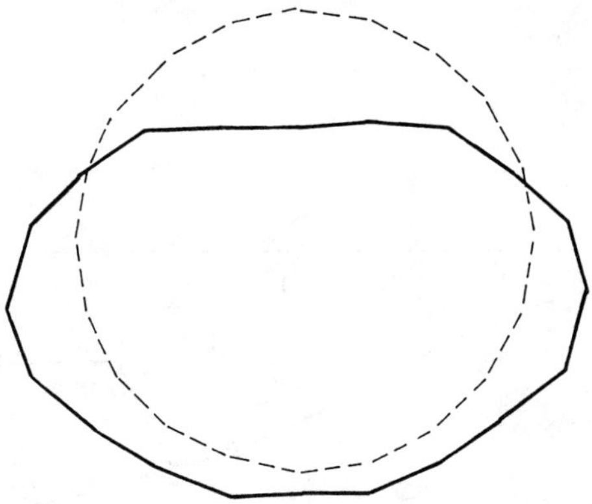

Pipe Installed Without Geofabric

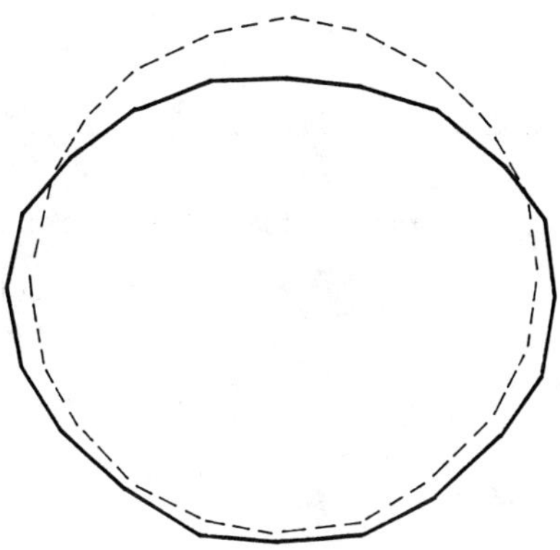

Pipe Installed With Geofabric 0.30 ft

Fig. 5 Deformations of the Pipe

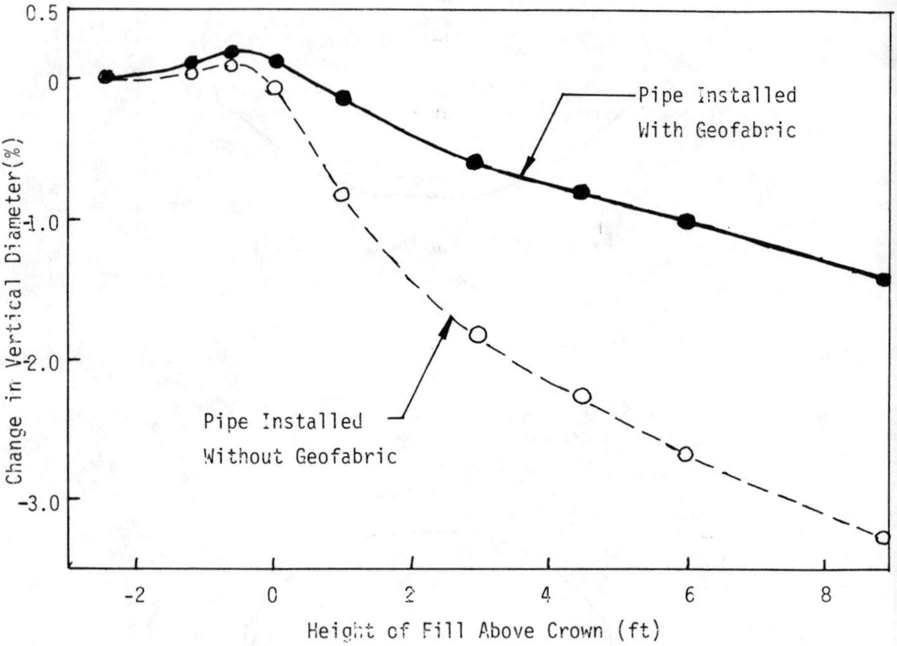

Fig.6 Initial Deflection Under Backfill Loads

backfill loads are shown in Fig. 4 for both installations with and without the use of geofabric. It is interesting to note that crown and haunch stresses are twice as high in the case of without geofabric when compared to that with geofabric.

Deflections

The deformed shapes of the pipe under backfill loads with and without the use of geofabric are shown in Fig. 5. The crown deflection at the end of backfilling for the installation without the geofabric is twice of that calculated with geofabric. The variation of the change in vertical diameter with the height of fill above the crown of the pipe for both, without and with geofabric in the installation is shown in Fig. 6.

Conclusions

This study of the behavior of a 60 inch diameter plastic pipe under backfill loads clearly indicates the potential use of geofabric to reduce the excessive crown deflection caused by poor native soils. It appears that the stresses in the pipe wall are also reduced significantly due to the lateral support strength added by the geofabric. The geofabric also causes a significant reduction in the bending moments and the loads on the pipe are carried more in the form of ring compression. This geofabric-soil-pipe interactive design would lead to a greater economy in the use of flexible sewer pipes.

Acknowledgments

This research was performed as part of an ongoing research program on soil-pipe interaction in the Department of Civil Engineering at Texas A&M University. Mrs. Joy Taylor typed the manuscript and James F. Dewey drafted the figures.

References

1. Chambers, R.E. and McGrath, T.J. (1981), "Structural Design of Buried Plastic Pipe," Paper presented at International Conference on Underground Plastic Pipe, New Orleans.
2. Duncan, J.M. and Chang, C.-Y. (1970), "Nonlinear Analysis of Stress and Strain in Soils," Journal of the Soil Mechanics and Foundations Division, ASCE, Vol. 96, No. SM5, Proceedings Paper 7513.
3. Duncan, J.M., (1976), "Finite Element Analysis of Buried Flexible Metal Culvert Structures," Laurits Bjerrum Memorial Volume, N. Janbu, F. Jorstad, and B. Kjoernsli, eds., Norwegian Geotechnical Institute, Oslo, pp. 213-222.
4. Duncan, J.M., (1979), "Soil-Culvert Interaction Method for Design of Metal Culverts," Transportation Research Board.
5. Duncan, J.M., Byrne, P.M., Wong, K.S., and Mabry, P.N., (1978), "Hyperbolic Volume Change Parameters for Nonlinear Finite Element Analyses of Stresses and Movements in Soil Masses," Geotechnical Engineering Report, University of California, Berkeley.

6. Duncan, J.M., (1979), "Behavior and Design of Long-Stem Metal Culverts," Journal of the Geotechnical Engineering Division, ASCE, Vol. 105, No. GT3, Proc. Paper 14429, pp. 399-418.
7. Duncan, J.M. and Jeyapalan, J.K., (1979), "Design Studies for Kaiser Aluminum Elliptical Culvert Structures Pinson Mounds, Jackson, Tennessee," Report for the State of Tennessee, Department of Transportation.
8. Duncan, J.M. and Jeyapalan, J.K. (1979), "Analyses of the Performance of Aluminum Box Culverts Subjected to 657B Scraper Loading," Report for Kaiser Aluminum and Chemical Sales, Inc.
9. Duncan, J.M., (1980), "Design Studies for Aluminum Structural Plate Box Culverts," Report for Kaiser Aluminum and Chemical Sales, Inc.
10. Duncan, J.M. and Jeyapalan, J.K., (1982), "Deflection of Flexible Culverts Due to Backfill Compaction," A paper presented at the 61st Annual Transportation Research Board Meeting, Washington, D.C.
11. Galili, N. et al. (1978), "Soil-Pipe Interaction, Stage I and Stage II, Final Report," Publ. No. 291, Agricultural Engineering Department, Technion City, Haifa, Israel.
12. Galili, N. and Shmulevich, I. (1981), "A Refined Elastic Model for Soil-Pipe Interaction," Paper presented at International Conference on Underground Plastic Pipe, New Orleans.
13. Gaube, E. and Mueller, W. (1981), "12 Years of Deformation Measurements on Sewer Pipes from Hostalen GM 5010," Paper presented at International Conference on Underground Plastic Pipe, New Orleans.
14. Hoeg, K. (1968), "Stresses against Underground Structural Cylinders," ASCE, No. SM4 6022.
15. Horn, W. (1979), "Field Tests of Plastic Pipe for Airport Drainage Systems," Report No. FAA-RD-79-86 WES TR GL-79-24.
16. Howard, A.K. (1981), "Diametral Elongation of Buried Flexible Pipe," Paper presented at International Conference on Underground Plastic Pipe, New Orleans.
17. Jeyapalan, J.K., Osequeda, F., and Horn, W.J., (1982), "Soil-Structure Interaction Analyses of Plastic Pipes," Preprint 82-511, American Society of Civil Engineers, Fall Convention, New Orleans, 14 pp.
18. Katona, M.G., Forrest, J.B., Odello, R.J., and Aligood, J.R., (1974), "Computer Design and Analysis of Pipe Culvert," Interim Technical Report 51-040, FHWA 3-1-1170, U.S. Naval Civil Engineering Lab., Port Hueneme.
19. Leonards, G.A. and Stetkar, R.E. (1978), "Performance of Buried Flexible Pipe," Report No. JHRP-78-24.
20. Molin, J. (1981), "Flexible Pipes Buried in Clay," Paper presented at International Conference on Underground Plastic Pipe, New Orleans.
21. Moser, A.P. (1981), "Strain as a Design Base for PVC Pipes?," Paper presented at International Conference on Underground Plastic Pipe, New Orleans.
22. Spangler, M.G. and Handy, R.L. (1982), Soil Engineering, 4th Ed., Intext Educational Publishers, New York.

A Soil-Structure Interaction Problem for a Group of Connected Flexible Pipelines

A.P.S. Selvadurai*

This paper examines the problem of flexure of a group of connected flexible pipelines which are deeply embedded in a soil mass. The group of pipelines is modelled as an infinitely long beam of finite width in which flexure takes place only in the longitudinal direction. An approximate analytical solution is developed for the problem in which the pipeline group is subjected to a concentrated load. Such concentrated loads can be induced at anchoring locations. The results presented in the paper illustrate the manner in which factors such as the thickness to diameter ratio of the pipe, the modular ratio of the soil-pipe system and the geometry of the pipeline group, influence the flexural moments developed in the buried pipeline group at the loaded location.

Introduction

Long distance buried pipelines are used quite extensively for the transport of materials such as oil and natural gas, coal slurries, mine tailing and water [1-4, 10-12]. In the construction of buried pipelines it is sometimes desirable to group or to connect several long distance pipelines to form a duct bank. This procedure enhances the lateral stability of the pipeline group, increases its overall stiffness and facilitates its economical construction. The composite action between the individual pipelines forming the group is achieved by providing metal or concrete connection blocks at desired intervals. Since the pipeline group in this configuration exhibits composite action it is necessary to prevent possible flotation of the group due to flooding action. This resistance to flotation is usually provided by mass concrete weights or ground anchors which are placed at salient locations. (Fig. 1). This paper examines the elastostatic soil-structure interaction problem related to such a long distance duct bank. For the purposes of analysis we assume that the duct bank can be idealized as a flat flexible beam of finite width and infinite length, which is embedded in bonded contact with an isotropic elastic soil mass. In this paper the analysis of the problem is restricted to the class of deeply embedded duct banks in which the width of the embedded pipe group b(=ND; where N is the number of pipes in the group and D is the outer diameter of a single pipe) is small in comparison to the depth of embedment H. Such a situation can occur in a vertical transition zone of a long distance duct bank. The analysis can also be extended to include shallow depths of embedment. The soil medium is assumed to be homogeneous and the effects of the excavation and backfill processes are ignored. The soil-structure interaction problem

*Professor and Chairman, Department of Civil Engineering, Carleton University, Ottawa, Ontario, Canada K1S 5B6

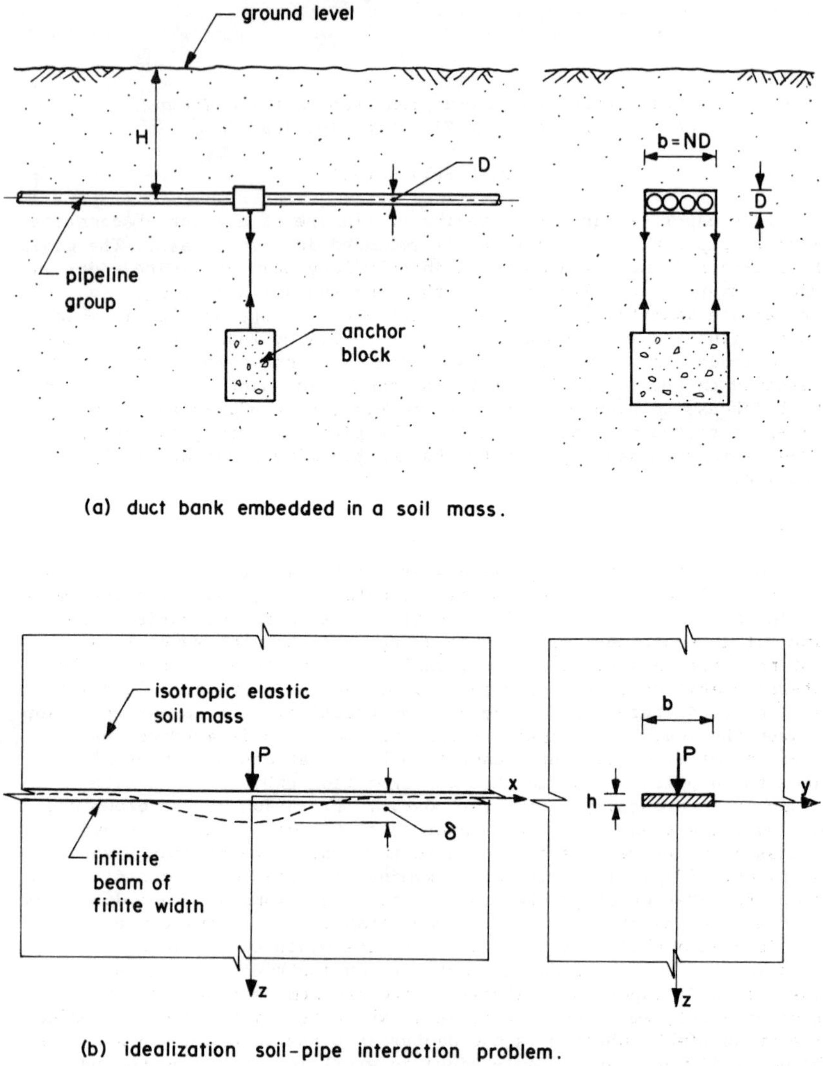

Figure 1. The soil-pipe interaction problem.

for the buried pipe group specifically investigates the flexural moments and flexural stresses that are developed in the duct bank due to a differential settlement which occurs at a section of the pipeline. Such differential settlements can be induced either by creep and consolidation settlements of the anchor block or the supression of flotation displacements at an anchoring location. This paper employs an analytical formulation to evaluate the flexural stresses developed in the pipeline group at the restrained location. The numerical results presented in the paper illustrate the manner in which the dimensions of the pipe, the number of pipes in the duct bank, the relative flexibility of the soil-pipe system influence the flexural stresses that are developed at the anchored location.

Analysis

We consider the problem of a duct bank of width b and thickness h which is embedded in an isotropic elastic soil mass. The depth of embedment is assumed to be such that the flexural interaction between the duct bank and the soil mass is relatively uninfluenced by the free surface of the soil mass. The total width of the duct bank b=ND and the thickness h of the duct bank is set equal to D. (D is the external diameter of a single pipe.) The pipes are grouped and connected together in such a way that flexure of the duct bank takes place only in the longitudinal direction. The duct bank essentially behaves like a flexible Bernoulli-Euler beam of flexural rigidity E*I* where E* is the elastic modulus of the pipe material and I* is the flexural rigidity of the pipe group. The duct bank is subjected to a differential settlement δ which is caused by the settlement at an anchored or restrained location. This differential settlement induces contact stresses at the soil-duct bank interface, which maintains complete bonded contact throughout the interaction process for all choices of the differential settlement δ. The idealized soil-duct bank interface occupies both the surfaces $z = \pm D/2$ and $y = \pm ND/2$, for $x \in (-\infty,\infty)$. In order to make the problem mathematically tractable we shall ignore the contributions from the vertical planes of the soil-duct bank interface and consider only the contact stresses developed on the planes $z = \pm D/2$. Also, for convenience we shall assume that the contact stresses effectively act on the planes $z = \pm 0$ of the elastic soil mass. We shall assume that the differential settlement at the anchor location induces a localized force P in the anchor rod. When the duct bank experiences flexure only in the x-z plane its deflection is defined by $w(x,y) = w(x)$. The embedded duct bank is effectively subjected to a loading $\tilde{p}(x,y)$ which causes symmetric flexure in the x-z plane. The contact stresses at the duct bank - soil mass interfaces are denoted by $\tilde{q}^{+}(x,y)$ and $\tilde{q}^{-}(x,y)$. The +ve and -ve superscripts signify the components of the contact stresses which act at the halfspace regions $z \geq 0$ and $z \leq 0$ respectively. For longitudinal flexure of the duct bank, the differential equation governing flexure can be written in the form

$$E^*I^* \frac{d^4 w(x)}{dx^4} + \int_{-b/2}^{b/2} \{\tilde{q}^{+}(x,y) - \tilde{q}^{-}(x,y)\}dy = \int_{-b/2}^{b/2} \tilde{p}(x,y)dy. \quad (1)$$

Considering the one-dimensional nature of the duct bank it is convenient to introduce a Fourier integral development of the variables in the soil-structure interaction problem. The deflection of the duct bank, the contact stresses at the interfaces and the external loads are expressed as averages taken over the width of the duct bank. i.e.

$$[w(x);\ bq(x);\ bp(x)] = \int_0^\infty [\bar{w}(m);\ \bar{q}(m);\ \bar{p}(m)]\cos(mx/a)\,dm \tag{2}$$

where

$$[\bar{w}(m);\ \bar{q}(m);\ \bar{p}(m)] = \int_{-b/2}^{b/2} [\frac{w(m,y)}{b};\ \tilde{q}(m,y);\ \tilde{p}(m,y)]\,dy \tag{3}$$

where a is a typical length parameter of the problem. Considering the Fourier representations, the governing differential equation (1) can be written in the form

$$E^*I^*\frac{d^4w}{dx^4} + b\{q^+(x)-q^-(x)\} = bp(x) \ . \tag{4}$$

In order to complete the description of the problem it is necessary to define the kernel function which relates the contact stresses $q^{\pm}(x)$ to the deflection $w(x)$. From the spatial symmetry of the problem it is evident that $q^+(x) = -q^-(x)$. The kernel function can be defined as a relationship between the transformed variables of $w(x)$ and $q(x)$, i.e. $\bar{w}(m)$ and $\bar{q}(m)$ respectively. Accordingly we can establish a relationship between $\bar{w}(m)$ and $\bar{q}(m)$ in the following form

$$\bar{w}(m) = \frac{\bar{q}(m)}{K(m)} \ . \tag{5}$$

Selvadurai [9] has presented an extensive account of the various techniques that can be adopted for the estimation of the kernel functions $K(m)$. Using solutions of the equations of elasticity for the loading of the soil-duct bank interface region ($z = \pm 0$; $-(ND/2) \le y \le (ND/2)$, $x \in (-\infty,\infty)$) it can be shown that several estimates can be developed by approximately satisfying the condition which requires constancy of displacement across the width of the duct bank. For example when the displacement across the width of the duct bank is approximately uniform it can be shown [5-9] that

$$K(m) = \frac{4G_s mND(1-\nu_s)\Psi(\beta)}{Ca(3-4\nu_s)} \tag{6}$$

where $1 < C < 1.13$; G_s and ν_s are the elastic constants of the soil mass; $\beta = NDm/2a$ and $\Psi(\beta)$ is a function which can be expressed as follows:

β =	0.1	0.5	1	3	8	∞
$\Psi(\beta)$ =	4.80	1.90	1.42	1.13	1.04	1.00

and for $\beta < 0.1$, $\Psi(\beta) \simeq (2/\pi\beta)\{\ln(1/\beta)+0.923\}^{-1}$.

Similarly, alternative expressions for $K(m)$ are [5-9];

$$K(m) = \frac{4G_s(1-\nu_s)(a+NDm)}{(3-4\nu_s)a} \tag{7}$$

and

$$K(m) = \frac{4G_s(1-\nu_s)}{(3-4\nu_s)} \{(2\beta + \tfrac{1}{2})\operatorname{erf}(\sqrt{2\beta} + (\tfrac{2\beta}{\pi})^{1/2} e^{-2\beta} - \beta\} \tag{8}$$

where

$$\operatorname{erf}(x) = \frac{2}{\sqrt{\pi}} \int_0^x e^{-t^2} dt \ . \tag{9}$$

The result (7) is accurate for small values of β and the result (8) is applicable for large values of β.

Concentrated Loading of the Duct Bank

The results presented in the preceding section can be used to develop a solution to the problem of the infinitely long duct bank which is subjected to a concentrated force P. Substituting (2) and (3) into (1) we obtain

$$E^*I^* \frac{d^4}{dx^4} \int_0^\infty \overline{w}(m)\cos(\frac{mx}{a})\,dm + 2\int_0^\infty \overline{w}(m)K(m)\cos(\frac{mx}{a})\,dm$$

$$= \int_0^\infty \overline{p}(m)\cos(\frac{mx}{a})\,dm \ . \tag{10}$$

For the particular case of the loading of the duct bank by a concentrated force, a solution of (10) gives

$$w(x) = \frac{Pa^3}{\pi E^*I^*} \int_0^\infty \frac{\cos(mx/a)\,dm}{[m^4+\Omega(m)]} \tag{11}$$

where

$$\Omega(m) = \frac{2a^4K(m)}{E^*I^*} \ . \tag{12}$$

The flexural moment in the duct bank is given by

$$M(x) = \frac{Pa}{\pi} \int_0^\infty \frac{m^2\cos(mx/a)\,dm}{[m^4+\Omega(m)]} \ . \tag{13}$$

These expressions for $w(x)$ and $M(x)$ can be numerically evaluated by employing the various approximations for K(m) defined by (6), (7) and (8). For purposes of numerical evaluation we shall consider the approximation for K(m) defined by (6). The maximum flexural moment in the duct bank is given by M(0). To numerically evaluate M(0) we set the length parameter a = D. The flexural rigidity of the duct bank

$$E^*I^* = \frac{\pi E^* D^4 N}{8} \; [\xi\{1 - 3\xi + 4\xi^2 - 2\xi^3\}] \tag{14}$$

where $\xi = t/D$. From (13) we obtain

$$M(0) = \frac{PD}{\pi} \int_0^\infty \frac{m^2 dm}{[m^4 + \tilde{\Omega}(m)]} \tag{15}$$

where

$$\tilde{\Omega}(m) = \frac{\beta \Psi(\beta)}{CR\phi S}$$

$$R = \frac{E^*}{E_s} \; ; \quad \phi = \frac{\pi(1+\nu_s)(3-4\nu_s)}{64(1-\nu_s)} \; ; \quad \beta = \frac{Nm}{2}$$

$$S = N\xi\{1 - 3\xi + 4\xi^2 - 2\xi^3\} \; ; \quad \xi = \frac{t}{D} \; . \tag{16}$$

Numerical Results and Conclusions

The integral (15) can be evaluated by using a Gaussian quadrature technique. The Figures 2-7 present values of $\bar{M}_0 = M(0)/(PD/\pi)$ derived from the integral expression (13). As is evident the maximum flexural moment in the duct bank, due to the load P, is influenced quite considerably by the relative modular ratio of the pipe material to surrounding soil mass (R). Other parameters such as the number of pipes in the duct bank (N) and the thickness to diameter ratio of the pipe (t/D in the range 0.020 to 0.005) only have a marginal influence on the maximum flexural moment. It may be noted that the factor ϕ yields the same result when ν_s is 0 or 0.5. Consequently the values of M(0) derived for $\nu_s = 0$ are also applicable to the situation in which the soil mass exhibits undrained behaviour. The numerical results presented in the paper for the maximum flexural moment indicate that for a range of relative rigidities (R) and pipe dimensions (t/D) of practical interest, the maximum flexural moment developed in a duct bank which is subjected to a localized anchor load P can be conveniently estimated from the approximate result

$$M_0 \simeq \frac{PD}{\pi} \left\{ \frac{\pi(3-4\nu_s)(1+\nu_s)}{48(1-\nu_s)} \; \frac{E^*}{E_s} \; \frac{t}{D}\left(1 - \frac{3t}{D} + \frac{4t^2}{D^2} - \frac{2t^3}{D^3}\right) N \right\}^{1/4} . \tag{17}$$

For a duct bank in which $t/D \ll 1$, the maximum bending stress in a single pipe at the point of application of the anchor load P is given by

$$\sigma_0^{max} \simeq \frac{P}{\pi D^2} \left\{ \frac{\pi(3-4\nu_s)(1+\nu_s)}{64(1-\nu_s)N^3} \; \frac{E^*}{E_s} \frac{D^3}{t^3} \right\}^{1/4} . \tag{18}$$

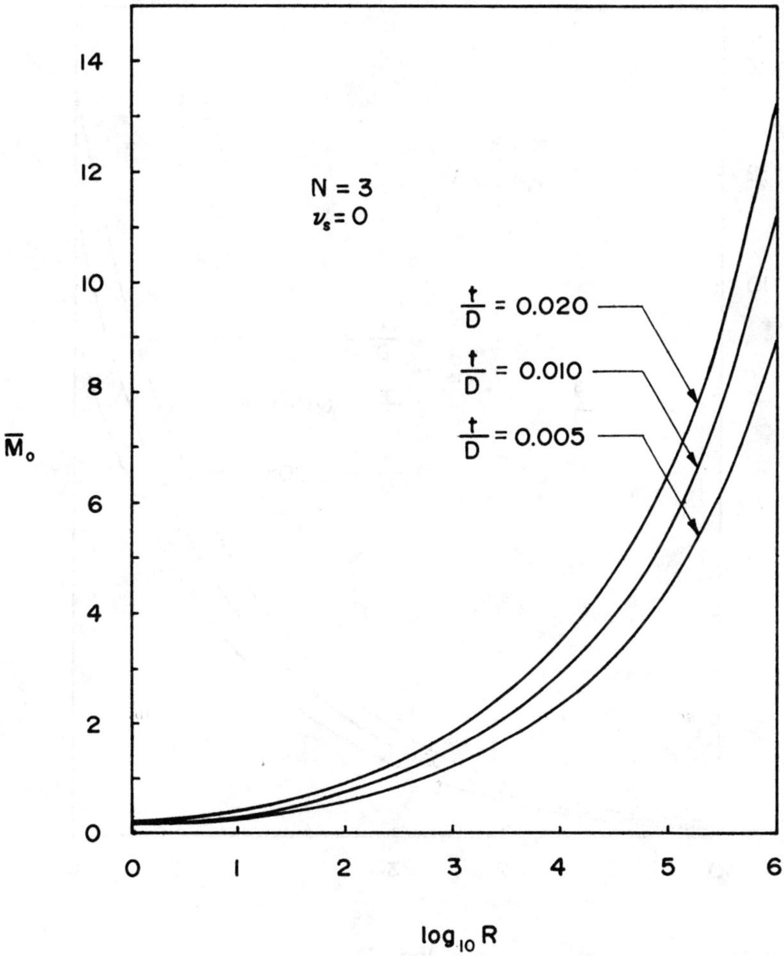

Figure 2. Flexural moment in the duct bank due to the action of a localized anchor force.

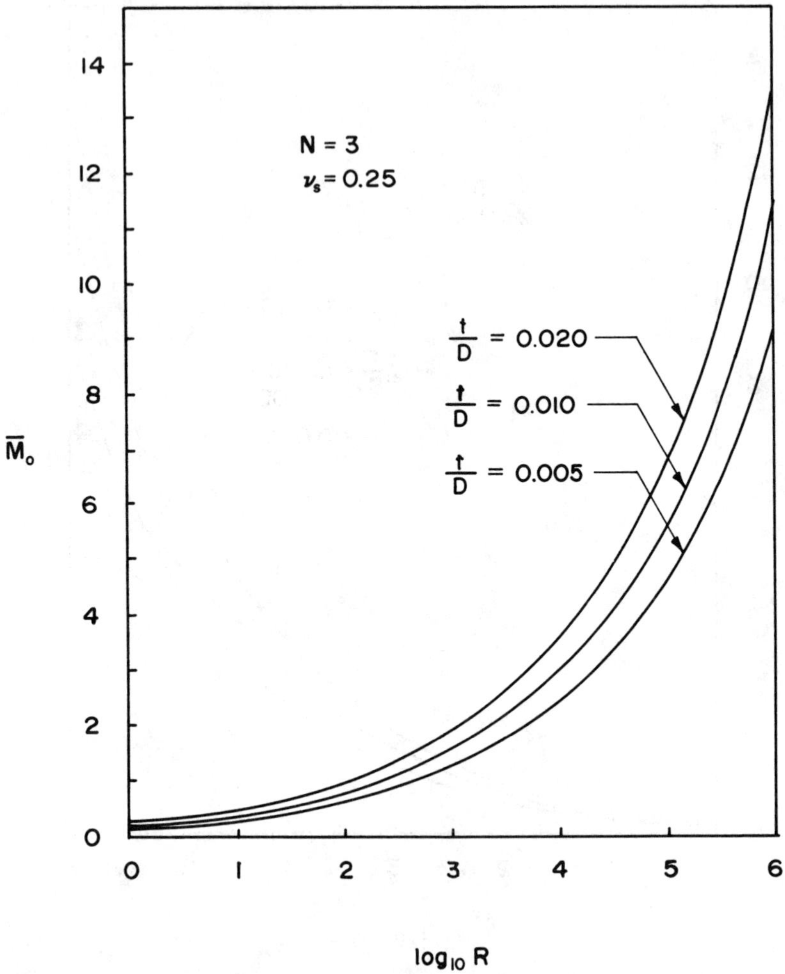

Figure 3. Flexural moment in the duct bank due to the action of a localized anchor force.

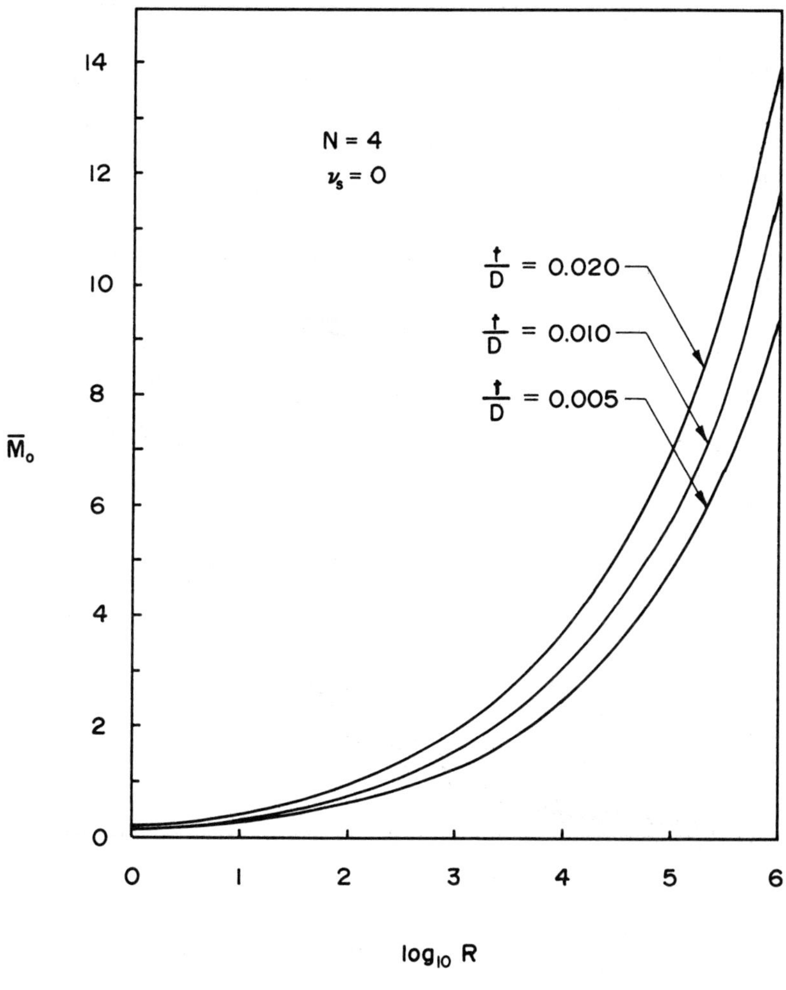

Figure 4. Flexural moment in the duct bank due to the action of a localized anchor force.

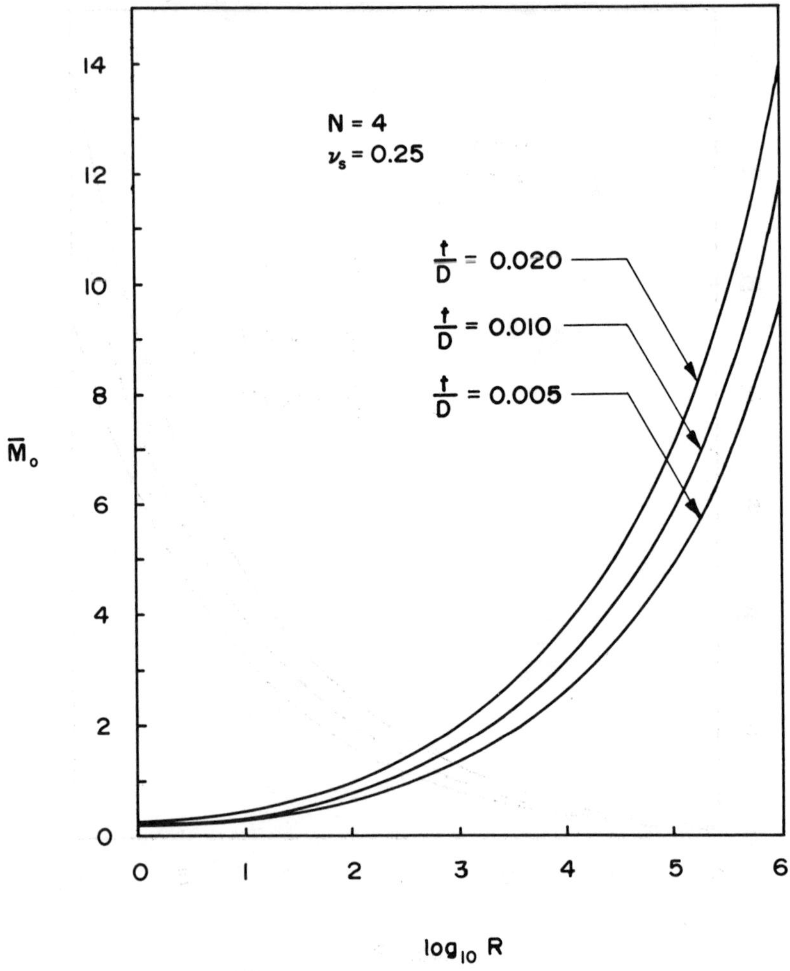

Figure 5. Flexural moment in the duct bank due to the action of a localized anchor force.

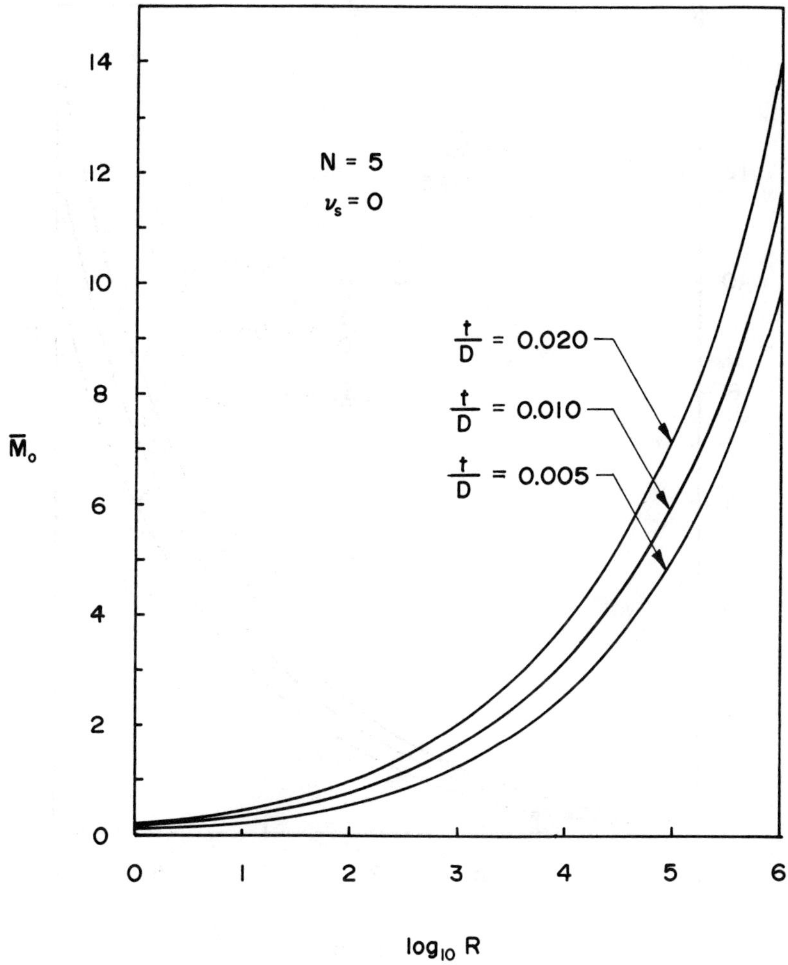

Figure 6. Flexural moment in the duct bank due to the action of a localized anchor force.

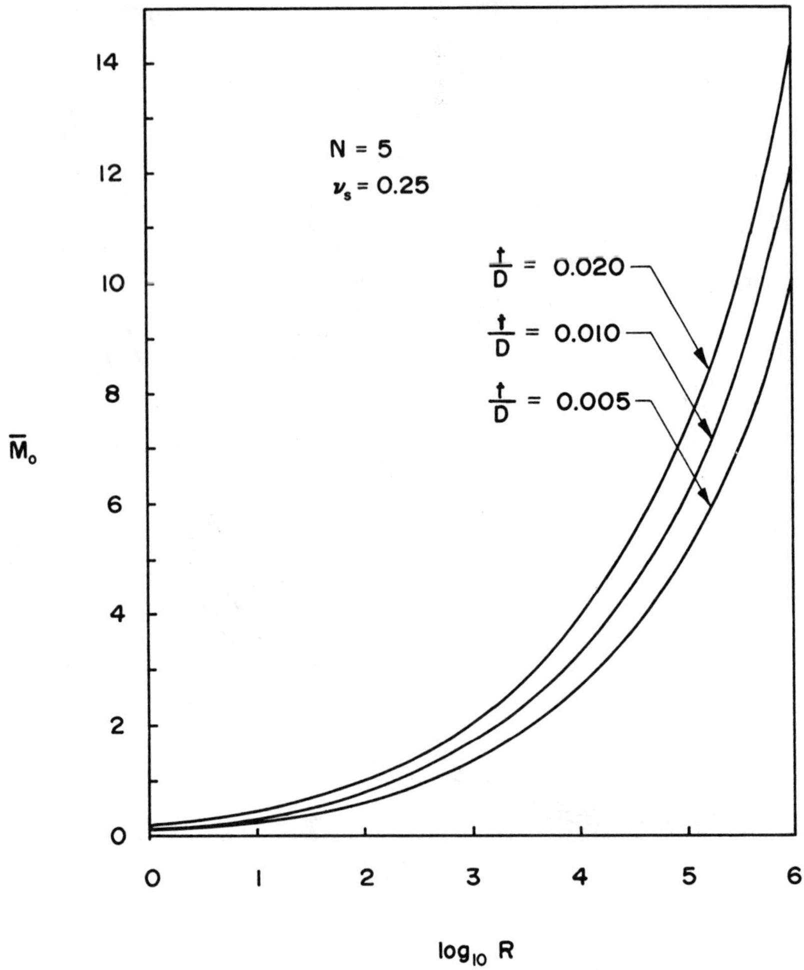

Figure 7. Flexural moment in the duct bank due to the action of a localized anchor force.

References

[1] Ariman, T., Liu, S.C. and Nickell, R.E., "Lifeline Earthquake Engineering - Buried Pipelines, Seismic Risk and Instrumentation", PVP 34, American Society of Mechanical Engineers, 1979.

[2] A.S.C.E., Journal of the Pipeline Division, Vol. 85, PL3, Proceedings Paper 2205, 1959.

[3] A.S.C.E., Journal of the Pipeline Division, Vol. 93, PL2, Proceedings Paper 5355, 1967.

[4] A.S.C.E., List of Source and Reference Material on Pipeline Design, Task Committee Report, 1980.

[5] Biot, M.A., "Bending of an Infinite Beam on an Elastic Foundation", Journal of Applied Mechanics, Vol. 4, A1-A7, 1937.

[6] Lekkerkerker, J.G., "Bending of an Infinite Beam Resting on an Elastic Halfspace", Proceedings of the Koningkl. Ned. Akad. Wetsch. Ser. B, Vol. 63, 484-497, 1960.

[7] Rvachev, V.L., "On the Bending of an Infinite Beam on an Elastic Halfspace", Journal of Applied Mathematics and Mechanics (PMM), Vol. 22, 984-988, 1958.

[8] Selvadurai, A.P.S., Elastic Analysis of Soil-Foundation Interaction, Developments in Geotechnical Engineering, Volume 17, Elsevier Scientific Publishing Co., Amsterdam, The Netherlands.

[9] Selvadurai, A.P.S., "Flexure of an Infinite Beam of Finite Width Embedded in an Isotropic Elastic Medium of Infinite Extent", International Journal for Numerical and Analytical Methods in Geomechanics (in press).

[10] Selvadurai, A.P.S. and Lee, J.J., Soil Resistance Models for the Stress Analysis of Buried Fuel Pipelines, Interim Report, Technical Research and Development Project, Research and Engineering, Bechtel Group Inc., San Francisco, December 1981.

[11] Selvadurai, A.P.S. and Lee, J.J., Soil Resistance Models for the Stress Analysis of Buried Fuel Pipelines, Final Report, Technical Research and Development Project, Research and Engineering, Bechtel Group Inc., San Francisco, December 1982.

[12] Shibata, H., Katayama, H. and Ariman, T., "Recent Advances in Lifeline Earthquake Engineering in Japan", PVP 43, American Society of Mechanical Engineers, 1980.

POLICIES FOR PIPELINE SAFETY NEAR URBAN AREAS

R. J. Allman, P.Eng. C.S.C.E.*
J. D. Dilay, P.Eng.**

ABSTRACT

Extensive oil and gas development in Alberta has resulted in the installation of approximately 200 000 kilometres (124 300 miles) of pipelines ranging from small diameter plastic lines for rural and urban gas distribution to medium and large diameter steel lines for the gathering and transmission of sweet gas, sour gas, crude oil, and refinery products. Two of the more critical types of pipelines, namely those carrying sour gas and ones transmitting high vapour pressure products, (ethane, ethylene, propane, and butane) require special attention because of the potential hazards to public safety, particularly in urban areas. Gas containing hydrogen sulphide (sour gas) is lethal in very low concentrations, and high vapour products form explosive mixtures with air. Both products can inundate large areas when a pipeline carrying them is ruptured. This paper discusses the basis for new policies concerning sour gas pipelines in proximity to expanding population centres and the recent developments taking place in the transportation of high vapour pressure products through urban areas, together with observations of their significance and impact.

INTRODUCTION

The Energy Resources Conservation Board (ERCB) is responsible for ensuring the economic, efficient and orderly development of energy resources in the Province of Alberta in Canada. One of its mandates is to see that energy-related pipelines are not only built to recognized standards, but that they operate safely under all conditions. In this regard, new and substantive control measures have been brought in to deal with the problem of urban development encroachment around existing sour gas fields. Foremost among the steps taken have been minimum separation distances between sour gas facilities and areas of public use. Similarly, once-distant refineries now close to population centres are served by numerous

---

* R. J. Allman, Senior Operations Engineer, Pipeline Department, Energy Resources Conservation Board, 640 - Fifth Avenue S.W., Calgary, Alberta, Canada, T2P 3G4

** J. D. Dilay, Assistant Manager, Pipeline Department, Energy Resources Conservation Board, 640 - Fifth Ave. S.W., Calgary Alberta, Canada, T2P 3G4

pipelines carrying high vapour pressure products. In this case the potential for disaster has led to revised land-use planning, upgraded pipeline standards, and new legislation designed to reduce the incidence of pipeline damage by outside forces. This paper will discuss in some detail the developments that have taken place.

THE PIPELINE-URBAN ENCROACHMENT PROBLEM

The pipeline-urban encroachment problem centres around the relationship of three inter-related factors, namely pipeline safety, urban development pressure, and land-use conflicts (26). Many interests are represented in this situation such as municipalities, planners, developers, pipeline operators, and regulatory agencies. Each group has its own point of view, and each must now recognize the effects of actions that can no longer be taken in isolation.

Pipeline safety has long been a concern of the pipeline industry. The design, installation, and operation of pipelines have always been undertaken with safety in mind, and this is supported by the numerous engineering codes and standards devoted to pipeline matters, augmented by continual research and testing of better materials, safety devices, and surveillance techniques. Why then, with all this engineering background and these resources does there appear to be a problem of transporting products, hazardous though they may be, through urban areas?

Damage by Outside Forces

The problem of pipeline safety in these areas is in most cases essentially one of protecting the pipeline from external damage by outside activities. Such damage can result either in instantaneous failure or long-delayed failure, depending upon the type and severity of the damage to the pipeline. The former can be serious enough in its scope and destructive force but the latter is much more insidious, starting out perhaps as perceived minor damage but later erupting into a full-scale disaster. Pipeline operators, therefore, are rightly concerned about the increasing frequency of damage to their pipelines by outside forces, and are looking beyond the traditional right of way protection methods for solutions.

Land-use Conflicts (27)

In response to population growth and the demand for housing and other amenities, the subdivision and development of land has become an increasing part of our social pattern. In many instances this activity is planned for locations adjacent to or around existing pipelines; on the other hand, pipeline economics may dictate a new route through an area that is either scheduled for urban development or where development already exists. Either way, there is an interface between pipelines and urban development that has potential for a number of problems. The situation is aggravated, of course, when the pipelines carry hazardous products.

A number of examples may be cited to show the kinds of land-use conflicts that may develop. A common occurrence is for municipalities to allow various types of development to encroach on pipeline rights of way, and there have been cases where residential lots have been planned directly over the pipeline (Figure 1). This of course ubstantially increases the risk of damage by outside forces to the pipeline, and generally means increased surveillance by pipeline company personnel.

Figure 1 - Residential Lots Over High Pressure Gas Pipeline

Alternatively, pipeline companies with existing rights of way through areas that are being planned, or have developed as established subdivisions, may wish to install a second or third pipeline within those rights of way to carry hazardous products --a not uncommon occurrence these days. This immediately changes the status quo from the residents' perceived safety point of view, and often it brings the pipeline company in direct conflict with developers and planning authorities.

A related area of potential conflict arises when an existing pipeline right of way literally bisects a planned development, and planners must either try to accommodate the right of way in the overall plan, or somehow try to have the pipeline relocated. The latter can be extremely costly, while the former may require special land useage not originally contemplated by the municipality or developers.

All of these conflicts are, of course, made worse when hazardous products are being transported through the pipelines. Although pipelines are inherently the safest mode of transportation for moving potentially dangerous petroleum products, accidents unfortunately do happen, and in 1979 one such accident took place in an Edmonton subdivision known as Mill Woods. It illustrates many of the problems still being faced by proponents involved in the pipeline-urban encroachment problem, and some of the details are presented here.

HIGH VAPOUR PRESSURE (HVP) PIPELINES

The Mill Woods Incident

At approximately 11:55 a.m. on Friday, 2 March 1979, a pipeline 219.1 mm (8.76 in) in diameter, licensed to Rimbey Pipe Line Co. Ltd. and operated by Gulf Canada Limited, ruptured in the Mill Woods area of Edmonton. Liquid propane at more than 7600 kPa (1102 psi) pressure escaped from a 448 mm (17.9 in) long split in the pipe wall and formed a cloud of propane vapour that drifted northward (Figure 2). At about noon a commercial vehicle entered the drifting vapour and ignited it. The driver was severely burned and his vehicle was destroyed.

Coincident with the formation of propane vapour, liquid propane entered a nearby storm sewer catch basin through an opening at street level and through fissures in the ground caused by the explosion. The propane spread into adjacent sewer lines, and explosive mixtures were subsequently detected over a wide area within the sewer system. As a result, over 19 000 residents were evacuated from their homes, while efforts went on throughout the ensuing hours to eliminate the explosive mixture from the sewer lines. A combination of water flushing and nitrogen gas blanketing successfully eliminated the danger, and some 23 hours after the initial pipeline failure, at 10:45 a.m. on Saturday, 3 March, residents were allowed to return to their homes.

In accordance with its mandate under the legislation, the ERCB called a public inquiry to determine the circumstances of the break, the emergency response taken to protect people in the area, and the longer-term issues of public safety. All major participants in the failure incident were required to attend, together with any interested parties that wished to be represented and make their views known. The investigation showed that the pipeline had been previously damaged in three places by a person or persons unknown, causing propane to escape from a pipeline rupture (Figure 3), and subsequently to enter the sewer system.

Following the inquiry, the ERCB issued a Decision Report (16), containing a total of 15 recommendations. These ranged from updated control-room operating procedures, education of construction personnel, and new legislation to control surface activities, to the establishment of contingency plans and facilities capable of handling pipeline emergencies of all types. These recommendations were followed up a year later with a second report (17), setting out in detail the overall progress made during the year, and an assessment by the ERCB of the current status of the recommendations.

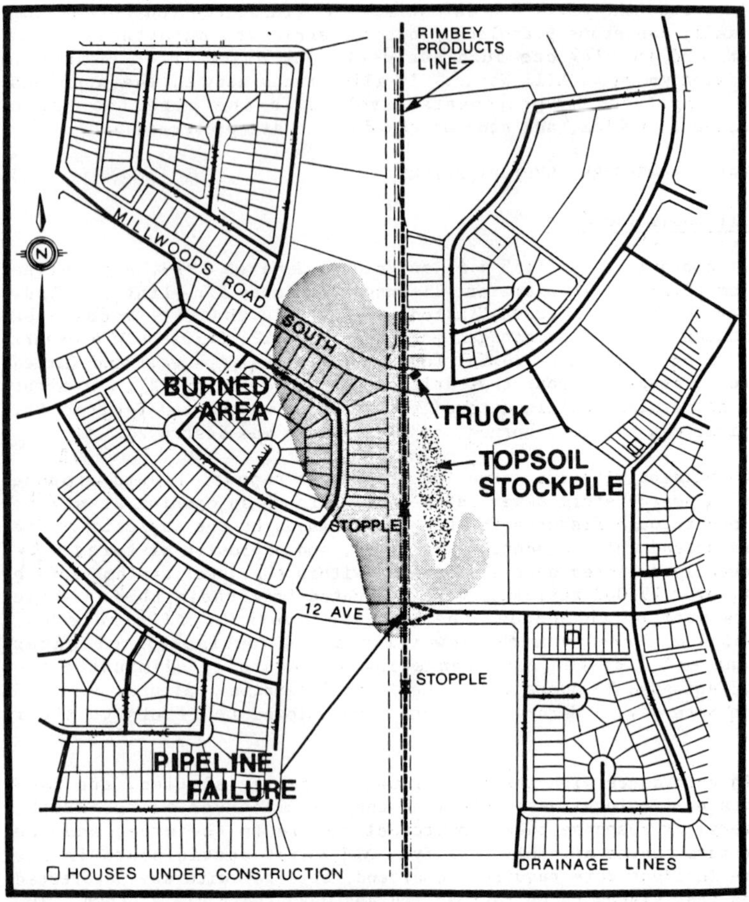

Figure 2 - Area of Propane Dispersion

The Mill Woods case pointed out the many problems inherent in the transportation of HVP products by pipeline through an existing urban area, and as such it was the catalyst for enlightened and updated action for the prevention of similar occurrences. The lessons learned and the remedial measures since taken will undoubtedly be relected in a greater awareness of pipeline-urban development problems, and some of the solutions available for coping with them.

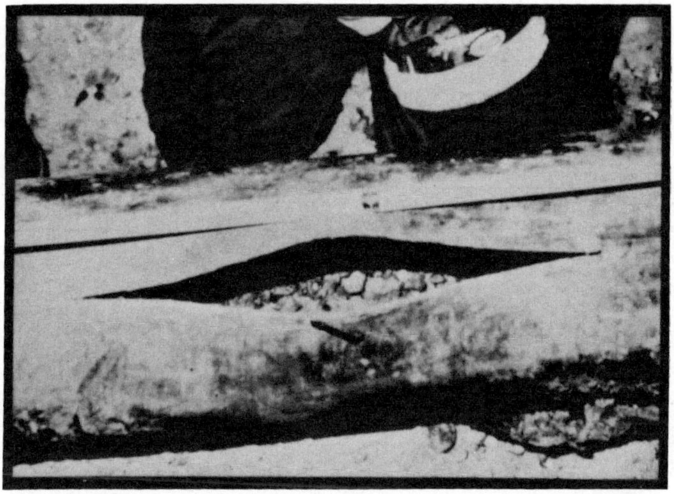

Figure 3 - The Ruptured Pipe

Transportation and Utility Corridors

One of the more recent and interesting planning developments in Alberta that has directly affected the pipeline-urban development problem is the establishment of transportation and utility corridors around the major cities of Edmonton and Calgary. They have been incorporated into the Government's Restricted Development Area plans, and as such form an integral part of the system around the two cities.

The corridor concept (29) is not a new one, and in recent years throughout North America many studies have been done and much attention paid to the idea of placing various modes of transportation in common rights of way. Such corridors can accommodate a variety of facilities for the conveyance of people, energy, information, or materials, and this is the situation in the case of the corridors around Edmonton and Calgary. They are planned specifically for highways, pipelines, communications circuits, and power lines; railways are not included (Figure 4). The figure shows "ideal" cross-sections which form the basis for planning the actual corridors; in practice, specific problem areas at different locations along the corridors have required variations in the cross-section configurations.

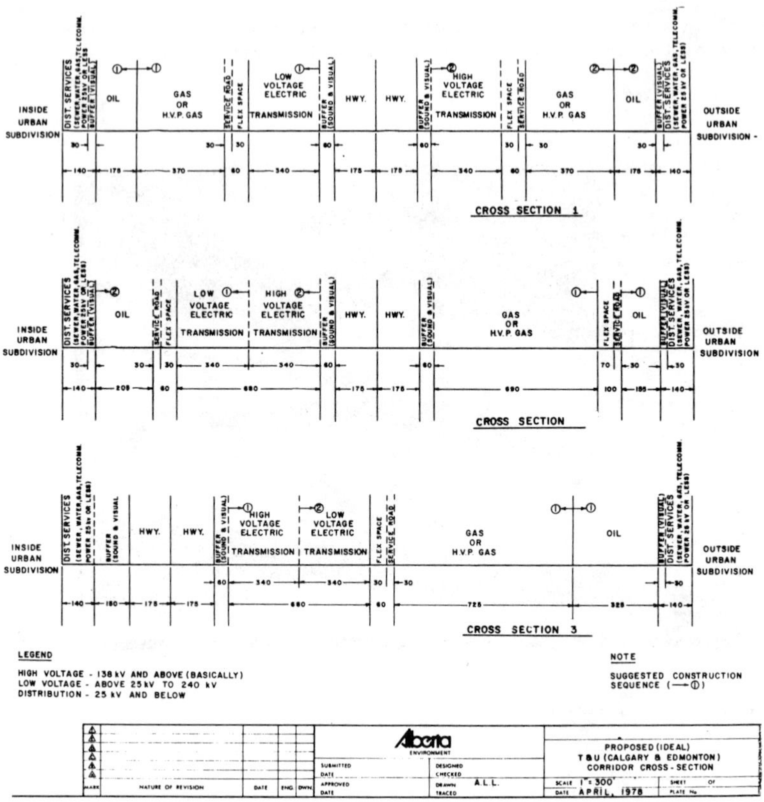

Figure 4 - Transportation and Utility Corridor Cross-Sections

Transportation and utility corridors are a particularly suitable solution to some of the previously cited problems. By dedicating a strip of land exclusively to the use of pipelines there is a more concentrated, orderly, and efficient land use, and considerably less fragmentation of urban areas by multiple rights of way. This in turn reduces the total environmental disturbance and restricts it to a limited area. Benefits may also result from this grouping of pipelines in a single, dedicated right of way by having common inspection and surveillance procedures in place, and this can be especially useful where there are HVP pipelines. A tangible safety benefit is also gained by the mere fact that having a number of pipelines properly marked and identified in a known right of way serves to make their presence known to others, and considerably reduces the possibility of damage by outside forces.

Notwithstanding the obvious advantages of placing pipelines in a single or multi-use corridor, there are some potential disadvantages which, if properly addressed at the time, can be minimized or eliminated entirely. For example, reconciliation of conflicting interests with other modes of transportation in such matters as engineering design (e.g. electrical interference problems between pipelines and power lines), timing of construction, legal matters, and insurance must all be addressed at an early stage and cannot be ignored. There is also the somewhat ironic problem that, by grouping the pipelines for the purpose of safety from outside forces, a major catastrophe in the area could damage or incapacitate a number of lines and services at one time.

In spite of the potential problems that may be encountered with a multi-use transportation and utility corridor, there are none that cannot be resolved with planning and foresight. Pipelines within such a corridor can obviously enjoy a number of advantages, not the least of which is to mitigate the problem of routing through urban and fringe-urban areas (7), and to minimize construction activity around existing pipelines in fully developed areas. Since transportation and utility corridors are still in the early development stages in Alberta, some of the long-term benefits of locating HVP pipelines within such corridors are still to be realized. However, even at this time, some of the tangible benefits are beginning to be recognized.

Other On-Going Developments

Transportation and utility corridors are but one way to overcome the problem of improving the integrity and safety of pipelines carrying HVP products through urban areas. People in many disciplines have studied the matter, and certain methods have evolved which time and experience have proven to be successful.

Reduction of Land-use Conflicts

The first and perhaps most obvious way is to make the pipeline rights of way an integral part of the land development plan. At first glance this may sound paradoxical --to plan subdivisions around potentially dangerous pipelines. As it turns out, this can be one of the most effective ways of eliminating pipeline damage by outside forces and controlling surface activities that might lead to pipeline failures. There are many examples of pipeline rights of way acting as green areas, parks, buffer zones and other low-density uses; the main criterion is imaginative planning.

Another way to avoid potential conflicts is for pipeline companies to consult with municipalities and planning authorities before a pipeline route is chosen, and to take land-use planning and future community growth into account in developing a pipeline project. The ERCB has made this one of the requirements for a pipeline permit, and applicants must now furnish proof that planned pipelines will not conflict with urban development.

One of the recommendations arising from the Mill Woods incident was to impose a setback distance for buildings from hazardous product pipelines in order to reduce the risk of external damage from surface activities, and to minimize the potential hazard to residents and structures adjacent to the lines. Such setback distances also allow space for the movement of vehicles or equipment without encroaching on pipeline rights of way, and discourage the practice of allowing building lots to be planned over the pipelines themselves.

An option that is always available is the relocation of existing pipelines from built-up or planned areas of population. This solution has obvious benefits and drawbacks depending upon the proponents involved, and each case requires close study and analysis. Relocation costs can be very high, and the question of payment can sometimes be very contentious. Offsetting such high costs can be the potentially increased value of vacated lands and the use to which such lands may be put. Notwithstanding the complex issues involved, relocation of hazardous product pipelines can be a viable solution under the right circumstances.

The One-call System

One of the ideas being considered in Alberta is the communication method known as the one-call system. This type of system has received wide acceptance in other countries, particularly the United States, and wherever it has been tried the results have been impressive. A one-call system provides a single toll-free telephone number for use by contractors, utilities, public agencies, and private citizens whereby owners of underground facilities can be notified by a central agency of the intent to excavate, tunnel, bore or otherwise disturb the ground surface. Owners or operators of pipelines or other underground facilities can then locate and mark such facilities prior to the start of any work. Construction activities, crossing agreements, and excavation permits can be checked at the same time.

One-call centres have proven extremely effective in reducing pipeline damage by outside forces, increasing public safety, and preventing loss of service. The reduction in costs has also been impressive. Other advantages have been increased public awareness of utilities and the consequences of damage, simplification of procedures for excavators, and immediate response from facility owners.

The advantages of such a system for HVP pipelines in urban areas are obvious. Alberta has embarked upon a study and possible implementation of a computerized one-call system (1), and indications to date are that it will be started in the near future. The benefits could well exceed the most optimistic predictions.

Changes in Pipeline Standards

Pipeline standards have always played a major role in pipeline safety, and their evolvement has been a measure of the changes within the industry itself. Recognizing the unusual properties of high vapour pressure petroleum products and the safety considerations that must be taken into account when moving such products through pipelines, CSA Standard Z183-M1982 (10) now includes upgraded requirements for HVP pipelines. The basic changes from previous editions are shown in Table 1.

Because of the varying population densities that can occur along the route of HVP pipelines and the inherent dangers associated with a pipeline failure, the concept of "Zones" was introduced; all other changes are related to this concept. With these changes, pipeline designers must now recognize the existence of urban areas along the routes of HVP pipelines, and incorporate the various factors into key elements of their pipeline design. In a sense, this begins to take on the rudiments of the design for gas pipelines where such design, based on population density, has been a long-standing principle.

Table 1 - Old and New Standards for HVP Pipelines

ITEM	OLD	NEW
ZONES	NONE	ZONE 1 & ZONE 2
DEPTH OF COVER	75 cm	ZONE 1 - 90 cm ZONE 2 - 120 cm
VALVE SPACING	30 km MAXIMUM	AS APPROPRIATE IN ZONE 1 15 km MAXIMUM IN ZONE 2 TRANSITION FROM ZONE 1 TO ZONE 2
DESIGN FACTOR	0.72	ZONE 1 — 0.72, ZONE 2 - 0.576
JOINTS	SLEEVED, COUPLED & PATENTED JOINTS ALLOWED WITH PROVISOS	WELDED JOINTS ONLY
TEST PRESSURE	125% MAX. OPERATING PRESSURE	ZONE 1 - 125% MAX. OPERATING PRESSURE ZONE 2 - 156% MAX. OPERATING PRESSURE

ZONE DEFINITIONS

ZONE 1 — A ZONE 1 LOCATION IS AN AREA EXTENDING 200 m ON EITHER SIDE OF THE CENTRE LINE OF ANY CONTINUOUS 1 km LENGTH OF A PIPELINE THAT CONTAINS 5 OR FEWER DWELLING UNITS INTENDED FOR HUMAN OCCUPANCY.

ZONE 2 — A ZONE 2 LOCATION IS AN AREA AS DEFINED FOR ZONE 1 THAT CONTAINS MORE THAN 5 DWELLING UNITS INTENDED FOR HUMAN OCCUPANCY, OR IS A FACILITY THAT CONTAINS 20 OR MORE PERSONS DURING NORMAL USE.

GENERAL — EACH DWELLING UNIT IN A MULTIPLE DWELLING UNIT BUILDING SHALL BE COUNTED AS A SEPARATE UNIT.

WHEN A FACILITY CREATES A ZONE 2 LOCATION REQUIREMENT, THE ZONE 2 AREA SHALL EXTEND AT LEAST 200 m FROM THAT FACILITY.

Believing that it is also in the interest of public safety to require operators of older HVP pipelines to meet safety standards applicable to similar new facilities, the ERCB now requires companies that operate such lines in urban areas to upgrade them retroactively, in accordance with or equivalent to CSA Standard Z183-M1982. Similarly, where the classification of an area changes to urban (Zone 2), an existing HVP pipeline in the newly-designated area must be upgraded to the latest design requirements within two years of the classification change.

The standards also reflect substantial changes in pipeline design that are not apparent from Table 1. Specifications can now be tailored to produce pipe for nearly every condition and climate, and special qualities have been built in that give additional safety margins for pipelines in critical areas. Along with the better pipe has come improved standards of welding, both in the materials used and the methods employed. These, together with newer inspection techniques, mean that the quality of welds now match and sometimes surpass the quality of the pipe. Even the standards for handling and installing pipe have markedly improved over the years, and systems can now be put into place with far less potential for problems later on.

Responding to the changing conditions of urban development and environmentally sensitive areas, the pipeline industry in Alberta has started to independently develop higher standards for leak detection for operating pipelines. Athough the present standards now include HVP pipelines, the newer ones will add greatly to the safety of people and property near such lines. They will also eliminate much of the uncertainty for pipeline designers concerning the levels of leak detection and prevention that should be maintained.

Advances in Pipeline Technology

In recent years, a number of significant technological developments have taken place in the field of pipeline transportation. These developments have contributed a great deal to the safety and integrity of pipelines carrying a variety of hazardous products, and continuing research is advancing the science even further.

Electro-magnetic survey instruments or "smart pigs" that can be inserted into pipelines to monitor the condition of the pipe walls have been developed to a high degree of sophistication. Originally intended to monitor internal corrosion, such instruments can now detect a wide range of flaws in the pipe, and store such information for detailed analysis at the end of each "run". This type of survey is invaluable for confirming the integrity of operating pipelines, and is fast becoming an established procedure with pipeline companies.

A variety of electronic and sonic detection systems have also been developed for monitoring small leaks in pipelines, and such devices are critical to the quick response time needed to prevent a leak from becoming a disaster of major proportions. Unfortunately, the effectiveness of these types of leak detectors for HVP pipelines is somewhat limited by the nature of the products being transported, and development is continuing to remove some of the shortcomings. This is not to say, however, that such devices should be discounted. Leak detection systems are an effective form of insurance, and the constant research going on is a valuable contribution to public well-being and safety.

In the authors' opinion, an equal, and in some ways more encouraging technological breakthrough for detecting leaks and breaks on HVP pipelines has been in the area of control room capabilities (18). Supervisory systems now have the ability to monitor a variety of operating data in far less time than was formerly possible, to do it with greater accuracy, and to repeat the data acquisition over very short intervals. Equally important, a modern control room supervisory system can predict and plot trends of critical operating parameters by computer simulation, and in doing so provide the operator with an advance means of detecting potential leaks or breaks along the pipeline. For HVP lines through urban areas, these are matters of prime importance.

Legislation to Reduce Pipeline Damage

Despite the not inconsiderable efforts of all parties to prevent damage by outside forces, the number of incidents has been gradually increasing. As one of the key agencies involved in the regulation of oil and gas pipelines in Alberta, the ERCB has been aware of the problem for some time, and in an attempt to reduce (if not eliminate) the number of incidents, has developed some special and in some ways unique legislation (22) (23).

The basis of this new legislation is to involve, for the first time, limited control of parties other than those in the pipeline industry. Before the start of construction activities around pipelines, anyone planning such work must first determine from available records whether pipelines exist within a specified distance of the work area. If any such lines are identified, the pipeline company must be contacted and the location of the pipelines marked in the field. Where construction work will take place within a pipeline right of way, written permission must first be received, and the lines must be accurately located on the site for position and alignment. Under certain circumstances lines must also be hand-exposed for positive identification. Temporary fencing of the pipeline right of way to limit access of heavy equipment may also be useful, and must be provided where such need is identified. Severe monetary penalties may be incurred for non-compliance with all these requirements.

As mentioned earlier, one of the most potentially dangerous situations that can arise is when a pipeline is struck by construction equipment resulting in a gouge, dent, crack, or similar type of damage to the pipe wall. Such damage, if not reported and repaired immediately, can lead to subsequent pipe failure, often with disastrous results. The Mill Woods accident in Edmonton was a good example of this type of situation. The new legislation now addresses this problem, and all such damage must be immediately reported to the pipeline company and construction activities stopped. Heavy penalties can be invoked if the law is ignored.

Other legislative changes now permit ERCB inspectors to suspend or shut down construction activities where it appears that the safety of a pipeline may be compromised, and an inquiry may be initiated if the matter is serious enough.

SOUR GAS PIPELINES

Sour Gas Reserves in Alberta

Some 22 per cent of Alberta's remaining established reserves of marketable gas of 1865 billion cubic metres (66.2 trillion cubic feet), or about 410 billion cubic metres (14.6 trillion cubic feet) derive from raw gas which is sour (20). That is, the raw gas contains deadly toxic and corrosive hydrogen sulphide (H_2S) as a component which must be removed by processing to render the gas saleable. Table 2 shows the distribution of Alberta's gas reserves by H_2S concentration. This distribution is expected to change with the long-range prospect that sweet and low-sulphur gas may predominate in reserve additions (28).

Table 2 Distribution of Alberta Sour Gas Reserves by H_2S Concentration

H_2S Concentration (%)	Initial Marketable Gas Reserve Basis (%)	Remaining Marketable Gas Reserve Basis(%)
less than 1	6.3	5.8
1-5	5.8	4.5
greater than 5	12.7	11.1
	24.8	21.4

The concentration of H_2S in raw gas in Alberta varies widely from pool to pool, from several grains per cubic metre to as much as 92 per cent in some unusual accumulations. It can also vary substantially from one part of the same pool to another, although this kind of variation is not common. As can be seen from Table 2, about half of the remaining sour gas reserves or about 11 per cent of the province's total (sour and sweet) gas reserves contain H_2S in concentrations greater than five per cent. About one quarter of the sour gas reserves are in the range of one to five per cent and the remaining quarter is at or less than one per cent. Generally speaking, H_2S concentration increases across the province from east to west, and with the depth of the formation.

One of the most dangerous of natural and industrial gases, H_2S has a toxicity to humans which has been compared to hydrogen cyanide (30). A concentration of 1000 parts per million (ppm) in the air will kill a person virtually instantly; at 700 ppm it will kill in a matter of minutes. Table 3 shows the physical effects of H_2S on humans at various concentrations (8). H_2S is so corrosive to steel that the Canadian Standards Association (CSA) requires special metallurgical properties for line pipe where the concentration of H_2S in the gas to be carried results in an H_2S partial pressure of 0.35 kPa (0.05 psi) or greater.

Sales gas specifications in Alberta for maximum H_2S content range from 0.25 to 1 grain per 100 cubic feet depending on the market area to be served with the gas. Thus, the sour raw gas from the wells must be gathered and transported in pipelines under pressure to a central processing facility for, among other processing requirements, the removal of H_2S.

Table 3 - Physical Effects of H_2S Reported by the API

Concentration (ppm)	Physical Effects
10	Obvious and unpleasant odor
20	Safe for 8 hours exposure
100	Kills smell in 3 to 15 minutes; may sting eyes and throat
200	Kills smell shortly; stings eyes and throat
500	Dizziness; breathing ceases in a few minutes; needs prompt artificial respiration
700	Unconscious quickly; death will result if not rescued promptly
1000	Unconscious at once; followed by death within minutes

Background to the Present Set-Back Requirements

The Alberta Petroleum Industry/Government Environmental Committee (APIGEC) is a committee of senior industry and government staff which was formed in 1972 to provide advice to the Minister of Environment on matters pertaining to the oil and gas industry and the environment. In 1974, APIGEC reported that there were over 100 sour oil and gas fields in Alberta, with 75 located within 16 kilometres (10 miles) of a city, town, or village (3). These figures have undoubtedly increased since 1974 to the present as a result of new gas discoveries and the development of existing fields, the expansion of urban centre boundaries, and the incorporation of settlements which have grown to reach village size. Figure 5 shows the location of sour oil and gas fields in relation to populated areas (the northern part of the province is excluded from the figure since the vast majority of the sour fields and population are located in the southern two-thirds).

APIGEC concluded that, notwithstanding a history of only a few uncontrolled H_2S releases from well blowouts, pipeline ruptures, or processing plant upsets, such releases could be a hazard to large numbers of people since many populated settlements were encroaching on sour gas operations, particularly in the area between Calgary and Edmonton (the province's two largest cities). For example, large fields involving H_2S concentrations of 30 to 35 per cent exist very near the large city of Calgary (pop. 623 000) (9). It concluded that little information was available on the encroachment problem and that inadequate controls existed. The report recommended that appropriate legislation be enacted to restrict residential development

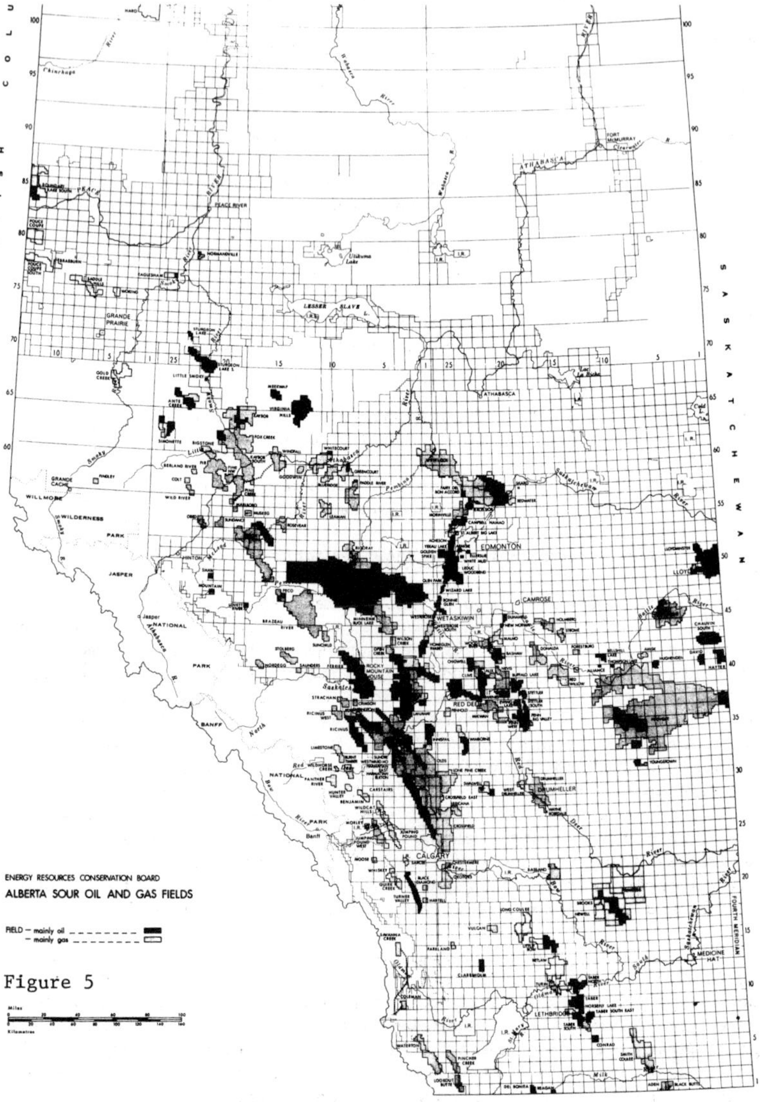
ENERGY RESOURCES CONSERVATION BOARD
ALBERTA SOUR OIL AND GAS FIELDS
FIELD — mainly oil
— mainly gas
BRITISH COLUMBIA
SASKATCHEWAN
EDMONTON
CALGARY
RED DEER
LETHBRIDGE
MEDICINE HAT
DRUMHELLER
WETASKIWIN
CAMROSE
JASPER NATIONAL PARK
BANFF NATIONAL PARK
GRANDE PRAIRIE
FOURTH MERIDIAN

Figure 5

near existing sour facilities, in particular prohibiting development within a 100-ppm isopleth of any sour facility, and that development applications be referred to the ERCB for comment.

An explanation of the term isopleth is required at this point. It is a line or contour of a constant H_2S concentration. The area enveloped by a 100-ppm isopleth, for example, would be subjected to H_2S concentrations of 100 ppm or more in the event of an uncontrolled release of H_2S, while the area outside it would have concentrations lower than 100 ppm. The concept of isopleths is a method of estimating the limits of gas dispersion and potential hazardous area following an accidental release of sour gas.

The recommendations were adopted by the ERCB (14) to restrict the encroachment of new sour facilities on existing residential development, and conversely, the authorities controlling new residential development adopted them (25) to restrict new residential development from encroaching on existing sour facilities. The ERCB requirements prevented new sour facilities to be located such that the 100-ppm isopleth would encroach on an existing urban centre or public facility or that the 500-ppm isopleth would encroach on any existing permanent residence.

Isopleth Calculation and Risk Assessment

The 1974 APIGEC report referred to the use of a Gaussian plume (steady-state) diffusion model to predict H_2S concentrations. It developed a family of curves for worst-case conditions of flat terrain, "moderately stable" atmospheric conditions, and a wind speed of 4.4 mph to show ground-level H_2S concentration as a function of downwind distance and source flow rate. While the results were developed based on constant source flow rate, such as for a well blowout, a method was suggested to predict approximate average concentrations from short-duration non-continuous releases, such as from pipeline ruptures.

The technique was applied to two existing sour gas developments, those in the Okotoks Field and also the Crossfield Field. The results of the study for the Crossfield case indicated that the 100-ppm isopleth would encroach across the city limits of Calgary. Indeed, in one area, it infringed on existing residential development.

Further study of the Crossfield Field indicated that the entire Town of Crossfield (pop. 1200) (9) would be within the 100-ppm isopleth. This fact plus the knowledge that the Town wished to expand toward the sour facilities indicated the need to study the matter in detail. The result was that a nearby sour gas well was abandoned and a main sour gas gathering pipeline was re-routed to ensure the safety of the residents of the Town and to have regard for the Town's expansion plans (13).

The above action involved the evolution of several combined steady-state/transient models to predict ground-level H_2S concentrations. But the results of the models differed by factors as large as

6 to 10. Consequently, sensitivity studies (14) were conducted on the models, and after some adjustments to them the conclusion was reached that, although some difference in results would continue, the models would generally agree if used properly. The sensitivity studies showed that assumptions on how the release rate occurs and the effect of plume rise are most critical factors. Additionally, some small-scale pipe-burst field tests were conducted to examine the behaviour of gas escaping from a ruptured pipeline and to test the validity of the assumption that a vertical sonic jet release would result. The results (5) showed, rather than a vertical sonic jet, the releases ranged from large volume, low level releases to angular jets oriented along the pipeline right of way.

The use of the models continued to result in very large isopleth distances and therefore a level of public safety that was high in the extreme. Consequently, studies were undertaken to identify and quantify the factors and events associated with the hazard imposed on the public by sour gas facilities. One of the results is shown in Figure 6 which is a plot of average urban individual risk versus separation distance for sour facilities for high, medium, and low H_2S releases, along with average individual risks associated with certain events (6). It shows that living in an urban area within 0.5 km of a sour gas pipeline is less risky than some natural disasters such as tornadoes.

Further Developments in Requirements

Following some experience with the new requirements, and considering the results of several studies which had been conducted to investigate further the potential hazards associated with sour gas operations, it was concluded that potential H_2S isopleths did not properly define the risk associated with a sour gas facility since they represent a worst possible and highly-improbable event. The isopleth policy was therefore considered to be overly restrictive and in need of revision. New requirements were then adopted (15)(19) which recognized the low probability of sour gas facility failures and which incorporated a number of simplifications in establishing acceptable separation distances between sour gas operations and other developments. Similarly, the legislation respecting encroachment of residential development on existing sour gas facilities was amended to require setbacks similar to the ERCB's separation distances (25). These are the requirements which are in place in Alberta today.

The separation distances specified, summarized in Table 4, increase as the potential H_2S release volume from the facility increases and as population density increases. There are four categories or levels of sour gas facility depending on the potential H_2S release rate or volume involved. For sour gas wells, a potential H_2S release rate determines the "level" of the well. The release rate for a well is calculated using the maximum wellhead deliverability that can be attained at any time through the tubing against zero wellhead pressure. For other facilities such as pipelines and processing plants which in the event of a failure can be blocked in with automatic valves operating on pre-set high- and low-pressure

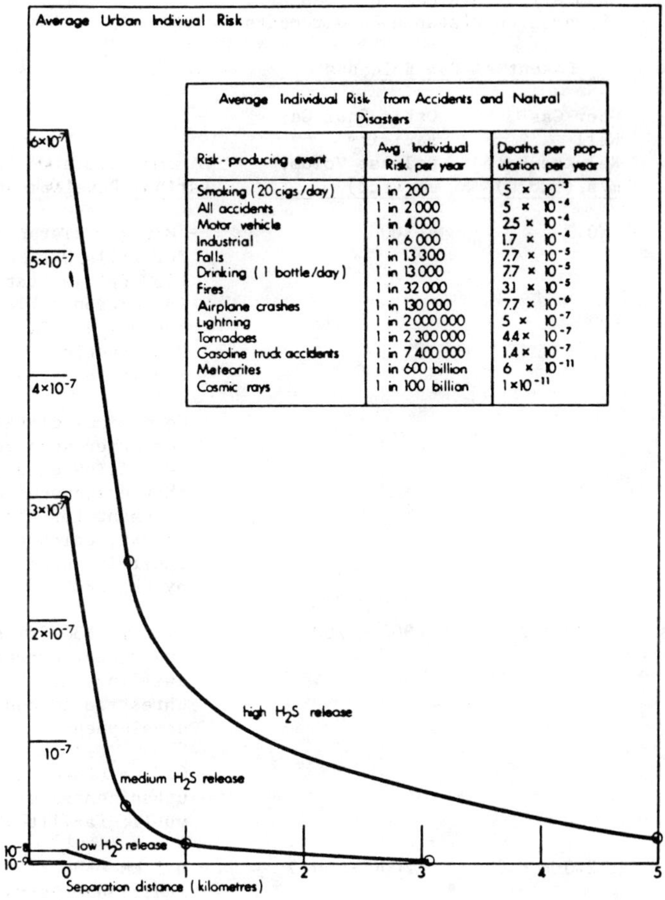

Average Individual Risk from Accidents and Natural Disasters		
Risk-producing event	Avg. Individual Risk per year	Deaths per population per year
Smoking (20 cigs/day)	1 in 200	5×10^{-3}
All accidents	1 in 2 000	5×10^{-4}
Motor vehicle	1 in 4 000	2.5×10^{-4}
Industrial	1 in 6 000	1.7×10^{-4}
Falls	1 in 13 300	7.7×10^{-5}
Drinking (1 bottle/day)	1 in 13 000	7.7×10^{-5}
Fires	1 in 32 000	3.1×10^{-5}
Airplane crashes	1 in 130 000	7.7×10^{-6}
Lightning	1 in 2 000 000	5×10^{-7}
Tornadoes	1 in 2 300 000	4.4×10^{-7}
Gasoline truck accidents	1 in 7 400 000	1.4×10^{-7}
Meteorites	1 in 600 billion	6×10^{-11}
Cosmic rays	1 in 100 billion	1×10^{-11}

FIGURE 6 AVERAGE URBAN INDIVIDUAL RISK vs. SEPARATION DISTANCE FOR SOUR FACILITIES.

Table 4 Summary of Minimum Sour Gas Separation Distance Requirements

	Potential H_2S Releases		
Level of Sour Gas Facility	Sour Gas Well Release Rate m/s (ft^3/s)	Other Sour Gas Facility Release Volume m (ft^3)	Minimum Distance to Various Developments
1	<0.3	<300	- Minimum distance for wells 0.1 km (330 ft) as stated in Section 2.110 of the Oil and Gas Conservation Regulations (24). - No minimum distance for other sour gas facilities except that associated with easement for right of way, unless otherwise specified by the ERCB.
2	0.3 - 2.0	300 - 2000	- 0.1 km (330 ft) to individual permanent dwelling and unrestricted country development. - 0.5 km (1640 ft) to urban centre or public facility.
3	2.0 - 6.0	2000 - 6000	- 0.1 km (330 ft) to individual permanent dwellings up to 8 dwellings per quarter section. - 0.5 km (1640 ft) to an unrestricted country development. - 1.5 km (4920 ft) to an urban centre or public facility.
4	>6.0	>6000	- As specified by the ERCB but not less than Level 3.

signal limits, the potential H_2S release total volume at standard pressure and temperature conditions determines the level of the facility and is calculated based on maximum operating pressure and the assumption that any automatic block valves close instantaneously upon failure of the facility.

For the purposes of the set-back requirements, public facility is defined as a recreational area such as a campground, or a public building such as a rural school or hospital, situated out of urban centres and any similar development the ERCB designates as a public facility. Unrestricted country development is defined as any collection of permanent dwellings situated out of urban centres and having more than eight permanent dwellings per quarter section.

The Implications of the Sour Gas Separation Distance Requirements

It has long been recognized by oil and gas operators, land developers, regulatory bodies, and the public that there needs to be separation between sour gas operations and the public. What has not been so clear is the appropriate distances for separation. The earlier requirements involving isopleth calculations and risk assessment were fraught with difficulties arising out of differences in assumptions and techniques for the calculations, and therefore, differences in results, the dearth of empirical data to substantiate the theoretical work, and the "worst-case" nature of the results. The current requirements, which specify separation distances that depend on the magnitude of the potential H_2S release and population density, are aimed at appropriate separation to provide acceptable risk to the public, and incorporate some simplifications not available under the earlier requirements.

The present separation distance requirements have enabled a more straight-forward consideration of most applications for new sour gas facilities but are not totally without difficulty themselves. In enunciating the new requirements, the ERCB stated that it recognized special circumstances may require a detailed review of a situation which could indicate a different H_2S release level classification for individual special cases than that resulting from the simplified calculations.

A recent example of a highly-contentious application that resulted in a lengthy public hearing, and ultimately denial of the application, was a proposal by Shell Canada Resources Limited for permits to construct Level 2 and 3 pipelines and associated facilities to gather sour gas from wells in the Moose and Whiskey Fields and transport it to an existing processing plant in the area (21). The H_2S concentration of the raw gas involved ranged from 7 to 40.4 per cent for the Moose Field wells and 10 per cent for the Whiskey Field well. Figure 7 shows the proposed routing of the pipelines and certain major features of the area, including Kananaskis Country and the 1988 Winter Olympic Games site. Kananaskis Country is an area in reasonable proximity to the City of Calgary, and therefore is already a popular recreation area. It has been established by the government of Alberta as a prime protection area to maximize its recreational

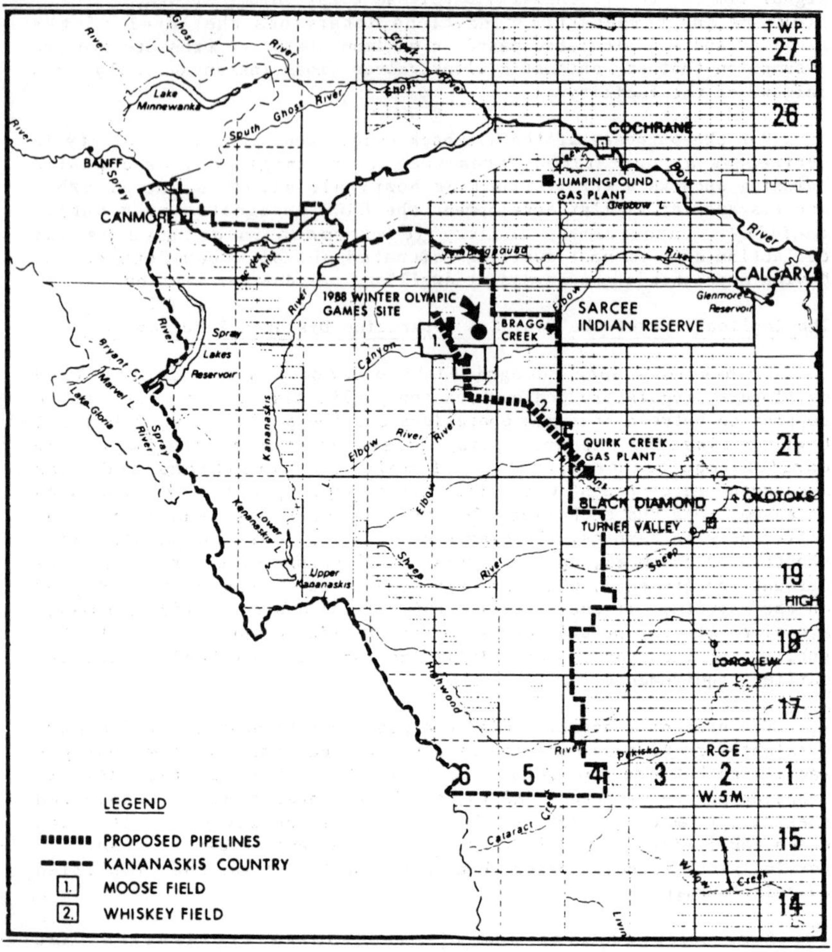

FIGURE 7 PROPOSED MOOSE AND WHISKEY FIELD DEVELOPMENTS

ENERGY RESOURCES CONSERVATION BOARD
ALBERTA, CANADA

value and to ensure protection of the environment and wildlife population. Rigid developmental controls have been imposed.

On considering the evidence presented at the public hearing, the ERCB concluded that, while there was a need for pipelines to transport the gas to processing facilities and that processing the gas at the plant proposed to be tied in to was the best alternative, the proposed pipelines were, for public safety reasons incompatible with specific development plans adopted for Kananaskis Country and should not be approved. These plans included educational and recreational programs involving day-use sites, parking lots, and bringing bus loads of school children into the area to see the unique geological and other features of the area such as ice caves. The ERCB concluded that, since the pipeline was proposed to be located in a narrow steep-walled canyon, the dispersion of any gas emitted would likely be restricted. The ERCB treated the specific development plans in the area for Kananaskis Country as a public facility and noted that the specified set-back distances would not be met by the proposed pipeline. Additionally, it concluded that production from certain of the wells would also be incompatible with these plans and that the wells should be abandoned if the Kananaskis Country plans were proceeded with.

Commenting on the significant implications of its decision, the ERCB said that the applicant, the province, and the public would forego revenues in the order of $500 million, but that notwithstanding, it was not prepared to sacrifice public safety. At the time of writing of this paper, Shell was in the process of amending its application.

The application of the separation distance requirements can have and have had serious implications on oil and gas operators, land developers, regulatory agencies, and the public alike. Of necessity, and notwithstanding efforts to simplify the process, application information requirements and the background work required to provide it have increased, in some cases, to the point where external consultants need to be retained. In addition, the manpower and expertise of the regulatory agencies which must consider the information has had to be increased. Considering that a proposal for new residential or other public-oriented development in or near a sour gas area could effectively preclude the installation of sour gas pipeline and other facilities, thus preventing the recovery of a very valuable resource (or that the converse may be true), developers whether they are for land or oil and gas should expect increasing opposition to development proposals in sour gas areas. To make matters worse, many of the largest and most sour gas reserves are located in close proximity to and even underlie some of most attractive areas for surface development, particularly in the Edmonton-Calgary corridor and in the prime recreation-potential areas of the Rocky Mountains Foothills. While such conflicts could theoretically be resolved through co-ordination of the planning, including timing, of both oil and gas and surface developments in a particular area by the proponents themselves, in practice, most cases require a formal process such as a public hearing of an application. Although not necessarily true in all cases, such hearings tend to be extremely adversarial in nature, and as a consequence, can create additional information requirements, procedural difficulties, and in general substantial lengthening of the regulatory process. Overall, this results in high costs for both applicants and interveners, and the regulatory agencies, the need to provide much more lead time in making applications, and in some cases, lost construction seasons. In extreme cases, the conflicts result in the modification to or denial of applications.

CONCLUSION

Pipelines are inherently safe, and the transportation of hazardous products by pipeline is demonstrably the safest and most efficient way to move them. The special problems that arise when these pipelines interface with urban areas are no reason to believe that the safety record must be compromised. However, it should be recognized that the present pace of urban development makes it essential that planners, developers, pipeline operators, and regulatory agencies work out and implement as quickly as possible the necessary policies and procedures to ensure compatability of land use and pipeline development.

REFERENCES

1 Alberta One-Call System Committee, 1980. Alberta One-Call System Feasibility Report.

2 Allman, R. J., Sharp, K. G., 1977. Board Staff Submission to Energy Resources Conservation Board Hearing into: Application Number 770422, Calgary, Alberta.

3 Alberta Petroleum Industry/Government Environmental Committee, 1974. Guidelines for Urban Development in Relation to the Sour Gas Industry.

4 Alberta Petroleum Industry/Government Environmental Committee, 1978. H_2S Isopleth Prediction, Phase I: Model Sensitivity Study.

5 Alberta Petroleum Industry/Government Environmental Committee, 1979. Hydrogen Sulphide Isopleth Prediction, Phase II: Pipe Burst Study.

6 Andrews, W. B., Rhoads, R. E., Franklin, A. L., Battelle Pacific Northwest Laboratories; Atwell, L. D., Schnitzler, W. J., Energy Resources Conservation Board, 1979. Risk Assessment of Sour Gas Facilities.

7 Associated-Pullman Kellogg Ltd., 1980. Relocation of Pipelines in the Mill Woods Area of Edmonton to Restricted Development Areas. 1, Calgary, Alberta.

8 Bander, T. J., Battelle Pacific Northwest Laboratories 1979. Literature Applicable to Descriptions of Diffusion of Uncontrolled Natural Gas Releases Containing Hydrogen Sulphide. Richland, Washington.

9 Bince, I. D., 1983. The Alberta List of Official Personnel in Federal, Provincial and Municipal Governments in the Province of Alberta. The Alberta List. Edmonton, Alberta.

10 Canadian Standards Association, 1982. Oil Pipeline Transportation Systems. CSA Standard Z183-M1982. Rexdale, Ontario, 182 p.

11 Choukalos, M., 1980. A Computer Model of the Risks from Gas Pipeline Ruptures. Air Quality Control Branch, Pollution Control Division, Alberta Environment. Edmonton, Alberta.

12 Energy Resources Conservation Board; Alberta Environment 1976. A Simplified Conservative Method for Calculating the Location of the 100 ppm H_2S Isopleth Around Sour Production Facilities.

13 Energy Resources Conservation Board, 1978. Decision on the Request by Amoco Canada Petroleum Company Ltd. for a Permit to Construct a Pipeline in the Crossfield Area. Decision 78-15. Calgary, Alberta.

14 Energy Resources Conservation Board, 1978. Encroachment of Sour Oil and Gas Operations Upon Urban Centres and Public Facilities. Interim Directive ID-OG 78-1. Calgary, Alberta.

15 Energy Resources Conservation Board, 1979. Minimum Distance Requirements Separating New Sour Gas Facilities from Residential and Other Developments. Interim Directive ID 79-2. Calgary, Alberta.

16 Energy Resources Conservation Board, 1979, Pipeline Failure Inquiry, Mill Woods Area, Edmonton, Decision Report 790161, Calgary, Alberta.

17 Energy Resources Conservation Board, 1980. Pipeline Failure Inquiry, Mill Woods Area, Edmonton. Post-inquiry Progress Review. Calgary, Alberta.

18 Energy Resources Conservation Board, 1980. Standard for Control-system Capabilities in Alberta. ERCB 80-F. Calgary, Alberta.

19 Energy Resources Conservation Board, 1981. Minimum Distance Requirements Separating New Sour Gas Facilities from Residential and Other Developments. Interim Directive ID 81-3. Calgary, Alberta.

20 Energy Resources Conservation Board, 1982. Alberta's Reserves of Crude Oil, Gas, Natural Gas Liquids and Sulphur. ERCB Report 82-18. Calgary, Alberta.

21 Energy Resources Conservation Board, 1982. Shell Proposed Moose and Whiskey Pipelines and Related Facilities, Kananaskis Area. ERCB Report 82-E. Calgary, Alberta.

22 Government of the Province of Alberta, 1981. Pipeline Amendment Act, 1981, Chapter 30. Edmonton, Alberta.

23 Government of the Province of Alberta, 1981. Pipeline Amendment Regulation, 1981, and the Pipeline Regulations (Alta. Reg. 298/75).

24 Government of the Province of Alberta. Oil and Gas Conservation Regulations Under the Oil and Gas Conservation Act. Alberta Regulation 151/71 as Amended.

25 Government of the Province of Alberta. The Subdivision Regulations Under the Planning Act. Alberta Regulation 132/78.

26 Murchie, G., 1981. The Pipeline-Urban Encroachment Problem: a Land-use Planning Perspective. Inter-Agency Planning Branch, Alberta Municipal Affairs. Edmonton, Alberta.

27 Strom, N., 1980. Developer-Pipeline Conflicts. Proceedings of the 6th Regional Conference on Energy Arteries - Lifelines for the Future. Edmonton, Alberta. pp. 151-156.

28 Warne, G. A., Webber, H. J., 1979. Sulphur Emissions - Alberta Energy Industries. Calgary, Alberta.

29 Weir, C. H., 1981. The Corridor Concept. Right of Way (USPS 466-080) 28(4), pp. 30-32.

30 Whittaker, J. D., 1980. Public Safety Aspects of Sour Gas Facilities. Air Quality Control Branch, Pollution Control Division, Alberta Environment. Edmonton, Alberta.

31 Wilson, D. J., 1979. The Release and Dispersion of Gas from Pipeline Ruptures, Alberta Environment. Edmonton, Alberta.

32 Wilson, D. J., 1981. Expansion and Plume Rise of Gas Jets from High Pressure Pipeline Ruptures. Pollution Control Division, Alberta Environment. Edmonton, Alberta.

33 Wilson, D. J., 1982. Predicting Risk of Exposure to Peak Concentrations in Fluctuating Plumes. Pollution Control Division, Alberta Environment. Edmonton, Alberta.

Polymer Modified Glass Fibre Reinforced Cement (PGRC) as a Protective Buffer Coating over Thin Film Anti-Corrosion Coatings

P.A. White C. Chem., F.R.S.C. *

SUMMARY

The search for energy sources in previously inaccessible areas requires new technology; drilling for oil and gas has required a research and development programme comparable to the United States space programme. The hostile environment, the drilling depths, the high temperature of extracted oil meant that previous technology was inadequate. Standard forms of pipeline protection such as coal tar enamel and bitumen were completely unacceptable when considering the temperature of the recovered oil.

The advent and development of thin film anti-corrosion coatings, typified by the fusion bonded epoxy coatings, have been successful but there have been difficulties. Whereas epoxy coating of 350 microns thickness have proved to be capable of corrosion prevention under normal conditions of usage, it has been found that under conditions of rough handling the thin epoxy coatings are in danger of being damaged. Thickening of the coating yields some benefit in this respect, but generally as the coating gets thicker, its physical properties change.

If a thicker, cheaper buffer coating could be laid over the thin film anti-corrosion coating, it would make possible a confident and wider use of such protected pipelines for heavy duties.

Such a product is available; known as a polymer modified glass fibre reinforced cement (PGRC), it was first developed as an innovative building material.

1.0 THE POLYMER MODIFIED GLASS FIBRE REINFORCED CEMENT OR PGRC SYSTEM

1.1 Introduction

Concrete, as we all know, is a composite of sand and course aggregates bound by cement and when reinforced with steel features prominently in the construction industry.

However, by adding other ingredients, mechanical and physical properties such as tensile and flexural strength, deformation capacity, porosity, drying shrinkage, freeze-thaw resistance, permeability, thermal movements etc. can be considerably improved.

Forton, B.V., Manchester, ENGLAND

Many years of research have proved that the incorporation of specialised polymers and glass fibre can contribute considerably to the improvement of such characteristics; the properties of the end product being determined by the composition and type of components, production techniques and curing conditions.

1.2 Glass Fibre Addition

The development of alkali-resistant fibre (AR-glass), some ten years ago now, has given a new impetus to the application of glass fibre reinforced concrete. The incentive for developing an alkali-resistant fibre material started from the assumption that the poor durability of concrete reinforced with non-alkali resistant glass fibres (E-glass) was due chiefly to these fibres being liable to attack in the alkaline environment of the cement matrix. The use of alkali-resistant glass fibre brought about a dramatic improvement. At the same time, however, it was found that AR-glass reinforced cement also deteriorated in the course of time, having an effect on such properties as tensile strength, deformability and impact resistance.

Fig. 1 shows a comparison of data on young and aged AR-glass and E-glass fibre reinforced concrete.

	AR Glass 5% (m/m) OPC			E glass 7%
	28 d	>10 yrs natural weathering		>3 yrs natural weathering
$\sigma_{t\,lop}$	9 - 10 (1)	6 - 8 (1)	MPa	5 - 7 (4)
σ_{uts}	14 - 17 (1)	6 - 8 (1)	MPa	6 - 7 (4)
ε_{ult}	10 - 15 (4)	0.2 - 0.3 (4)	o/oo	0.2-0.3 (4)
$\sigma_{b\,lop}$	14 - 17 (1)	12 - 15 (2)	MPa	12 - 14 (4)
σ_{mor}	35 - 50 (1)	12 - 15 (2)	MPa	12 - 14 (4)
ε_{u}	7 - 10 (4)	0.5 (3)	o/oo	0.5 (4)
Izod/ Charpy	17 - 31	2 (3)		1.5 (4)

Fig. 1
Properties of grc

It has been found that in spite of a considerable decrease in ageing rate little improvement in final strength is achieved by using AR-glass. The reasons for this are thought to be:

- the increase of the stress transfer between cement matrix and fibres;
- the increase of the physical incompatability between the glass fibres and the hardening matrix.

The first cause is illustrated in Fig. 2, which shows that in the aged concrete the glass fibres have become firmly embedded in the cement matrix.

Bartos (Ref.5), of the University of Paisley, in Scotland, has demonstrated that as a result of increasing stress transfer the deformability of a composite of fibres and a brittle matrix decreases considerably and may eventually become equal to the deformability of the matrix itself.

The physical incompatibility of the matrix and glass fibres is apparent, for instance, from the occurence of bending stresses and transverse forces in the fibres.

Fig. 3 shows the effect of bending that causes stresses in the glass fibres in a a sample under load. This effect is observable in the first place from a lowering of the effectiveness of the fibre material, and hence in a lowering of the strength of the composite.

Stress concentration in glass fibre
caused by sharp bend

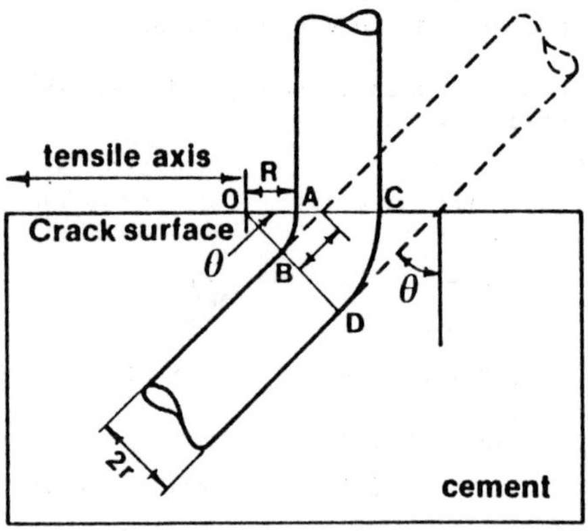

Fig. 3

An improvement of the properties of aged GRC can be attained only if all three causes of the deterioration of the quality of GRC can be removed or, at least, substantially reduced.

Consequently, for a GRC system to be effective it should satisfy the following requirements;

- the stress transfer between fibres and matrix should be controlled;
- the compatibility of fibres and matrix should be increased or separation between matrix and fibres should be effected;
- finally, the alkaline attack of the glass fibres should be suppressed.

1.3 Polymer Addition

For more than fifty years now, the addition of water-dispersed polymers to cement mortars with the object of modifying the cement matrix has been a known expedient. Dingley (Ref.6) has summarised the literature published on the subject up to 1974 and has thoroughly studied the properties of cement modified by a polymer dispersion in relation to the polymer content. His principal conclusions are that the polymer modified matrix has a higher deformability and a lower E Modulus.

In addition, with polymer-modified mortars the use of a lower cement/water ratio is possible, the matrix shows better cohesion, rapid drying of the matrix is prevented, and finally the formation of crystalline Ca(OH)2 is largely suppressed. The first two effects are shown in Figs. 4 and 5 (borrowed from Dingley, Ref. 6).

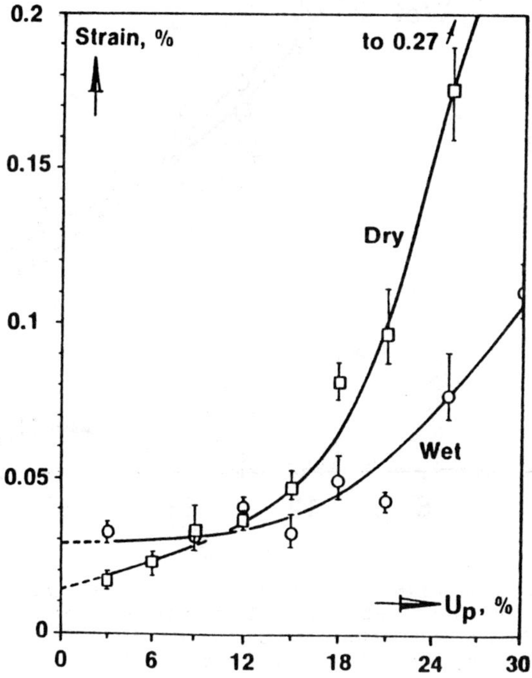

Fig. 4

Axial strain at failure for samples tested under tension plotted as a function of polymer volume fraction Up. O = 28 days old wet, □ = 28 days old, dry. Bar marks indicate range of values obtained.

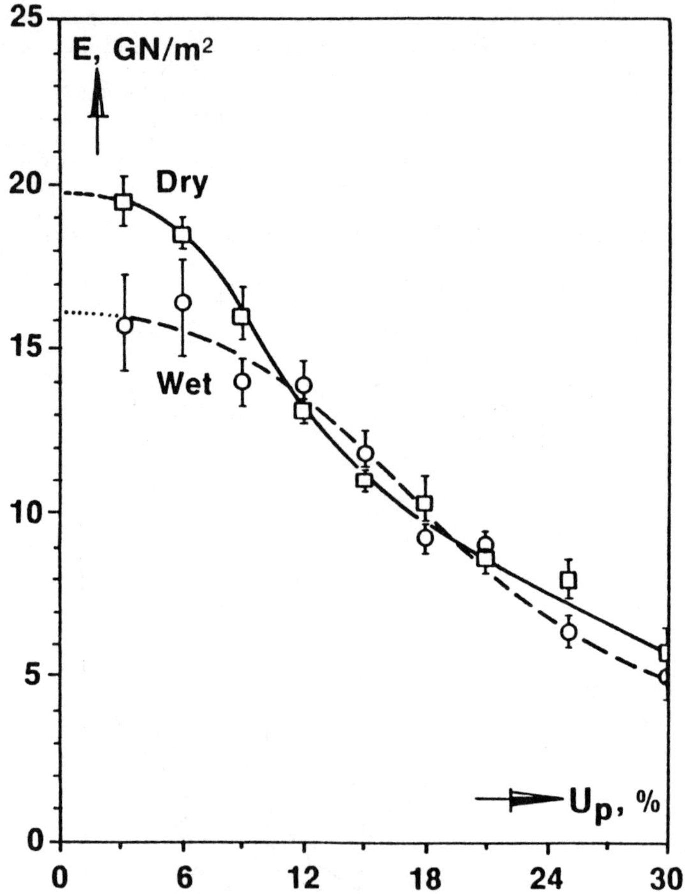

Fig. 5

Young's modulus E, determined from tensile tests plotted as a function of polymer volume fraction Up.O = 28 days hydration, tested wet, □ , 28 days hydration, tested after drying to 50% RH. Bar marks indicate range of values obtained.

Further, a thin coating of polymer modified cement that has been properly modified with the correct polymers will adhere tenaciously to old concrete and brick - also to wood and even to metal, plastic and glass and other difficult substrates, once considered off limit to cement mortars.

Test results also substantiate dramatic improvements in impact strength and abrasion resistance of cement finishes upon the addition of specialised polymers. Outdoors, they have the tensile strength and adhesion to resist freeze-thaw lifting, as well as the durability to withstand weathering and ultra-violet degradation; one typical use being as a maintenance coating for structural steel.

The dramatic performance improvements are obtained by choosing the correct polymer modification and retaining the water required for hydration - through the conventional mechanism of film formation. That is, when water first evaporates from the surface of the mortar, a film of coalesced polymer spheres forms around the sand and cement particles, and acts as a barrier to prevent further evaporation. In essence, the polymers act as an 'internal curing membrane', thus eliminating any need for further water curing of the cement.

It would appear, therefore, that a combination of glass fibre and specialised polymer would create the ideal system.

1.4 The FORTON Material System

(Combination of E-glass and specialised polymers - Ref. DSM published literature - FORTON PGRC, A Manysided Construction Material, written by M.J.N. Jacobs, October 1981).

It was an important discovery to find that the processes taking place during the mixing of a polymer modified cement mortar with glass fibre resulted both in the attainment of the desired conditions necessary for good ageing resistance of the product and the retention of their individual properties described previously.

The glass fibre used for reinforcement of cement consists of loose fibre bundles which are composed of hundreds of individual filaments (Fig. 6).

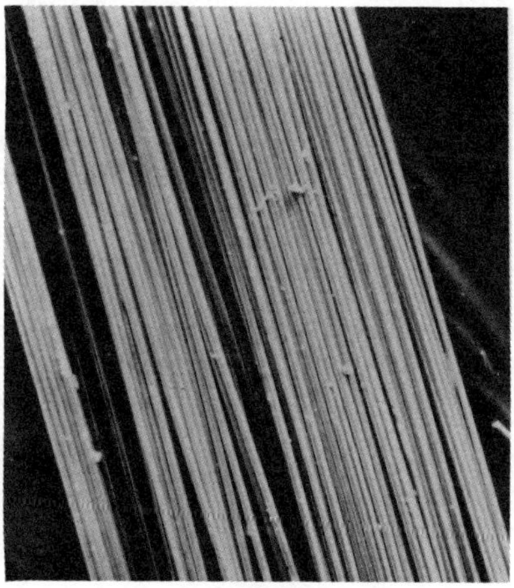

Fig. 6
Glass Fibre Strand

The diameter of the fibres is about 10 microns; the gaps between them having a width of a few microns. The diameter of the cement particles is on average 30 microns, which means that these particles cannot pass through the gaps (Fig. 7).

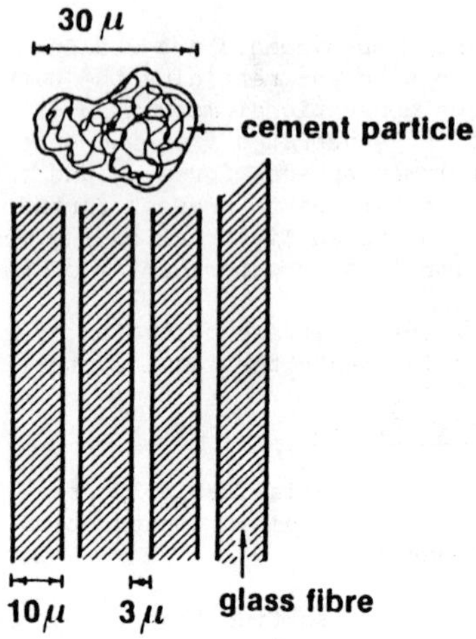

Fig. 7

Relative sizes of cement particles
and interfilamentary voids

The polymer particles in the Forton-System (now patented) measure some tenths of a micron, allowing them to penetrate the inter-filamentary gaps. During the mixing of the fibres with the polymer dispersion, the loose glass fibre bundles take up water, due to the capillary forces acting in the gaps. This water carries the polymer particles into the fibre bundles (Fig. 8). When the mortar has dried sufficiently, through the evaporation/hydration of the cement, the polymer particles adhere to each other, resulting in a coherent polymer film which uniformly spreads over the glass fibres. Fig.9 shows the polymer film after the cured matrix has been dissolved.

The comparatively soft film isolates the glass fibres from the hard brittle cement matrix. The effect of the polymer is supplemented by active acid groups, reacting with the $Ca(OH)2$ groups released during hydration of the cement. This reaction considerably decreases the formation of the $Ca(OH)2$ crystals.

The conditions for good ageing behaviour in PGRC result from the mechanism described above:

- reduction of stress transfer between the matrix and the glass fibres

- isolation of the fibres from the highly alkaline cement matrix and hard crystalline hydration products.

Fig.8
Polymer particles between glass fibres

Fig. 9
Polymer film on glass fibre

The properties of the end product are determined by the composition and type of components and can be altered to meet individual requirements.

The essential features of a PGRC coating:

- adhesion to steel, plastic and concrete
- flexibility
- impact resistance
- weather resistance
- anti-corrosion properties
- compressive strength
- ease and quickness of processing
- low cost

make it an ideal material for use as a buffer coating for the mechanical protection of thin film anti-corrosion coatings on pipelines.

2.0 EXPERIENCE OF TENNESSEE GAS PIPELINE COMPANY (Sokol Ref. 8)

2.1 Background

Fusion bonded powder epoxies have been in common use on gas pipelines in the United States since the mid-1960's. Although the service performance of this genre of line pipe coating is generally considered superior to other types, it does have several drawbacks. One of these is the necessity for closely controlling application parameters; the other is susceptibility to mechanical damage due to the thinness of the coating.

2.2 The Problem - Damage

Such mechanical damage falls into four categories:
(a) Loading and transit damage
(b) Handling damage at the job site
(c) Scoring and bruising when making field (cold bends with unlined dies

And in the case of offshore work:
(d) Damage caused by concrete application.

Handling and transit damage can be minimised by appropriate inspection and loading precautions, but such precautions are all too often inadequate. The use of lined bending dies can eliminate bending damage, but such dies are not always available and when they are, are costly and must be kept clean. Concrete damage had, it was thought, been eliminated by use of the compression method of concreting over 0.014 in (350 microns) thick epoxy (Tennessee Gas standard thickness) or by impacting concrete on 0.025 in. (625 microns) thick epoxy. During 1978-1979, field problems with concreting damage to epoxies were brought to their attention and investigation followed. For clarity it must be noted that no existing United States installation offers both fusion bonded powder epoxy and concrete coatings. Therefore, more handling and movement of pipe must be performed there than, for example, in Europe.

The conclusions of the investigation showed that:

(a) Transit damage to pipe coatings on rail cars was not infrequent, even when rubber separators were used.

(b) Severe handling damage was commonplace at the concrete yard.

(c) Scoring of the coating (and the usual handling damage) also occured at compressive concrete coating yards, unless the process was closely monitored.

(d) Damage caused by concrete impacting was far worse than anyone had suspected; in many cases, the holiday detector showed over 200 jeeps per joint, excluding innumerable chipped and gouged areas which had less than 0.008 in. (200 microns) epoxy left, but did not jeep, as shown in Fig. 24.

2.3 The Investigation of the Problem

Working under the assumption that damage from the first three areas could be minimised, the parameters of concrete impacting were investigated. The following resulted:

(a) Both 190 lb/cu.ft. (3060 kg/cu.m.) and 140 lb/cu.ft. (2255 kg/cu.m) concrete application produced intolerable damage to 0.25 in. (625 microns) thick epoxy coatings.

(b) Epoxy coatings up through 0.0285 in. (712 microns) thickness were severely damaged (0.35 jeeps/sq.ft. [3.50 jeeps/sq.m.] plus chipping/scoring which did not jeep). Data extrapolation indicated that an epoxy coating thickness of approximately 0.045 in. (1125 microns) would be necessary to assure that the remaining coating thickness at any chipped area remained above 0.012 in. (300 microns), the Tennessee Gas specified minimum.

(c) The tangential velocity of the applicator rolls and brushes ranged from 58 to 120+ mph (93 to 193 km/h), but applicators claimed that high velocities were necessary to achieve optimum physical properties.

(d) Since impact energy is proportional to the cube of the particle size, the full scale tests with sieved aggregate were run. Damage was lessened but not eliminated. In addition, the percentage of 'heavies' (Ilmenite or Hematite particles not passing a number 4 sieve) varied from two to ten per cent from batch to batch; this was reflected in higher damage rates for larger aggregate.

2.4 Possible Solution - A Buffer Coating

Due to the lack of enthusiasm on the part of concrete applicators to solve the problem by changing parameters or application methods, the search began for a suitable buffer coating to prevent handling, transit and concrete application damage. Although the compressive method can avoid concrete application damage, competitive concreting systems were to be maintained, and hence a solution to the impact damage problem was necessary. Such a buffer coating had to meet the following criteria:

(a) Cost should be comparable to the extra expense of increasing average thickness of 0.014 in. (350 microns) to 0.025 in. (625 microns). Higher costs would justify altering concrete application methods, such as by first applying 0.5 in. (12.7 mm) of grout, followed by the heavy weight mix.

(b) Cathodic protection must not be shielded (i.e. the buffer coating must have a resistivity of less than 1000 ohm-cm).

(c) The buffer coating must adhere well to the fusion bonded coating and not permit slippage to occur either between the concrete and buffer or between the buffer and fusion bonded epoxy.

(d) Concrete impact resistance must be sufficient to reduce damage to an acceptable level (a 0.012 in. [300 micron] epoxy thickness must be maintained).

The low resistivity requirement ruled out butyl rubber, polyethylene and foamed polyurethane. Inadequate bonding eliminated foamed polyurethane. Inadequate adhesive bonding eliminated foamed polyurethane, ethyl silicate and certain gypsum/mica coatings. Poor impact resistance and cost eliminated a cement-impregnated tape.

2.5 The Choice of Buffer Coating

Within the limited time available to find a solution, the only buffer coating found to meet all the criteria was a coating based on an adaptation of the PGRC material, described previously, with preliminary properties as below.

Resistivity in sea water in the order of 900 ohm-cm. Adhesion to the epoxy substrate exceeded 400 psi (2.8 kg/cm^2). Preliminary tests on small panels, using an impact concrete applicator simulator with 80 mph (129 km/h) tangential brush velocity and 1/4 to 3/8 in. (6.35mm to 9.52mm) diameter Ilmenite aggregate produced a maximum penetration of 0.035 in. (875 microns) into the PGRC. Most damage was 0.020 in. (500 microns) deep. No damage occured to the fusion bonded epoxy coating.

2.6 Field Testing

The next step was to conduct a full scale test using two joints of 12 3/4 in. (323.85mm) O.D. pipe coated with 0.014 in. (350 microns) of epoxy. One was coated with oil (as a release agent) and both joints were then coated with PGRC. Half of the oiled joint was then stripped of PGRC as a test control area, and both joints were subjected to impact by standard (unsieved) 190 lb/cu.ft. (3060 kg/cu.m.) Ilmenite/PEA gravel aggregate with a tangential brush velocity of slightly over 100 mph (170 km/h). Distance from brush tangent contact to the pipe being 19 in. (482.60mm).

The control (unprotected) area of the oiled pipe, even at voltages as low as 1400 V, jeeped continuously. The PGRC half was stripped and showed absolutely no impact damage. The second joint (without release agent) showed no exposure of epoxy after concrete application, see Fig. 29. During simulated handling damage (gouging with bare forks of a forklift and dropping) some epoxy was exposed but no damage to the epoxy occured.

2.7 Operational Commitment

As a result of these successful tests, a commitment was made to use a PGRC buffer coating over 14 mils (350 microns) epoxy on two offshore jobs in the Gulf of Mexico. The two jobs encompassed 15 miles (24.13 km) of 36 in. (914.40mm) and 11 miles (17.70km) of 26 in. (660.40mm) line pipe. Work commenced in May, 1980, at Reading and Bates Construction Company's Houston Plant, and pipelaying began late Summer. (See Figs. 26 and 27).

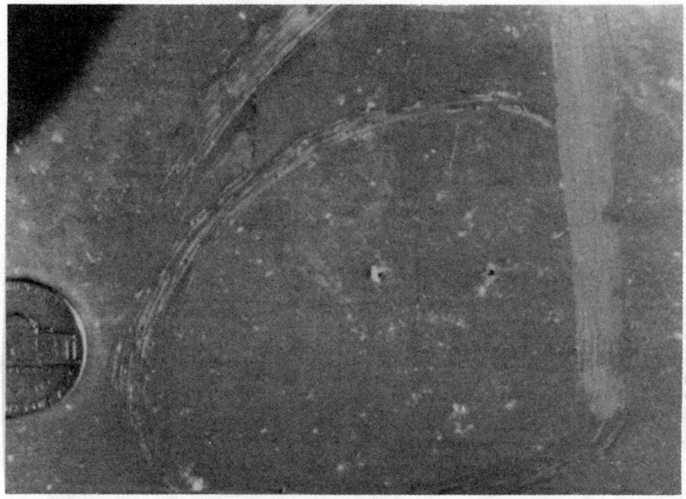

Fig. 24

Typical concrete impact damage to fusion bonded epoxy powder. The two central areas show full penetration of the 0.025 in. (625 microns)

Fig. 25

Typical appearance of PGRC overcoating after concrete application and hosing. Note rough texture and lack of significant damage from concrete impact.

Fig. 26

Application of unpigmented PGRC at Reading & Bates Construction Company - Pipe Coating Division's Houston Plant

Fig. 27

Removal of overcoated pipe from carriage. This coating was applied at ambient Summer temperatures and required over 24 hours for cure.

2.8 Conclusions of Tennessee Gas Pipeline Company

2.8.1 PGRC as a Buffer Coating for Mechanical Protection

Over the course of the overcoating, handling, barge movement, concrete application and transporting to the lay barge of this 36 miles (57 km) of pipe, only one instance of damage to the epoxy substrate was reported and that was due to a direct hit with a forklift. Not only were the original aims of preventing transit and impact damage achieved, but damage by aggregate thrown onto the concrete applicator carriage wheels (which otherwise would have been ground into the epoxy) was totally prevented and also damage by wire cable slings which were used to convey pipe from supply barges to the lay barges, was eliminated.

2.8.2 PGRC as a Coating to Prevent Slippage

The 36 in. (914.40mm) pipe was laid to a depth of 260 ft. (79.25 m) with a tension of 60,000 lbs. (27,216 kgs). The 26 in. (660.40 mm) pipe went into 330 ft. (100.59 m) of water with a maximum tension of 80,000 lbs. (36,288 kgs). Not a single instance of concrete slippage occured, though laying continued in seas of up to 12 ft. (3.66 m). It was noted that concrete adherance was superior to anything experienced before - there was absolutely no sagging at the lower quadrant. Normally, without the buffer coating, there would have been a 1/16 to 1/18 in. (1.58 mm to 1.41 mm) gap along the bottom quarter of the circumference.

The first jobs used a 0.090 in. (2.28 mm) nominal thickness. In the course of additional full-scale tests carried out during these jobs, it was determined that a 0.075 in. (1.90 mm) nominal thickness ($\pm$ 0.015 in. [0.38 mm]) was adequate protection.

During Autumn and Winter of 1980, an additional 31 miles (49.90 km) of large diameter (16 - 36 in. [406.40 mm - 914.40 mm]) pipe was coated in this low millage with successful results.

Since 1980, more lines have been laid both in North America and Europe using this type of buffer coating system.

3.0 FURTHER DEVELOPMENT OF THE PGRC BUFFER COATING

To date, experience in North America and Europe with PGRC buffer coatings has been restricted to its use as a protective membrane between thin film anti-corrosion coatings and the concrete weight coating for negative buoyancy. (For typical properties of this type of PGRC coating, see Fig. 28.

FORTON 2000 CRETESHIELD
Protection against concrete impingement and slippage

PHYSICAL PROPERTIES
(Tested at 23°C [73°F])

TEST	METHOD	VALUE
Density (brush impingement)	ASTM E12-70	> 2013 kg/m^3 (>125 lb/ft^3)
Compressive Strength	ASTM C42-77	> 7 MPa (> 1000 psi)
Impact Resistance	ASTM C14-77	> 18 J (> 160 in lb)
Indentation Resistance	Carver Press 1¼ in Chisel bevelled 30°	> 11365 kg (> 25,000 lbs)
Adhesion	Elcometer	≥ 2.8 MPa (> 400 psi) Failure cohesive
Electrical Resistivity	5% NaCl 5V 48 hr	< 1000 ohm cm
Flexibility	4 point modular bending jig	> 1.5° per pipe diameter. as per ANSI B31.8 Pipe Code Requirement
Resistance Freeze-Thaw	DIN 105	Good
Resistance UV Radiation	Weather-O-Meter	Good

(Data established under laboratory conditions for FORTON 2000 CRETESHIELD over fused bonded epoxy coated steel panels)

Preheat: Epoxy coated panels preheated to 60°C

Postcure: FORTON 2000 CRETESHIELD 1.8mm (0.072 in.)

Normalisation: Overcoated panels normalised for 24 hours at room temperature

(All information given in good faith but without warranty)

Fig. 28

Over the past 12 months research by Forton has lead to new formulation producing a cold applied PGRC coating with greater flexibility. This reformulated coating has been developed with a view to its utilisation as a Rockshield for on and offshore pipelines.

The following is a summary of the mechanical and physical conditions to which a Rockshield coating would be subjected during handling, laying, transportation and the operation service life of a pipeline and the corresponding properties required of such a coating.

Mechanical/Physical Conditions	Required Properties
In the coating yard:	
- load out - storage in stacks - ties - slings - discontinuous supports	- early compressive strength - thickness of coating - indentation depth
During loading and unloading operations:	
- narrow ropes or slings - puncturing by discontinuous supports	- indentation resistance - point loading
Alongside the ditch on supports or on the ground:	
- support distance - bending long welded sections under own weight - temperature variations - expansion or shrinkage - rubbing along the ground - field bending	- bending capacity - tensile strength - indentation resistance - coating thickness - deformation capacity - crack behaviour - tensile strength of wrap
Backfilling and pipe testing:	
- soil settlement and movement changes in volume of the soil (wet/dry) - soil loading stress (dead weight) - circumferential deformation during field testing	- resistance to wrinkling - point load resistance - deformation capacity

The material properties and performance data obtained by testing the reformulated PGRC coating under standard laboratory conditions are shown in Figs. 29 and 30.

MATERIAL PROPERTIES OF FORTON 2000 (PGRC) ROCKSHIELD

1. Apparent Density (dry): (water displacement specific gravity)	1.84 - 1.89 kg/m^3 (115 - 118 lbs/cu ft)
2. Compressive Strength:	$\geqslant$ 45 N/mm^2 (46 - 55) ($\geqslant$ 6428 psi)
3. Tensile Strength:	6 - 8 N/mm^2 (850 - 1140 psi)
E Modulus	12 - 15 GPa
4. Bending Strength:	12 - 16 N/mm^2 (1700 - 2280 psi)
5. Strain Capacity:	1.5 - 2.0%
6. Drying Shrinkage:	final value 1.2%
7. Thermal Expansion	$12 \times 10^{-6}/^oK$
Value λ	0.7 W/m^oK
8. Moist Behaviour:	
- water absorption	$0.9 - 1.3 \times 10^{-3} kg/m^2 \sec^{0.5}$
- water penetration	$0.04 \times 10^{-3} m/\sec^{0.5}$
- water vapour permeability coefficient	$\delta = 420 \times 10^{-15}$ sec
Value μ	= 620
9. Electrical Resistance: (Volume Resistance)	dry: 1.3×10^{13} ohm cm 10°/65% RH: 6×10^7 ohm cm
ASTM D257	wet fresh water: $1 - 4 \times 10^6$ ohm cm
Impedance Method	wet, 5% salt water: $0.5 - 1 \times 10^5$ ohm cm

(All information given in good faith but without warranty)

Fig. 29

PERFORMANCE DATA OF FORTON 2000 (PGRC) ROCKSHIELD

1. Adhesive Strength:	
a) Erichsen Adhesion Tester:	2.5-4.0 N/mm^2 (360 - 570 psi)
b) Slip out test (shear): Adhesion bond to FBE	0.5-1 N/mm^2 (70 - 140 psi)
2. Impact Test:	
a) Acc. ASTM G14 (Erichsen Impact Tester) 8 in (203 mm) dia weight = 2.0 lbs (0.91 kg)	2.2 kg m (160 in lbs) Slight surface indentation
b) Falling Ball 2.5 in (63.5 mm) dia weight = 2.2 lbs (1kg)	3.0 kg m (220 in lbs) Slight surface indentation
c) Concrete Tile dimensions = 12 x 12 x 3in (305 x 305 x 76 mm) weight = 38.5 lbs (17.5kg) contact point (sharp edge)	Sharp indentation without cracking or spalling off 17 kg m (> 1520 in lbs) at 4-6 mm (0.16 - 0.23 in)
3. Indentation Test:	
a) Chisel 30.1 cm (11.8 in) 30°, 1mm (0.04 in) dia Failure: crack	Indentation depth: 1.5 - 1.9 mm (0.059 - 0.074 in) Indentation resistance: 176 N/mm^2 (25,000 psi)
b) Cone 90°, 2mm (0.08 in) dia Failure: crack	Indentation depth: 2.4 - 2.9 mm (0.094 - 0.11 mm) P max load: 7100 N
c) Ball 30 mm (1.18 in) dia Failure: crack	Indentation depth: 1.3 - 1.4 mm (0.050 - 0.054 in) P max load: 14000 N
4. Bending Capacity:	2.58° per pipe dia
5. Elongation:	2.2%

(All information given in good faith but without warranty)

Fig. 30

The laboratory test pipes were prepared by hand spray up of FORTON 2000 mix over fusion bonded epoxy to a coating thickness varying from 0.19 in - 0.39 in. (5mm to 10mm), with an outer reinforcement. The pipe length was 24 in. (609.60 mm), and the diameter 8.0 in (203.20 mm). The coating was cured under ambient laboratory conditions (23°C (73.4°F] ± 2°C [35.50°F] and 65% relative humidity).

With the development of a reformulated flexible PGRC coating grade, a comprehensive system for the mechanical protection of thin film anti-corrosion coating on pipelines is now available, to ensure confident use of such protected pipelines in the severest environmental conditions.

APPENDIX I

REFERENCES

1. P. Bartos: Analysis of Pull-Out Tests on Fibres Embedded in Brittle Matrices. Journal of Materials Science 15 (1980) 3122 - 2128

2. BRE IP 36/79 Nov. 1979 Properties of GRC: Ten Years Results

3. R.G. Dingley: The Structure and Properties of Hydraulic Cement Partes Modified by Polymer Latex. Thesis, University of Southampton, Dept. of Mechanical Engineering, April 1974.

4. K.L. Litherland, D.R. Oakley, B.A. Proctor: The Use of Accelerated Ageing Procedures to Predict the Long Term Behaviour of GRC-Composites. Boston, Nov. 1980 Fi.

5. A.J. Majumdar: Properties of GRC Concrete International 80, London. Fibrous concrete pp. 48-68.

6. D.R. Oakley: Presentation of: Fibre Reinforced Cement, Mortar and Concrete. University of Salford, May 1981 (UK)

7. Present work.

8. D.R. Sokol, P.A. White, M.E. Rohn: A Protective Membrane for Thin Film Anti-Corrosion Coatings. Fourth International Conference on the Internal and External Protection of Pipes (BHRA), Netherlands, September 1981.

Lateral Soil Resistance in Soil-Pipe Interaction

A.P.S. Selvadurai*, J.J. Lee[+], R.A.A. Todeschini[+] and N.F. Somes[+]

This paper presents a summary of certain elementary soil models that can be used to represent the lateral soil resistance in the analysis of a soil-pipe interaction problem. The models have elastic-plastic, bi-linear or hyperbolic forms and the parameters characterizing these models depend on the linear elastic parameters and the shear strength parameters of the soil surrounding the pipe.

Introduction

Buried pipelines are used extensively for the bulk transportation of energy resources such as heated oil, slurried coal and natural gas (Ariman et al. [1], ASCE [2-4], Selvadurai and Lee [18,19], Shibata et al. [21]). Unlike above ground pipelines, the structural design of buried pipelines should take into consideration the mutual interaction between the pipeline and the surrounding soil. Such interaction can be induced by a variety of factors including thermal- and pressure-induced expansion, ground movement during earthquakes, effects of surcharge loads such as landfill and traffic, frost heave, soil swelling and pipe flotation during flooding. The modelling of the soil resistance is a crucial component in the geotechnical study of the soil-pipe interaction problem. The soil resistance is influenced by a number of factors including the variability of the soil properties along the length of the buried pipeline, differences in the properties between the natural ground and the compacted backfill, the condition at the soil-pipe interface and the relative flexibility of the soil-pipe system. In view of the complexities associated with the soil-pipe interaction problem, it is desirable to introduce simple geomechanical models which would attempt to duplicate the soil resistance at a pipe section. The soil resistance at a pipe section is characterized by the load-displacement behavior of the embedded pipe section subjected to lateral (horizontal displacement), vertical (upward and downward displacements), axial (displacements along the axis of the pipe) and torsional (rotation about the pipe axis) loads. This paper presents a brief summary of the methodologies currently employed in the estimation of the lateral soil resistance at a pipe section. The lateral soil resistance is usually developed at bends and horizontal transition sections of buried pipelines.

* Professor and Chairman, Department of Civil Engineering, Carleton University, Ottawa, Ontario, Canada; Formerly Visiting Researcher, Applications Engineering Group, Research and Engineering, Bechtel Group, Inc.
+ Applications Engineering Group, Research & Engineering, Bechtel Group Inc., 50 Beale Street, San Francisco, CA, U.S.A.

Methods for Soil-Pipe Interaction Analysis

In the context of this paper, the modelling of soil resistance is concerned with establishing representative mathematical or physical models for the mechanical behavior of the soil mass surrounding the buried pipeline. Such modelling is an essential first step in any soil-pipe interaction analysis. The mathematical models which describe the soil resistance must take into account the complexity of behavior of both naturally occurring and compacted soils. The stress-strain behavior of soils is influenced by a number of mechanical and physical factors including the particle size distribution, the density, void ratio, moisture content, the in situ state of stress and the stress history. These factors give rise to soils which display markedly non-linear, irreversible and time-dependent stress-strain phenomena and to soil masses which exhibit anisotropic and nonhomogeneous material properties (Desai and Christian [9], Vermeer and Luger [24]). The elastic-plastic idealization of soil behavior is a convenient first approximation for the stress-strain response of soils which are subjected to static or quasi-static loads and which display little or no time-dependence in the stress-strain characteristics. The pipeline itself can be represented by a structural element which exhibits stiffness characteristics in the extensional, flexural and torsional modes. The interaction between the soil and the buried pipeline can be studied by using the following methods of analysis: (a) analytical methods, (b) finite element techniques and (c) simplified methods.

Analytical Methods - The analytical methods involve solution of the three-dimensional equations of elastic-plastic equilibrium in the context of a soil-pipeline interaction problem. Solution of these equations is an extremely complicated task and to date there have been no attempts made to solve soil-pipe interaction problems by using rigorous analytical schemes. It is important to note that, despite the simplifications offered by the elastic-plastic idealization of the soil, the geometry of the pipeline is not at all simple and usually has a three-dimensional form. Vertical and horizontal bends, transition curves etc., may be encountered on a single length of pipeline. Under these circumstances, the analytical methods, which incorporate inelasticity in the soil properties, offer little prospects for the analysis of soil-pipeline interaction problems. Some limited progress can be made in the application of three dimensional elasticity techniques for the study of certain classes of soil-pipeline interaction problems (Selvadurai [20]).

Finite Element Methods - Finite element techniques are applied quite extensively for the analysis of soil-foundation interaction problems (see for example, Desai and Abel [8], Desai and Christian [9], Gudehus [10], Wittke [25] and Selvadurai [15]). Methods of analysis currently in use incorporate very sophisticated models of soil behavior and have capability to handle fully three-dimensional problems. However, these numerical techniques are both time consuming and expensive. Furthermore, the use of such techniques is warranted only in situations where the stress-strain parameters for the soil are determined either from laboratory tests or in situ tests. With reference to the buried pipeline problem, such information is rarely available at any design stage.

Simplified Methods - Simplified modelling of the soil-pipe interaction problem is an attempt to alleviate the deficiencies of analytical and finite element techniques. In a simplified analysis, the soil-pipe interaction is studied by idealizing the 'overall behavior' of the soil surrounding the pipe by a one-dimensional mathematical or mechanical model. Such an idealization takes into consideration both the stress-strain characteristics of the soil and the geometry of the soil-pipe system. The simplified modelling of soil-pipe interaction concentrates on the development of a mathematical relationship for the lateral load (referring to Figure 1; this load is the induced transverse load P_T) vs. lateral displacement (transverse displacement δ_T) behavior of a buried pipeline at any particular section along its length. The displacement δ_T is caused by thermal gradients, internal pressure and other effects. The load P_T corresponds to the resultant of surface tractions which act at the soil-pipe interface along the horizontal direction. It is assumed that the load-displacement relationship can be represented by elastic-plastic, hyperbolic and bi-linear approximations in which the soil stiffness (K_T) is governed by the elementary elastic properties (Young's Modulus E; Poisson's ratio ν) of the soil and the pipe dimensions while the limiting soil resistance (P_{LT}) is governed by the shear strength characteristics of the soil or the soil-pipe interface. These idealized constants K_T and P_{LT} can be affected by factors such as flexibility of the pipe wall, bending flexibility of the pipe, depth of burial of the pipe, density and compaction of the back-fill and the load/unload behavior of the buried pipe. The influences of these factors are discussed in detail by Selvadurai and Lee [18].

Existing Procedures for Determining the Lateral Load-Displacement Response

The procedures currently documented in the literature for simplified analysis of soil-pipe interaction use the concept of an effective lateral load (P_T) vs. lateral displacement (δ_T) relationship for the embedded pipe at any cross-section. The cross-section of the pipeline is considered to be rigid and the effective P_T - δ_T relationship is developed by considering a two-dimensional plane strain model of the soil-pipe configuration. Two particular forms of the load-displacement relationship can be found in the literature (Audibert and Nyman [5], Kennedy et al. [12], Peng [14], Luscher et al. [13], Audibert et al. [6]). These are a hyperbolic form of the P_T - δ_T relationship which is largely based on experimental observation and an elastic-plastic form of the P_T - δ_T relationship which is based purely on empirical considerations.

The Hyperbolic P_T - δ_T Relationship - The experimental investigations reported by Audibert and Nyman [5] and Das and Seeley [7] are concerned with the measurement of the load-displacement relationship for rigid pipes and rigid anchor plates embedded in cohesionless soils. The experimental results indicate that, for the soils tested, the load-displacement behavior could be approximated by a hyperbolic relationship of the form

$$\bar{P}_T = \frac{\bar{\delta}_T}{a+b\bar{\delta}_T} \tag{1}$$

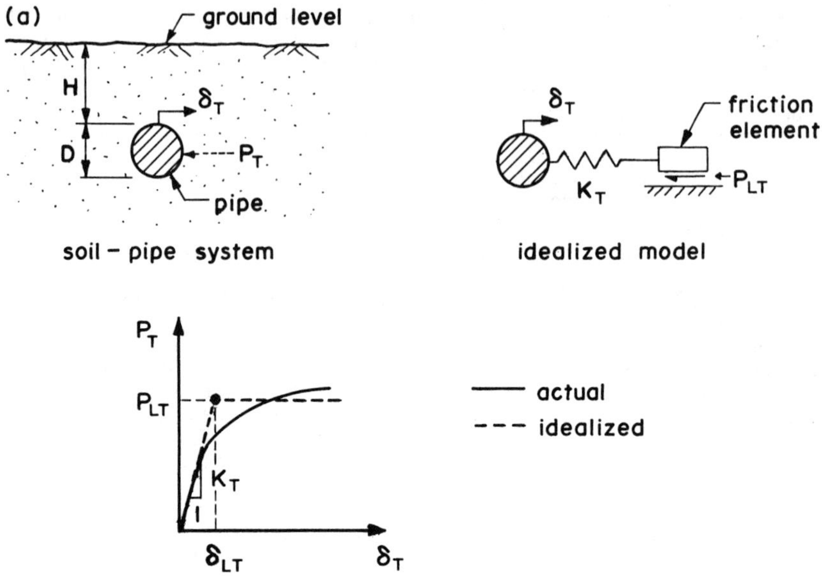

Fig 1 Simplified Lateral Load – Displacement Response of the Buried Pipe

where $\bar{P}_T = P_T/P_{LT}$; $\bar{\delta}_T = \delta_T/\delta_{LT}$; δ_{LT} is the pipe displacement at which P_{LT} is reached and a and b are constants. The values of a, b, P_{LT} and δ_{LT} are estimated from the experimental results. The hyperbolic form of the P_T - δ_T (or $\bar{P}_T$ - $\bar{\delta}_T$) relationship attempts to duplicate more closely the non-linear behavior observed in the tests. As discussed in the previous section, the transverse load-displacement relationship for a buried pipe is affected by a number of factors. The reported experimental investigations examine only a limited number of variables. On the basis of the experimental investigations, the load-displacement relationship (1) can be recast in the specific form

$$\bar{P}_T = \frac{\bar{\delta}_T}{0.15+0.85\bar{\delta}_T} \quad (2)$$

Two further results are necessary to make the representation (2) explicit. The value of P_{LT} is calculated by making use of the results derived for the horizontal bearing capacity of a footing of width D embedded at a depth H_0 (= H + D/2) in the soil (see e.g. Hansen [11]). For a cohesionless soil, considering a pipe element of length s,

$$P_{LT} = \gamma H_0 D s N_q^* \quad (3)$$

where N_q^* is the bearing capacity factor (Figure 2). The variation of N_q^* with depth ratio (H_0/D) and ϕ is shown in Figure 2a. It may be noted that for a cohesionless soil, the limiting soil resistance is derived primarily from the self weight of the soil (γ) and the angle of internal friction (ϕ). For a purely cohesive soil the limiting soil resistance is given by

$$P_{LT} = cDsN_c^* \quad (4)$$

where N_c^* is the bearing capacity factor for a vertical foundation embedded at a depth H_0. The variation of N_c^* with depth is shown in Figure 2b. In equation (4) c refers to the undrained cohesion.

The ultimate lateral displacement δ_{LT} at which the cohesionless soil derives its maximum resistance depends on the soil density, the pipe diameter and the effective depth of embedment H_0. The experimental results given by Audibert and Nyman [5] and Das and Seeley [7] indicate that in loose sand δ_{LT}/H_0 varies from about 6% for a 1 in. diameter pipe to about 3% for a 4.5 in. diameter pipe, and tends to an asymptotic limit of 2% for pipes of larger diameter. Similarly, for dense sand, δ_{LT}/H_0 decreases from 3.5% to an asymptotic value of 1.5%. On the basis of these observations δ_{LT}/H_0 is assigned the range of values 1.5% to 2.0%. Normalized ultimate displacement (δ_{LT}/H_0) at which a purely cohesive soil derives its maximum resistance is estimated at 3% to 5% (Audibert et al. [6]).

Empirical P_T - δ_T Relationship - The empirical relationship for the lateral load-displacement behavior of the buried pipe (Peng [14])assumes the elastic-perfectly plastic form discussed previously. The limiting soil resistance is estimated from relationships of the form given in the equations (3) and (4). The empiricism in the method stems from the fact that the soil stiffness K_T is derived by stipulating a priori the limiting displacement required to mobilize P_{LT}. In particular, it is

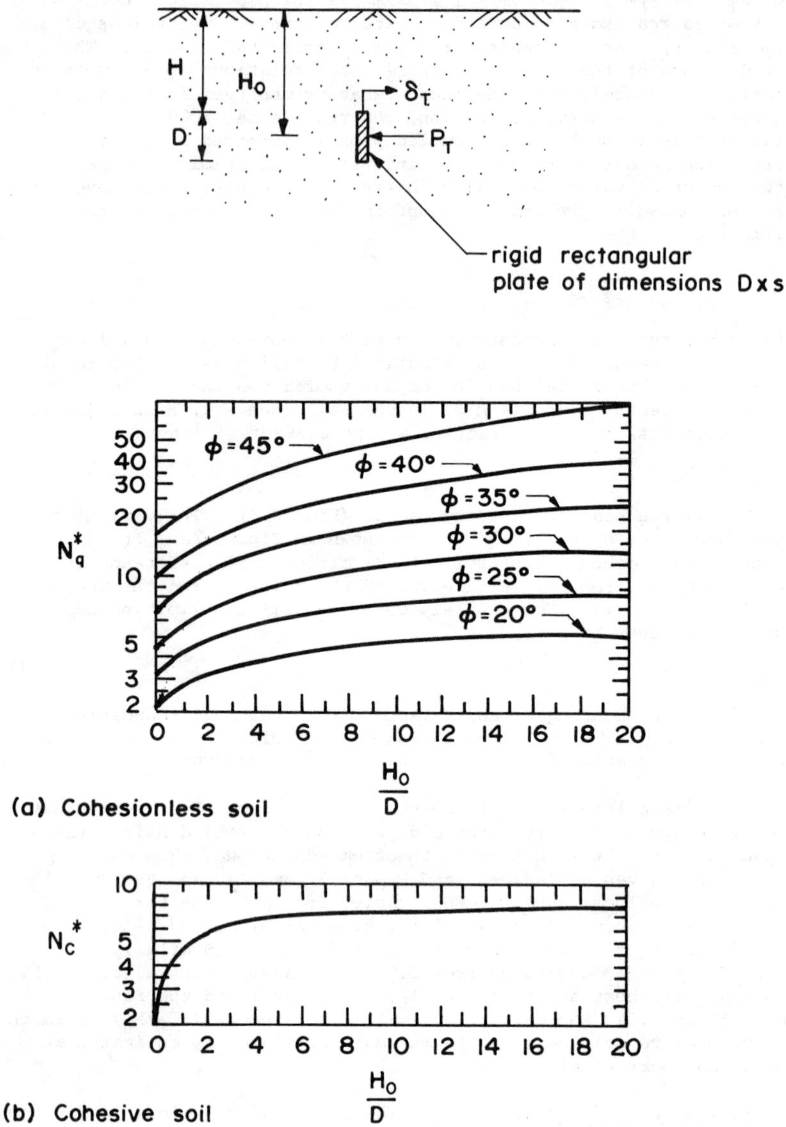

Fig 2 Bearing Capacity Factor for the Embedded Anchor Plates (Reference II)

assumed that P_{LT} is mobilized at approximately the same displacement ratios (δ_{LT}/H_0) discussed previously in connection with the hyperbolic load-displacement response.

A summary of P_T - δ_T relationships currently (1983) employed in soil-pipe interaction analysis is presented in Table 1. The hyperbolic load-displacement relationship attempts to duplicate more closely the non-linear character of the experimental curve. In the opinion of the authors, there is insufficient experimental evidence on full-scale behavior of buried pipes to justify the use of hyperbolic P_T - δ_T relationship for the full range of backfill density. As the relative density of the backfill or the surrounding soil mass reaches its optimum, the load-displacement curve for the buried pipe is expected to deviate considerably from the assumed hyperbolic form in the pre- and post-peak ranges (Figure 3) at shallow depths of burial ($H < 3D$). The discrepancy should become less significant as the backfill density approaches that of a loose soil or when the depth of burial becomes greater than 6D.

The purely empirical technique of expressing the value of K_T in terms of δ_{LT} and P_{LT} always underestimates the value of K_T even for a loose granular soil (Figure 4). Both techniques currently available for the estimation of the P_T - δ_T relationship are dependent upon the estimation of P_{LT} and δ_{LT}. The magnitude of P_{LT} can be assessed reasonably accurately from a knowledge of the shear strength data (c and ϕ). The value of δ_{LT} is particularly sensitive to the relative density of the soil, the depth of the pipe and the shear strength parameters c and ϕ. The available empirical estimates primarily take into account the influence of backfill density.

Proposed Modelling of Lateral Load-Displacement Response

The proposed modelling of the P_T - δ_T relationship for the buried pipe is based on the elastic-perfect plastic idealization discussed previously. In contrast to the empirical approach, which determines the value of K_T by prescribing P_{LT} and δ_{LT}, here an attempt is made to derive K_T in terms of the deformability parameters (E,ν) of the surrounding medium. Also the limiting soil resistance P_{LT} is estimated by using certain elementary "plastic limit equilibrium" solutions. The latter results are derived from the analysis of plausible failure mechanisms induced in the soil by movements of the pipe. To aid the development of these results it will be assumed that (i) the pipe is embedded in "adhering contact" with the surrounding soil, (ii) the soil exhibits isotropic linear elastic or perfectly plastic behavior, (iii) the 3-dimensional behavior of the pipe as it bends about an axis can be represented by the simpler plane strain approximation, (iv) the cross-section of the pipe is rigid, and (v) the backfill is compacted to the same state as the natural soil such that there is no material non-homogeneity.

Elastic Behavior - In order to develop the elastic stiffness K_T for the buried pipe the following plane strain problem is examined. The pipe which is embedded in bonded contact with the elastic soil mass is subjected to a lateral displacement δ_T. The state of deformation

Table I

EXISTING LOAD-DISPLACEMENT RELATIONSHIPS
FOR
BURIED PIPES

Hyperbolic Relationship

Soil Type	Compaction State	δ_{LT}	P_{LT}	$P_T - \delta_T$ Relationship
Sand	Dense	$0.015\ H_0$	$P_{LT} = \gamma H_0 D\ s\ N_q^*$	$P_T = \frac{P_{LT}\delta_T}{0.15\,\delta_{LT} + 0.85\,\delta_T}$
	Loose	$0.020\ H_0$		
Clay	Dense	$0.030\ H_0$	$P_{LT} = cD\ s\ N_c^*$	
	Loose	$0.050\ H_0$		

Empirical Relationship

Soil Type	Compaction State	δ_{LT}	P_{LT}	$P_T - \delta_T$ Relationship
Sand	Dense	$0.015\ H_0$	$P_{LT} = \gamma H_0 D\ s\ N_q^*$	$P_T = \frac{P_{LT}\delta_T}{\delta_{LT}}$; $\delta_T < \delta_{LT}$
	Loose	$0.020\ H_0$		
Clay	Dense	$0.030\ H_0$	$P_{LT} = cD\ s\ N_c^*$	$P_T = P_{LT}$; $\delta_T > \delta_{LT}$
	Loose	$0.050\ H_0$		

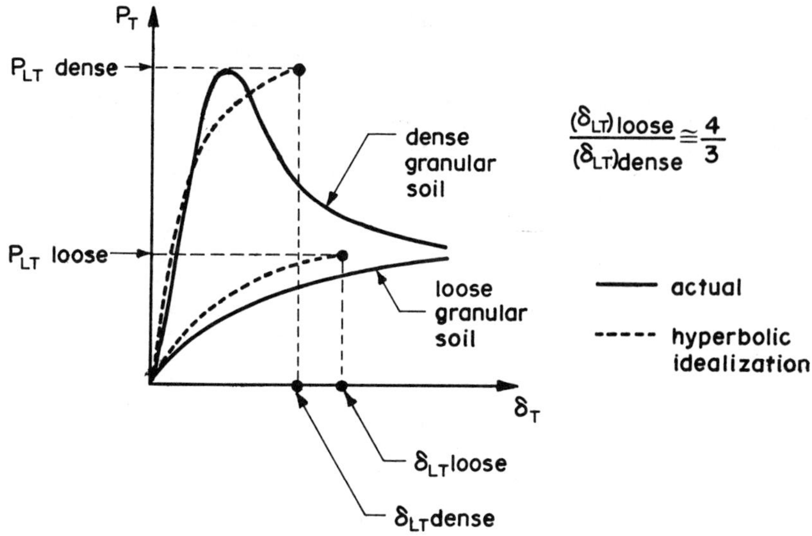

Fig 3 Hyperbolic Representation of the $P_T - \delta_T$ Relationship

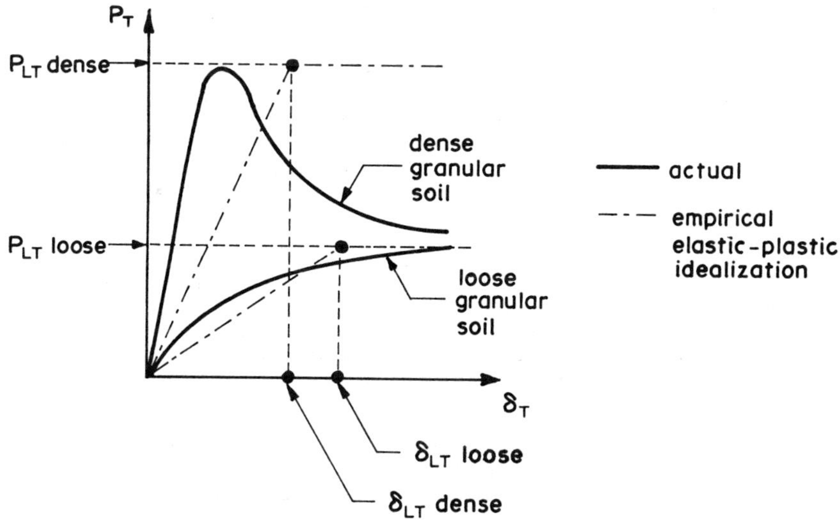

Fig 4 Empirical Elastic-Plastic Representation of the $P_T - \delta_T$ Relationship

in the soil mass is assumed to be such that the displacements reduce to zero at a cylindrical boundary which is located at $r = H_0$ from the centre of the buried pipe, Fig. 5. The stress analysis of the resulting 2-dimensional problem in the classical theory of elasticity can be examined by using an Airy stress function approach (see, e.g., Timoshenko and Goodier [22]). Details of the analytical procedures are given by Selvadurai and Lee [18] and will not be repeated here. It is sufficient to record here the final expression for the lateral soil stiffness:

$$K_T = \frac{12\pi Es}{[9\ln(\rho)-\{\frac{\rho^2-1}{\rho^2+1}\}]} \tag{5}$$

where $\rho = 2H_0/D$. It may be noted that this expression for K_T incorporates the depth of embedment of the pipe. Unfortunately, limits such as $H_0 \to 0$ and $H_0 \to \infty$ cannot be recovered as special cases of (5), without encountering degenerate terms. This is a deficiency in any plane strain model associated with an elastic medium of infinite extent (which is the case as $H_0 \to \infty$). For other finite values of H_0/D, (5) will yield finite estimates for K_T. Within the range of depths of embedment encountered in practice (i.e. $3D < H_0 < 5D$) the soil stiffness can be approximated by the relationship

$$K_T = \pi Es\{0.60 \text{ to } 0.80\} \tag{6}$$

The equation (6) represents the elastic lateral stiffness of an embedded pipe section which maintains complete contact with the surrounding isotropic linearly elastic soil medium. The assumed linear behavior will be valid only for soils which are subjected to small strains. Consequently the elastic stiffness of the buried pipe defined by (6) will be accurate for small pipe displacements. It can be shown (Selvadurai and Lee [18]) that when the embedment ratio H_0/D is approximately 3, the normalized limiting pipe displacement (δ/H_0) elastic limit is approximately $0.5\ \varepsilon_\ell^*$ where ε_ℓ^* is the strain at which deviation from linear elastic behavior is observed in a triaxial test specimen of the backfill/undisturbed soil. The cell pressure in the triaxial test is taken as γH_0 where γ is the unit weight of the backfill/natural soil.

Limiting Soil Resistance - During the lateral movement of the pipeline, the surrounding soil is subjected to a complex state of stress. When this stress state satisfies a defined failure criterion the soil will experience contained or uncontained plastic flow. Physically, the 'failure of the soil' results in either the generation of well defined discrete slip planes along which the limiting soil strength is reached and/or the development of plastic flow in an entire region. The manner in which the soil fails around the pipe defines the limiting soil resistance P_T. The mode of failure of the soil in the vicinity of the buried pipe can be influenced by factors such as depth of embedment of the pipe, the direction of movement of the pipe, the basic soil type (granular, cohesive or combinations thereof), the density and compaction state of the backfill and natural material and the characteristics of the soil-pipe interface.

For shallow depths of burial, the planes or zones of plastic flow usually extend to the ground surface (Figure 6). As a consequence the limit load exhibits a distinct peak value. With loose cohesionless

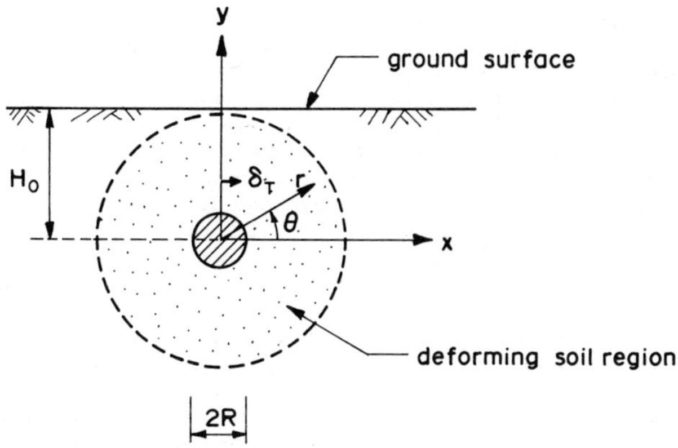

Fig 5 Plane Strain Model of the Pipe Embedded in an Isotropic Elastic Soil Mass

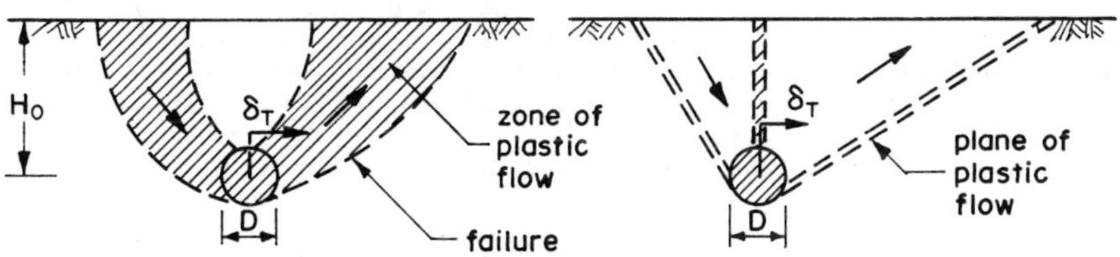

(a) Modes of Failure — Shallow Embedment

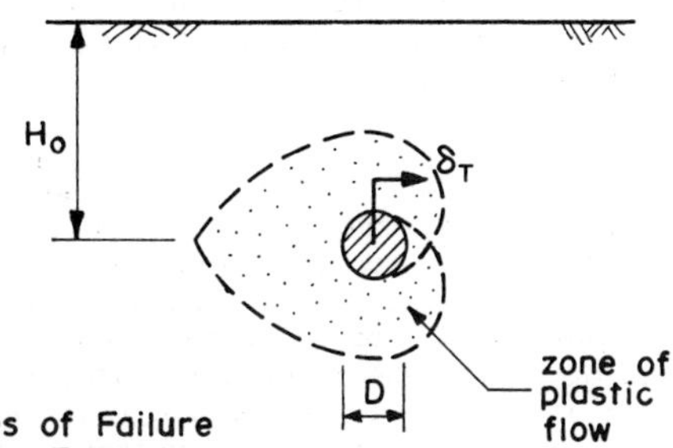

(b) Modes of Failure - Deep Embedment

Fig 6 Typical Failure Modes for the Soil Surrounding the Buried Pipe

soils the mode of failure consists of the generation of plastic zones with large strains whereas with dense cohesionless soils(or soils which are compacted or precompressed) the limit load is reached by the occurrence of distinct planes of shear failure. The extent of the failure zone will depend on factors such as ϕ , the relative density of the soil and the characteristics of the soil-pipe interface (rough or smooth). When the depth of burial becomes large (H > 6D), the zone of plastic flow is contained within the soil mass (Figure 6). As a result, the P_T - δ_T curve does not exhibit a distinct peak. Due to the "sinkage action" the load P_T can continue to increase beyond the initial mobilization of plastic flow.

With cohesive soils (i.e. normally consolidated clays) the load-displacement behavior of the buried pipe follows a form similar to that which is observed in a loose granular soil. As the depth of burial increases the load-displacement curve can exhibit the progressive increase in the limiting soil resistance discussed previously. Similar effects have been observed in deep plate load tests conducted on purely cohesive soils (Selvadurai et al. [16], Selvadurai and Nicholas [17]).

In order to estimate the limiting soil resistance P_{LT} it will be assumed that (i) the pipe movement induces a state of plane strain in the soil over a finite length s, (ii) the rigid pipe can be replaced by a rigid anchor plate with a width equal to the pipe diameter, (iii) the soil-pipe interface is either smooth or completely adhering, and (iv) the failure criterion corresponds to the classical Mohr-Coulomb postulate, $\tau = c + \sigma\tan\phi$, where τ is the shearing stress on the failure plane; c is the cohesion and ϕ is the angle of internal friction.

The article by Selvadurai and Lee [18] gives a complete account of the various estimates for P_{LT} for pipes embedded in granular and cohesive soil media. For example, by using a Rankine analysis for wedge failure (Figure 7) (i.e. planes ab and cd in Figure 7 are assumed to be smooth) it can be shown that the limiting soil resistance of a pipe embedded in a granular soil takes the form

$$P_{LT} = \gamma H_0 Ds \left\{\frac{4\tan\phi}{\cos\phi}\right\} \tag{7}$$

The assumption of frictionless conditions on the planes ab and cd (Figure 7) provides a lower estimate for the upper bound of the limiting soil resistance. When friction is incorporated along these planes the mode of failure no longer corresponds to the simplified Rankine state. The failure mode consists of radial shear zones (Figure 8; regions abg and cde) in which plastic energy is dissipated and rigid zones (Figure 8; regions hbg and cfe) in which plastic energy is dissipated along the boundary planes. The failure limit load for this mode can be examined by recourse to numerical techniques. The results cited in equation (3) corresponds to this particular failure mode, i.e.

$$P_{LT} = \gamma H_0 Ds\, N_q^{*}\left(\frac{H_0}{D}, \phi\right) \tag{8}$$

The Rankine analysis discussed previously can also be applied to derive an elementary estimate for the limiting soil resistance P_{LT} developed in a cohesive soil due to lateral movement of the pipe. With cohesive

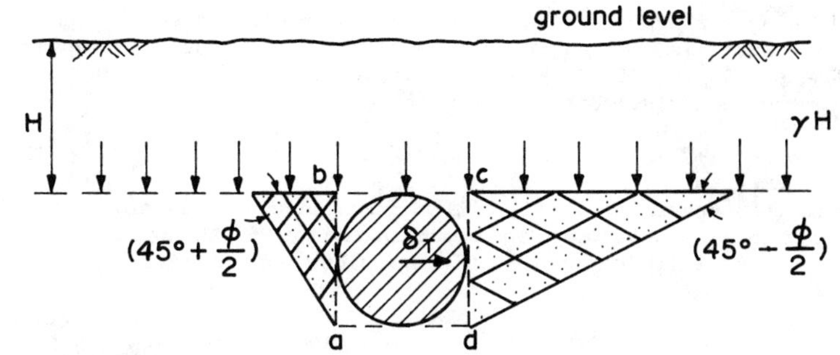

Fig 7 Failure Modes for the Lateral Displacement of the Pipe Embedded in a Granular Soil -Smooth Interfaces ab and cd

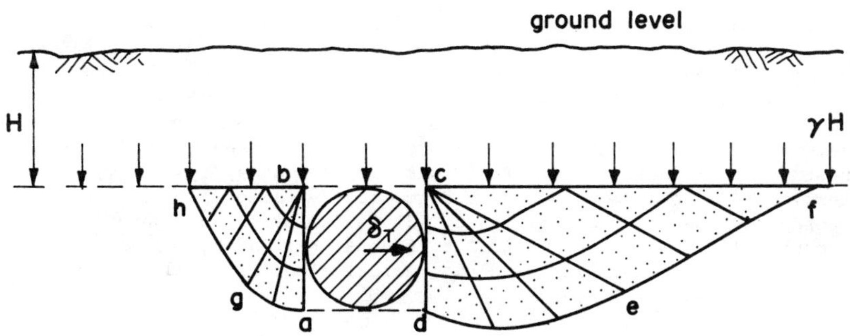

Fig 8 Failure Mode for the Lateral Displacement of the Pipe Embedded in a Granular Soil -Rough Interfaces ab and cd

soils the active earth pressure component can be neglected since cohesive soils contain sufficient strength to support a vertical cut to a depth H_c of at least $2c/\gamma$. If $H_c > H_0$, the active earth pressure can be omitted from the anlysis. For this condition

$$P_{LT} = 2c\ sD\ \{1+ \frac{\gamma H_0}{2c}\} \tag{9}$$

For large depths, the failure mode will correspond to the development of a failure mechanism similar to that shown in Figure 9. An application of the upper bound limit theorem yields the following estimate for P_{LT}

$$P_{LT} = 2c\ sD\ \{\frac{7\pi}{8} + \frac{1}{2} + \frac{1}{2\sqrt{2}}\} \simeq 2c\ sD\ \{3.5\} \tag{10}$$

It may be noted that results of tests on horizontal anchors embedded in saturated cohesive soils (see, e.g., Tschebotarioff [23]) give $P_{LT} \simeq 2csD\{4.7\}$. Other plausible failure mechanisms such as those shown in Figure 10 and 11 can be used to improve the upper bound estimate. Finally, the result cited in equation (4) gives a further estimate for the limiting soil resistance developed due to the lateral movement of the pipe

$$P_{LT} = c\ sD\ N_c^*\ (\frac{H_0}{D}) \tag{11}$$

Bilinear Representation of the Lateral Soil Resistance

As observed by Audibert and Nyman [5] and others, the lateral load-displacement relationship can be approximated by a hyperbolic relationship of the type (2). The lateral soil resistance model proposed here attempts to describe this hyperbolic relationship by an equivalent bi-linear (or tri-linear) form. The Figure 12 indicates, typically, the lateral load-displacement behavior described by equation (2). The proposed bi-linear representation of (2) is indicated on Figure 12 by the broken line OBC. The initial stiffness K_{T1} of the bi-linear model is chosen to coincide with the initial slope of the hyperbolic representation (2). Hence

$$K_{T1} = \left\{\frac{d}{d\delta_T}\ [\frac{K_T\delta_T}{1+\lambda\delta_T}]\right\}_{\delta_T\to 0} = K_T \tag{12}$$

where $\lambda = K_T/P_{LT}$.

In order to define the secondary stiffness modulus K_{T2} (see Figure 12), it is assumed that departure from the initial load-displacement behavior (characterized by the slope K_{T1}) occurs at an initial yield which is defined by P^*_{LT}. This load can be expressed as a proportion of the ultimate lateral soil resistance P_{LT}, i.e.

$$P^*_{LT} = \beta\ P_{LT} \tag{13}$$

where β is a constant. The secondary modulus K_{T2} is defined as the slope of the line which passes through the location B (see Figure 12) and is tangential to the hyperbolic curve. The point of intersection of the line BC and the P_T axis is D. The load level $\tilde{P}_{LT}$ corresponding to the location D is expressed as a proportion of P_{LT}, i.e.

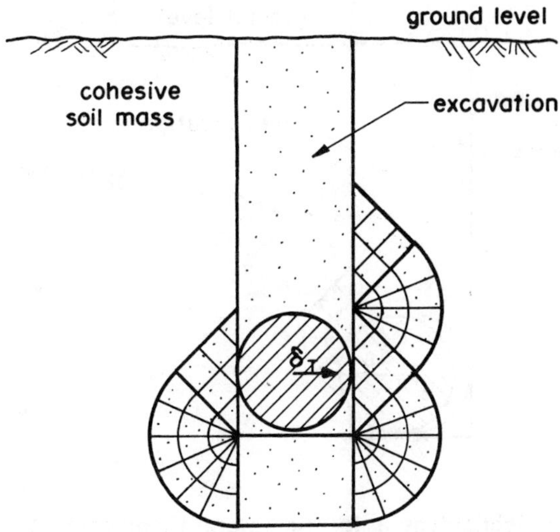

Fig 9 Failure Mechanism 1 for the Lateral Displacement of the Pipe Embedded in a Cohesive Soil

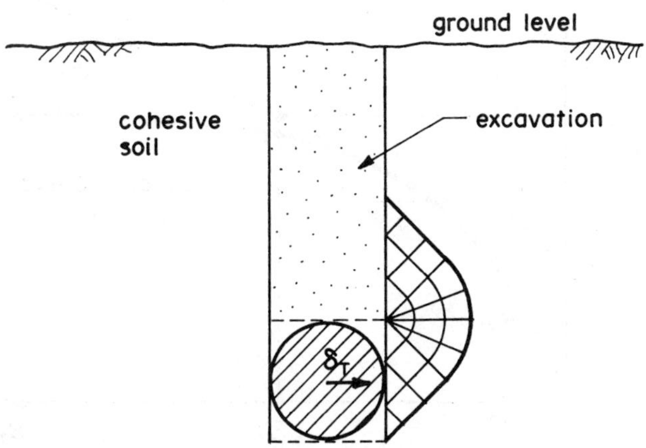

Fig 10 Failure Mechanism 2 for the Lateral Displacement of a Pipe Embedded in a Cohesive Soil

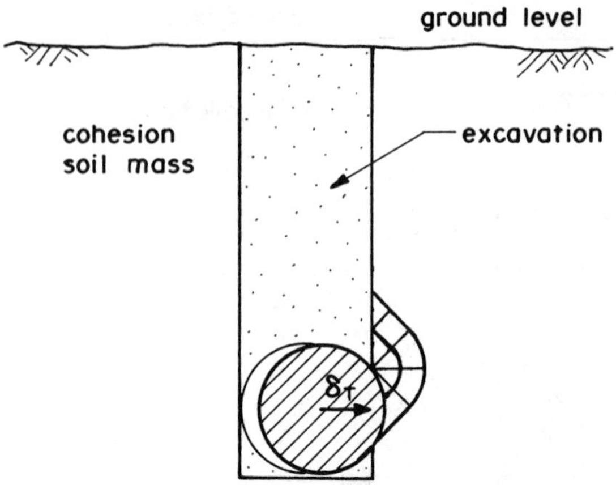

Fig 11 Failure Mechanism 3 for the Lateral Displacement of a Pipe Embedded in a Cohesive Soil

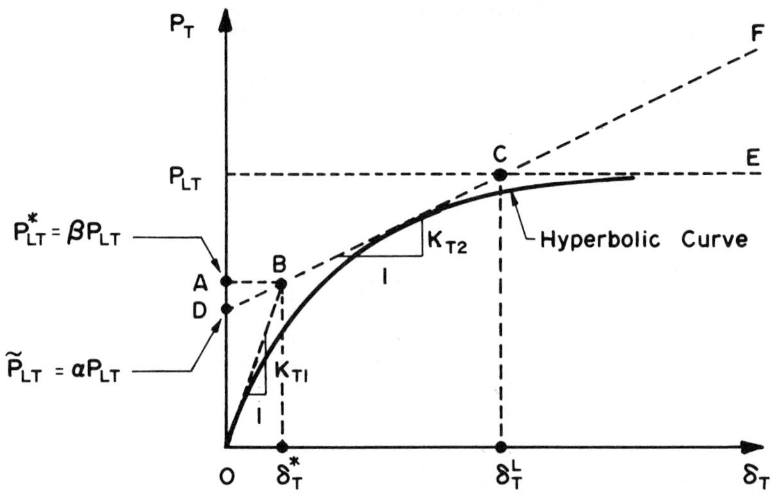

Fig 12 Bilinear Modeling of Lateral Load — Displacement Response

$$\tilde{P}_{LT} = \alpha P_{LT} \tag{14}$$

where α is a constant.
Avoiding details of calculation it can be shown that

$$K_{T1} = K_T \tag{15}$$

$$K_{T2} = K_T \ \Phi(\alpha) \tag{16}$$

$$\beta = \frac{\alpha+\sqrt{\alpha}}{2-\alpha+\sqrt{\alpha}} \tag{17}$$

where

$$\Phi(\alpha) = [1-\sqrt{\alpha}]^2 \tag{18}$$

In order to define the bi-linear representation of the hyperbolic lateral load-displacement relationship, three parameters are necessary. These are K_T, P_{LT} and α. The parameters K_T and P_{LT} have been defined previously. The constant α has to be specified in any analysis. The value of α may be chosen as the load level at which a departure from the initial linear-elastic behavior is to be expected in the transverse load-displacement response. It may be assumed that the load level $\tilde{P}_{LT}$ corresponds to a working load obtained by reducing P_{LT} by a judiciously chosen factor of safety. In most geotechnical applications $\alpha \simeq 1/3$.

Values for the parameter $\beta(\alpha)$ may be obtained directly from the curve given by Figure 13. A similar curve for $\Phi(\alpha)$ is shown in Figure 14. It may be noted that

$$\delta_T^* = \frac{1}{\lambda}\left[\frac{\alpha+\sqrt{\alpha}}{2-\alpha+\sqrt{\alpha}}\right] \tag{19}$$

and

$$\delta_T^L = \frac{1}{\lambda}\left[\frac{1+\alpha+2\sqrt{\alpha}}{1-\sqrt{\alpha}}\right] \tag{20}$$

The extension of the model to the tri-linear domain is quite straightforward. The slope K_{T2} is assumed to be valid up to the load level P_{LT} (i.e. the location C in Figure 12). For the tri-linear representation, beyond the location C, the load-displacement response is essentially rigid-plastic.

Conclusions

This paper contains a review of the procedures currently used to describe the lateral load (P_T)-displacement (δ_T) response of a pipe embedded in a soil mass. Such lateral deformations can be induced by pressure- or temperature-induced expansion of a buried pipeline at regions near bends or transition zones. The load-displacement relationship is evaluated by examining a plane strain model of the buried pipe. In current literature, this load displacement behavior is approximated by either a hyperbolic or an elastic-rigid plastic relationship between P_T and δ_T. Both experimental and empirical results are used to substantiate these formulations. This paper provides

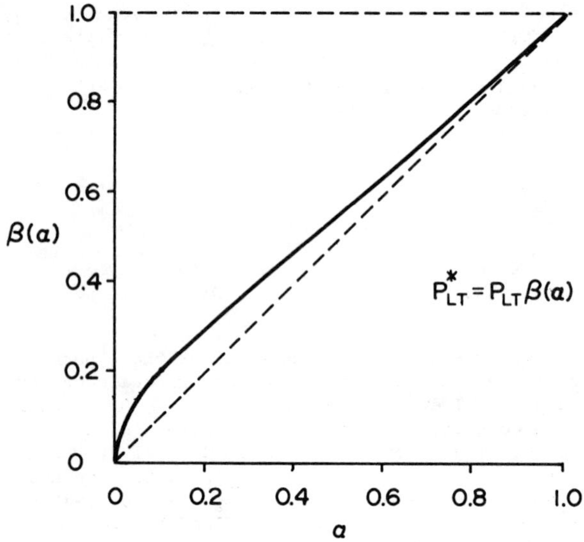

Fig 13 Variation of Parameter $\beta(a)$

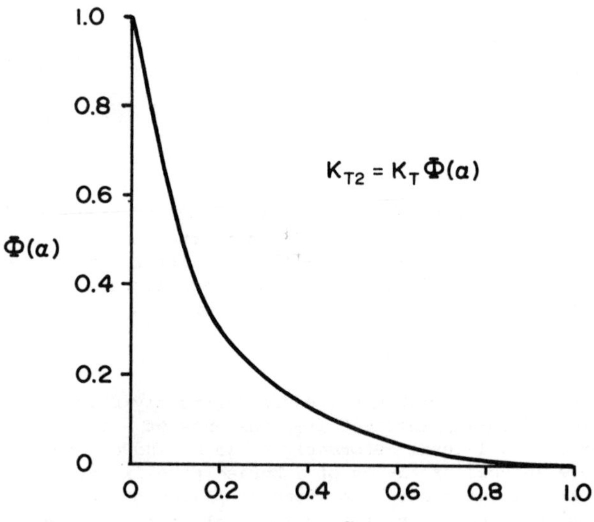

Fig 14 Variation of Parameter $\Phi(a)$

a methodology for evaluating the parameters which govern these elementary hyperbolic and elastic-plastic soil models. It is shown that the initial slope of the load-displacement curve is governed by the linear elastic deformability characteristics of the soil and the geometry of the soil-pipe configuration. The ultimate soil resistance is governed by the shear strength parameters of the soil. The paper also presents a bi-linear representation which attempts to model more closely the hyperbolic load-displacement response.

References

[1] Ariman, T., Liu, S.C. and Nickell, R.E., "Lifeline Earthquake Engineering - Buried Pipelines, Seismic Risk and Instrumentation", PVP 34, American Society of Mechanical Engineers, 1979.
[2] A.S.C.E., Journal of the Pipeline Division, Vol. 85, PL3, Proceedings Paper 2205. 1959
[3] A.S.C.E., Journal of the Pipeline Division, Vol. 93, PL2, Proceedings Paper 5355, 1967
[4] A.S.C.E., List of Source and Reference Material on Pipeline Design, Task Committee Report. 1980
[5] Audibert, J.M.E. and Nyman, K.J., "Soil Restraint Against Horizontal Motion of Pipes", Journal of the Geotechnical Engineering Division, Proc. A.S.C.E., Vol. 103, October 1977, pp. 1119-1142.
[6] Audibert, J.M.E., Lai, N.W. and Bea, R.G., "Designing Pipelines to Resist Instabilities - Part 3", Pipeline and Gas Journal, Vol. 207, June 1980, pp. 49-53.
[7] Das, B.M. and Seeley, G.R.,"Breakout Resistance of Shallow Horizontal Anchors", Journal of the Geotechnical Division, Proc. A.S.C.E. Vol. 101, September 1975, pp. 999-1003.
[8] Desai, C.S. and Abel, J.F., Introduction to the Finite Element Method, Van Nostrand, New York, 1972.
[9] Desai, C.S. and Christian, J.T., Numerical Methods in Geotechnical Engineering, McGraw-Hill, New York, 1976.
[10] Gudehus, G., (Ed.) Finite Elements in Geomechanics, John Wiley, New York, 1977.
[11] Hansen, J.B., Earth Pressure Calculation, Danish Technical Press, Copenhagen, Denmark, 1953.
[12] Kennedy, R.P., Chow, A.W. and Williamson, R.A., "Fault Movement on Buried Oil Pipeline", Transportation Engineering Journal, Proc. A.S.C.E., Vol. 103, September 1977, pp. 617-633.
[13] Luscher, U., Black, W.T. and Nair, K., "Geotechnical Aspects of Trans-Alaska Pipeline", Transportation Engineering Journal, Proc. A.S.C.E., Vol. 101, November 1975, pp. 669-680.
[14] Peng, L.-C., "Stress Analysis Methods for Underground Pipelines", -Part 2, Soil-Pipe Interaction, Vol. 47, May 1978, pp. 65-74.
[15] Selvadurai, A.P.S., Elastic Analysis of Soil-Foundation Interaction, Developments in Geotechnical Engineering, Vol. 17, Elsevier Scientific Publishing Co., Amsterdam, The Netherlands, 1979.
[16] Selvadurai, A.P.S., Bauer, G.E. and Nicholas, T.J., "Screw Plate Testing of a Soft Clay", Canadian Geotechncial Journal, Vol. 17, September 1980, pp. 465-472.
[17] Selvadurai, A.P.S. and Nicholas, T.J.,"Evaluation of Soft Clay Properties by the Screw Plate Test", Proc. 10th International Conference in Soil Mechanics and Foundation Engineering, Stockholm,

Vol. 2, June 1981, pp. 567-571.

[18] Selvadurai, A.P.S. and Lee, J.J., Soil Resistance Models for the Stress Analysis of Buried Fuel Pipelines, Interim Report, Technical Research and Development Project, Research and Engineering, Bechtel Group Inc., San Francisco, December 1981.

[19] Selvadurai, A.P.S. and Lee, J.J., Soil Resistance Models for the Stress Analysis of Buried Fuel Pipelines, Final Report, Technical Research and Development Project, Research and Engineering, Bechtel Group Inc., San Francisco, December 1982.

[20] Selvadurai, A.P.S., "Flexure of an Infinite Beam of Finite Width Embedded in an Isotropic Elastic Medium of Infinite Extent", International Journal for Numerical and Analytical Methods in Geomechanics, (in press).

[21] Shibata, H., Katayama, H. and Ariman, T., "Recent Advances in Lifeline Earthquake Engineering in Japan", PVP 43, American Society of Mechanical Engineers, 1980.

[22] Timoshenko, S.P. and Goodier, J.N., Theory of Elasticity, McGraw-Hill, New York, 1970.

[23] Tschebotarioff, G.P., Foundations, Retaining and Earth Structures, McGraw-Hill, New York, 1973.

[24] Vermeer, P.A. and Luger, H.J., Deformation and Failure of Granular Materials", Proceedings of the I.U.T.A.M. Symposium, Delft, Balkema Press, Rotterdam, The Netherlands, 1982.

[25] Wittke, W., (Ed.) Numerical Methods in Geomechanics, Aachen 1979, Proceedings of the 3rd International Conference on Numerical Methods in Geomechanics, Vol. 1-4, A.A. Balkema, Rotterdam, The Netherlands 1979.

STEEL PIPELINE PROVIDES NEW WATER SUPPLY FOR JORDAN

by: Robert L. Thoem[1] Fellow, ASCE;
Charles L. Meyer[2], Member, ASCE;
Mark C. Boner[3]

INTRODUCTION

Background

The Hashemite Kingdom of Jordan is located in the Middle East and is bordered by Saudi Arabia, Israel, Syria, and Iraq. The climate is arid; the terrain is rolling to mountainous with elevations ranging from 400 meters (1,300 feet) below sea level in the Jordan Valley and Dead Sea area to points exceeding 1,500 meters (5,000 feet). Much of the area is rocky with shallow soil cover. Temperatures can vary significantly with extremely high temperatures occurring in the Jordan Valley area.

The region has in the past relied on well and spring systems to provide potable water supply. However, rapid economic and population growth have outpaced the capacity of existing groundwater resources. The capital city of Amman, with a population of about 1,000,000, has experienced severe water shortages with some parts of the city receiving water service only once or twice a week. Nearly all buildings have roof storage tanks to provide water when the distribution system does not meet the demand.

Regional Water Projects

As a result of the critical water resource need, Jordanian government agencies have proceeded agressively to explore and develop other potential water supplies. Two major projects are currently underway.

A northern regional supply, which taps a recently discovered major artesian aquifer, was recently designed and is currently in the contract bidding stage. This project will supply several cities and villages in the northern part of the country. A second artesian source nearby remains to be developed.

The southern regional supply, which is farther along in development than the northern project, will pump excess irrigation water from a canal system in the Jordan Valley to the city of Amman and surrounding

1) Associate Chief Environmental Engineer, Stanley Consultants, Inc., Muscatine, IA, USA
2) Chief Civil Engineer, Stanley Consultants, Inc., Muscatine, IA, USA
3) Resident Engineer, Stanley Consultants, Inc., Amman, Jordan

metropolitan region (Figure 1). This paper deals only with the southern region project and specifically the pipeline planning, design, and construction.

Southern Project Features

Engineering studies for this project started in July 1979, and the design was completed ten months later in May 1980. Project reviews and arrangements for financing from several international funding sources required over a year, with bids being received in late 1981. Contract agreements were finalized in May 1982.

The project consists of an intake pump station, five high lift booster pump stations, a water treatment plant, and a major reservoir (see Figure 2). Water in the irrigation canal at 230 meters (755 feet) below sea level is pumped to the terminal reservoir 1,040 meters (3,412 feet) above sea level, a total lift of 1,270 meters (4,167 feet). Distance from the intake canal to the reservoir near Amman is approximately 37 kilometers (23 miles). Design capacity of the system is 45 million cubic meters per year (22,500 gpm or about 32 mgd).

PIPELINE PLANNING

Initial planning studies for the southern project assessed several factors to define the design and construction features. These included pipe materials, optimum diameter, corridors for routing the pipeline, and techniques for installing the pipe.

Pipe Materials

The pipeline construction over a major portion of the route is located in rugged terrain. In the six-kilometer (3.7-mile) Jordan Valley escarpment area, for example, the pipeline is laid at an average slope of 14 percent with several stretches much steeper. The area is seismically active and has several active faults which need to be crossed by the piepline. In addition, design pressures over 35 kilograms per square centimeter (500 psi) are required at several locations in the in the pumping system.

During the preliminary project planning, a decision was made to utilize steel pipe to meet the severe pressure, seismic, and construction demands on the pipeline. Large pipe bending equipment was proposed to accommodate the many vertical and horizontal curves required in the rugged topography.

Pipe Diameter

The desirable size of pipe was established through life-cycle cost analyses. Construction costs were estimated for various pipe diameters capable of handling the design flow and compared with the cost of energy associated with friction losses. The annual power cost required for the various pipe sizes was converted to a present worth value using a 10 percent discount rate over a 50-year period. Static head was not

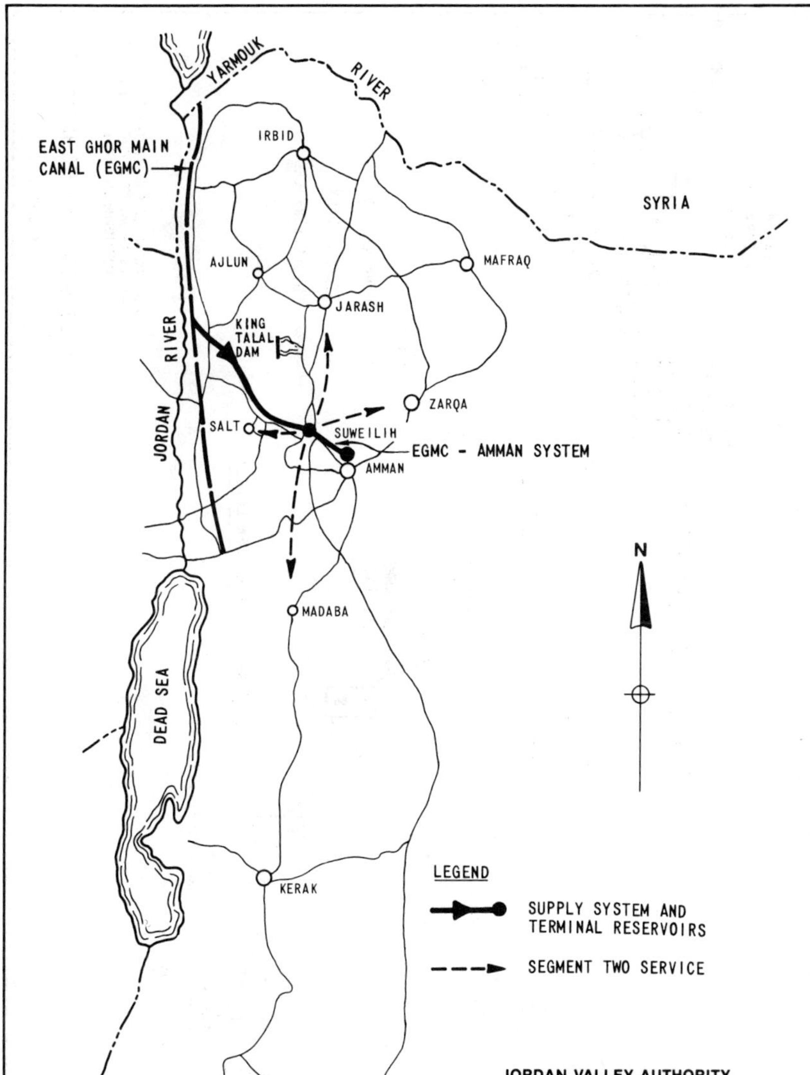

TOTAL SYSTEM LAYOUT

FIGURE 1

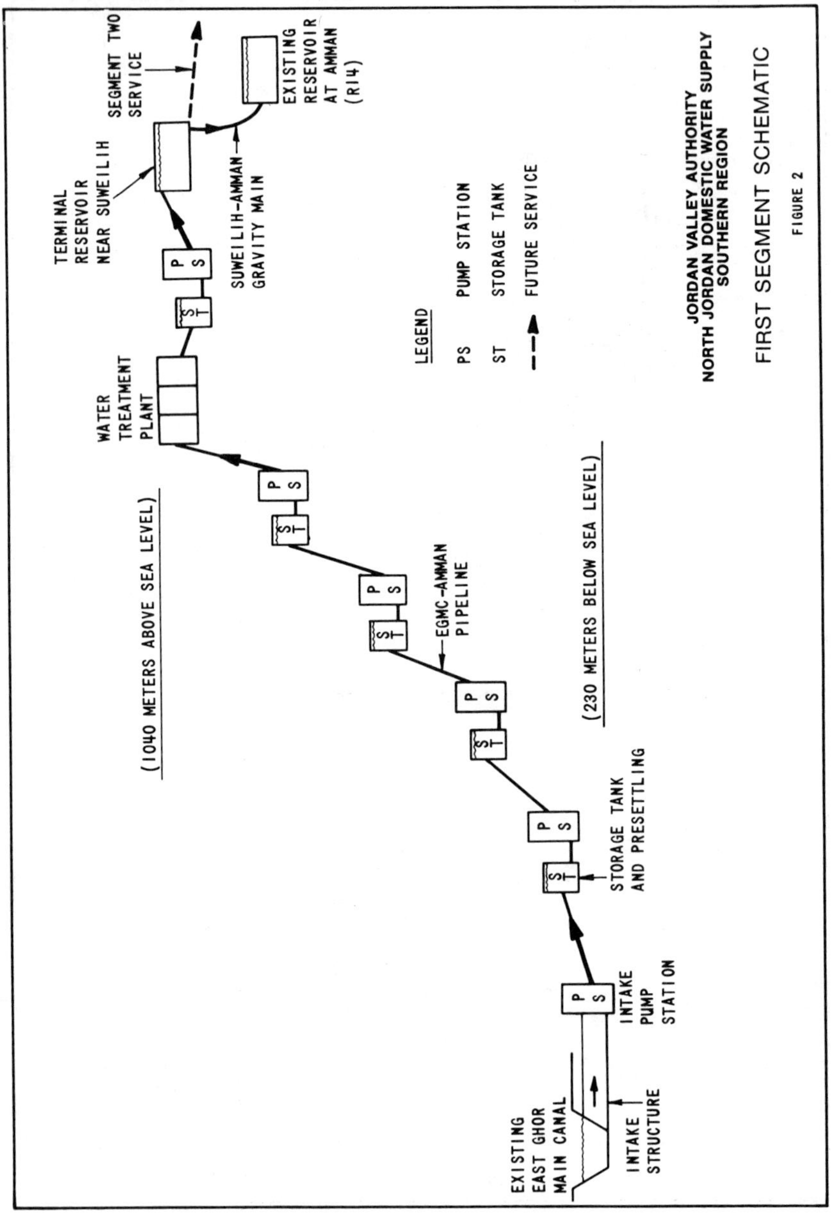

FIRST SEGMENT SCHEMATIC

FIGURE 2

considered since it is the same regardless of pipe size and maintenance costs were not included since they are low in comparison with the construction and power cost.

The life-cycle analyses yield an optimum diameter for given flow conditions with a tradeoff between the high power and low construction costs for small diameters and vice versa for the large diameters. Conclusions from these evaluations were further tested through sensitivity analyses. One test assumed electric power rates would not escalate during the economic life and a second assumed they would grow at an annual rate of 8 percent. These analyses yielded a range of optimum diameters and the selected 1,200-mm (48-inch) diameter pipeline was within this bracketed range.

Route Alignment

Reconnaissance surveys were made of the potential corridors for the pipeline from the Jordan Valley to the Amman area. The reconnaissance evaluations narrowed the choices to three route alignment alternatives. Data gathered from the environmental assessment work on the entire project as well as construction cost estimates were utilized in assessing desirability to the three routes. The main selection criteria utilized included the following:

1. Slope of land
2. Distance
3. Land elevation
4. Intermittent stream crossings
5. Land use
6. Potential growth patterns
7. Right-of-way
8. Transportation access
9. Utilities
10. Soil conditions

Another important feature in the route alignment study was the ability to locate the various pumping stations so the same size pumps and motors could be utilized at each facility. This factor required several iterations in selection of the three candidate route options.

The selected corridor was a central route (Figure 3) compared to the other two options of one to the north and another to the south. Cost comparisons of the northern and central alignment were so close that the selection was based on intangible factors such as those listed above. The most outstanding advantage for the central route was the potential of serving a nearby small community, a recreational park, and developing areas located along existing roads.

The cost evaluations also indicated that for a major pipeline such as that required on this project, the shortest distance is extremely important. Slightly longer routes which took advantage of easier construction terrain were more than offset by the added pipe and installation cost.

Installation Technique

During the planning study, several different installation options were considered for the transmission pipeline including a buried pipe, an aboveground supported pipe, an at-grade installation, or a combination.

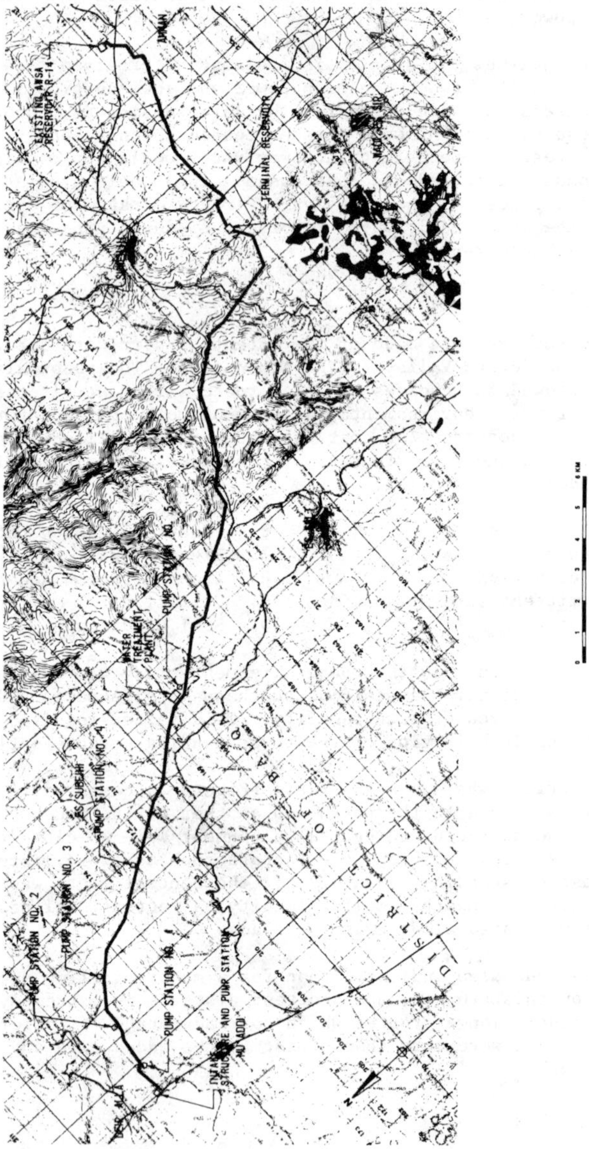

JORDAN VALLEY AUTHORITY
NORTH JORDAN DOMESTIC WATER SUPPLY
SOUTHERN REGION

PIPELINE ROUTE

FIGURE 3

After evaluating these installation alternatives, a totally buried pipe was selected. Large temperature variations which would affect a supported pipe design are avoided with the buried pipe. The underground installation also avoids anchorage requirements since the soil load on the pipe holds it in place and resists thrust forces. Another major advantage of the buried pipe is that it provides no interference to land use and is more aesthetically pleasing. Pipe damage due to vandalism or other activities is also minimized.

The primary advantage of the supported pipe was in areas where the cost of rock excavation was anticipated to be significant. Maintenance of the aboveground facility was anticipated to be more extensive.

The "at-grade" approach is an arrangement where the pipe would be partially buried to about the middiameter point. This option would avoid some of the more extensive excavation costs for the buried alternative. It has a major disadvantage of requiring two different types of coating for the portion underground and the part exposed to the atmosphere. In addition, specially designed anchors at various points would be needed to restrain pipe movement since soil load would not be available to serve that purpose. The partially buried installation also limits land use and movement.

PIPELINE DESIGN

The combination of extreme topography, severe seismic conditions and major faults, wide ranging soil conditions, and high working pressures presented significant obstacles to the design team.

Internal Pressure

Design pressures required for the pumping system range from below 17.6 kilograms per square centimeter (kg/sq cm) (250 psi) to 39.5 kg/sq cm (560 psi). The design pressures include normal working pressures plus those experienced during a transient condition under pump or electric failure. Transient conditions were controlled through a combination of pump discharge control valves, hydropneumatic surge tanks, and air/vacuum valving.

The pipeline is divided into three different pressure classes for the pipe material and wall thickness as follows:

Class 1 - 0 to 17.6 kg/sq cm (0 to 250 psi)
Class 2 - 17.6 to 29.6 kg/sq cm (250 to 420 psi)
Class 3 - 29.6 to 39.5 kg/sq cm (420 to 560 psi)

The number of pressure classes was determined through investigation of the hydraulic requirements along the pipeline profile. Various pressure groupings ranging from two to five classes were considered. Lengths of pipe required for different pressure groupings were determined. Estimates of pipe costs for various pressure class combinations were made and compared. A system consisting of three classes was selected as being the most practical considering the pipe costs and management of different pipe classes during construction.

Wall Thickness

Wall thicknesses for differing steel grades were determined for the selected three pressure classes. All pipe was specified to be manufactured and supplied in accordance with American Petroleum Institute (API) 5LX specifications. Pipe costs for various grades of steel and wall thicknesses were compared to assess potential advantages of increasing the API X grade of steel and reducing the wall thickness.

Three different steel grades were selected, X42, X46, and X52 for Class 1 through 3 pressure ranges, respectively. Minimum wall thickness for Classes 1 through 3 is 9.53 mm (0.375 inch), 13.21 mm (0.520 inch), and 14.45 mm (0.569 inch), respectively. The minimum wall thickness included consideration of handling, buckling, collapse under complete vacuum, and deflection under earth load.

Bedding and Backfill

Select bedding material with a maximum particle size of 12.5 mm (0.5 inch) is being placed entirely around the pipe in the trench. Typically, a minimum of 150-mm (6-inch) bedding is required under the pipe and 300 mm (12 inch) above the pipe. All bedding is placed by hand and compacted prior to backfilling the remainder of the trench. Backfill material for the remainder of the trench is restricted to soil and rocks smaller than 150 mm (6 inch) in size.

In general, the pipeline was planned for a minimum of 1-meter (3.3-feet) cover. Adequacy of this minimum cover was checked using a mathematical model (PIPLIN) to determine the pipe-soil interaction. Analyses were made of pipe-soil reaction at side bends, over bends, and sag bends along the pipeline profile. Results of these analyses indicated normal bedding and backfill requirements will be sufficient to resist pipe movement.

Fault Crossings

The geotechnical investigations revealed that the entire Jordan Valley area is subject to seismic activity and that three other locations along the pipeline are potentially active fault zones. The valley area has the potential to develop both horizontal and vertical displacements while the other fault crossings are susceptible to only vertical movement. Limited investigation of the seismic exposure for the pipeline route indicated the maximum surface fault displacement for a 100-year event earthquake to be 0.3 meters (1 foot) for both vertical and horizontal movement. Modeling was employed to evaluate the potential reaction of the pipeline for these two types and magnitudes of fault crossings. These analyses utilized various steel grades, wall thicknesses, internal pressure, temperature differential, and soil types to determine the fault-pipe interaction.

Results of the fault investigation involving only vertical slip displacement determined the impact of a seismic event would be minimized by providing an unanchored section of pipe approximately 340 meters (1,100 feet) on each side of the fault zone. The unanchored section has a bedding with less compaction than a normal trench and also a backfill with particle sizes not greater than 75 mm (3 inch). The bedding and backfill requirements would enable the pipe to displace vertically through the backfill during a seismic event, thus minimizing the strain and deformation on the pipe.

In the Jordan Valley area, where the pipe could be subject to both horizontal and vertical movement, a trapezoidal trench section, approximately 426 meters (1,400 feet) on each side of the fault zone is provided with one-to-one side slopes and the same bedding and backfill as previously described for the vertical displacement faults. The trapezoidal section allows the pipe to displace through the soil not only vertically, but horizonally, under a seismic event.

Corrosion Control

Fusion-bonded epoxy exterior coating with a thickness from 0.305 to 0.407 mm (0.012 to 0.016 inches) is being provided to control corrosion of the steel pipe. A coal tar epoxy lining with total thickness of 0.381 to 0.635 mm (0.015 to 0.025 inches) will protect the interior of the pipe. An impressed current cathodic protection system is also provided along the pipeline with rectifiers and anode beds located at each pump station site.

Appurtenances

A series of air release, vacuum, and drain valves are being installed to provide satisfactory pipeline operation during normal operation and to allow draining for maintenance purposes. Access manholes are also provided approximately every kilometer to permit entry for maintenance and inspection.

PIPELINE CONSTRUCTION

Cost and Schedule

The southern region project is separated into five contracts with a total construction cost of approximately $U.S. 110,000,000. The pipeline contract constitutes approximately 40 percent of this total. A Lebanese contractor is responsible for the pipeline construction.

The entire project is scheduled for completion in the second half of 1984. The pipeline contract is progressing more rapidly than all other contracts and is expected to be finished this year. Average production for the pipeline construction has been about 215 meters (700 feet) per calendar day.

Clearing and Trenching

In general, all construction operations started from the Jordan Valley and proceeded uphill towards Amman; however, clearing and trenching started from uphill sections and worked downward. Slopes from the toe of the escarpment to Pump Station 2 were as steep as 50 percent. The right-of-way in this stretch was overcut to lessen the grade for equipment operation. This area was cleared using a D9 Caterpillar.

The initial slope was so steep the D9 had to use the blade and ripper to work itself upfill. Fuel trucks were positioned at Pump Station 2 and at the toe of the escarpment. Figures 4, 5, and 6 show the pipeline construction activities. Because of delays in initiating blasting operations, the trench was opened with a D-9 ripper in the rocky areas.

In general the clearing consisted of a 20-meter (66-foot) pathway. Restoration requirements will reduce the pathway to about 5 to 15 meters in width, depending upon side slopes. Trenching has consisted for the most part of vertical wall trench (see Figure 5) except in the seismic zones as previously described.

Rock blasting operations were difficult due to numerous villages along the pipeline route and due to governmental restrictions. Blasting was accomplished using ammonia nitrate and electrical detonators; time delay detonators were not allowed. Blasting operations required continued surveillance by the military and local police.

Pipe Laying

All pipe was manufactured in France and supplied in 12-meter (40-feet) sections. It was shipped to the port of Aqaba and transported overland to a pipe yard located along the right-of-way between the treatment plant and Amman. Stringing of the pipe was usually accomplished in single length sections due to the steep winding roads leading to the right-of-way. However, a double jointing operation was set up at the pipe yard to fill in time when the welding crews were held up along the pipeline.

For most of the route the pipe was strung along the trench. At the escarpment and other very steep areas the pipe was double jointed at the toe of the escarpment. Each of the double jointed pipe sections was welded in the trench at an average of three to four welds per day. This operation took about one month for installing one-half kilometer of pipe.

The pipe bending machine, which weighed 65 tons, had to be transported to the toe of the escarpment by a longer route due to the narrow winding roads. Sections of pipes requiring bends were made at the toe of the escarpment, then transported by boom and Caterpillar uphill to its appropriate location. The bending machine was transported along a goat path reconstructed by bulldozers for bending past Pump Station 3. About 90 percent of the pipes had some degree of bend between Pump Stations 2 and 4. For long bends the pipe was typically double jointed.

Welding operations, on the average, consisted on about 12 welders accomplishing approximately 15 joints per day on the thin wall pipe and about 12 joints a day on the thick wall pipe. The X52 pipe required about nine passes per joint whereas the X46 needed eight passes and the X42 pipe needed seven passes. The stringer bead was made using Fleetweld rods at 4-mm (5/32-inch) diameter. The hot pass and filler and cap welds were accomplished using Shield Arc rods at 4-mm (5/32-inch) and 4.76-mm (3/16-inch) diameters. One hundred percent of the joints were inspected using gamma radiography performed by the Royal Scientific Society of Jordan.

FIGURE 4 - Construction of escarpment near Pump Station 2.

FIGURE 5 - Trenching at top of Jordan Valley escarpment.

FIGURE 6 - Pipe laying operation in progress.

Completed external pipe joints were mechanically cleaned and Raychem shrink sleeves were applied. Interior joints were also mechanically cleaned and were painted with a primer and several coats of coal tar epoxy.

The bedding material was hydraulically flooded to obtain the compaction requirements. In flat areas, wheel loaders and backhoes were used to spread bedding material for subsequent mechanical compaction. On the escarpment and in the steep areas a 977 Caterpillar crawler loader and a 235 Caterpillar crawler with clamshell attachment were used to distribute and apply bedding material for subsequent compaction. Nuclear moisture density gauges were used to check bedding and backfill compaction.

SUMMARY

The southern region water supply project represents a very significant undertaking for the Hashemite Kingdom of Jordan and numerous features make it a unique water system. Completion of the project next year will enable continued growth and development in Jordan.

A UNIFIED GEOTECHNICAL ENGINEERING SYSTEMS APPROACH APPLIED TO ARCTIC PIPELINE PLANNING AND DESIGN

By Charles L. Vita[1], M.ASCE and James W. Rooney[2], M.ASCE

ABSTRACT

Presentation is made of an applied systems approach that identifies, defines, and integrates into the overall project implementation process those geotechnical engineering functions, processes and data valuable to the planning and design of complex pipeline projects in geotechnically variable arctic terrain. System objectives, components, attributes, and internal/external functional relationships are identified and explained from a practical perspective. Geotechnical route characterization is put into a landform-based, systems format. The systems approach is presented from a project perspective, starting with definition/identification of needed geotechnical input/output, and progressing through route selection, preliminary design, final design, and construction application. Criteria for system development and application are identified as they relate to arctic pipeline projects.

INTRODUCTION

The purpose of this paper is to conceptually identify and outline a systematic approach to geotechnical engineering for complex arctic pipelines--a principal element of the arctic pipeline design process. Figure 1 illustrates important system relationships.

The fundamental geotechnical difficulty inherent in the design of arctic pipelines is the great variability and influence of earth materials (of various compositions and thermal states) encountered along the alignment. Earth material (rock and soil) variability will occur in ways similar to earth material in temperate climates with the additional major concern for variability in thermal state: from completely frozen to unfrozen, from freezing to thawing, and states in between. The structure of frozen soils (permafrost) can be particularly complex: ground ice composition and spatial characteristics can lead to sharp changes in conditions over short distances. Permafrost can range from "ice-poor" to "ice-rich"; from "ice-bonded" to not ice-bonded. The degree of ice-bonding can be strongly temperature dependent over narrow temperature ranges (below 32°)--particularly in marine or saline environments. In general, the mechanical behavior of ice-bonded soil is temperature and time dependent: as frozen temperature or loading time increases strength and stiffness decrease and compressibility increases; these effects are most pronounced where soil moisture is undergoing solid/liquid phase change, at or below 32°F. If thawed, ice-poor permafrost can be "thaw stable", (exhibiting little volume reduction on thawing under overburden loads) and ice-rich permafrost "thaw unstable" (exhibiting large volume reduction on thawing). Thaw-stable implies no distress to any supported structure; thaw-unstable implies potential distress to supported structures--the

---

[1] Vice President, R&M Consultants, Inc., Seattle, Washington
[2] President, R&M Consultants, Inc., Anchorage, Alaska

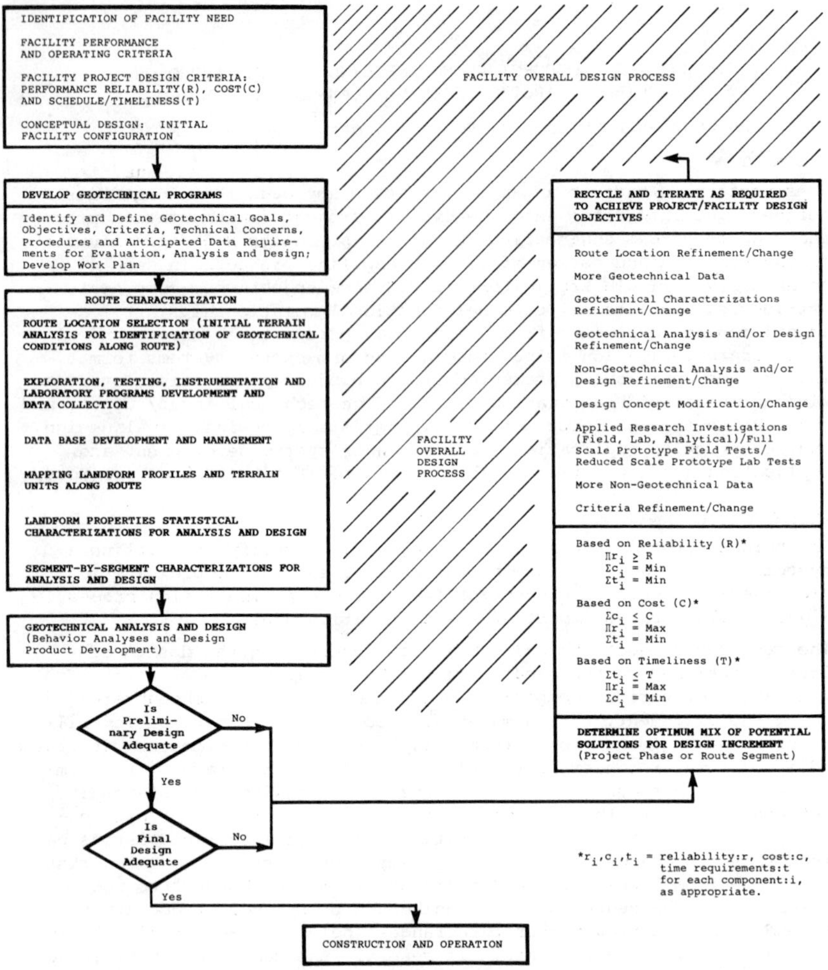

FIGURE 1
GEOTECHNICAL ENGINEERING SYSTEM
(Within The Facility Overall Design Process)

terms are relative and structure dependent: what is thaw-stable for one structure, e.g., a pipeline, could be thaw-unstable for another, e.g., a precision turbine foundation.

Because geotechnical conditions in the arctic are complex and highly variable, it is commonly impractical to achieve exploration programs of sufficient magnitude to characterize soil conditions with certainty along an entire alignment. Therefore geotechnical evaluations must be done with limited field data. In practice, the tendency has been toward making apparent conservative interpretations of soil conditions; a major limitation has been the lack of systematic control of design conservatism: some segments are characterized more or less conservatively than necessary to achieve a given level of performance reliability.

The route geotechnical engineering process presented here for project planning, analysis, design and construction application addresses these concerns. The process is iterative based on successive refinement and incremental improvement with several interacting elements occurring to varying extents simultaneously in time. These are briefly summarized in logical succession: (1) geotechnical program development; (2) terrain conditions along the (proposed) route(s) are identified using available information to provide an initial interpretation of subsurface conditions; (3) route(s) are selected and refined throughout the planning and design process based on both geotechnical and non-geotechnical considerations; (4) surface and subsurface exploration, testing, and instrumentation programs are designed and executed to obtain needed information; (5) resulting field and laboratory data are used to characterize geotechnical properties both on a site-specific route basis and on a landform statistical basis as needed for analysis and design; (6) analysis and design is conducted first on a preliminary basis and then on a final design basis. The adequacy, sufficiency, and mix of geotechnical field, laboratory, experimental, and analytical information as required in the design process are controlled by criteria of facility performance reliability (R), cost (C), and design time (T) constraints.

GEOTECHNICAL PROGRAM DEVELOPMENT: SYSTEM DESIGN

This element--which can be most influential in actually achieving a timely, cost-effective and technically-effective design--consists of identifying, defining, and planning the scope, purpose and form of the geotechnical program: its goals, objectives, procedures, criteria and project-integrated work plan needed to realize an acceptable facility design. This includes identifying geotechnical concerns and the information needed to address those concerns: the assessment approaches, assumptions, definitions, necessary field/lab data, analytical procedures, and the performance criteria used to establish the sufficiency of the information (i.e., its measure of effectiveness) compatible with higher level facility criteria.

The purpose and scope of the geotechnical engineering function is to support route selection/refinement and facility design through detailed identification and assessment of geotechnical and geological conditions pertinent to design, construction or operation including recommended control or mitigation of hazards or problems along the alignment which can potentially impact facility construction or operation, adjacent facility integrity, or environmental concerns. This encompasses all efforts to identify, obtain, analyze, evaluate, synthesize and apply geotechnical and geological data in a time-, technically-, and cost-effective manner and at a scale of effort compatible with project needs.

These efforts can be systematically formulated and applied to support project objectives by identifying actual and potential geotechnical effects of concern based on experience with geotechnical behavior and its interaction with constructed facilities, and then, in an iterative process of successive refinement: (1) determining data needs, (2) obtaining needed data, (3) assessing the data and geotechnically characterizing the route for (4) engineering analysis and synthesis of the identified geotechnical concerns, followed by (5) design consideration and incorporation into the project design.

ROUTE CHARACTERIZATION

All geotechnical engineering is ultimately based on evaluation of field conditions. The route characterization process establishes mile-by-mile geotechnical data necessary for evaluating each analysis/design consideration along the route, including: mechanical and thermal stability, soil/structure mechanical and thermal interaction analysis/design, foundation selection/analysis/design, and earthwork considerations.

Several factors influence the characterization: geotechnical variability; constraints/limits imposed by the available data base--due to limited field exploration and laboratory testing, and data bias effects; details of how the data is used in analysis and design; facility design sensitivity; and level of required reliability.

Characterization parameter values are established by analysis and synthesis of available data. Geomechanical parameters (e.g., soil strength and compressibility parameters) and geothermal parameters (thermal conductivity, heat capacity, latent heat and ground surface/ climatic thermal factors) are determined by: field exploration and testing; laboratory measurements; calculations using stochastic correlations based on limited testing of representative samples obtained from drilling programs; and correlations based on semiempirical formulas, literature information, or engineering judgement.

Systematic route characterization [4] would include interactive elements for exploration (geotechnical and geological data collection), data base development and management, and geotechnical characterizations for route location selection, analysis, design, and construction. Development criteria for route characterization would require the process be user-oriented and capable of:

- Producing segment-by-segment soil profiles and associated properties appropriate for engineering analysis and design using the available data base, analytical techniques, time and budget;
- Systematically accounting for constraints/limits of: the analysis methodology, available data base, design performance sensitivity, construction considerations and environmental impacts;
- Being updated by the addition of data in a timely manner;
- Being understood and efficiently incorporated into the project design process.

Exploration: Geotechnical and Geological Data Collection

Exploration programs are designed to obtain the field and associated laboratory data needed to address the actual and potential geotechnical concerns identified in the Geotechnical Plan. Such programs can include: (1) remote sensing (non-ground contact) exploration: aerial photography, aerial reconnaissance, geophysical techniques; (2) field investigations: ground reconnaissance, soil/rock boring and sampling programs, in-situ testing, geophysical surveys, field design prototype testing, observations of field performance of similar existing facilities, field instrumentation of all types; and (3) laboratory testing of representative samples obtained from field programs, and small scale prototype testing. Data/information from exploration are placed in a data bank (computerized and hard copy) for subsequent processing.

These programs have three general, interrelated purposes: (1) to supply site-specific subsurface information to meet all the identified facility design needs (geotechnical and non-geotechnical) segment-by-segment along the entire alignment, (2) to provide data for statistical characterization of landforms for the purpose of minimizing the amount of site-specific data otherwise required for item (1) to meet the facility performance reliability criteria, and (3) to provide necessary information for the identification and mapping of route terrain features (landforms and associated details) used in route characterization (particular emphasis must be placed on delineating boundaries and transitions between terrain features of engineering significance). Programs should be designed and executed to optimize these purposes in combination in a way that cost-effectively supplies the data needed to meet facility design needs, including performance reliability, cost and time considerations.

Data Base Development and Management

User-oriented computer systems are necessary where very large amounts of geotechnical data must be stored, accessed and processed in a cost-effective and timely manner. This element represents the computer hardware and software tools for route characterization and analysis, including: geographic information system (GIS) data base development, management and processing systems, remotely sensed image processing, interactive graphics, and interactive mathematical analyses (statistical data processing and engineering modeling) software. These tools, which have been and are continuing to undergo significant, rapid development, must be effectively integrated into a working system which supports the geotechnical and non-geotechnical needs.

Route Selection

The first stage of route characterization is the initial identification of geotechnical conditions along candidate route(s) for the purpose of route selection. This is done using terrain analysis techniques to identify visible features of the ground surface that can be correlated with geotechnical behavior and subsurface conditions. A geotechnical model of proposed or candidate route(s) can be developed by mapping observed landforms and associated terrain features in plan and profile along the entire route(s) or corridor. In profile, the route centerline can be represented by a "landform profile" showing the landforms occurring from the ground surface to a limiting depth (beyond which there is generally little affect of soil or rock conditions on facility performance); in plan, "terrain units" can be used to map landform profiles along a route corridor (or any area). Landform identification provides a basis for making general engineering predictions of geotechnical behavior through correlation between landform type and geotechnical properties. Given the landforms and terrain features occurring along the proposed route(s), general evaluations and predictions (qualitative and statistical) of engineering behavior can be made and compared against the geotechnical route selection criteria to ascertain the most effective route (after considering all higher level non-geotechnical project/facility constraints).

Segment-By-Segment Landform Characterization

Subsurface conditions inferred by terrain analysis and using available information from field investigations and laboratory testing programs form the principal data source for segment-by-segment landform identification and characterization along a route.

After initial route selection, terrain analysis techniques and available data are used to improve and refine maps of landform types along the route. Kreig and Reger [2] elucidate many of the pertinent considerations in the development of terrain unit maps and landform-type profiles for the Trans-Alaska Pipeline System (TAPS) project. For the TAPS project, the terrain unit/landform profile concept offered qualitative guidance to the interpretation and extrapolation of borehole data and proved a highly useful tool for: allocating exploration funding and efforts; selecting pipe construction modes along the alignment; locating material sources and disposal sites; evaluating reroute alternatives; evaluating slope stability; conducting resistivity studies for establishing cathodic protection procedures; determining workpad thicknesses; earthwork materials management; estimating ditching and pile installation-drilling rates for construction scheduling; establishing oil spill contingency plans; and establishing erosion control procedures.

Landform data for route characterization is available at three geographic scales; in order of decreasing size or areal extent but increasing specificity these are (1) for all occurrences of a landform or landform group (of similar landforms), (2) for occurrences of a landform or landform group in a particular local region or physiographic province or subprovince, and (3) for segment/site-specific occurrences of the landform in the region. The segment/site-specific scale can be subdivided into subsegment/sites as desired to the meaningful limit supported by available data. The quantity of available data for landform characterization follows the geographic scale order, decreasing from (1) all occurrences, to (2) local occurrences, to (3) segment/site-specific occurrences.

In a typical design segment data will be scarce; often, based only on segment data, a meaningful characterization cannot be achieved. The working principal for predicting landform soil properties wherever the landform occurs along the route is to utilize all available representative data obtained for the given landform. Combining landform data from different scales and locations requires data "updating" techniques. Bayesian updating techniques [3,4] can be used to provide a rational and systematic basis for combining landform data obtained from different scales and locations (geographic updating).

Each level of characterization has certain uses and limitations, depending on landform property variability and design requirements. Route characterization based on landform identification only, accomplished with the least amount of total effort and/or available data, may be useful for feasibility studies including route selection. However, unless landform property distributions are not location dependent, it will not support final design for a geotechnically sensitive facility without significant risk of multiple failures or, alternatively, compensatingly high levels of design conservatism--although, for a geotechnically tolerant or high risk facility, it may be adequate. Characterization based on landform and region may support preliminary stages of design for geotechnically sensitive facilities in variable terrain or may be adequate for final design in more uniform terrain and/or for certain geotechnically tolerant facilities. Final design of geotechnically sensitive facilities in statistically heterogeneous terrain requires more detailed and conservative segment/site-specific route characterization.

In short, the geographic specificity/conservatism in the route characterization required to support a given level of performance reliability must become greater in proportion to (1) the variability of the landforms, and (2) the geotechnical sensitivity of the design. A facility having a route traversing both variable and uniform landforms and having a suite of design modes available (each having greater or lesser geotechnical sensitivity or performance capability), can be designed such that the geographic specificity/conservatism of the characterization and the design modes are matched (on a mile-by-mile basis) with the terrain conditions to achieve the most cost-effective mix meeting project requirements [5]. The matching process requires mile-by-mile design decisions based on multi-technical, economic, and schedule considerations. In all cases, the principals of route characterization remain the same.

Segment-by-Segment Soil Profile Development

Soil profiles are engineering constructs used to idealize soil stratigraphy and associated material properties within a specific site area or segment length for engineering analysis and design purposes. Segment-by-segment soil profile development conceptually starts with identification of landform profiles, which are then refined as needed using available site-specific data [3,4,5].

A segment is defined as a reach of alignment which can be adequately characterized for a given (set of or single) engineering purpose(s) by a single soil profile. Each point of a route can belong to more than one soil profile (and associated segment) where significantly different characterizations are required for the various specific analysis and design purposes--e.g., within the limits of available data, segment-by-segment soil profiles characterized for slope/embankment stability may be different than soil profiles characterized for frost heave, bearing capacity, or erosion, etc.

The primary objective of site/segment soil profile development is to adequately represent--as simply as possible--the pertinent geological and geotechnical details that could potentially affect the conditions or structures of concern. Soil profile detail has practical meaning only if it influences geotechnical behavior and facility performance; unnecessary details include refinements beyond the resolution of analysis or judgement to predict meaningful differences in geotechnical responses or, moreover, refinements which either do not or could not affect design. Structures sensitive to geotechnical behavior tend to need more detailed soil profiles than less sensitive structures, for a given level of acceptable risk (the trade-offs are between design economies and economies of greater geotechnical exploration and analysis). Soil profile development is dependent on: (1) engineering behavior(s) of concern and design performance sensitivity(s), (2) prediction capability(s), (3) available data/information, and (4) man power/budget/schedule constraints. For a given level of acceptable risk, conservatism must increase in proportion to the geotechnical variability, data limitations, and design performance sensitivity of the facility to nonconservative error in the geotechnical input data.

GEOTECHNICAL ANALYSIS AND DESIGN

Geotechnical analysis and design efforts required to achieve facility performance/reliability criteria consists of two interrelated areas: (1) Analysis necessary for evaluating geotechnical behavior and performance of the route and its interaction with facility design, construction and operation, including: settlement, heave, slope/embankment stability, liquefaction, faulting, erosion, soil/structure static-, dynamic- and earthquake-induced interactions of all kinds (mechanical and thermal) and (2) Design product development and support for: pipeline and related facilities foundation designs; hazard control (mitigation and elimination) design recommendations; construction scheduling, techniques, and control of hazards; and earthwork suitability and design, including grading and excavations. Geotechnical analysis and design is accomplished first on a preliminary design basis and then on a segment-by-segment, final design basis; it is an iterative process of successive refinement (as are all elements of the system) integrated with the allied, non-geotechnical aspects of the facility design process.

Preliminary geotechnical design/response analyses are conservatively developed based on generalized characterizations of pertinent conditions influencing facility design (e.g., as conceptually outlined in [1] for frost heave). In this way, a limited number of analyses can be used to assess the route on a segment-by-segment basis. Segments found to be of potential concern are subject to detailed re-analysis. Preliminary design analyses are developed by: (1) examining the state of conditions along the alignment pertinent to design (topography, geotechnical mechanical and thermal properties, groundwater conditions, permafrost states and ground ice conditions, climatic conditions, design earthquake ground motion parameters, and design configurations including construction zone geometry, etc.); (2) grouping and characterizing similar conditions into a limited number of representative configurations or "typicals" based on conservative, idealized conditions; (3) analyzing typicals for structural/mechanical and thermal responses and geotechnical/structural interactions, including construction phase and operations phase affects; (4) applying the results of (3) to preliminary facility design.

Final design includes detailed engineering of those site-specific sections of the alignment requiring further analytical or field qualification or case-specific design solutions to provide acceptable performance and reliability. The most appropriate design solution (e.g., pipe mode design change, construction zone change, reroutes or minor alignment adjustments, or mitigative measures) will vary from segment to segment depending on geotechnical factors, design and construction considerations, including environmental and drainage concerns. The selection of the most appropriate design solution and the details of the solution is an iterative process: a trial solution is chosen, one which is also satisfactory to the non-geotechnical considerations, and analyzed for acceptability. The cost-effective solution which provides satisfactory performance for the particular segment of concern is the basis of selection.

Design solutions based on mitigation may be used to achieve satisfactory performance wherever changes in design mode/construction zone geometry or alignment cannot be implemented or are excessively costly. Mitigation techniques can include: pipe support below hazardous soils; insulation to minimize thaw degradation or frost heave; above ground support; grading to flatten slopes; control of subsurface drainage conditions; soil densification; excavation and replacement of hazardous soil; grouting or chemical stabilization; ground freezing; slope buttressing; construction scheduling; use of overfill in thaw subsidence areas.

Each segment analysis and design basis should be documented so that during subsequent design, construction, and operation, documentation can be reviewed as appropriate to see if a significant changed condition has occurred which could require segment redesign. Changed conditions can be associated with changes in project criteria, design methodology, data base, design concepts, alignment, construction zone geometry, pipe mode, or construction procedures and scheduling.

The role of synthesis in the integrated segment-by-segment geotechnical characterization, analysis and design process is essential in arctic and sub-arctic regions, particularly where large diameter pipelines are involved. Analyses must be examined in overview, segment-by-segment, integrating the various "parts" of the process, in-total and in-light of the engineering and geological experience, understandings, analytical techniques, judgements and assumptions used in analysis and design, checking that: all pertinent factors have been identified and adequately considered; assumptions are reasonable for each analysis; results of each analysis are consistent and reasonable; and after being integrated, results and recommendations are reasonable, and compatible with other segment results. The elements of the process are iterated as needed to accomplish these objectives.

Results of the various geotechnical efforts are combined with non-geotechnical design information in the design process to produce an acceptable facility design. The necessary design integration of geotechnical and non-geotechnical information to formulate an acceptable project design is also an iterative, interactive process--one which requires ongoing geotechnical engineering involvement, including necessary re-assessment of pertinent conditions as the facility design evolves and progresses.

APPENDIX I.--REFERENCES

1. Davison, Bruce E., Nottingham, Dennis, Rooney, James W., and Vita, Charles L., "Chilled Pipeline Frost Heave Mitigation Concepts," Proceedings, Pipelines in Adverse Environments (Vol. I), ASCE, New Orleans, LA, January 15-17, 1979.
2. Kreig, Raymond A. and Reger, Richard D., "Preconstruction Terrain Evaluation for the Trans-Alaska Pipeline Project," Geomorphology and Engineering, Donald R. Coates, Ed., Dowden, Hutchinson, and Ross, 1976, pp. 55-76.
3. Vita, Charles L., "A Landform-Based Probabilistic Methodology for Site Characterization," Proceedings Nineteenth Annual Symposium on Engineering Geology and Soils Engineering, Idaho State University, April, 1982.
4. Vita, Charles L., "The Systematic Use of Geotechnical Data for Characterization of Transportation Routes," Proceedings, Twentieth Annual Symposium on Engineering Geology and Soils Engineering, Idaho Department of Transportation, April 6-8, 1983.
5. Vita, Charles L., "Geotechnical Characterization and Analysis of Transportation Facility Routes," R&M Consultants, Inc., unpublished manuscript (in progress).

Thaw Settlement Analysis
For Buried Pipelines in Permafrost

Kenneth J. Nyman[1], A.M. ASCE

Within the past decade, analyses for soil support and stress in buried pipelines has evolved from simple approximations using closed form solutions to use of sophisticated analytical methods which incorporate plastic analysis of steel and nonlinear soil behavior. Reasons for this relatively quick evolution in pipe analysis methodology are primarily due to more stringent restrictions to license projects and the need to develop pipelines in adverse environments. More specifically, with regard to the nuclear power industry, buried cooling water circulation lines are analyzed and designed to maintain structural integrity under soil loadings associated with earthquakes; marine pipelines are designed to withstand mud slides; and buried arctic pipelines are designed for adequate restraint against thermal pipeline loadings as well as to maintain integrity against thaw settlement and frost heave. This paper focuses upon determination of allowable thaw settlement for the design of cross country pipelines in permafrost areas. The purpose of the thaw settlement evaluation is to aid in the selection of the proper pipeline construction mode which will ensure pipeline integrity and minimize impact on the environment and adjacent structures.

---

[1]Senior Geotechnical Engineer, ARCO Pipeline Co., Alaskan Division, 515 South Flower Street (BOA-3222), Los Angeles, CA 90071

For cross country design, the potential thaw settlement of each pipeline segment is calculated in accordance with the project design criteria and is compared against the allowable design values. The appropriate construction mode to mitigate adverse settlement is selected for segments where the potential thaw settlement exceeds the allowable design values. Based on soils information obtained from investigations along the pipeline route, geothermal-geotechnical analyses are performed to predict thaw depth over the design life of the system and resulting cumulative thaw settlement potential. The effectiveness of this method is dependent upon a well defined soil properties data base along the alignment. Design values for allowable thaw settlement are determined from stress analysis for appropriate combinations of pipeline settlement configurations and soil conditions.

It is anticipated that pipeline settlements can occur at transitions to and within permafrost areas where the thaw front penetrates supporting soils below the pipe bottom. The amount of soil settlement (below the pipe) is controlled by the thaw depth and thaw strain properties of the soil contained within the thaw zone below the pipe. Disruption of the natural heat balance between the atmosphere and the ground caused by construction disturbance, application of a gravel workpad at the ground surface, and surface water infiltration into the ditch will increase the thaw depth beyond the seasonal active layer. In addition, buried pipelines operating at temperatures above the initial permafrost ground temperature will induce additional thaw penetration below the pipe bottom during the operating life of the system.

Although buried hot oil pipelines have an inherent thaw settlement potential due to the positive heat flux radiating from the pipe, chilled gas pipelines also have a thaw settlement potential if the negative heat flux from the pipe is not sufficient to counteract the effects of surface disturbance from construction and surface water accumulation within the ditch profile.

Surface disturbance, thaw strain properties of the permafrost and water convection may vary along a design segment thus resulting in a potential for differential or non-uniform settlement of the supporting soils below the pipe. The differential settlements should not exceed allowable limits associated with overstressing of the pipeline segments. Allowable pipeline settlements can be evaluated using numerical techniques to solve the differential thaw settlement soil-structure interaction problem.

DIFFERENTIAL THAW SETTLEMENT CONFIGURATIONS

Differential thaw settlement configurations are classified as Transition Zones or Span Zones. A "transition" is defined as the boundary of a thaw settlement zone. For long settlement zones the maximum pipe stresses are located in the vicinity of the transition boundary. As the settlement zone decreases in length, the pipe stiffness plays a more active role in transmitting overburden load to the transition zone, relocating maximum pipe stress toward the center of the "span" across the settlement zone. For both cases, differential thaw settlement is promoted by a non-uniform thaw depth and/or variation in thaw settlement soil properties.

Possible differential thaw settlement configurations for Transition Zones and Span Zones are presented in Figure 1 through Figure 3 and are briefly described as follows:

Figure 1 — Transition Zones

a) Non-Uniform Thaw Front

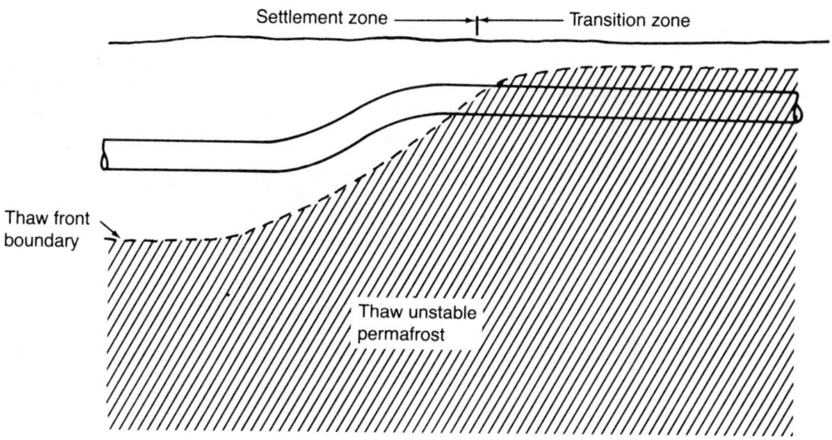

b) Non-Uniform Thaw Settlement Soil Properties

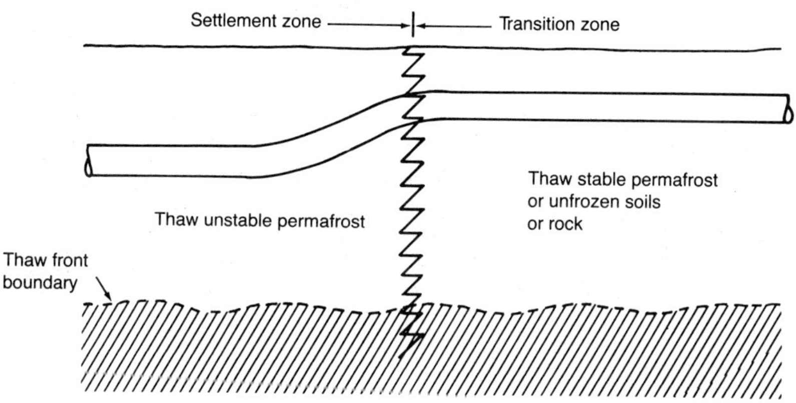

Figure 2 — Span Zones (Concave)

a) Non-Uniform Thaw Front

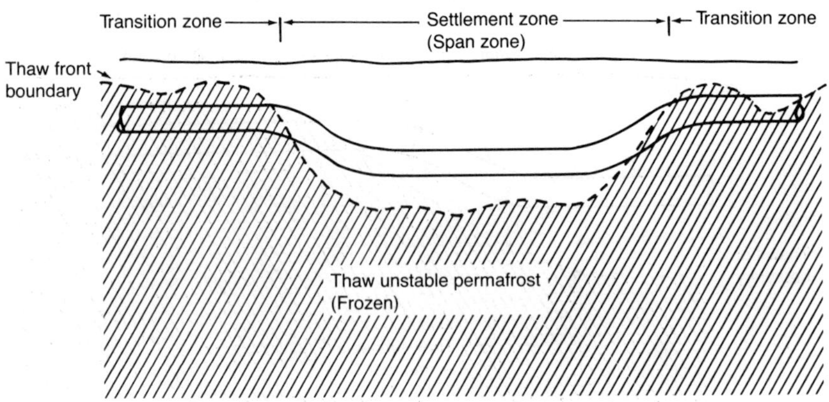

b) Non-Uniform Thaw Settlement Soil Properties

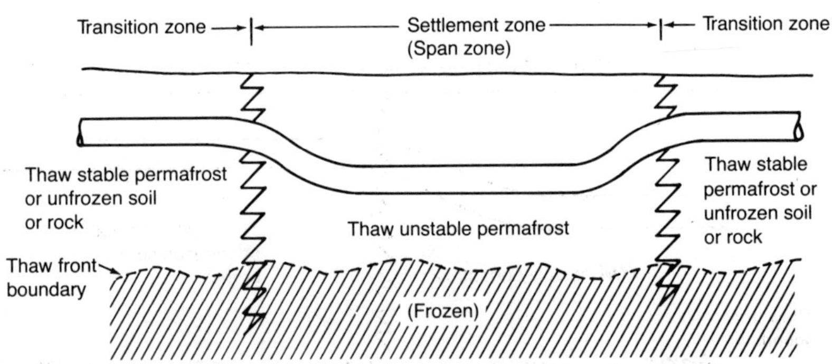

Figure 3 — Span Zones (Convex)

a) Non-Uniform Thaw Front

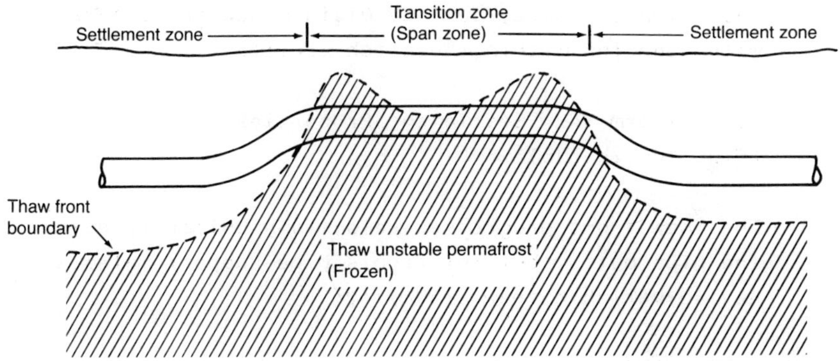

b) Non-Uniform Thaw Settlement Soil Properties

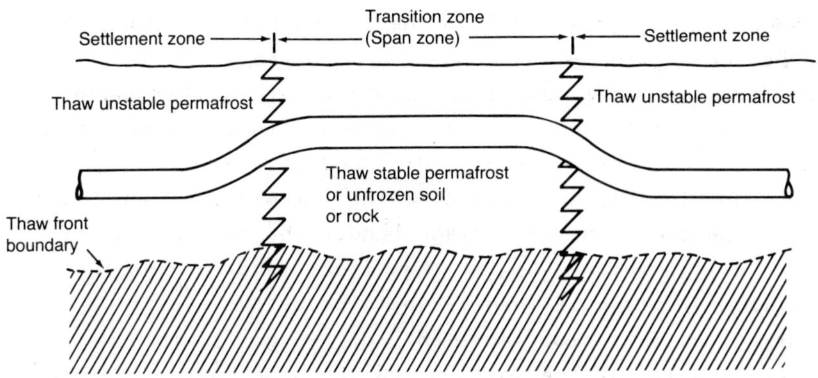

Non-Uniform Thaw Depth (Fig. 1a, 2a, 3a)

The thaw front varies at a shallow depth along the pipe profile. This may be promoted by natural variation in soil heat transfer properties or radiation exposure, surface disturbance or surface water ponding within the ditch profile.

Non-Uniform Thaw Settlement Properties (Fig. 1b, 2b, 3b)

The thaw front progresses relatively uniformly below the pipe bottom but intercepts pipe support materials having variable thaw stability.

THAW SETTLEMENT MECHANISM

It is necessary to understand the thaw settlement mechanism in order to evaluate the appropriate soil load-displacement functions required as input for stress analysis. A schematic representation of a non-uniform thaw front progressing from the ground surface along a pipeline segment will be used to illustrate the differential thaw settlement mechanism (Fig. 4 through Fig. 7). As the thaw front progresses to the pipe bottom during time increments dt_1, dt_2, and dt_3 (Fig. 4 through Fig. 6) the soils contained within the thaw front envelope consolidate in stages under the self weight of the in-situ soils. As the thaw front moves beyond the horizontal plane defined by the pipe bottom elevation during the next time increment, dt_4 (Fig. 7), thaw consolidation of the soils below the pipe bottom promote a loss of support for both the pipe and overlaying soils. This produces a net downward load on the pipe due to the weight of the pipe, and forces developed from overburden soils acting on the surface of the pipe. As a result, the pipe and overburden soils tend to move downward to conform

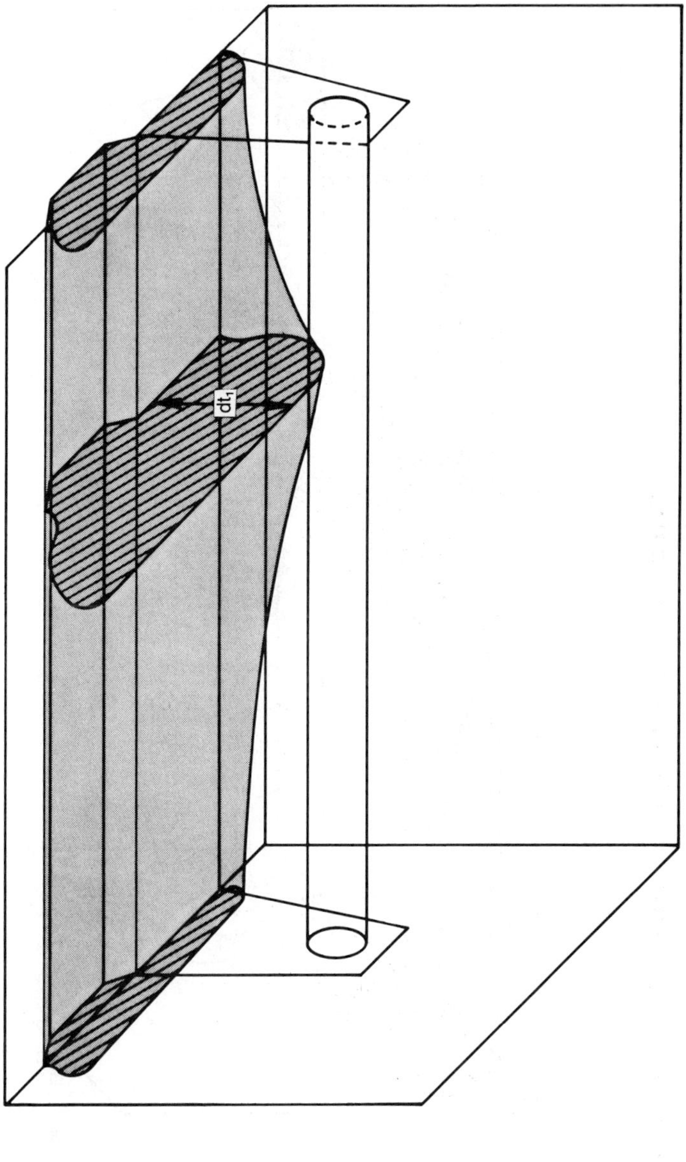

Figure 4 — Non-Uniform Thaw Front — Time Increment = dt_1

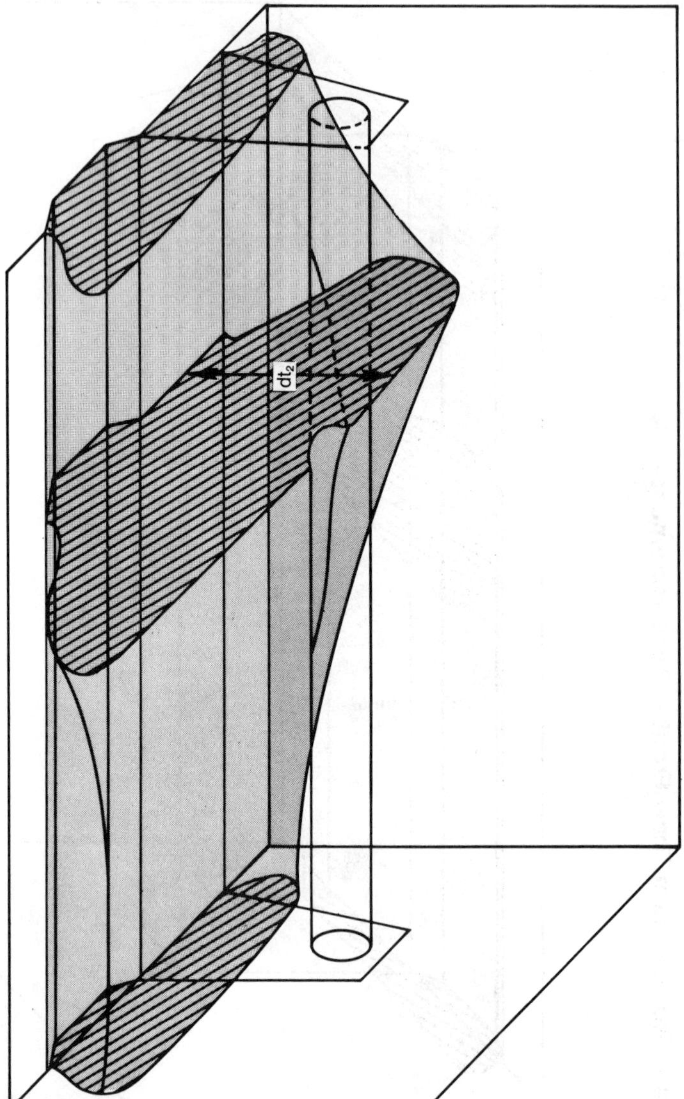

Figure 5 — Non-Uniform Thaw Front — Time Increment = dt_2

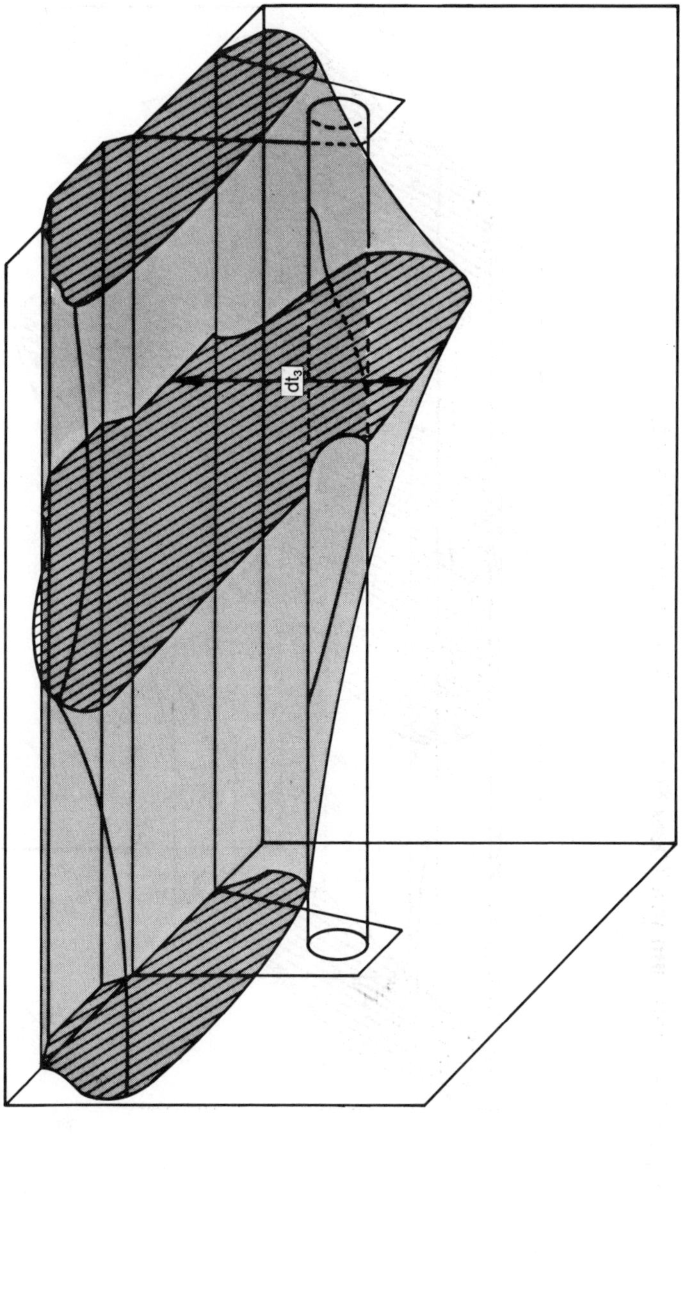

Figure 6 — Non-Uniform Thaw Front — Time Increment = dt_3

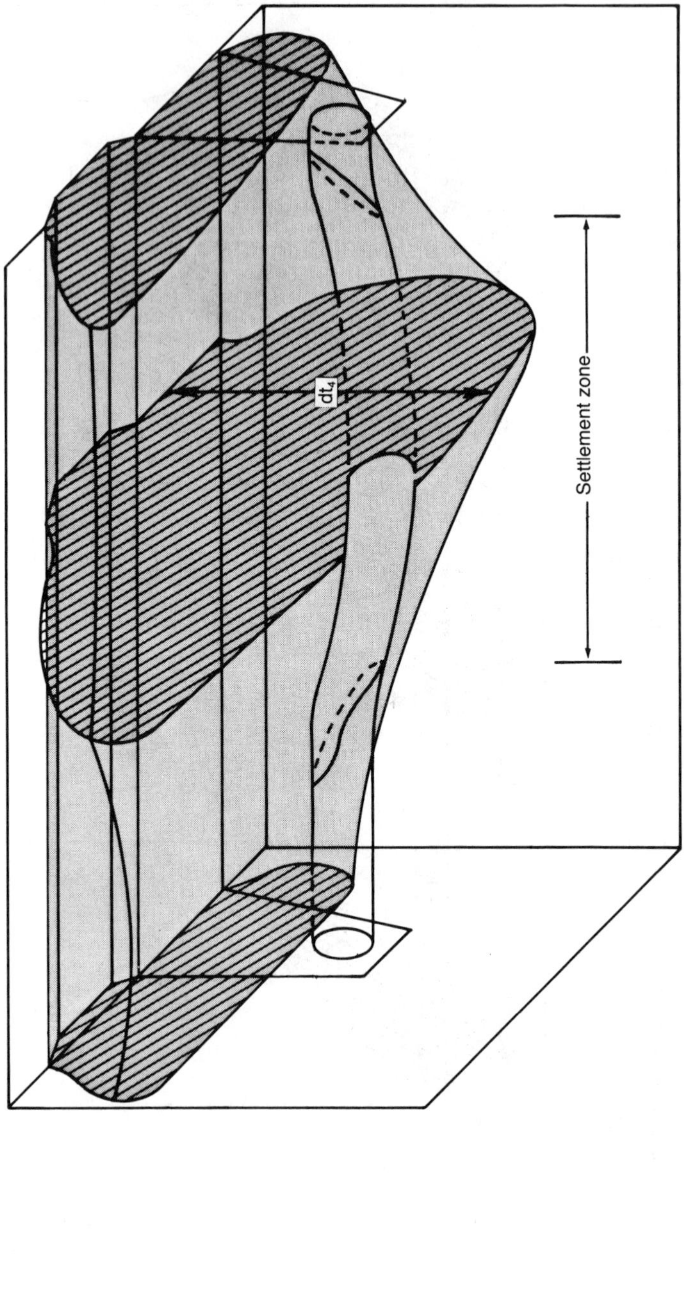

Figure 7 — Non-Uniform Thaw Front — Time Increment = dt_4

with the settlement profile. The downward thrust developed against the pipe by the overburden soils within the settlement zone is illustrated in Fig. 8 and is similar to the soil-pipeline interaction model proposed by Marston for an embankment constructed over a positive projecting conduit (9). The thaw front at time increment dt_1 (Fig. 8a) has not intercepted the pipe profile resulting in no unbalanced forces on the pipe from thaw consolidation of the overburden soils. The vertical in-situ soil stresses at the pipe location are probably distributed away from the pipe surface due to the presence of a frozen "soil arch" around the pipe. During time increment dt_2 the thaw consolidated soil mass from increment dt_1 displaces downward, δ_{soil}, due to thaw consolidation of the frozen soils encompassed by thaw front increments dt_1 and dt_2 (Fig. 8b). As the thaw front intercepts the pipe profile, the effects of a frozen arch around the pipe diminish, and the weight of the overburden soil, W, is applied to the pipe surface. In addition, shearing stresses, s, occur along vertical planes developing above the pipe. This is caused by variation relative settlements of the consolidating soils directly above the pipe, and soils adjacent to the sides of the pipe. Similarly, as the thaw front approaches the pipe bottom during time increment dt_3 (Fig. 8c), the pipe is subjected to the weight of overburden soils and the downdrag force mobilized by soils moving downward relative to the pipe (downward pipe movement is negligible since the pipe is supported by frozen soils at the end of time increment dt_3). During time increment dt_4, the soils beneath the pipe thaw consolidate under the soil forces developed on the surface of the pipe , and the pipe weight (Fig. 8d). As a result, the consolidated soil mass from previous time steps and the pipe contained within the soil mass move downward. As resistance to pipe settlement develops as soil bearing capacity within the transition

Figure 8 — Soil/Pipe Interaction Resulting from Progressing Thaw Front

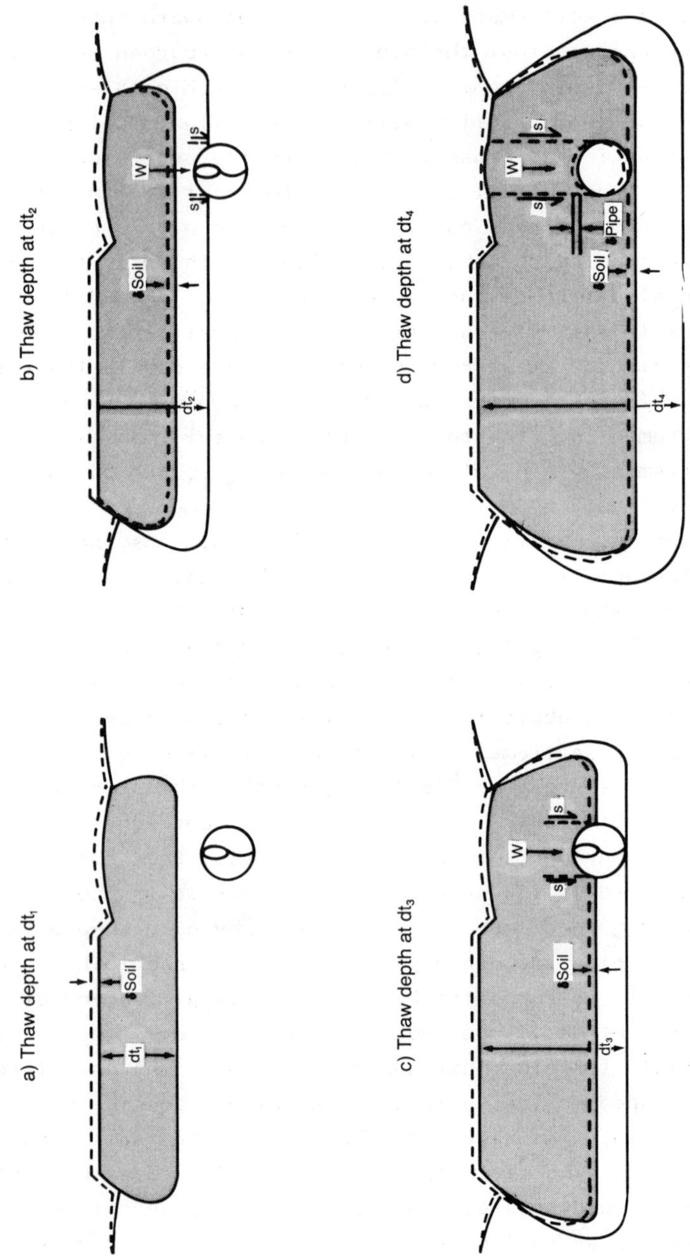

zone adjacent to the settlement zone, relative settlements between the pipe and surrounding soil mobilizes additional shear (downdrag) along the central prism of soil above the pipe. Large thaw settlements may mobilize ultimate uplift capacity of the prism as shown in Fig. 8d.

ANALYTIC MODEL FOR THAW SETTLEMENT

The beam-on-elastic foundation model used for pipeline thaw settlement is presented in Fig. 9. It is pointed out that this analytical model is "non-generic" in nature and could represent other soil-pipeline differential movement problems such as mud slides or fault crossings (1). The soil loading and support conditions are represented as "soil springs" attached to the pipe; each spring having a load-displacement relationship (P-Y) characteristic of the soil response for each direction of loading. The thaw settlement problem requires values of soil restraint associated with vertical-uplift, downward-bearing; and longitudinal-axial directions of pipeline movement. Thaw settlement of soils below the pipe can be simulated by imposing a uniform base displacement, δ_{soil}, of the springs within the settlement zone. The soil loading conditions are provided by the springs attached to the top of the pipe within the settlement zone, and represent the downward thrust equivalent to the uplift resistance of the soil overburden. Restraint against pipe settlement is provided by springs along the transition zones (axial, bearing, and uplift), and is potentially provided by soil springs beneath the pipe (bearing) within the settlement zone wherever the pipe conforms to the settlement profile and attains equilibrium against the consolidated soils.

Figure 9 — Analytic Thaw Settlement Model

a) Prior to Settlement Beneath Pipe

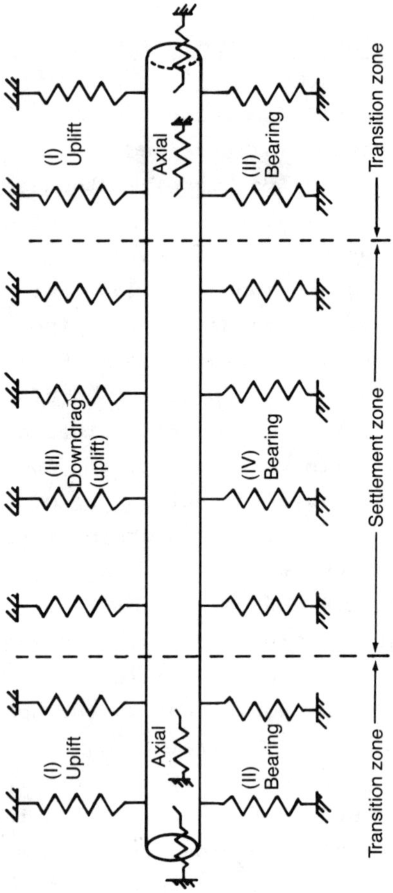

b) Following Settlement Beneath Pipe

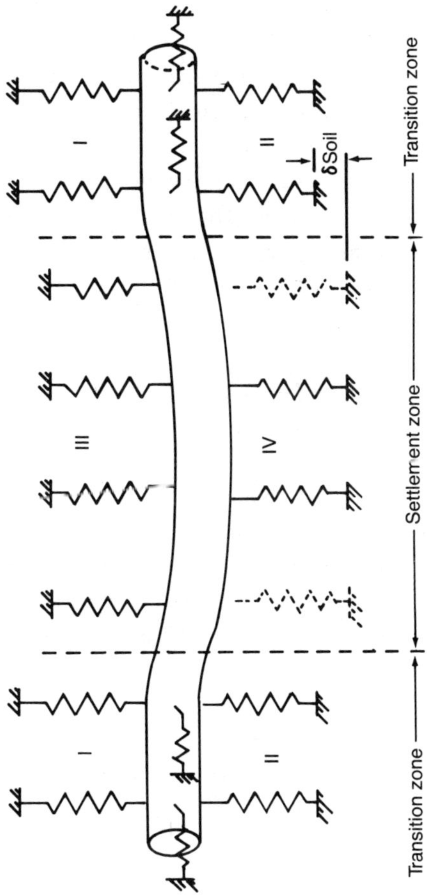

LOAD DISPLACEMENT FUNCTIONS

Load-displacement functions describing the behavior of the soil springs can be represented as bilinear approximations of the actual soil response against pipeline displacement similar to those used for the analysis for the restraint of bends (1, 2, 3, 5, 6, 7). The bilinear functions are typically defined by an ultimate value of soil restraint, P_u, and corresponding yield displacement, Y_u, as presented in Fig. 10. For the thaw settlement analysis, it is conservative to assume upper bound soil response for the soils acting downward on the pipe within the settlement Zone III and for soils below the pipe transition zone II (fig 9). Stiffer soils in these areas will produce higher pipe stresses during thaw settlement. Conversely, lower bound soil-response for soils beneath the pipe within the settlement zone IV will maximize pipe stresses during thaw settlement.

Expressions for ultimate soil restraint and yield displacement are obtained from published research information, and are based on results from prototype and model tests on buried circular pipe piles, anchors, and strip footings. The soils considered for pipeline thaw settlement analysis are frictional granular materials.

Thaw Settlement Zone: The soil overburden loading on the pipe within the settlement zone III can be modeled as two components of unfrozen-thaw consolidated soil response (Fig. 8d):

1) Weight of soil prism above the pipe.

2) Frictional downdrag forces acting on each side of the soil prism.

Figure 10 — Bilinear Representation of Soil Response

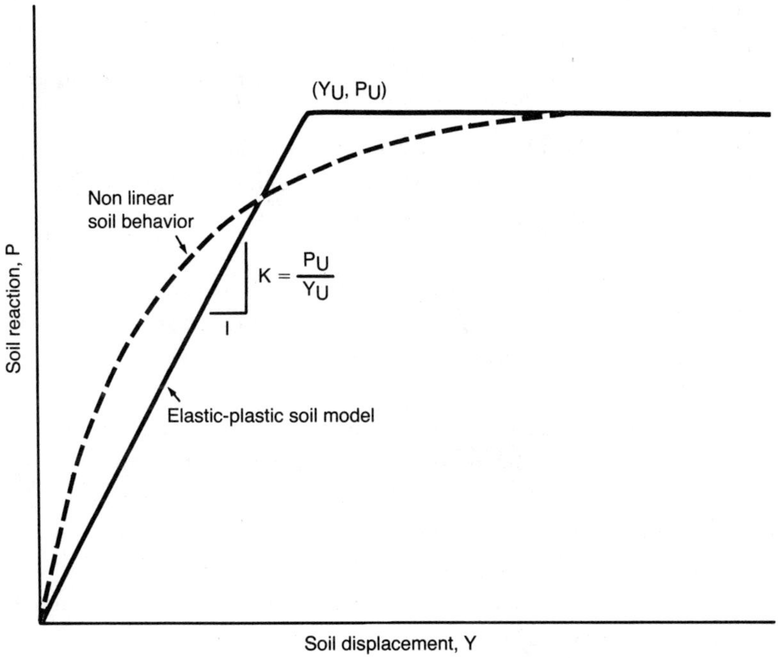

The soil response is similar to that associated with uplift resistance of horizontal anchors or the downward force of an embankment constructed over a positive projecting conduit. An estimate of ultimate resistance for unfrozen soils, P_u (uplift-unfrozen) can be calculated from the following limit equilibrium solution (5):

$$P_{u(uplift-unfrozen)} = \gamma DC + \gamma (C+.5D)^2 \, K_o \, \frac{4}{3} \, Tan\, \phi \qquad \text{(Eq. 1)}$$

Where:
- γ = effective unit weight of soil
- D = O.D. of pipe
- C = depth of cover from top of pipe to ground surface
- K_o = Coefficient of earth pressure at rest = 0.5
- ϕ = effective angle of internal friction for the soil above the pipe

The first component of Eq. 1, the soil weight, is developed against the pipe prior to thaw of soils beneath the pipe as illustrated earlier.

The yield displacement at which both components of uplift resistance are fully mobilized is dependent upon the depth of pipe burial. Investigators report that the yield displacement associated with pullout of horizontal anchor plates ranges from approximately 1 to 5 percent of the depth of cover. A value of 1 percent of the cover depth to the pipe springline is recommended for the thaw settlement analysis to mathematically represent an upper bound stiffness of the soils thrusting on the pipe.

$$Y_{u\ (uplift-unfrozen)} = 0.01\ (C+0.5D) \quad . \ . \ . \ . \ . \ . \ . \ . \quad \text{(Eq. 2)}$$

The overburden thrust given by Eq. 1 is valid for pipe embedment depths less than approximately six equivalent pipe diameters of soil cover over the pipe.

Potential support of the pipe within the settlement zone is provided by bearing capacity of the thaw consolidated soils beneath the pipe bottom, and can be estimated from the following equation (4):

$$P_{u\ (bearing\text{-}unfrozen)} = (C+D)\,\gamma D\,(Nq'-1) + 1/2\,\gamma D^2 N_\gamma' \quad \text{(Eq. 3)}$$

Where Nq', N_γ' = bearing capacity factors for local shear conditions (4).

The yield displacement required to develop ultimate bearing capacity is a function of the pipeline diameter and ranges between 5 to 20 percent of the effective bearing area of circular pipe. A value of 20 percent is recommended to represent a lower bound stiffness of soils beneath the pipe within the settlement zone:

$$Y_{u\ (bearing\text{-}unfrozen)} = 0.20D \quad . \ . \ . \ . \ . \ . \ . \ . \ . \ . \quad \text{(Eq. 4)}$$

The soils in the settlement zone contained within the thaw envelope are considered to be unfrozen thaw-consolidated materials. Soil properties used in Eq. 1 to represent soils between the pipe bottom and ground surface should be upper bound values consistent with those associated with the backfill materials.

Similarly, the soil parameters assigned to the unfrozen materials below the pipe bottom should represent lower bound thaw consolidated properties.

Transition Zone: The soils adjacent to the settlement zone provide axial, uplift, and bearing restraint against pipe settlement. These soils may be frozen or unfrozen, depending upon the configuration of the progressing thaw front. Frozen soils or rock support adjacent to the settlement zone will produce the maximum pipe stress condition.

The optimum support condition is where the soil is unfrozen above the pipe bottom overlying frozen soils or rock beneath the pipe. The nominal soil support condition is where the soils above and below the pipe are unfrozen thaw consolidated.

The axial load-displacement parameters for unfrozen soils have been well researched for buried circular columns such as piles or shafts. These load displacement relationships are applicable to buried horizontal pipes. The value for ultimate load transfer can be approximated from the following expression (1):

$$P_{u\ (axial\text{-}unfrozen)} = \frac{\pi D}{2} \gamma (C+0.5D)(1 + k_o) \tan\beta \quad \text{(Eq. 5)}$$

Where: β = angle of friction between the backfill and pipe coating

Based on past research, it has been commonly accepted that the maximum load transfer occurs at an ultimate displacement ranging from 0.1 to 0.2 inches.

$$Y_{u\ (axial\text{-}unfrozen)} = 0.1 \text{ to } 0.2 \text{ inches} \quad \ldots\ldots \quad \text{(Eq. 6)}$$

The load-displacement parameters for uplift resistance of unfrozen soils within the settlement zone III are applicable to unfrozen soils along the transition zone I (Eq. 1 and Eq. 2).

An upper bound estimate of bearing capacity for unfrozen soils is given from the following equation:

P_u (bearing-unfrozen) = $(C+D)\ D\ \gamma\ Nq + 1/2\ \gamma D^2 N_\gamma$ (Eq. 7)

Where: Nq, N_γ = bearing capacity factors for general shear conditions (4).

As stated previously, the yield displacement required to mobilize ultimate bearing capacity of unfrozen soils ranges from 5 to 20 percent of the pipe diameter. A value of 5 percent is recommended for the thaw settlement analysis to represent an upper bound stiffness of the bearing soils along the transition zone.

Y_u (bearing-unfrozen) = 0.05 D . . . . (Eq. 8)

Frozen soils and unfrozen rock beneath the pipe along the transition zone are assumed to have the same relative magnitude of restraint properties against short term loading, due to the high shear strength of these materials. The ultimate bearing capacity of these materials is estimated from the following equation adopted from classical soil mechanics theory:

P_u (bearing-frozen) = $N_c\ D\ S_u$ . . . . (Eq. 9)

Where: N_c = bearing capacity factor = 5
S_u = shear strength of rock or frozen soil
= 50 kip/ft^2 to 1000 kip/ft^2 (7)

The yield displacement required to mobilize ultimate bearing capacity of frozen soil or rock is estimated to be approximately 0.1 inch in accordance with correlations from laboratory stress-strain data proposed by Skempton (8):

Y_u (bearing-frozen or rock) = 0.1 in. . . . . . . (Eq. 10)

SOIL SUPPORT CONDITIONS FOR THE THAW SETTLEMENT ANALYSIS

Depending upon the most predominant thermal state or soil condition anticipated for the pipeline embedment, allowable thaw settlements may be evaluated for two soil support conditions along the transition zone:

Optimum Condition - Unfrozen soil above pipe bottom and frozen soils or rock below the pipe bottom.

Nominal Condition - Unfrozen soils above and below the pipe.

For all conditions, the soils within the thaw settlement zone are assumed to be unfrozen-thaw consolidated.

COMPUTATION OF ALLOWABLE THAW SETTLEMENT FOR CROSS COUNTRY DESIGN

A rigorous assessment of most pipeline restraint problems requires computerized solution techniques due to the complex nature of the soil-structure interaction mechanism. Numerous public domain programs which can mathematically model the pipe as a beam-on-elastic foundation can be utilized to achieve a satisfactory approximation for allowable settlement. However, specialized mathematical models for buried pipelines are available, and can simulate effects of differential pipe temperatures and internal operating pressures.

A symetrical pipe settlement profile along the soil settlement zone can be assumed as first approximation for the allowable settlement used for design. Due to symetry, only half the settlement span is modeled using

Figure 11 — Allowable Soil Settlement versus Settlement Zone Length

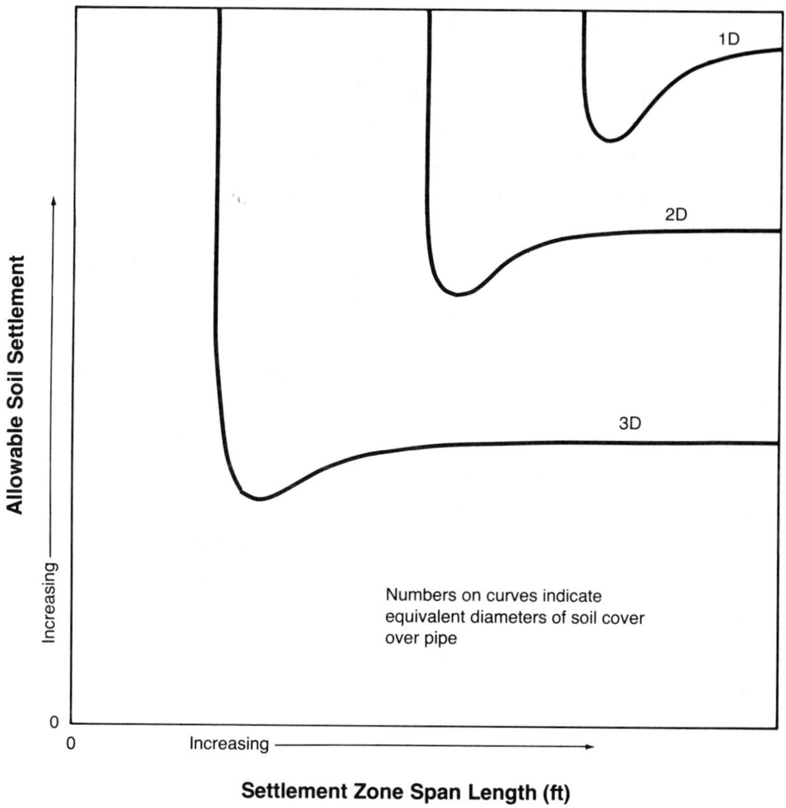

appropriate end restraints. Typical relationships between allowable soil settlement, settlement zone length and embedment depth for a pipeline subjected to abrupt settlement are illustrated in Fig. 11. For a given pipe embedment depth, the pipe is able to bridge "short" settlement zones (infinite soil settlement beneath pipe) and support the soil overburden without reaching the maximum bending stress criteria. The apparent stiffness against bending for "long" settlement zones diminishes, thus allowing the pipe to conform to the soil profile to achieve a consistent allowable settlement which is independent of the settlement zone length. For "intermediate" settlement zone lengths, the relative pipe span stiffness approaches a critical condition where adverse bending stresses are developed prior to bridging or conforming to the soil settlement profile. For this condition, the allowable settlement minimizes and is associated with the critical span for the pipe embedment geometry. The allowable settlement and critical settlement span length decrease with increasing pipe embedment depths. The allowable settlement associated with the critical span is used for design, since it should be assumed that the critical span could potentially occur along the pipeline alignment as the worst condition for thaw settlement.

REFERENCES

1) Audibert, J. M., LAI, N.W., Bea, R. G., "Design of Pipeline to Resist Sea Floor Instabilities and Hydrodynamic Forces", ASME Publication 78-PET-37, 1978

2) Audibert, J. M. E., Nyman, K. J., "Coefficients of Subgrade Reaction for the Design of Buried Piping", Proc., 2nd ASCE Specialty Conference, Structural Design of Nuclear Power Plant Facilities, New Orleans, Vol. IA, Dec. 1975, pages 109-141

3) Audibert, J. M. E., Nyman, K. J., "Soil Restraint Against the Horizontal Motion of Pipes", Journ., Geotechnical Eng. Div., ASCE, GT10, Oct. 1977, pages 1119-1142

4) Bowles, J. E., Foundation Analysis and Design, McGraw-Hill Book Co., 2nd Ed., 1977, p. 116

5) Luscher, V., Thomas, H. P., Maple, J. A., "Pipe-Soil Interaction Trans-Alaska Pipeline", Proc., ASCE Specialty Conference, Pipelines in Adverse Environments, New Orleans, 1979, Vol. 2, pages 436-502

6) Nyman, K. J., "Soil Response Against the Horizontal-Vertical Motion of Pipes", Preprint 82-534, Presented at ASCE Fall Convention, 1982, New Orleans

7) Nyman, K. J., Reese, L. C. "Field Load Tests of Instrumented Drilled Shafts in Coral Limestone", Masters Thesis, University of Texas at Austin, Dept. of Civil Engineering, May 1980

8) Skempton, A. W., "The Bearing Capacity of Clays", Proceedings, Building Research Congress, London 1951, pages 180-189

9) Spangler, M. G., Soil Engineering, International Textbook Co., 2nd Ed., 1963, pages 403-408

Pipeline Supported Structures on Driven Piles in Permafrost: State of the Art

Dennis Nottingham[1], M.ASCE and Alan B. Christopherson[2], M.ASCE

ABSTRACT: Pile driving techniques in permafrost have developed rapidly over the last 5 years; the driven pile method using a thermally modified drilled pilot hole now offers a fast and economical solution to many frozen foundation problems. Most types of piles, including H-shape, pipe, and sheet piles now can be driven, assuming that there is a knowledge of soil conditions and that correct procedures are followed. This paper explores the latest research, testing and production pile driving developments in Alaska. Current proposed pile design approaches will be discussed as related to short- and long-term loading (creep), frost jacking, and lateral loading. The scope of this paper is based largely on the authors' actual practical experience with approximately 5,000 piles driven in cold and warm permafrost soils as well as recent laboratory work and field testing experience on a large number of driven piles. The proposed criteria presented in this paper are primarily addressed to the practicing design engineer, including design and construction considerations. As more research and experience accumulate, factors presented in this paper may be subject to change.

Introduction

In the past, piles used as structural support in permafrost have consisted primarily of the drilled and slurry backfill type. Early pile driving efforts in Alaska by the U.S. Army Corps of Engineers, the State of Alaska, oil companies and others have had varying success. Although piles had been driven in permafrost, the methods used were not entirely reliable or economical. Conversely, development of slurry backfill pile placement has advanced through extensive use on the Trans-Alaska Pipeline and North Slope oilfield projects.

Slurry backfill piles require a number of operations, including large diameter drilling equipment, pile placement and support equipment, slurry preparation, haul and placement. Loads cannot be placed until freezeback occurs, which can result in lengthy delays.

---

[1]President, Peratrovich, Nottingham & Drage, Inc., 1506 West 36th Avenue, Suite 101, Anchorage, Alaska 99503

[2]Senior Engineer, Peratrovich, Nottingham & Drage, Inc., 1506 West 36th Avenue, Suite 101, Anchorage, Alaska 99503

Driven pile placement requires less equipment and support. When properly planned, installation placement rates of possibly 2 to 3 times the rate for slurry piles can be achieved and with a reduced crew size. Principal limitations of driven piles are location tolerances and present lack of refined driving equipment, although many good components do exist. Driven piles can be loaded after placement much sooner than slurry backfill piles.

Pile driving in permafrost using thermally modified pre-drilled pilot holes has now developed to the point where predictability, reliability, and economy make it a viable method in most applications. Current research efforts have helped define pile driving characteristics associated with the general soil types encountered. Contractors also are becoming aware of techniques and advantages, particularly after recent experience with more than 5,000 piles driven into permafrost on the North Slope and other locations in Alaska.

Arctic and sub-arctic soil conditions present unique foundation problems for pipeline design and construction that differ from the contiguous United States, primarily because of both continuous and intermittent permafrost. Intermittent permafrost regions are probably the most difficult from a design standpoint unless pipeline support methods are used which become less sensitive to in situ ground conditions. Examples of typical soil conditions which may be encountered in arctic and sub-arctic regions during pile driving are shown in Figure 1. Based on these examples, it is evident that driven pile foundations have significant advantage over drill and slurry pile installation methods, particularly where groundwater conditions present problems for the drill and slurry method.

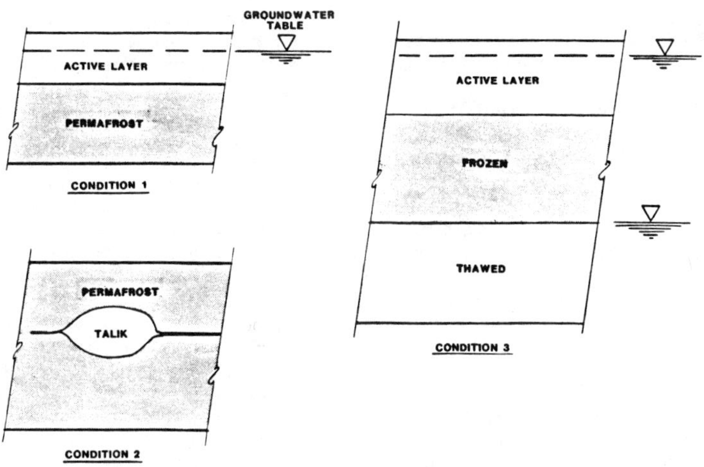

FIG. 1.—Typical Permafrost Ground Conditions

Also it appears that design problems can be best solved in some cases by the use of above-ground pipeline foundation systems specifically designed to attain economics by repetitive construction procedures, minimized field labor, and efficient component design which considers construction methods and tolerances.

This paper is directed primarily toward design engineers who must apply results of knowledge gained to date in a practical manner. A design concept is presented that uses short-term loading criteria to define maximum adfreeze limits under any condition including long-term, followed by long-term loading to establish pile adfreeze limits based on creep deflection. This approach clarifies a past area of confusion to many engineers concerning the appropriate values to use for long-term strength.

Driven Pile Placement Methods

Piles, including H-shape, pipe and sheet can be driven with impact and vibratory hammers and the more sophisticated sonic hammers (depending on soil conditions). Where extremely hard driving is encountered, an impact hammer and cast-steel pile driving tips are necessary. Experience has shown that even with relatively easy driving, pile tips should be used with an impact hammer to prevent tip damage. Pipe piles are particularly subject to tip ovaling and flattening during impact driving into pilot holes. Pile tips on pipe piles should be of the open, flush exterior type, and H-pile tips should be flush on the exterior. Vibratory hammers are particularly good in fine-grained saturated thawed soils or weak frozen soils, such as those produced by the thermally modified pilot hole method. They have been used for slow driving in warm frozen silts without the use of pilot holes. Vibratory hammers may have difficulty driving piles into cold dense frozen soils or where there is a predominance of coarse gravels and cobbles, or hard layers. Pile tips should be considered when installing vibratory driven piles into dense granular soil conditions with thermally modified pilot holes. Typical examples of pile tip accessories for different hammer types are shown in Figure 2.

It should be noted that piles made of mild steel (i.e. A36, A252 etc.) have not been observed to fracture while driving with impact hammers in extremely cold environments; however, they may fail in various modes from improper design or driving methods. To more clearly identify various hammer types suitable for use in cold weather and permafrost, the following discussions are presented.

Impact Hammers.-These hammers rely on falling mass to produce energy and have many forms and types. Practice in Alaska now centers primarily on diesels, with some air and hydraulic hammers also in use. Diesel hammers work well if kept warm, and if they meet with sufficient driving resistance to assure fuel ignition. When used with pilot hole thermal modification and short piles, driving is often too easy for efficient diesel hammer operation. Air hammers offer very controllable driving, but during cold weather periods line deicers or heaters may be needed to prevent freezeup. Hydraulic impact hammers are small, fast-hitting devices that offer tremendous potential for driving small piles at remote building foundation sites. Mounted on tracked vehicles, they are highly efficient and mobile machines. As with all hydraulic systems in cold weather, attention must be given to the use of proper fluids and keeping components warm and protected from the environment.

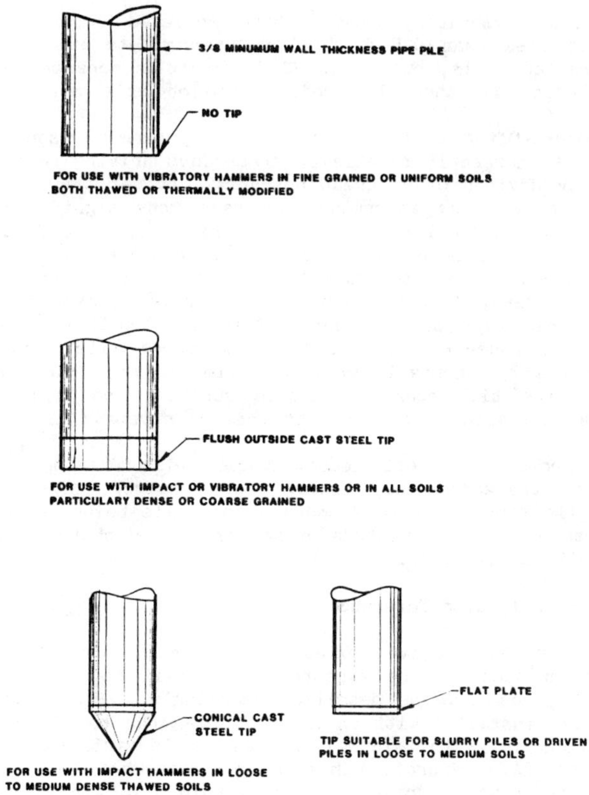

FIG. 2.—Pile Tip Accessories

Typical driving rates in permafrost for properly sized impact hammers are 1 ft (300 mm) per minute in warm fine-grained soils without a pilot hole, 1 ft (300 mm) per minute in dense warm granular soils with the use of a pilot hole, and up to 5 ft (1.5 m) per minute by use of thermally modified pilot holes in most soil types and temperatures.

Vibratory Hammers.-These hammers are either hydraulic or electric and operate on a principle which uses two counter-rotating eccentric weights. The largest vibratory hammers used on the North Slope typically have only the driving energy equivalent to a small impact hammer, and will perform the same should difficult driving be encountered. They are particularly good in fine-grained saturated soils or under conditions where soil particles can be easily displaced. As a result, vibratory hammers are highly efficient when used with thermally modified pilot holes.

Properly sized vibratory hammers have achieved driving rates in permafrost of less than 0.5 ft (150 mm) per minute at best in some warm fine-grained soils, but up to 20 ft (6 m) or more per minute in most soils when the thermally modified pilot hole method is used properly.

Sonic Hammers.-Often confused with vibratory hammers, sonic hammers and drills are inherently capable of tremendous driving rates. These high frequency devices offer great potential, but at the present time they are expensive, few in number and have many significant operational problems, particularly in cold weather. In most frozen fine-grained soils without pilot holes, driving rates comparable to vibratory hammers using thermally modified pilot holes have been achieved. To date, frozen granular soils have presented difficult driving for this type pile hammer. Thermally modified pilot holes have been used under these conditions to speed pile installation. Without the use of thermally modified pilot holes, voids have been noted around the pile near the ground surface, and typically pile embedment is increased to account for loss of strength in these upper sections.

Figure 3 shows a typical impact hammer pile driving operation, complete with tracked crane, leads, diesel hammer, pipe piles and cast-steel pile tips. Figure 4 shows a test pile being driven with a vibratory hammer. Both methods shown in these photos utilized a thermally modified pilot hole.

Driven Pile Installation Tolerances

Designers specifying driven piles must recognize that less precise placement tolerances may be expected over drill and slurry installations, and plans must be detailed accordingly. Horizontal tolerances of piles installed with an impact hammer can be ±2 in (±50 mm) with an extreme of ±3 in (±76 mm) in plan, while variation from plumb may be up to 2%. Vibratory hammers are somewhat better in this regard, and can usually drive piles to within a 0.5-in (12-mm) horizontal tolerance and virtually plumb with some effort. This is because piles can be vibrated up and down the thawed pilot hole until desired tolerances are achieved. Piles driven by impact hammers cannot be adjusted in this manner. An important factor in achieving specified design tolerances when using pilot holes is to drill an accurate plumb hole, since the pile may tend to follow hole alignment during driving. At times it may be desirable to drill the pilot hole 1-3 ft (0.5-1 m) deeper than the desired pile tip elevation, particularly if driving to close vertical tolerances.

Recent developmental work with driven piles in permafrost has shown that piles can be installed in almost any soils, but final placement tolerances, plumbness and plan location may not be achieved as easily as with drill and slurry placement methods. Construction tolerances are still however, within reach of sound structural design.

Pile/Cap Plate Connection.-This is probably the most important single aspect from a cost and method standpoint. Many of the problems experienced on previous jobs have directly resulted from labor contract stipulations and not necessarily from physical conditions.

The best method to connect steel components, from a cost, time and strength standpoint, is by welding. The question then arises when deciding where welding should be done--in the shop or field.

FIG. 3.—Impact Hammer Pile Driving Operation

For driven piles with less precise placement tolerances, field welded connections between the driven pile and cap plate are best if welding methods, such as automatic or semi-automatic, can be devised to quickly and efficiently perform the job. Not only is it necessary to technically solve this problem, but field labor must be organized to minimized the presence of non-essential personnel. Piles can be effectively driven with pre-welded cap plates or bolted attachment plates, but this creates the need for very close tolerance control, thus increasing production costs. Figure 5 presents different connection details.

Pile Cross Member.-The Trans-Alaska pipeline project and several North Slope pipeline projects have illustrated the extreme costs of above-ground support designs. Typically, pipeline support designs are of such size that full scale tests can be performed to supplement theory which may result in significant savings when thousands of components are involved. Conventional use of built-up wide flange beam sections may lead to extremely high welding costs and tend to dictate pile cap plate location tolerances. Cross beam shapes such as structural steel tubing and pipe require less than half the steel of a comparable wide flange. Other advantages include surface preparation and painting. The surface area requires a third less preparation and paint than wide flanges. Figures 6 and 7 show several schematic pipeline to cross member connection details.

FIG. 4.—Vibratory Hammer Pile Driving Test

To reduce potential accummulated soil and water pressures within driven pipe piles, placement of a small diameter weep hole just above final ground line elevation prior to driving is recommended. From the authors' experience, this hole need not be greater than 1 in (25 mm) in diameter. It has been noted on several driving jobs that water will spray out of these weep holes several feet as the pile nears grade; when a welded cap plate is attached to the pile.

Experimental Testing

Split-Spoon Frozen Soil Penetration.-Common methods of soil sampling often include standard penetration tests (SPT) utilizing a 2 in

(51 mm) outside diameter, 1.4 in (36 mm) inside diameter split spoon, with a 140-lb (64-Kg) hammer dropped 30 in (762 mm). Blow counts are recorded for 18 in (457 mm) in 6-inch (152-mm) increments. The blows per foot over the last 12 in (305 mm) are used in establishing relative density estimates (blows per foot). In the past, permafrost has been difficult to test due to high blow counts required to drive the sampler (typically 100+). Thus, little relationship could be observed between standard penetration tests and actual pile driving or soil conditions.

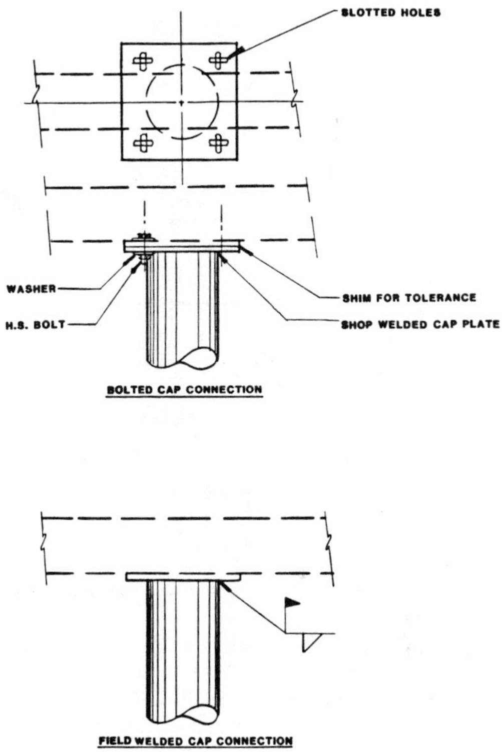

FIG. 5.—Pile/Cap Connections

From past experience, H-piles have been successfully driven in warm permafrost (usually silts) without the use of pilot holes. Driving rates have been on the order of 1 ft (305 mm) per minute. Similar rates have been achieved using pilot holes in warm sandy and gravelly soils. Warm permafrost is assumed as a general term for perennially frozen soils that are above 30.2°F (-1°C), yet remain in a bonded frozen state. Cold permafrost includes frozen bonded soils with a temperature lower than 30.2°F (-1°C).

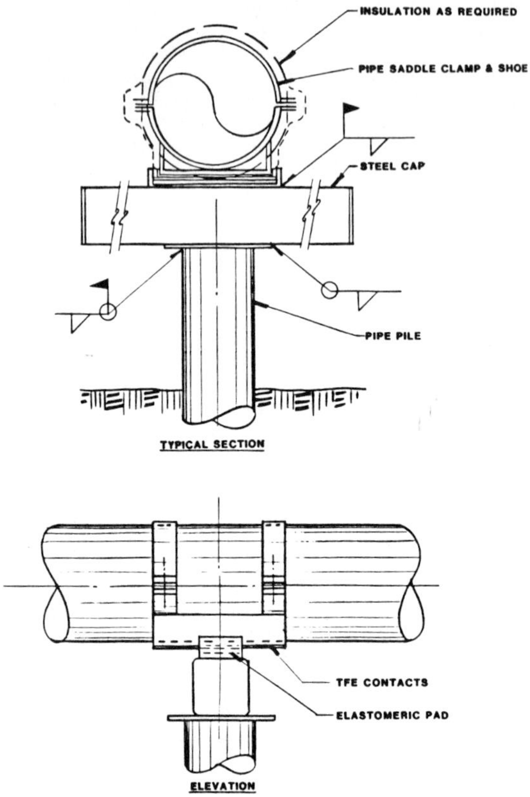

FIG. 6.—Typical Longitudinal Expansion Support

Driving refusal has been observed while attempting to drive piles into pilot holes drilled in cold gravelly soils, and driving has also been difficult in sandy silty cold permafrost. Many instances of severe pile damage have been recorded while attempting to drive into cold permafrost.

To establish a meaningful method of evaluating driving resistance in permafrost, the standard split-spoon driving test has been employed, but only for a 6-in (152-mm) penetration. In actual field work, blow counts for this penetration should be started only when it is apparent that the split spoon is seated in frozen soil and not penetrating loose cuttings.

Results from experiments (7) using SPT in frozen ground are shown in Figure 8. Analysis of these data and subjective comparison to past pile driving and soil sampling experience shows that suitably designed piles probably can be driven in fine-grained soils as cold as 26.6°F (-3°C) and in coarse-grained soils down to 30.2°F (-1°C), using pilot holes. Piles probably can not be driven efficiently without pilot holes in frozen soils much colder than 31.2-30.2°F (-0.5 to -1.0°C), depending on soil type. With this information, parameters are established for potential driven pile foundations in permafrost. Obviously, in permafrost colder than these temperatures, some method such as pilot hole thermal modification must be used to achieve suitable soil temperature in the immediate pile area during driving.

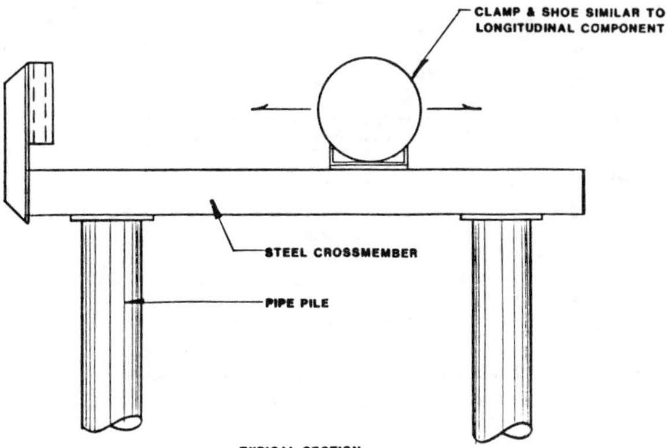

FIG. 7.—Typical Expansion Loop Slide Support

Pilot Hole Thermal Modification.-The technique of modifying permafrost temperature by use of a small pilot hole and non-circulated hot water allows easier pile driving where previously it seemed impossible. A pilot hole is a drilled hole generally extending at least to the desired pile tip depth. Pilot hole thermal modification has proven to be a more reliable and controllable method than other methods of thawing, including steaming. If used properly, it affects the in situ soil thermal regime significantly less than drill and slurry methods.

The small amount of controlled heat introduced into permafrost surrounding a pilot hole dissipates into the cold North Slope ground environment soon after driving, thus allowing soil freezeback to occur very quickly. The process of using hot water rather than steam (contrary to Russian and other efforts), results in a more uniform radial heat exchange along the length of the pilot hole, rather than spot thawing due to localized steaming. Two recent examples of contractors using steam thawing instead of hot water occurred at Bethel and near Prudhoe Bay, Alaska. At Bethel the use of steam thawing around sheet piles resulted in large cavernous voids and a larger thaw zone than necessary. This may be significant due to the warm thermally sensitive permafrost found in this region. At the offshore gravel island called Seal (Prudhoe Bay), steam thawing has been tried to improve the installation of a sheet pile cell around the perimeter of the island. The steam thawing resulted in significant ground settlements adjacent to the wall. These depressions were later filled with gravel at considerable expense.

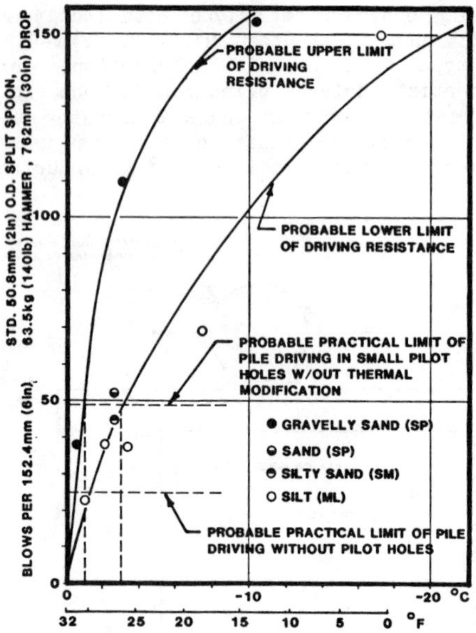

FIG. 8.—Split-Spoon Soil Sampling Resistance Versus Frozen Soil Temperature

Use of water-filled pilot holes in permafrost may have other important implications. Water offers a non-compressible media. When subject to shock, it tends to transfer vibrations, causing soil particles to temporarily loosen then densify against the pile, thereby promoting good adfreeze bond between the pile and soil. Water also fills all voids in the soil around the pile, thus assuring a strong pile/soil/ice bond. Regardless of the exact mechanism, water-filled pilot holes in general allow piles to be driven easily and improves placement accuracy.

Drilled pilot hole diameter can be determined for any size pile by equating the pilot hole diameter to the pile diameter and subtracting 2d. Where d is the desired isotherm distance from the edge of the pilot hole in inches (mm). For H-piles, an equivalent diameter slightly larger than the section depth may be appropriate. By this logic, a good starting point for pile driving in most soils would be as shown in Table 1:

Table 1.-Pile Type versus Pilot Hole Size

Pile Type	Pilot Hole Diameter
8 5/8 in (219 mm) dia.	4 in (102 mm)
10 3/4 in (273 mm) dia.	6 in (152 mm)
12 3/4 in (323 mm) dia.	6 in (152 mm)

Table 1.-Pile Type versus Pilot Hole Size (continued)

Pile Type	Pilot Hole Diameter
16 in (406 mm) dia.	8 in (203 mm)
18 in (457 mm) dia.	8-10 in (203-254 mm)
HP 10	4-6 in (102-152 mm)
HP 12	6 in (152 mm)
HP 14	6 in (152 mm)
Sheet piles or large diameter piles	4 in (102 mm) dia. @ 12-18 in (305-457 mm) O.C.

Using concepts taken from Nottingham (6) concerning early discoveries relating to pile driving with water-filled pilot holes, and curve fitting techniques, the following approximate relationship was established:

$$d = k(T)^{1/2} \qquad (1)$$

Where: d = isotherm distance from the pilot hole edge in inches (mm)
k = constant for various soil types and isotherm desired
T = time in minutes

Considering a 26.6°F (-3°C) isotherm for silty soils and a 30.2°F (-1°C) isotherm for gravelly soils, k will be approximately 0.3-0.5 for most conditions, using British units. For a thaw time of 60 minutes, d from Equation 1 will be approximately 2 to 4 in (50-101 mm).

In practice, vibratory pile driving tends to cause soil to be vibrated from the sides of the hole and be displaced to the pile tip. This will cause the pilot hole water to rise along or in the pile, depending on type and results in the following:

1. Soil refreezes near the pile tip faster because of lower heat requirements, as verified by field measurements.
2. The water-soil mixture created by driving action acts to effectively slurry the pile into place.

To date, field observations in 23°F (-5°C) soils indicate freezeback in less than 2 days and in 19.4°F (-7°C) soils in about 1 day. Structural strength for most load applications is achieved after this period. To the authors' knowledge, no significant frost jacking during the refreeze process has been observed or measured in the Prudhoe Bay area.

Based on laboratory tests and field experience, initial temperature of pilot hole water does not appear to be critical. Water temperatures between 60-80°F (15-27°C) may be suitable for warmer permafrost, while water temperatures between 150-212°F (65-100°C) appear to be appropriate for cold permafrost. In actual installations water has been placed in pilot holes 40-60 minutes prior to pile driving with good success.

Pile Loading Criteria

Presented on Figure 9 is an idealized mode of pile action in ice-rich permafrost under constant loading. After a load increment is applied, for a period of a few hours to a few days and depending on pile length, a load adjustment period will be required for stresses to be uniformly distributed over the pile surface. This period is often described in literature as primary creep. Steady state creep is of interest for the low long-term adfreeze stresses normally used in design, and is often referred to as secondary creep. For most structural applications, it is usually necessary to limit long-term creep of piles to less than 0.5-1 in (12.7-25 mm); thus pile failure within the tertiary creep region of Figure 9 is of less interest to the design engineer.

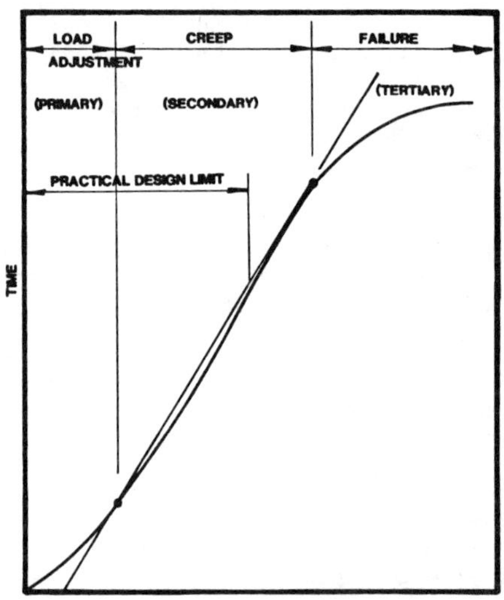

FIG. 9.—Idealized Modes of Pile Settlement in Ice-Rich Permafrost

Four specific conditions are of importance to the design engineer:

1. short-term vertical loading
2. long-term vertical loading
3. frost jacking loading
4. lateral loading

Short-Term Vertical Loading.-Short-term pile tests in cold permafrost have demonstrated tremendous adfreeze resistance values, but values rapidly decrease near 32°F (0°C). Short-term has been conservatively taken in this paper to be loads of generally less than 5 hours' duration. This loading group contains the following categories:

1. building live loads (other than permanent loads such as furniture, files, etc.)
2. wind loads
3. earthquake loads
4. moving vehicle loads
5. ice impact forces
6. other loads applied for short duration

By accurately assessing these loads, the design engineer can reduce pile lengths where short-term loads are a significant factor because of potentially higher allowable adfreeze stress, and where creep is not an important factor. Figure 10 depicts the authors' current recommended design adfreeze stresses with an approximate safety factor of 3 for the short-term condition. It is important to note that for driven piles the adfreeze strength of coarse-grained soils to steel is much lower than for fine-grained soils (11). The thermally modified pilot hole approach helps to assure a slurry bond; however, high allowable adfreeze stresses are not advisable in coarse-grained soils. Tests on slurry piles indicate little difference in strength compared to similar size driven piles in fine-grained soils, but slurry piles exhibit greater short-term strength in coarse-grained soils.

Design charts shown in this paper do not account for end bearing, which may be significant in some conditions, such as for gravel. However, end bearing is ignored in favor of a more conservative approach at this time.

When designing piles, engineers should disregard pile embedment in the active zone as contributing to pile strength.

The graphs used here do not account for saline soil conditions. The design engineer confronted with this situation should perform additional tests, which are beyond the scope of this paper.

Long-Term Vertical Loading.-Piles in permafrost are subject to creep-related settlement when loads are of a sustained nature. This cannot be overemphasized in designing significant structures such as water tanks, heavy machinery supports, or other critical structures. The phenomenon is not unlike creep, which engineers commonly consider for design of concrete or timber structures.

Conversely, failure of the designer to separate short-term loads from long-term loads may lead to uneconomical foundation solutions.

Long-term creep is sensitive primarily to these aspects:

1. sustained loading
2. soil temperature
3. pile diameter
4. soil type and moisture content
5. vibrations
6. adfreeze stress

Long-term deformation of ice and ice-rich soils may be approximately represented by steady-state creep where the flow law for ice provides the upper limit for ice-rich soil. Limited numbers of long-term tests on driven piles in various soil types and temperatures have produced a series of approximate creep rates for various adfreeze values (3).

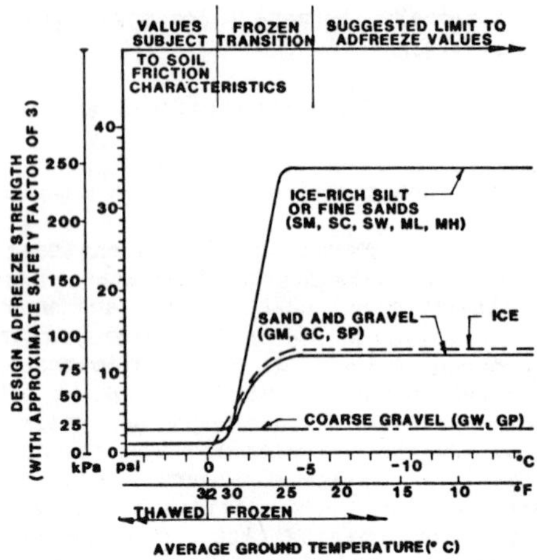

FIG. 10.—Proposed Maximum Short-Term Design Adfreeze to Steel Piles

Pile displacement rate is very dependent upon the applied shaft shear stress. However, the effect of temperature is also clearly important, with pile displacement rates changing by as much as one order of magnitude for soil temperature changes of a few degrees.

The effect of pile diameter is also important, as noted by Nixon and McRoberts (5), and Weaver (12). Increasing pile diameter appears to lower allowable stress for equal creep rates. In other words, under equal adfreeze stress, a large diameter pile will settle faster than a smaller diameter pile.

For piles supporting sustained loads long-term deformation is highly sensitive to adfreeze stress, and design loads must be held to low levels for supports which are critical with respect to deflection.

Figure 11 was constructed from the best data currently available for ice-rich soils. The chart is based mainly on soil temperature and pile size, and only generally considers soil type and moisture content. This chart can be used conservatively for most soil types, including granular soils, but pure ice will have greater creep rates. Soils with ice contents greater than 50% should be viewed with consideration given to potentially greater creep rates and reduced adfreeze strength.

As an example, the following design analysis is presented for an 18 inch diameter pile:

Where: U = allowable vertical pile movement in inches (mm) per year
a = pile radius or equivalent radius in inches (mm)

Average soil temperature = 31.3°F (-0.4°C)
Soil type = frozen sand

a = 18/2 = 9 in (457/2 = 229 mm)
U = 2 in (51 mm) maximum settlement in 30 years
= 0.07 in (1.7 mm) per year

then $U/a = 0.07/9\ (1.7/229) = 0.0074$ in (0.0074 mm)
$= 7.4 \times 10^{-3}$ year^{-1}

Then from Figure 11, allowable long term adfreeze = 2.5 psi (17.2 KPa).

A check for allowable short term adfreeze on Figure 10 indicates thawed strength conditions should be considered; however, 2.5 psi (17.2 KPa) appears safe.

As more research is done, data reflected in Figures 10 and 11 may be subject to change. At this time, until additional data becomes available, it appears that piles subject to vibratory loads may experience settlements at least double those developed under static loading at the same adfreeze value. For warm frozen soils, thawed soil strengths should be considered.

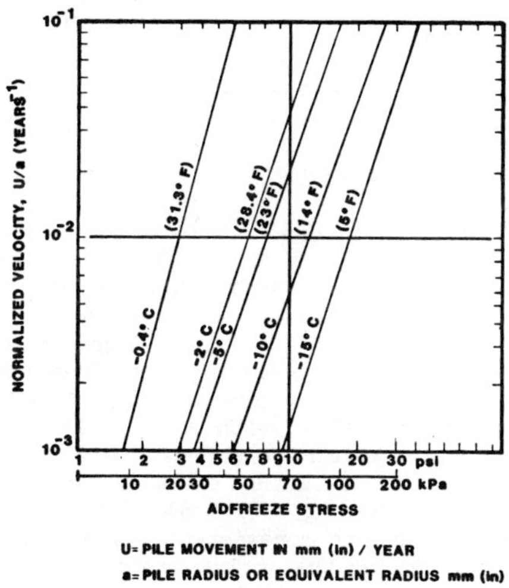

FIG. 11.—Proposed Design Pile Settlement for Ice-Rich Permafrost

Long-term creep calculations should consider average soil temperature over the pile length, excluding the active layer. Where average temperature varies with the season, it may be desirable to evaluate time increments to reflect these variations. Where unusual conditions such as heavily long-term loaded closely spaced piles are used, group creep action should be given special consideration; this aspect is beyond the scope of this paper.

Pile Frost Jacking.-Heave of structures and piles as a result of active layer freezing and ice lens formation, as illustrated on Figure 12, has been a reoccurring problem in northern regions. This figure must be studied in detail to visualize all the implications of frost heave. Often weak soils such as peat are common near the ground surface, which when frozen may exhibit low soil/pile adfreeze. To the contrary, fine-grained soils with strong adfreeze potential are also common in this area. Soil/pile adfreeze bond strength which gives rise to limiting values of jacking, is maximized since surface temperatures are usually very low during this period, while deeper resisting soil temperatures may be near their highest value. With all factors taken into account, an extremely complex and variable situation confronts the design engineer. Factors in this complex array that contribute to jacking forces include:

1. surface soil type
2. active layer depth
3. presence of free ground water
4. rate of active layer freeze
5. pile surface characteristics and adfreeze strength
6. soil temperature
7. soil shear strength

Failure conditions may consist of the following; the first two do not affect the integrity of the pile:

1. active layer soil/pile adfreeze failure
2. active layer soil shear failure
3. permafrost soil/pile adfreeze failure

In the laboratory, it is much more difficult to create frost heave than it would initially seem. Normally, only by selecting certain soils and slowly controlling the freezing front advance will significant soil volume change occur during the freezing process. Rapid soil freezing usually will not produce significant heave in soils. This observation is consistent with known pile jacking cases which seem to dominate in the warmer permafrost regions. Additionally, in warmer regions weaker resisting soils are more prevalent. Measurements to determine pile jacking of installed piles on the North Slope indicate movements are rare to nonexistent, even where piles are placed to depths of less than twice the active layer.

Some design engineers have had success with preventing pile heave by backfilling the top few feet around a pile with gravel. Its significance is apparent in the lower pile/soil adfreeze strengths for coarse soils, coupled with possible freezing front redirection, with resulting decrease in soil/ice expansion near the pile. Various active layer bond breakers also have been used with varying degrees of success.

A fairly rational but conservative approach can be taken by design engineers in solving this problem merely by resisting the maximum possible adfreeze force in the active layer with similar forces in the permafrost layer; plus any sustained loads. Obviously, if thawed soils are present, considerable embedment may be required to resist jacking. Present designs on the North Slope are very conservative and commonly use 40 psi (275.8 KPa) pile adfreeze jacking stress over a 3-ft (1-m) active layer, resisted by adfreeze values of about 12 psi (82.7 KPa). This assumes granular permafrost soils. Many areas are primarily silty sands, but sandy gravels do occur. Examples of pile jacking for properly designed piling on the North Slope are nonexistent to the authors' knowledge, while on the other hand there are several examples of pile creep.

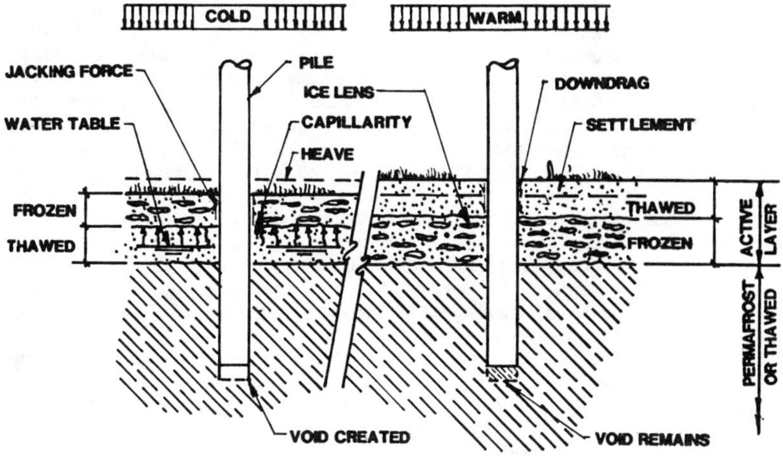

FIG. 12.—Frost Action and Pile Heave

Figure 10 suggests design values for short-term loading in various soils, which can be applied to the frost jacking problem. Use of these values with an additional factor of safety for resisting forces should result in a safe design. For example, under the following conditions a 35-ft (12-m) pile penetration should be sufficient to resist jacking:

- o 60 in (1.5 m) active layer of silt
- o frozen gravel below silt 23°F (-5°C) average
- o no sustained pile loading

then: $$\text{pile embedment} = t_a + \frac{f_j\ (t_a^*)\ \text{F.S.}}{fr} \qquad (2)$$

$$= 60 + \frac{35\ (60)2}{12} = 410 \text{ inches } (34.2 \text{ feet}) \quad (\text{say } 35 \text{ feet})$$

where: t_a = thickness of active layer in inches (mm)

t_a^* = thickness of equivalent active layer in inches (mm) - reduced to account for sustained vertical pile loads

f_j = factored jacking stress psi (KPa) (Figure 10)

f_r = factored resisting stress psi (KPa) (Figure 10)

F.S. = factor of safety (2 shown here, but should always be greater than 1)

Lateral Loading.-Common practice at this time is to assume pile fixity near the bottom of the active layer for permafrost soil conditions. Observations in cold permafrost indicate that laterally loaded piles will usually bend near the top of the frozen surface. Permafrost exhibits high strength and resistance to lateral pile movement for short-term loading conditions. Less is known about laterally loaded pile creep for long-term loads. This is an area deserving more research.

Design Limitations.-Design approaches, such as included in this paper, have limitations and conditions that must be reviewed. Due to limited background data at this time, methods shown are considered to be conservative, and should special conditions warrant, more thorough investigation is highly recommended.

Soil Grain Size.-There appears to be definitely lower adfreeze strength associated with clays or, on the other extreme, coarse granular soils. When clays are encountered, additional tests should be performed to establish design criteria. Coarse-grained soils, particularly those with low moisture contents, should be considered more like thawed soils. Pile adfreeze in coarse-grained soil may be low due to low soil/pile area contact; pile creep may also be low, but end-bearing generally is very high.

Soil Salinity.-Frozen soils, particularly near marine coasts, have been measured with high salt content and resulting freeze point depression. The high variability of possible adfreeze values under these conditions requires site specific criteria establishment which is not defined in this paper.

Thermal Change.-Construction and development will often cause soil thermal regime changes. Design of driven piles should consider performance over the project life; design engineers should make every effort to identify future potential problems.

Ice.-Massive ice is common in permafrost soils but presents little problem with pile driving for most methods. Frozen soils creep in a manner similar to ice, which is consistent, since ice partially forms the bonding and adfreeze mechanisms for soil. However, ice does creep faster than most soils and if it is to be used for long-term structural purposes, adfreeze values must be low and consistent with ice flow theories found in the literature. The authors' have considered soils with ice content greater than 50% for long-term constant load conditions.

Driving Methods.-The authors highly recommend the use of thermally modified pilot holes for the following reasons:

1. Foundation soil can be logged and examined for exact conditions at each pile.

2. Difficulties in achieving driving tolerances are reduced.

3. Driving stresses and pile installation time are reduced.

4. Soil slurry is developed around the pile thus assuring a more complete adfreeze bond.

5. Impact on the permafrost thermal regime around the pile is minimized, compared to slurry methods.

6. Discontinuous permafrost, taliks, and perched water tables present few construction problems when compared to slurry piles in similar conditions.

Pile Types.-Steel pipe piles generally offer the best structural shape for most foundation solutions. H-piles, or web-reinforced H-piles with tips, may offer a good solution under difficult driving conditions using impact hammers. Design adfreeze strength values shown in this paper should be assumed to act on the encompassed area of H-piles, and not on the total contact area. For standard H-piles shapes this area would be about 4 times the section size times embedment in permafrost. For creep calculations, a pipe pile with its perimeter equal to the encompassed perimeter of an H-Pile may be used. Structural design of piles for driving is critical, particularly for pile tips. Computer driving analysis and use of available pile accessories can assist the design engineer. Recent research by the authors' revealed untreated wood pile may rot in arctic environments, thus driven steel piles should be considered as an alternative for these materials.

Conclusions and Recommendations

Driven piles in permafrost can offer an attractive alternative to many foundation systems. Suitable design procedures recognizing permafrost pecularities must be used, with attention to loads, creep and frost jacking over the design life. Currently, driving procedures and techniques presented in this paper are being implemented during production driving programs on the North Slope.

Certainly one of the most obvious problems with driven piles in permafrost is the lack of reliable pile load test data literature. There is need for more full-scale field testing and half-scale, highly controlled laboratory testing in varied soils types and temperatures. These should be of significantly different nature than many past efforts. Weaver (12) concluded that many previously published test results have been inconclusive and are not suitable for creep comparisons. Small-scale laboratory tests may be subject to scale factor problems which yield erroneous results. Testing of half-scale models will allow duplication of a large number of conditions at much lower cost than full-scale field tests. Results will also be more uniform than with field tests, which are subject to many hard-to-control variables.

Pile frost heave (jacking) force is an area particularly requiring additional testing and research. Another area is the establishment of maximum allowable settlement values before piles begin to show loss of strength (failure or tertiary creep). Very little is known about this aspect as applied to practical design situations.

Additional reference data to support this work can be found in the paper by Manikian (3).

Acknowledgments

The authors are grateful to ARCO Oil and Gas Company for their cooperation and dedication to the further development of driven pile techniques in permafrost on the North Slope. Also the authors wish to thank Todd Nottingham for his technical assistance in experiments for this project.

Appendix.-References

1. Crory, F.E., 1973, "Installation of Driven Test Piles in Permafrost at Bethel Air Force Station Alaska," U.S. Army Cold Regions Research Engineering Laboratory, Technical Report 139.

2. Crory, F.E., 1975, "Bridge Foundations in Permafrost Areas," U.S. Army Cold Regions Research Engineering Laboratory, Technical Report 266.

3. Manikian, V., 1983, "Pile Driving and Load Tests in Permafrost for the Kuparuk Pipeline System," in Proceedings of the Fourth International Conference on Permafrost, V. 1: Fairbanks, Alaska USA, Polar Research Board, National Academy of Sciences.

4. Nixon, J.F., 1978, "Foundation Design Approaches in Permafrost Areas," Canadian Geotechnical Journal, V. 15(1), P. 96-112.

5. Nixon, J.F., and McRoberts, E.C., 1976, "A Design Approach for Pile Foundations in Permafrost," Canadian Geotechnical Journal, V. 13(1), P. 40-57.

6. Nottingham, D., 1981, "Method and H-pile Tip for Driving Piles in Permafrost," U.S. Patent 4,297,056.

7. Nottingham, D., and Christopherson, A.B., 1983, "Design Criteria for Driven Piles in Permafrost," Research Report, State of Alaska Department of Transportation and Public Facilities.

8. Parameswaran, V.R., 1978, "Adreeze Strength of Frozen Sand to Model Piles," Canadian Geotechnical Journal, V. 15(4), P. 494-500.

9. Parameswaran, V.R., 1981, "Adreeze Strength of Model Piles in Ice," Canadian Geotechnical Journal, V. 18(1), P. 8-16.

10. Rooney, J.W., Nottingham, D., and Davison, B.E., 1976, "Driven H-Pile Foundations in Frozen Sands and Gravels," Proceedings--Second International Symposium on Cold Region Engineering, Fairbanks, Alaska, pp. 169-188.

11. Tsytovich, N.A., 1975, "The Mechanics of Frozen Ground," McGraw-Hill Book Company, pages 157 to 162.

12. Weaver, J.S., 1979, "Pile Foundations in Permafrost," Doctoral Thesis submitted to Department of Civil Engineering, Edmonton, Alberta.

DUCTILE IRON PIPE
FOR UNDERWATER SERVICE

Ben C. Helton*

Introduction

Ductile iron pipe with ball and socket joints, either bolted or boltless, is an extremely versatile product for use in subaqueous construction in either freshwater or saltwater environments. Ductile iron, and its predecessor, gray iron pipe has been used in the United States in this type of service for more than 45 years. Instrumental to the efficacy of this product is not only the flexibility and restraint against joint separation provided by the ball and socket joint, but also the inherent strength, impact resistance and corrosion resistance of ductile iron.

Construction of underwater pipelines is at best a difficult undertaking, challenging not only the strength and versatility of the piping material, but also the skill and ingenuity of the installer. Because of the many difficulties of this type of construction, the piping material most suitable to the widest range of installation methods should be specified. The toughness, versatility and reliability of ductile iron ball and socket pipe have long made it a favorite of utilities, engineers and contractors for underwater construction.

This paper presents several general methods of construction that can be used in installing ball and socket pipe underwater. The methods presented are general, and can be varied to suit the requirements of the particular installation. The first section discusses the basic characteristics of the joints, followed by a discussion of various construction practices and the several factors that must be considered to successfully install subaqueous ball and socket joint ductile pipelines.

History and Characteristics of Ball Joint Pipe

The American Cast Iron Pipe Company was a pioneer in ball and socket gray iron pipe during the mid 1930's, during which time the Molox ball and socket gray iron pipe was developed. This product was marketed until as recently as 1972, when all production of gray underwater piping was discontinued in favor of ductile iron.

Initially, Molox pipe was manufactured in 4" - 42" sizes only, but the size range was later extended to include the 48" and 54"

* Manager of Technical Services, American Cast Iron Pipe Company, 1501 31st Avenue North, Birmingham, Alabama 35202

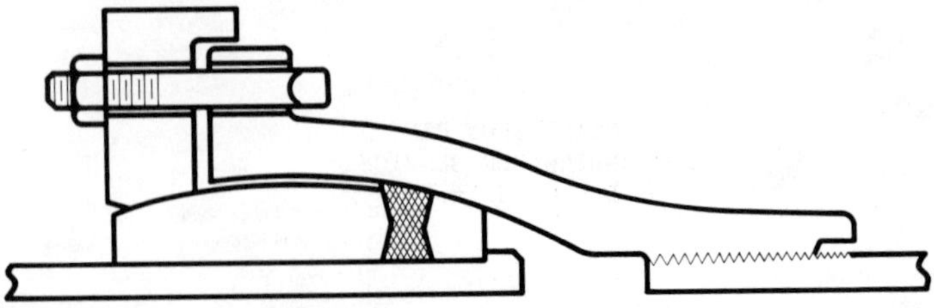

MOLOX BALL JOINT

sizes. It is now available only in 42" - 54" sizes. Features of the joint include separately-cast bell and ball ends, and a totally enclosed duck-reinforced gasket. The alloy steel gland incorporates a circumferential lip that overhangs the socket flange to provide extra strength in the full joint deflection of 12 $1/2^0$. Bolts normally furnished with this joint are stainless steel. In the available sizes, the Molox joint is rated for 150 psi water working pressure. Adequate safety factor (2X minimum) is provided to allow for the rigors of installation and unexpected surges and pressure increases after installation.

The Flex-Lok boltless ball and socket joint was developed in the 1960's and was initially produced in 4" - 24" sizes. The size range has now been extended to include 30", 36", 42" and 48" sizes and will ultimately include 54".

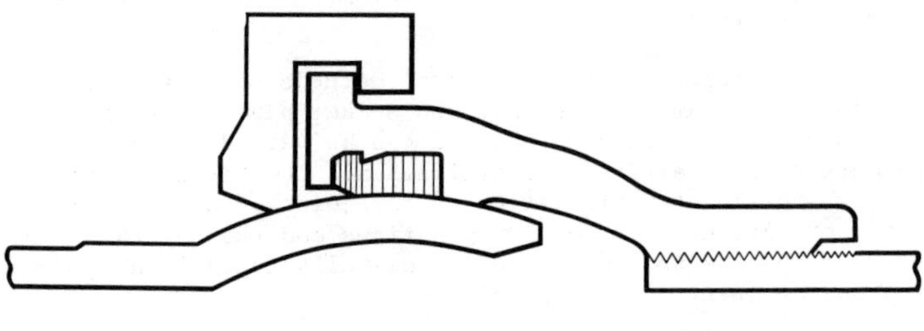

FLEX-LOK BALL JOINT

The Flex-Lok ball and socket pipe features an integrally-cast ball and a threaded-on spherical socket, all 60-42-10 grade ductile iron. The retaining gland in sizes through 24" is also 60-42-10 ductile iron. However in 30" - 48" sizes the gland is alloy cast steel for greater strength and safety factor. Available deflection in all sizes of the Flex-Lok joint is 15^0 and the pressure rating is 250 psi working pressure in all sizes with safety factor of 2X.

Construction Methods

There are three general categories of construction techniques that are employed in installing subaqueous ductile pipelines: the pipe can be assembled on a bank and then pulled into position from the opposite bank; the pipe can be lowered from a barge; or the water can be diverted and the pipe installed in a dry trench using normal good construction practice.

Installation from Bank: By far the most common method of installation of ductile iron ball and socket underwater pipelines is that of pulling from the bank. This is due to the economies that can be effected due to the ease and simplicity of this type of installation. The method consists of assembling the pipe on the bank, and alternately pulling the assembled lengths and adding more pipe until the installation is completed.

Several factors must be considered in determining whether the pulling method of installation is feasible.

1. Availability and suitability of the banks. On the assembly bank, there must be enough space to construct the assembly ramp, store the pipe, and accomodate the digging and pipe handling equipment. The bank also must be sufficiently stable to support excavating and pipe-handling equipment. It must also be graded to provide the proper angle of entry by the pipe into the water.

The opposite bank must provide access, space, and stability for the pulling equipment. It must also be graded to provide the proper exit angle for the pipe as it emerges from the water.

2. Length of line. The length of the line to be pulled is an important consideration so that the capacity of the pulling equipment or the published safe end pull of the pipe is not exceeded. However, even very long lines can be installed by pulling from the bank if the installation is properly engineered.

In the vast majority of jobs, the assembled pipe is dragged along the bottom in a previously prepared trench. In the case of small diameter lines, generally 20" or less, the pipe has negative buoyancy even if it contains no water and hence will not tend to float. In some cases, buoyancy must be added to the line to reduce the underwater weight, and thereby reduce to amount of force required to pull the line. In the case of larger diameter lines, generally 24" and larger, the pipe is lighter than the water it displaces and therefore it will float if it contains no water. In this case, weights have to be added to the line to keep it in position on the bottom.

Buoyancy and safe end pull data are shown in Table 1 below for both types of joints.

Table 1
Buoyancy and Safe End Pull Data
Flex-Lok and Molox Pipe

Pipe Size	Flex-Lok Pipe				Molox Pipe			
	Wt./Lgth. Under Water Full of air		Maximum Safe End Pull		Wt./Lgth. Under Water Full of Air		Maximum Safe End Pull	
(In)	(Lbs)	(Kg)	(Tons)	(KN)	(Lbs)	(Kg)	(Tons)	(KN)
4	267	121	21	186				
6	353	160	35	311				
8	384	174	49	435				
10	448	203	63	560				
12	576	261	76	676				
14	551	250	90	800				
16	464	210	100	889				
18	441	200	110	978				
20	392	178	119	1058				
24	-54*	-24*	130	1156				
30	-769*	-349*	225	2001				
36	-1653*	-750*	255	2268				
42	-678*	-308*	305	2713	131	59	293	2606
48	-354*	-161*	350	3113	-366*	-166*	376	3345
54					-2825*	-1281*	479	4261

*Negative underwater weight indicates pipe will float when it contains no water.

3. Condition of bottom. For a line to be pulled successfully across the bottom, reasonable smoothness and a lack of obstructions on the bottom are required. Any large rocks, stumps or other heavy debris that is present must be removed.

One variation of the bank installation that is sometimes used is to float the line into position. This can be a much more involved procedure than dragging across the bottom because the assembled pipeline must be lowered to the bottom in a controlled manner to avoid damage to the joints. Further problems can also be encountered due to wind, current, or boat traffic during construction.

When the pulling method is employed, a launching ramp is normally constructed on the assembly bank extending into the water. Such a ramp can be built with previously-used construction material, such as rail and crossties, or structural members, timbers, etc.. The ramp, which will be removed when the installation is completed, provides two particular advantages to the installer. It provides a convenient assembly area off the ground greatly facilitating joint assembly and it greatly reduces the amount of tensile load required to pull the line if it is properly constructed.

Photo 1 below shows such a ramp.

Photograph 1 - Entry Ramp

This ramp was constructed of steel sheet and rail, and was built to accomodate twin 30" ball and socket Flex-Lok lines. In designing and building the launching ramp, the length, size and weight of the line must be considered. If the crossing is short and the pipeline is of small or medium diameter, then probably no special ramp is required. The pipe joints can be assembled on timbers for cleanliness, the timbers removed, and the line then be pulled on the ground into the water.

Photograph No. 2 above shows 16" Flex-Lok pipe being installed from the bank without benefit of a launching ramp.

If, however, the line is long, and pulling load may become an important factor, then a well designed and constructed ramp is a wise investment, even though it is a temporary structure to be removed when the pipe installation is completed.

Photograph No. 3 - 30" Molox Lines on Entry Ramp

Photograph No. 3 above shows installation of twin 48" Molox lines. These lines were assembled and pulled simultaneously across the waterway. Note the support timbers, pulling sled and the well-designed rail ramp built on crossties to support the pipe during the pulling operation.

Installation from Barge: The installation of a ductile iron ball and socket subaqueous pipeline from a barge is another effective installation method, though it is generally more expensive than installation from a bank for two reasons:

1. Labor and equipment costs. The use of floating construction equipment, such as barge-mounted cranes and excavation equipment, and the specialized personnel necessary to move and operate that equipment can add materially to the cost. The use of divers is also not a negligible expense.

2. Speed of installation. Installation from a barge is necessarily slower than installation from a bank due to the increased problems of material handling as well as the extra time required to move the equipment.

Although this method is generally more expensive than installing from a bank, in some cases it is mandatory due to conditions of the line or construction site. Where the line is of exceptional length; where the bottom is rough or is unstable and a trench will not remain open long enough to complete an installation from the bank; or where the route of the line is not straight and cannot be pulled are examples where the use of a barge is necessary

Installation from a barge can be accomplished by either of two methods. The first method consists of using the barge as a work platform, and assembling all pipe joints on the barge. The barge must be equipped with a ramp extending to the bottom of the waterway. The pipe are assembled on the ramp, and as pipe are added, the barge is moved forward, and the assembled pipeline is allowed to slide down the ramp into the prepared trench. If the pipe is buoyant, then weights must be attached to it, or it must be maintained full of water as the installation progresses. It should also be noted that the ramp should be designed to support the pipe all the way to the bottom to prevent unnecessary beam loads from occurring on the pipe joints.

Photograph No. 4. Forty-Eight Inch Flex-Lok Pipe on Assembly Ramp

The second method of installation from a barge involves underwater joint assembly by divers. Either one pipe at at time or several pipe assembled on the barge and supported by a strong-back are lowered to the bottom to be received and assembled to the previously installed pipe by divers. This is the least used method. It is extremely slow due to the underwater jointing work required. Another disadvantage to this method is that contamination to the pipe by the surrounding water is unavoidable.

Photo No. 5 - Thirty Inch Molox
Pipe Being Lowered into Water
from Barge.

Dry Installation: The third type of installation of ductile iron ball and socket pipe is dry installation. When crossing small shallow stream beds, many times the water may be diverted, or coffer dams may be constructed and the pipe installed in a dry trench using normal good pipe laying practice. In this situation, the special features of ball and socket pipe may not be required unless there is an unstable soil condition, and other types of restrained or unrestrained joint ductile iron pipe may be used successfully.

Testing

Before the completed subaqueous line is connected to land lines, and before it is backfilled, pressure tests should be performed to insure the integrity of the joints and to insure that no damage occurred to the pipe during installation. If the line has negative buoyancy, either due to its own weight or due to attached weights, a low pressure air test of fifty psi or less is suggested to locate any possible leaks. The line should then be filled with water and hydrostatically tested, normally to 1.5 times the expected working pressure. If the line is buoyant when it contains no water, then the air test will have to be omitted. If at all possible, due to the difficulty of repairing leaks in a subaqueous line, periodic pressure tests should be made during construction, especially construction of a long difficult line.

Conclusion

Successful construction of the underwater pipeline is a function of careful prior planning; the use of sound construction methods and the choice of the proper piping material. Ductile iron ball and socket pipe is a tough, versatile product, adaptable to various installation methods, which provides a long, trouble-free service life.

PIPELINE RIVER CROSSINGS: OPERATIONAL AND MAINTENANCE ASPECTS

Walfred E. Hensala, M. ASCE [1]

Abstract: Advanced technology on river morphology, river mechanics, and pipeline river crossings is used to identify deficiencies in existing pipeline crossings of rivers and to define corrective measures. A program of surveillance of pipeline river crossings is incorporated into an operations and maintenance program to monitor river changes and to limit maintenance and corrective construction.

The growth of cross country pipelines as major transportation systems has resulted in an associated advancement in the science and engineering technology associated with pipelines. A very specialized branch of technology for the crossings of rivers and streams has also evolved to address unique problems of construction and to limit the possibilities of loss of service. Northwest Pipeline Corporation is utilizing that technology to address the operational and maintenance aspects of pipeline river crossings.

While logical methods and criteria do exist, and the design and construction of major crossings are usually addressed in detail using proper criteria, often constraints in the route selection process result in crossings that are less than optimum. These constraints may be occasioned by pipeline alignments to the crossing sites, land use, and environmental concerns. The use of inapplicable empirical approaches, the lack of historical records and adequate data, or improper interpretation of the data available can also contribute to river crossings of pipelines that are less than ideal.

As a result of these constraints and possibly in conjunction with changes in the natural process or changes brought about by the cumulative effect of man's activities, the integrity of the pipeline crossing may be threatened and corrective measures may become necessary.

1 Senior Engineer, Transmission Systems, Northwest Pipeline Corporation, 295 Chipeta Way, Salt Lake City, Utah

A program of surveillance of pipeline crossings, initiated early and coupled with knowledge of the river morphology and mechanics of the river, and based on an understanding of the design process and its supporting criteria, can determine possible deficiencies in the crossing. Properly conceived, the monitoring program will provide an historical record of changes in the river morphology, and when used in conjunction with hydrologic data on the river will provide a predictive model for metamorphosis of the river. It will define acceptable changes in the river morphology and determine when minimal corrective measures can be utilized to protect the integrity of the pipeline crossing.

Northwest Pipeline Corporation and its predecessor companies have monitored river crossings since beginning operations in 1957. The program has evolved from one which initially monitored only cover over the pipeline to one which incorporates river behavior and morphology and original design criteria. It now utilizes river mechanics and bank rehabilitation to protect pipeline crossings, as well as channel stabilization structures and alterations of the pipeline facility.

While the program is in use throughout the Northwest pipeline system, it is particularly useful in the area west of the Cascades in Washington and Oregon. In this area there are in excess of 200 river and stream crossings of which more than 100 are monitored on a scheduled basis on over 500 miles of pipeline right of way. More than half of the right of way has looped pipelines.

The streams crossed have large fluctuations in flow from high to low stages, but the periods of high stages are fairly predictable. In addition, most of the streams crossed contain valuable fisheries, including anadromous fish, so any alteration of the facilities must address impacts to the aquatic ecology.

The monitoring program looks beyond the location of the pipeline crossing and reviews the reach of the river which can affect the crossing. Aerial photographs are utilized as a data base and changes in the river behavior are noted.

Stream bed profiles at the crossing are surveyed on a schedule based on past experience, or following flood stage events, to check cover over the pipeline in the channel. Lateral channel migration is checked on the photographs, as are changes in bends and bars. Hard spots in the bank are noted as are flow constrictions. The effects of flood control structures and river regulation are defined. The program includes information on the type of bank and bed material, current velocities and depth, and the type and density of vegetation on the banks. The morphology of the stream and its floodplain are reviewed and the geometry of the bends and meanders determined. Water resources data and climatological data are obtained from federal and state agencies and used to determine approximate timing of higher stages and their duration.

Environmental aspects are reviewed with state and federal agencies when applicable and information on fisheries and timing of possible construction determined.

Land use and its effects on the pipeline and the river are determined and monitored. While land use practices outside of the pipeline right of way are beyond the control of the pipeline company, potential problems can be identified and addressed to avoid possible emergencies.

When protective bank structures have been utilized, adjacent structures in the same reach and those in adjacent reaches are observed to gain further information on successful structures, as well as those that fail. Type of structure and materials used, method of procurement, approximate age and condition and apparent maintenance are noted. The information is used in a continuing effort to develop more cost-effective and adequate structures.

When changes in the river behavior are noted, the possible use of natural methods of bank protection are considered. Where stream flow conditions are not severe and erosion has not progressed these minimal stabilization methods may avoid the use of major bank rehabilitation work, massive river training structures, or alteration of the pipeline facility.

These natural methods include shaping of the banks, land use practices, and the use of vegetation. They involve the consent and coordination of the adjacent landowners, and for the pipeline company, a possible source of continuing maintenance of areas outside of the control of the pipeline company. They can be effective when utilized properly, but are possibly most effective when combined with structural revetments. Where erosion has progressed, or extreme flow conditions exist, protection by structural revetments or river training structures appear to be more feasible.

If a pipeline crossing is threatened, either by the dynamics of the river system or through alteration of the hydrologic regimen by man's activities, major corrective measures become necessary.

An analysis of the original pipeline crossing criteria and design procedures, together with available historical data on river hydrology and morphology is used to determine cost effectiveness of corrective measures. Measures considered include channel revetments, river training structures, or alteration of the pipeline facility. Analysis of corrective measurements must consider the interaction with land use requirements, regulation by governmental agencies, societal and institutional constraints, as well as the requirements of the pipeline system and the dynamics of the river processes. Measures other than the alteration of the pipeline facility require continuing coordination with federal and state agencies having primary jurisdiction over the river, as other measures for the protection of the pipeline would be considered a secondary use.

River training methods include jetty systems, spurs, groins, vanes and guide banks. Most require extensive work on full reaches of the river and coordination with the agencies. They imply changes in the river system and are best utilized to allow the mechanics of the river to provide most of the work of change. The concept can be adopted to correct problems on smaller streams or make minor changes in larger streams. River training structures must be based on adequate design and valid data.

Bank stabilization through the use of revetments, designed and constructed using the proper criteria, is often most practical and cost effective. If extensive revetment work is required, esthetic values, recreational uses of the river, and environmental concerns must be considered to obtain the necessary approvals and proceed with construction.

All projects that will affect water resources or wetlands require permissions or approvals from governmental agencies. Permits are required to protect water resources, aquatic ecology, environmental values and on larger streams, navigable waters.

An understanding of the permit process and its requirements must be incorporated in the early planning for any work in rivers and streams. Agency concerns must be considered to avoid delays and excessive costs. It is prudent to discuss proposed work with agencies early in the planning stage, to allow the project to meet its objectives and still address regulatory requirements. The procedures to obtain authorization can be complex, time-consuming and, at times, frustrating; but can be used to provide benefits. To a large extent much of the documentation of design methods and their rationale is a result of regulatory requirements to obtain authorization for major projects.

Advances in the technology associated with pipelines and particularly the specialized area of river crossings have generally focused on design and construction. Northwest Pipeline Corporation is applying these advances in technology to the operational and maintenance aspects of pipeline river and stream crossings. A method of surveillance of river and stream crossings, which is at the present time used on a case-by-case basis, is being developed into a structured program.

Appendix - Bibliography

Fenwick, G.B., ed., "State of Knowledge of Channel Stabilization in Major Alluvial Rivers," Technical Report No. 7, Committee on Channel Stabilization, U.S. Army Corps of Engineers, Waterways Experiment Station, Vicksburg, Miss., Oct. 1969

Klingeman, P.E. and Bradley, J.B., "Willamette River Basin Streambank Stabilization by Natural Means", Water Resources Research Institute, Oregon State University, Corvallis, Ore., June 1976

Shen, H.W., ed., River Mechanics, 1st ed., Volumes 1 and 2, Water Resources Publications, Fort Collins, Colo. 1971

INSPECTION OF THE TANANA RIVER PIPELINE BRIDGE

Alan Christopherson[1], M.ASCE, William Swigart[2], M.ASCE, and Michael Fries[3], A.M. ASCE

ABSTRACT: A 5-year inspection program for the 1,200 ft Tanana River Pipeline Bridge was conducted during the fall of 1982. Presented in this paper are the planning, coordination, engineering methods and techniques developed by the pipeline operator, engineering consultant and contractor on a fast-track basis. In addition to the technical challenges, logistics and cold weather conditions were considered since the structure is located in sub-Arctic regions.

INTRODUCTION

The Tanana River Pipeline Bridge is located along the Alaska Highway about 90 miles south of Fairbanks, Alaska and about 160 miles south of the Arctic Circle (Fig. 1). The bridge is a suspension structure and supports the 48-in. diameter Alyeska Pipeline 1,200 ft across the Tanana River (Figs. 2 and 3). In 1982, after about five years of service, the bridge came due for a thorough inspection program. The owner of the bridge, Alyeska Pipeline Service Co., selected the team of Peratrovich, Nottingham & Drage, Inc. (Engineering Consultants) and Construction and Rigging, Inc. (Contractors) to perform the inspection. The object of the program was to inspect all cable nodes and structural steel connections, and other points of interest on the bridge; and to measure the tension in selected cables. The contract, awarded on September 15, 1982, stipulated that the inspection be completed by about October 15, 1982, in order to avoid working in bitter winter conditions.

---

[1] Senior Engineer, Peratrovich, Nottingham & Drage, 1506 West 36th Avenue, Suite 101, Anch., AK 99503
[2] Senior Engineer, Peratrovich, Nottingham & Drage, 1506 West 36th Avenue, Suite 101, Anch., AK 99503
[3] Structural Engineer, Alyeska Pipeline Service Company, 1835 South Bragaw Street, Anch., AK 99512

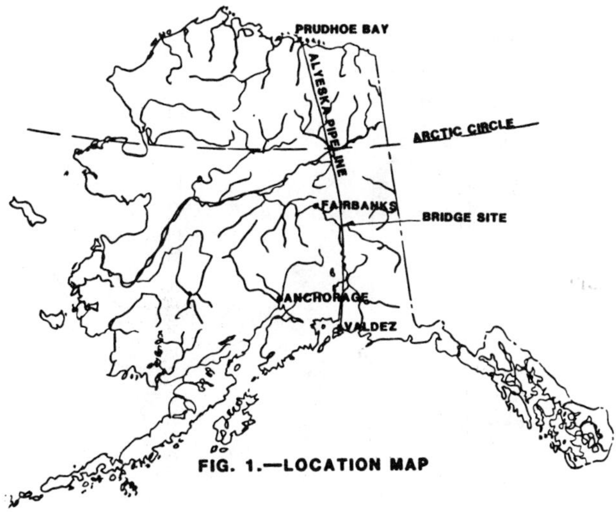

FIG. 1.—LOCATION MAP

PROJECT CONSTRAINTS

A number of constraints had to be accommodated during the inspection of the bridge.

1. The bridge has virtually no safe access for men or equipment. The towers have no ladders or other convenient means of ascending. There is no access across the bridge except by walking the top of the pipe, which is exceptionally slick and has very few safe handholds.

2. Much of the bridge is inaccessible to conventional cranes as about 900 ft of the bridge lies over the river and the towers are on the order of 160 ft high.

3. The project had to be completed within about one month of the notice to proceed. This included final design, fabrication of special equipment, transport of equipment and manpower to the site, mobilization on site, and actual performance of the inspection.

4. The system had to be workable in winter conditions. At 160 miles south of the Arctic Circle even the beginning of winter can have substantial effect.

METHODS AND EQUIPMENT

A combination of methods and equipment was selected for inspection of the various portions of the bridge within the constraints listed above (Fig. 4). The following discussion is presented for each method.

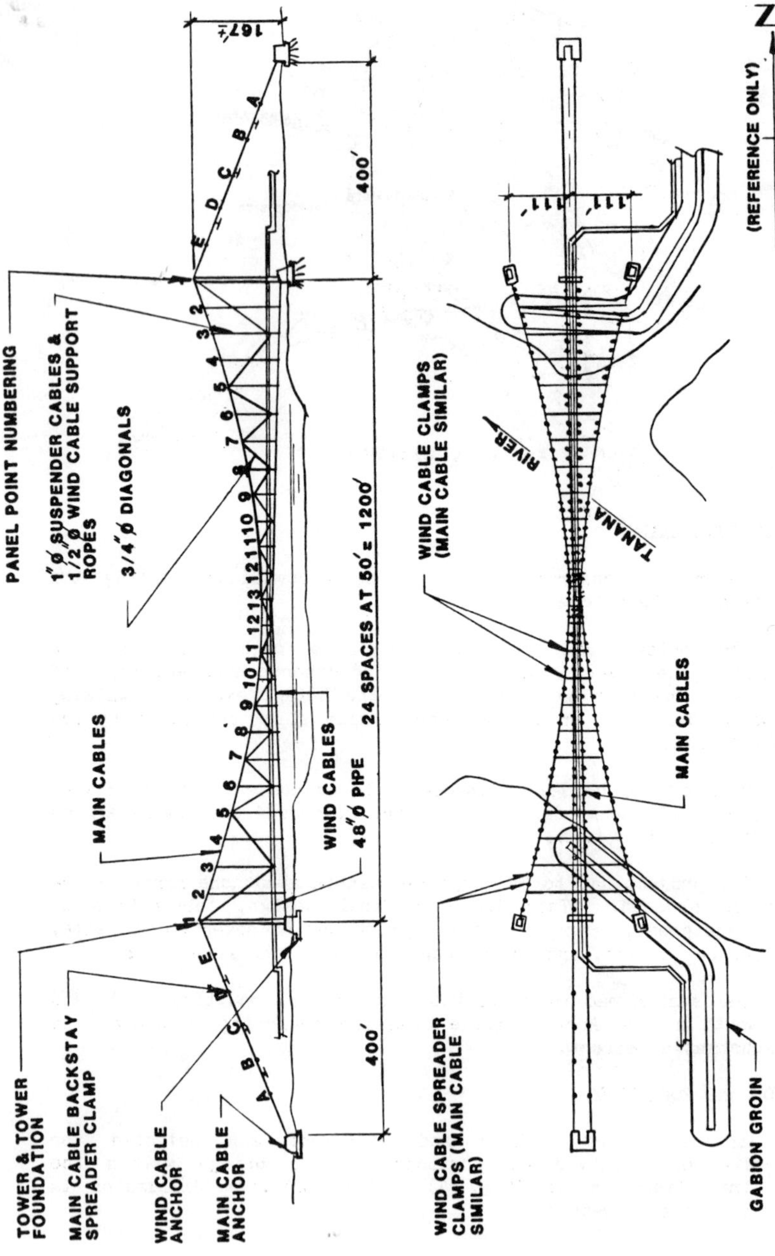

FIG. 2.—PLAN AND ELEVATION OF BRIDGE

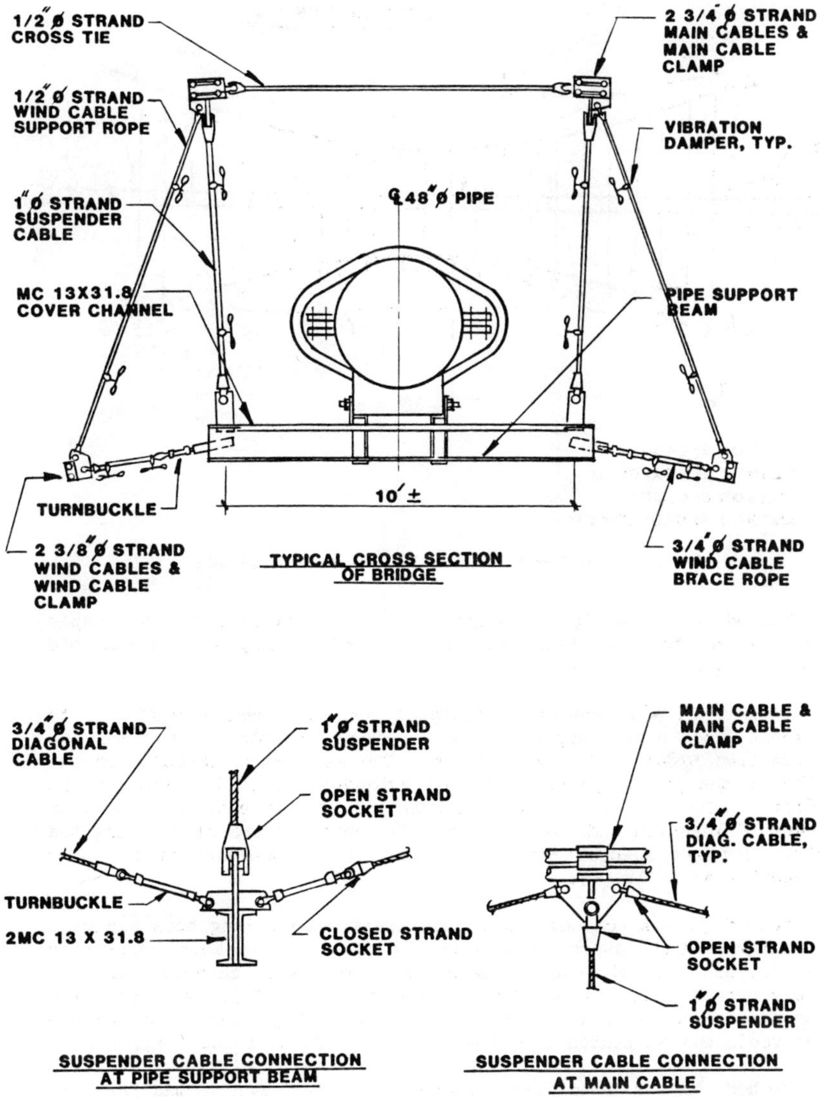

FIG. 3.—TYPICAL SECTIONS OF BRIDGE

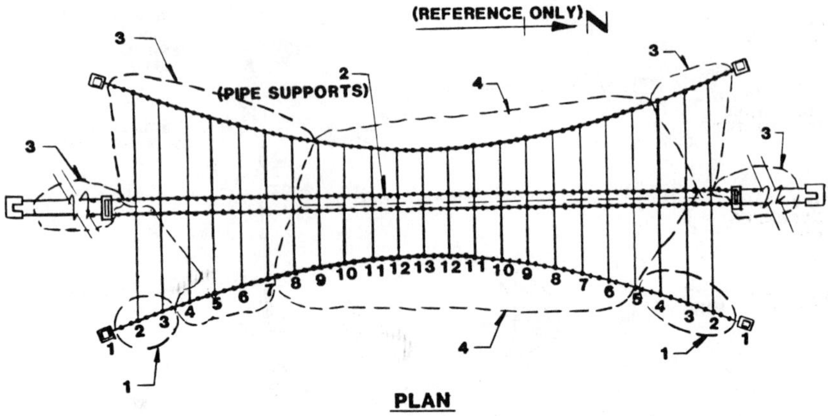

METHOD 1: LADDER
METHOD 2: SCAFFOLD
METHOD 3: CRANE
METHOD 4: CABLEWAY

FIG.4.—METHODS USED TO INSPECT BRIDGE

Method 1.- A 40-ft extension ladder was used to inspect cable anchors and those portions of the wind cables that were accessible from the ground (Fig. 5).

Method 2. - A specially designed scaffold system was used to inspect the pipe support beams. The scaffold hung from two galvanized 5/8-in. dia. wire ropes. The wire ropes passed over the tops of the pipe support beams and extended the full length of the bridge. The cables were counterweighted at one end, to allow for thermal expansion and contraction. The scaffold travelled along the wires and its special hangers allowed it to pass around the pipe support beams (Fig. 6).

Method 3. - A conventional truck crane with a long boom was used to inspect the towers, the upper reaches on the main suspension cables, and the main cable back stays. Two 40-ton capacity Linkbelt truck cranes with 130 ft of main boom and 50 ft of jib were selected for this portion of the job. A 4-ft by 4-ft platform for two men and tools was suspended from the load line of the crane (Fig. 7).

Method 4. - A specially designed cableway system was used to inspect the remainder of the main cables and wind cables over the river. The system consisted of a 40-ton capacity Linkbelt truck crane at each end of the bridge. The crane's load lines were joined together, and at the point of connection a two-man platform was suspended. The platform could travel horizontally by one crane taking up on its load line while the other let out. The platform could travel vertically by one crane taking up or letting out on its

FIGURE 5 INSPECTION METHOD 1 - LADDER

load line while the other held tight its load line. By moving horizontally and vertically along the axis of the cableway system the two-man platform could be positioned near any point on the bridge catenary. A grappling hook was used as necessary to position the platform up to 10 ft perpendicular to the cableway system. In order to access each of the two catenaries, two setups of the cableway system were required, one on each side of the pipe (Fig. 8).

FIGURE 6 INSPECTION METHOD 2 - SCAFFOLD SYSTEM

Cable Tension Measuring Device. - A Martin Decker UB2 Cable Tension Indicator was used to measure in place the tension in the diagonal bridge cables. The device functions by measuring the force required to deflect a cable a known distance between two points.

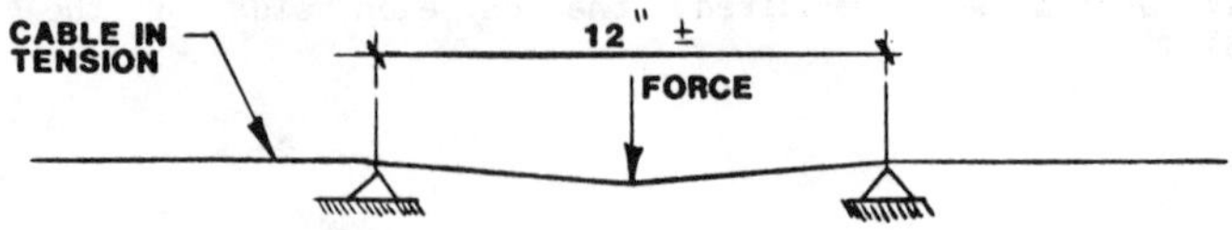

The force is correlated to a graph for the specific cable diameter and type to obtain the tension (Fig. 9).

FIGURE 7 INSPECTION METHOD 3 - CRANE

Safety Cable. - A galvanized 1/4-in. diameter wire rope safety cable was installed on top of the pipe and secured laterally at 50-ft intervals to provide a continuous handhold for men crossing the pipe. The cable was counterweighted rather than fixed at one end in order to allow for thermal expansion and contraction (Fig. 10).

DESIGN OF THE CABLEWAY SYSTEM

The analysis and design of the cableway system was complicated by a number of interrelated factors (Fig. 11).

1. The weight of the two-man platform and its contents is small compared to the weight of the cable which suspends it. This requires a complex analysis of the cable as a caternary rather than as straight lines.

FIGURE 8 INSPECTION METHOD 4 - CABLEWAY SYSTEM

2. The force in the cable varies depending on the horizontal and vertical position of the platform with respect to the tips of the two crane booms. That is, the closer the platform is, vertically or horizontally, to a boom tip, the higher the force in the cable becomes.

3. The longer the boom, the lower the forces in the cable while inspecting any given point on the bridge, because of the greater sag of the cableway catenary. But, the longer the boom the greater the tendency for the crane to tip. Also, the longer the boom the more difficult to set up the crane in the limited space available.

4. When one crane holds its load line while the other crane takes up or lets out on its load line, the vertical movement of the platform is much greater than the takeup or letout of the one crane. The closer the platform is vertically to the boom tips, the more exaggerated is this vertical movement of the platform. Also, any horizontal "bounce" of the boom tips would be greatly exaggerated as vertical movement of the platform.

FIGURE 9 CABLE TENSION INDICATOR

5. As cable forces increase, consideration must be given to failure of various system components, some of which are the cable, the stability of the cranes, the crane boom pendants, and the crane boom.

To reduce the complexity of this project a system was devised to "decomplicate" the above interrelated factors. First, all components of the system were made capable of resisting the stresses that would be produced by the greatest line pull that the crane drums were capable of exerting. Second, the crane booms were tied down in the front and in the rear to prevent overturning of the cranes and bouncing of the booms.

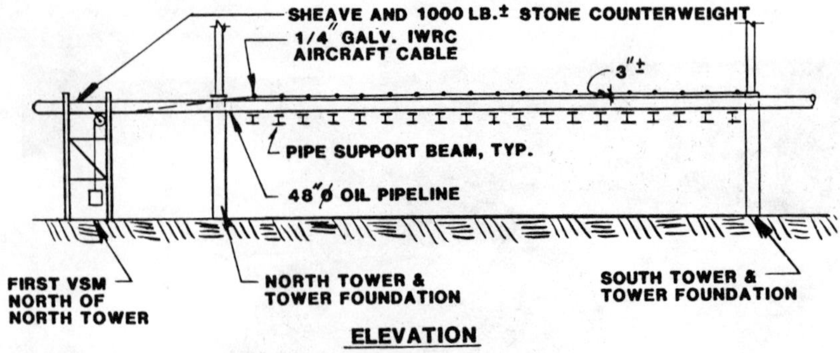

FIGURE 10 SAFETY LINE SYSTEM

With these two limitations and some ordinary care the system could be operated safely. The system would always run short of power before it ran short of strength.

WEATHER

Severe winter weather (-20°F) arrived a couple of weeks before anticipated. In spite of the weather, the project proceeded on schedule. Some of the adverse affects of the weather, however, are as follows:

> Snow on the pipe made for slow passage across the pipe. The cold caused the cable tension indicator to malfunction, which was corrected with a special hydraulic fluid. The cold also made the construction equipment difficult to start and to warm up. Ice fog produced by the river decreased visibility. Ice pans which formed in the river made operating the safety boat very difficult as well as increasing the difficulty of pulling the cableway wires across the river during setup. The weather also took its toll on the labor force, as inspection team members periodically had to spend time indoors getting warm (Fig. 12).

CONCLUSIONS

Through a novel approach, careful engineering, and thorough planning the Owner, Engineer and Contractor were able to perform a tough job in record time. In spite of bitter winter conditions, the time from notice to proceed to project completion was one month and one day. The methodology developed during this program makes it possible for future inspection programs and construction operations to be conducted in a similar manner.

A second conclusion may be drawn for the benefit of owners and designers of bridges. Bridges do require periodic inspection and maintenance. Wherever possible, permanent or removable fixtures should be included for easy access to various parts of the bridge.

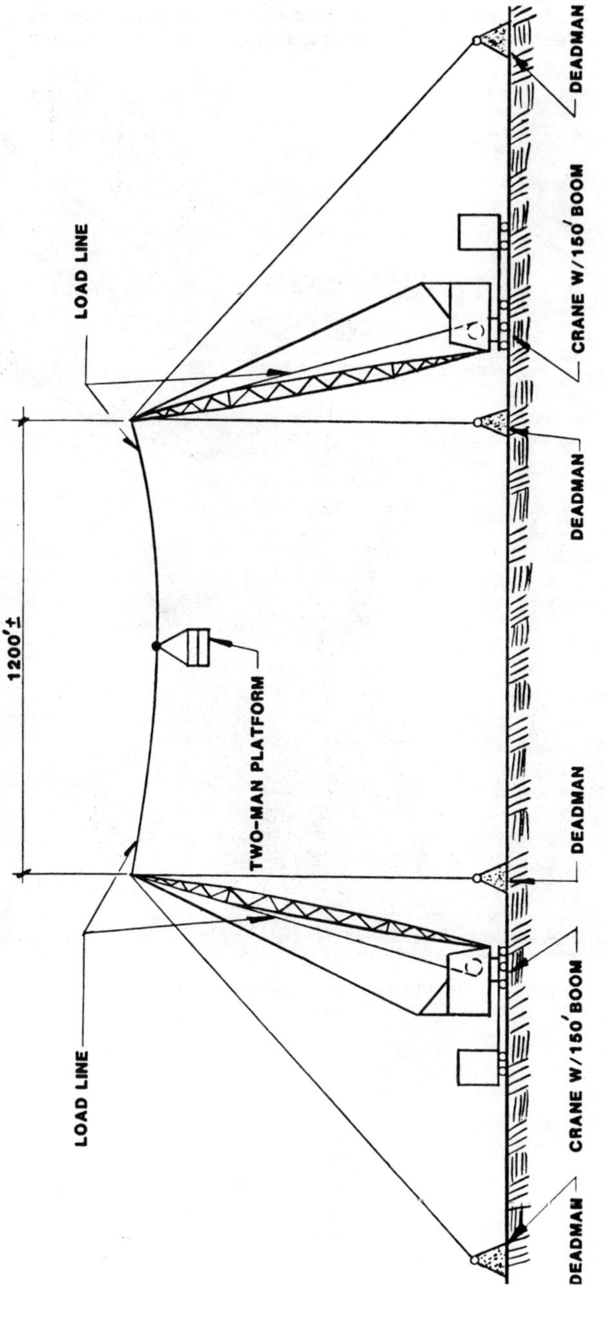

FIG. 11.—ELEVATION OF CABLEWAY SYSTEM

Good access would decrease cost of inspection and maintenance, while increasing effectiveness and encouraging greater frequency of inspection.

FIGURE 12 INSPECTION IN PROGRESS AT -10°F

DESIGN FOR PIPELINE WIND VIBRATION IN ARCTIC AREAS

By Yee Kwok Lee*, M. ASCE

ABSTRACT

In arctic areas, economic considerations dictate that the unsupported pipeline spans be greater then is generally used elsewhere. Consequently, they are more susceptible to the vortex shedding vibration caused by a steady wind. It can induce premature failure of the line because of low damping ratio of a steel pipeline and cause metal fatigue problems. Results of research into methods of remedy by others is described. A design of an add-on vortex spoiler has been evolved from that study for one of the arctic gathering systems. The vortex spoiler could be easily clamped on the pipeline if the line would encounter wind vibration problems during service. For a river crossing where the prevailing steady wind is known, the vortex spoiler also is preferred because a long clear span is needed for the passage of large ice sheets in spring break-up. The effectiveness of spoilers has been observed and is under further closer monitoring in the field.

INTRODUCTION

When a steady windstream passes a cylindrical obstruction alternating vortices are shed normal to the wind direction forming a Von Karman vortex trail (Figure 1), named after Theodore Von Karman who first investigated the vibration failure of the Tacoma Narrows Bridge in 1940 and theorized this oscillation due to wind (1). Although this phenomenon is well known by engineers involved in the design of suspension bridges, high towers, stacks, chimneys, electricity transmission lines, etc., it is much less frequently encountered in the pipeline industry. It is probably due to the fact that the slenderness ratios are relatively small for most of the aboveground supported pipelines; therefore, they are normally aerodynamically stable.

In arctic areas, the unsupported pipeline span, in general, is greater than those located elsewhere. This is an effort to minimize the high construction cost in such an adverse environment. Consequently, these

*Director of Civil Engineering, Williams Brothers Engineering Company, 6600 S. Yale Ave., Tulsa, OK 74177

pipelines are more susceptible to the vortex shedding vibration caused by a steady wind.

This phenomenon of pipeline oscillation due to wind can cause premature failure of the lines, because of low damping ratio of welded pipelines and metal fatigue of pipes. Thus, it is prudent to consider this dynamic aspect of wind load as well as direct wind pressure force on aboveground supported pipelines.

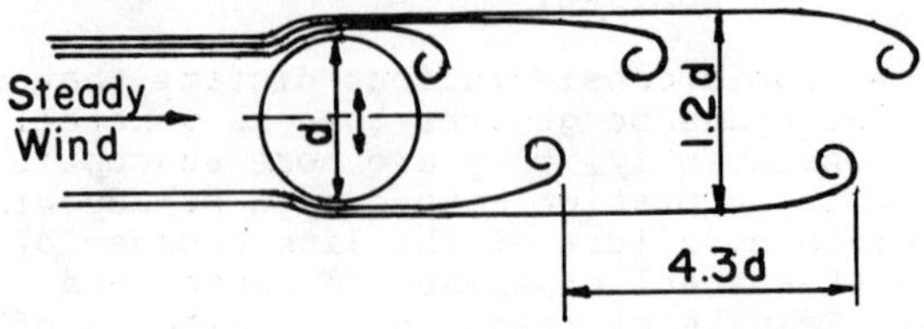

Figure I. VON KARMAN TRAIL

VORTEX SHEDDING FREQUENCY OF WIND

Experience with steel stacks and other circular cross section structures reveals that serious vibration might occur from steady winds of low velocities ranging from 10 to 50 mph. When wind velocity exceeds 50 mph, the natural turbulence generated in the airstream around the cylinder inhibits the periodic vortex formation (3,6). Therefore, wind velocity higher than 50 mph poses no threat on pipeline resonance with wind vortex shedding. There is a definite correlation between steady wind velocity, diameter of pipeline, frequency of vortex shedding and Reynolds Number (2). It may be expressed as follows:

$$\frac{fd}{v} = S \qquad (1)$$

in which

f = frequency of vortex shedding of wind, cps
d = outside diameter of pipeline, ft.
v = steady wind velocity, ft/sec.
S = Strouhal number.

Strouhal number is a function of type of fluid, pipe motion and Reynolds Number (RE). The value RE = 300,000 is known as the critical Reynolds Number. Values below and above this number are called subcritical and supercritical, respectively. The idealized and theorized vortex trail by Von Karman may be considered applicable and reasonably accurate, for six or seven diameters downstream, up to the critical Reynolds Number. At this critical number, there is a sharp drop in the drag coefficient and abrupt rise in the Strouhal numbers.

Beyond this point, the speed of rotation and intensity of shear associated with the vortices cause them to break down nearer and nearer the cylinder, with increase of Reynolds Number likewise the wake narrows with regular vortex formation no longer visible. If the Reynolds Number is known, the value of S can be obtained from test data (5). However, for low heights of aboveground supported pipelines in the arctic, the Reynolds Number cannot be evaluated because turbulence can modify its value. From experiments, a range of Strouhal Numbers from .18 to .28 should be considered in the design. Subsequently, a range of probable vortex shedding frequencies of wind may be computed and an envelop of critical frequencies developed (Figure 2), for a specific size of pipeline by using Equation 1.

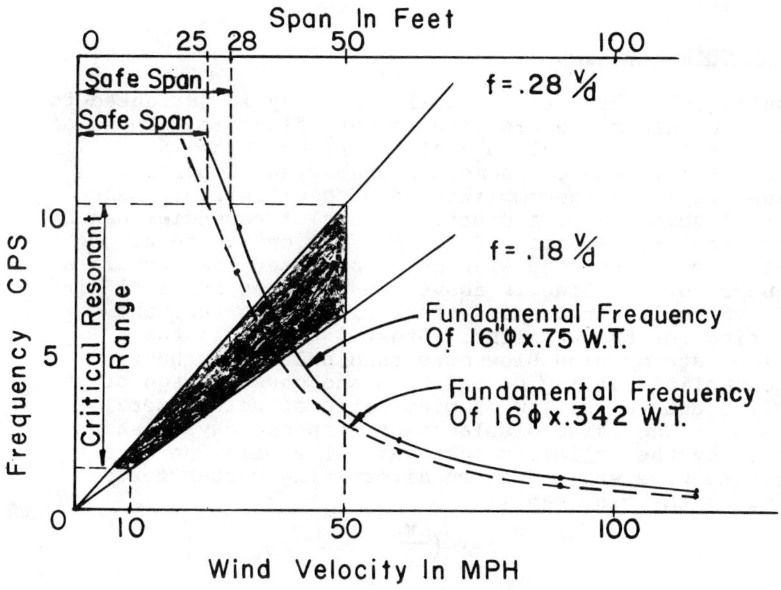

Figure 2. CRITICAL RESONANT FREQUENCY AND SAFE SPAN

NATURAL FREQUENCY OF PIPELINE

Values of natural frequency of a simply supported pipeline having a flexural rigidity EI and unit length weight W are as follows:

$$f_n = \frac{n^2 \times .743}{L^2} \left(\frac{EI}{W}\right)^{.5} \qquad (2)$$

where

n = Number of mode shape, n = 1 is the fundamental frequency.
L = Length of pipe between supports, ft.
E = Young's modulus of pipe, psi.
I = Moment of inertia of pipe, in^4
W = Total gravity load of pipe including self-weight, lb/ft.

RESONANCE EVALUATION

Normally only the first natural frequency is warranted to look into unless the pipeline is very flexible. Those of higher modes are a multiple of the first. The fundamental frequency is most probably and important because of its large amplititude of oscillation. Based on trial spacings of supports, natural frequencies can be calculated from equation 2 and plotted on Figure 2. Should the unsupported span be so selected that the first frequency of pipeline is above the range of critical wind vortex shedding frequency, there will be no resonant vibration for the pipeline. Otherwise, should the critical steady wind blow more than half of an hour, a large amplititude may be built up and cause damage of the pipeline due to the low damping ratio of welded metal pipeline. The large displacement response may be shown clearly by the following mathematical expression of a vibrating pipe excited by an alternating vortex force acting at the mid-span (4).

$$y = \frac{2gP}{WL} \sum_{n=1}^{\infty} \frac{\sin\left(\frac{n\pi}{2}\right)\sin\left(\frac{n\pi x}{L}\right)}{\omega_n^2 - \omega^2} \sin \omega t - \frac{2gP}{WL} \sum_{n=1}^{\infty} \frac{\sin\left(\frac{n\pi}{2}\right)\sin\left(\frac{n\pi x}{L}\right)}{\omega_n^2 - \omega^2} \sin \omega_n t \qquad (3)$$

where

P = Alternating vortex shedding force
W = Total gravity load of pipe including self-weight per unit length
L = Length of pipe between supports

t = Time
ω_n = Natural angular frequency of pipe
ω = Angular frequency of the disturbing force.

The first series is proportional to the term sin ωt which represents disturbing force due to wind, and the second series is proportional to sin ω_n which represents the free vibration of pipe.

If $\frac{\omega^2}{\omega_n^2} = \alpha^2$ then the first series can be written as

$$y = \frac{2PL^3}{\pi^4 EI}\left(\frac{\text{Sin}\frac{\pi x}{L}}{1-\alpha^2} - \frac{\text{Sin}\frac{3\pi x}{L}}{81-\alpha^2} + \frac{\text{Sin}\frac{5\pi x}{L}}{625-\alpha^2} - \cdots\right).$$

$$\text{Sin}\,\omega t \qquad (4)$$

For a small value of α^2, that is, for $\omega << \omega_n$ the first term alone of the series represents the steady state response with good accuracy. The ratio of dynamic to static deflection is $\frac{1}{1-\alpha^2}$. If ω is very close to any one of the natural frequencies, ω_n, then the first term in the series becomes very large and a condition of resonance occurs.

FATIGUE STRESS

In the case of known frequent steady prevailing winds crossing a pipeline, even calculation of frequencies rules out the possibility of resonance of pipeline with wind vortex shedding. A low amplitude vibration still may occur as shown in equation 4. Cyclic and reversal stress generated due to this alternating force of the vortex shedding should be evaluated and compared with safe metal fatigue stress to avoid the premature metal fatigue failure. The maximum alternate harmonically varying forces on a pipeline in a direction perpendicular to that of the steady windstream for the computation of pipe bending stress can be obtained from equation 5.

$$P = C_k\left(\frac{1}{2}\rho v^2 A\right) \text{Sin}\,\omega t \qquad (5)$$

in which

C_k = Karman force coefficient = 1 (See Ref. 2)
ρ = Mass density of air
v = Velocity of wind
A = Projected area
ω = Angular frequency of the disturbing force

METHOD OF STABILIZATION OF PIPELINE AGAINST WIND VIBRATION

When a shorter unsupported span for a pipeline is not cost effective, in order to eliminate the risk of damage from wind induced vibration, several devices to stabilize the pipeline can be considered. Most of these devices fall into the following two categories.

1) Vortex Spoilers - prevent formation of vortex shedding by modifying wind sensitivity parameters of a pipe. A comparison of various designs and concepts (Figure 3) of this group is shown in Table 1.

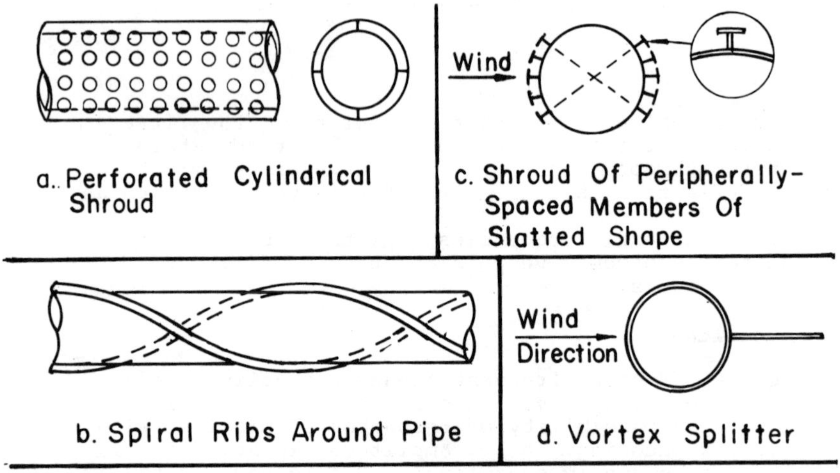

Figure 3. VARIOUS CONCEPTS OF VORTEX SHEDDING SPOILER

TABLE 1. COMPARISON OF VORTEX SPOILERS

Description of Device Inventor/Investigator	Function/Effectiveness/ Advantage
a. A perforated cylindrical shroud fitted around and spaced apart from the pipeline by P. Price of Stanford University in 1956. (1)	The flow of air through lattice openings is complex. The mechanism of vortex elimination not theorized; however it is effective in suppressing vortex excitation in subcritical, transition, and supercritical regions of Reynolds Numbers.
b. Spiral rib or strake around pipeline by C. Scruton and D.E.J. Walshe of National Lab., 1957. Further tested by L. Woodgate and J.F.M. Mabrey in 1959. (1,7,8)	Experiments indicated a proper combination of spirals, pitch and strake height suppressed vortex formation effectively in all ranges of Reynolds Numbers.
Spiral rib of circular shape by W. Weaver of MIT in 1959. (1)	Winding of 4 spirals of 12 pipe-diameter pitch showed effective in suppressing oscillation in subcritical and transition range of Reynolds Numbers, but not in supercritical Reynolds Number. The wind tunnel test revealed that the spoiler need not be applied over the total length of the member.
c. A shroud around at least part of the pipe periphery comprised of spaced slatted shapes by Henry H. Wong of Bearsoen, Scotland. (10)	The shroud weakens vortices and disrupts their periodic formation at their upstream source. Therefore, wake buffeting of the downstream structure will be diminished. An incomplete shroud can be effective if the flow direction is known.
d. Longitudinal fin at downwind side. Extended a distance of at least one pipe diameter. (5)	The fin or plate stabilized vortices at each side of fin, so they remain fixed in its position and the alternate shedding and associated side thrusts are stopped.

2) Vibration Damper - Increase the effective damping through the installation of mechanical devices to reduce the amplitude to a safe level.

Numerous devices were used for transmission lines and stacks. A typical Stockbridge damper clamped on transmission lines, consisting of a piece of stranded steel cable about 12 inches long, carrying cast iron weights of about 1 lb. Another common type is a friction damper, consisting of springs and dashpot, normally mounted in guy cables of a stack (2). However, these are unsatisfactory for stabilizing pipelines as they would require an elaborate auxillary damper system in order to dissipate large amounts of vibration energy especially in the fundamental model.

It is understood that Russian engineers developed a friction telescoping vibration damper for a Siberian pipeline in Arctic areas (9). This 20 and 28-inch gas pipeline from Messoyakha gas field to Norilsk is supported aboveground the entire length. Vibration induced by high winds has contributed to support failures and to pipe jumping off supports. To prevent wind induced vibrations, these devices were installed intermediately between pipe supports. This device consists of a strut composed of one piece of small diameter pipe inserted inside another. The outer pipe is slotted over a considerable length from the end into which the inner pipe is inserted. A heavy clamp is bolted around the slotted pipe to squeeze it against the inner one. The free end of one pipe is fastened to the pipeline through a universal joint and the free end of the other is anchored to the ground in a like manner. The exact length is easily adjusted through the telescoping connection after which the clamp is drawn up tightly. A sketch of this interesting device has been developed from the above description as shown on Figure 4.

DESIGN TO PROTECT A PIPELINE IN KUPARUK FIELD

The Kuparuk pipeline, 16 inch O.D. x .342 WT is supported above ground on piles at 60 feet spacing for the entire 30 miles. The pipeline span increases to 115 feet, to prevent jamming of ice floes in spring breakup, at the Kuparuk River crossing. The pipe wall thickness of this approximate one mile crossing also is increased to .750 inch to carry the longer span. In both on-shore and river crossing portions of the pipeline, the fundamental frequencies of pipeline fall within the range of critical resonant frequencies of vortex shedding induced by a postulated range of steady wind from 10 mph to 50 mph (Figure 2).

Spans would have to be reduced to about 25-28 feet to avoid the possible problem of vibration. Due to the lack of records of steady wind at the time of design, reasonable assumptions had to be made for a cost effective design. The probability of prevailing steady wind within 10 - 50 mph following the river flow normally is relatively higher than that of low height on-shore pipelines. A decision was made to protect the river crossing portion of pipeline with add-on vortex shedding spoilers.

The type of spoiler adapted for deployment was of the spiral strake kind. It was designed with three spiral strake of a 9 foot pitch, and 2 inches height (Figure 5) on a 2 half-circle cylindrical clamp of 10 gage steel fabricated in 9 foot segments for ease of handling and installation. The design possesses the following merits:

a. Effectiveness in suppressing vortex formation in all applicable ranges of Reynolds Number in the wind stream, independent of wind attack angle.

b. Reasonably effective for partial application on a span.

c. Relative ease in fabriation and installation.

To minimize cost, these spoilers were installed at quarter points of a span with 18 foot length of spoiler for each application. Inference from tests conducted for circular shape of spiral (1) indicates this arrangement possibly would reduce the vibration to about 2% of the resonant amplitude.

Away from the river crossing, due to low height of the pipeline, average 5 feet above grade, it is doubtful that steady wind of enough frequency and duration, to cause damage to the pipeline, would occur for the majority portion of the pipeline. Thus, the design philosophy here was to build the pipeline without vortex spoilers on it. Should vibration become a problem later at some locations, then field installation of spoilers could remedy this situation with ease. This did happen.

The construction of the pipeline was completed in 1981. In January of 1982, the line patrol witnessed a "bouncing" effect on the pipeline just east of river crossing. The effected area was approximately one mile in length. The maximum measurement taken at mid-span indicated a vibration amplitude of approximately one inch. Wind velocity and direction taken at two nearby locations registered from 8 to 20 mph, with directions generally at a small angle from the perpendicular direction of the pipeline. These small amplitudes induced negligible bending stress in the pipeline. However, 31 9-foot segments of spoiler were added on to the pipeline for precaution up to March of 1983. A field monitoring program in the meantime has been established to measure direction, frequency, duration of steady wind, and pipeline vibration amplitudes. This program is aimed to evaluate more accurately the effectiveness of the spoiler and to optimizing the usage of spoiler length for metal fatigue control of the pipeline.

ACKNOWLEDGEMENT

The Kuparuk Pipeline Project was for Atlantic Richfield Company. The assistance from M.J. Fatur and K.J. Nyman both of Arco is highly appreciated. In addition, credit is due to the following staff of Williams Brothers Engineering Company for their input and support in the design phase: R.E. Browning, J.P. Patel, J.N. Rau, and R.N. Thrinh (currently with C.F. Braun). Special thanks to J.N. Rau and J.R. Wall for their review of this paper.

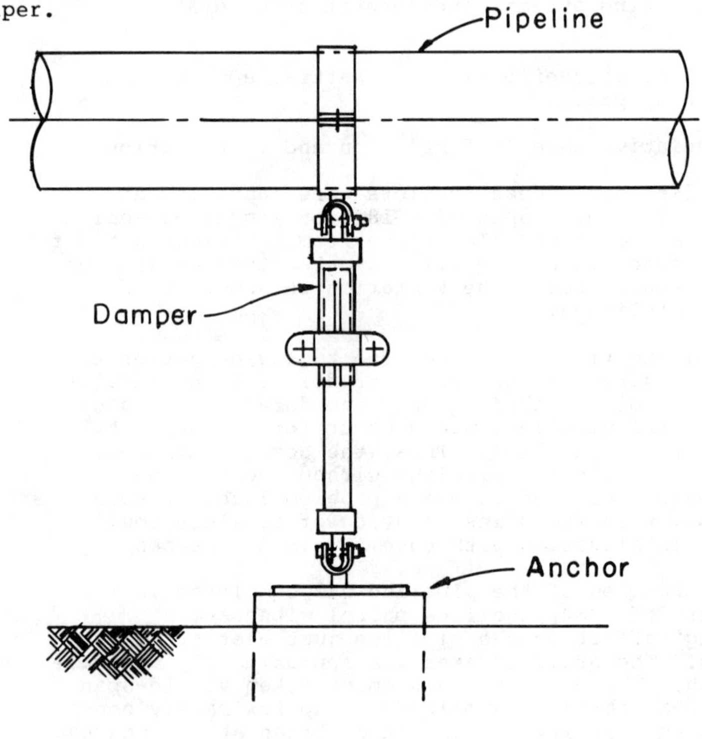

Figure 4. DAMPER USED IN RUSSIAN ARCTIC PIPELINE

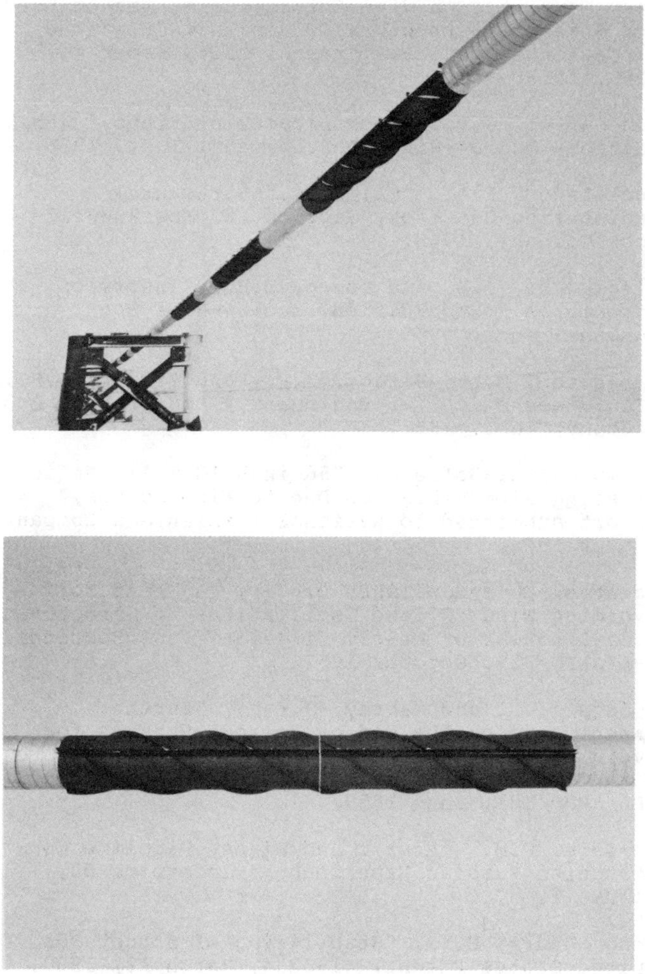

Figure 5
VORTEX SPOILER FOR KUPARUK RIVER PIPELINE CROSSING

REFERENCES

(1) Task Committee on Wind Forces, Committee on Loads and Stresses, Structural Division ASCE, "Wind Forces on Structures," trans. ASCE, Paper No. 3269, 1959.

(2) Den Hartog, J.P., "Mechanical Vibrations," 4th Edition, McGraw-Hill Book Company, Inc., 1956.

(3) Dockstader, E.A., "Chimneys," Structural Engineering Handbook, Section 26, McGraw-Hill Book Company, 1968.

(4) Timosugnko, S.P. and Young, D.H., "Theory of Structures," 2nd Edition, McGraw-Hill Book Company, Inc., 1965.

(5) Albertson, M.L., Barton, J.R., and Simons, D.B., "Fluid Mechanics for Engineers," McGraw-Hill Book Company, Inc., 1960.

(6) Snamprogetti-Sofregax, "56-inch IGAT II, Sefid Rud Crossing Pipe Vibration Due to Wind Forces," Report submitted to National Iranian Gas Company, August 1978.

(7) Scruton, C. and Walshe, D.E.J., "A Means for Avoiding Wind Excited Oscillations of Structures with Circular or Nearly Circular Cross Sections," NPL/AERO/335, October 1957.

(8) Woodgate, L. and Mabrey, J.F.M., "Further Experiments on the Use of Helical Strakes for Avoiding Wind Excited Oscillations of Structure with Circular or Near Circular Cross Section," NPL/AERO/381, June 1959.

(9) Sanders, W.M., "Arctic and Winter Pipeline Construction Russian Experience," Interpipe 80, Volume I.

(10) Wong, Henry, H.Y., "Stabilizing of Structures," United States Patent, 4193234, March 18, 1980.

EVALUATING EXTERNAL PIPELINE CORROSION

William T. Mori, A.M. ASCE,[1] and Gerald P. Duppong[2]

ABSTRACT

Two existing cast iron sewage pipelines were surveyed for external corrosion and the results of the survey analyzed. The surveys and analysis were used to determine the present condition and estimate and remaining useful life of the pipelines. Field testing and statistical analysis indicate that the two pipelines are being corroded and that selective dissolution of the iron is taking place. Recommendations were: monitor both pipelines, replace them in the near future (5 to 10 years), and provide cathodic protection for the replacement pipelines.

INTRODUCTION

Cast iron pipelines corrode with age, and failures can eventually occur. Surveys to determine the extent of pipeline corrosion can indicate the possibility of such failures before they occur. These surveys also allow an evaluation to be made to determine whether to continue use of the pipeline or replace it.

This paper describes the procedures that were used to perform an external pipeline corrosion survey of two existing sewage pipelines and the subsequent statistical analysis. The surveys and analysis allowed the existing condition of the pipelines to be determined and the useful life of the pipelines to be estimated.

BACKGROUND

The cases presented document the evaluation of the external corrosion of two cast iron sewage pipelines in Washington State. Figure 1 shows the locations of both pipelines.

Case A

The 7,000-foot-long (2,134 m), 20-inch-diameter (508 cm) cast iron pipeline was installed on railroad right-of-way along Sinclair Inlet in Bremerton, Washington, in 1949. Class 150 pipe with 3/16-inch (4.8 mm) cement mortar lining and bell and spigot joints was

---

[1] Sanitary Engineer, CH2M HILL, 1500-114th Avenue S.E., Bellevue, Washington 98004.

[2] Corrosion Engineer, CH2M HILL, 1500-114th Avenue S.E., Bellevue, Washington.

FIGURE 1
Pipeline Locations

specified. All joints were lead caulked. According to the record drawings, the pipe was buried with 1 foot (0.3 m) to 5 feet (1.5 m) of soil cover.

Case B

The cast iron pipeline was installed in the beach along Port Gardner Bay between the cities of Mukilteo and Everett, Washington, in 1963. The pipeline comprises 17,000 feet (5,181 m) of 12-inch (30-cm), 14-inch (35-cm), and 16-inch (40-cm) cast iron pipe. Class 100 pipe with 3/32-inch (2.40-mm) cement mortar lining and mechanical joints was specified. The pipeline joints were not electrically bonded. According to the record drawings, the pipe was buried with 1 foot (0.3 m) to 10 feet (3 m) of soil cover.

CORROSION OF CAST IRON

Cast iron is an alloy composed primarily of iron, carbon, and silicon. Selective dissolution of iron from a cast iron pipe is associated with two types of corrosion: general corrosion and pitting corrosion. General corrosion, which is caused by the corrosive action of the soils, results in a uniform reduction of pipe wall thickness with time. This general corrosion process is known as graphitization and is the most common form of corrosion on buried cast iron pipe.

During the process, selective dissolution of ferrite iron leaves behind graphite flakes. Graphitized cast iron has a low structural strength. However, the appearance of graphitized cast iron is deceiving and it may look as if it still has a sound structural integrity. To fully evaluate graphitized cast iron pipe, the graphite must be removed.

Pitting corrosion results in highly localized areas of metal loss. Pitting corrosion leads to perforation of the pipe wall in a relatively short period of time, even though general corrosion may reduce the overall pipe wall thickness only moderately. The selective dissolution caused by the two types of corrosion is illustrated in Figure 2.

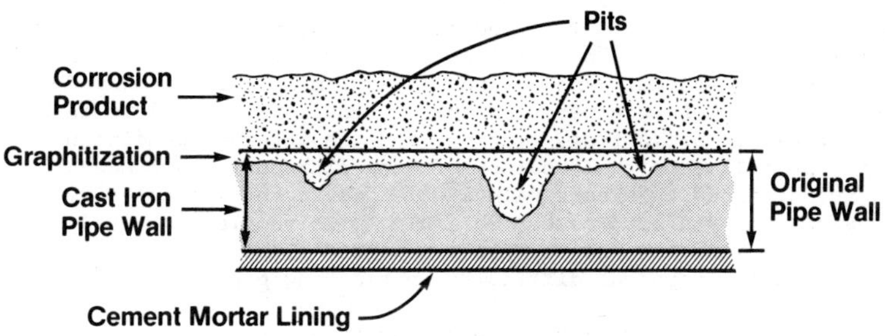

FIGURE 2
Corrosion of Cast Iron Pipe

The rate of either type of corrosion depends on the characteristics to which the cast iron pipe is exposed. The primary soil characteristics that affect the corrosion rate of buried cast iron are electrical resistivity, pH (acidity), soil aeration (oxygen content), and chloride and sulfate salt concentrations.

FIELD TESTING

Visual Inspection of Excavated Pipelines

Both pipelines were exposed by excavation for inspection. The soil conditions in Case A were gravelly sands. In Case B, sands with clay and silt were encountered. The pipeline in Case B was covered with salt water periodically because of tidal fluctuations. The pipeline in Case A was never submerged in salt water. The pipeline in Case A was covered with thin, tightly adhering red and black corrosion products. The Case B pipeline was covered with thick red and black corrosion products at all test locations. The corrosion products in both cases were easily removed by a high-velocity jet of water. The pipe surfaces beneath the corrosion product were generally graphitized. Probing the pipe surface with a screwdriver revealed moderate localized corrosion (pitting) in Case A and significant pitting in Case B. The odor of hydrogen sulfide was observed in Case A and the sound of escaping gas and bubbling on the pipe surface was noted in Case B.

Pipe Wall Thickness. The pipe wall thickness of both pipelines was measured using ultrasonic (high-frequency sound) test methods to determine the amount of metal loss caused by general corrosion. The measured pipe wall thickness in Case A ranged from 472 mils (11.99 mm) to 616 mils (15.65 mm), and the measured pipe wall thickness in Case B ranged from 490 mils (12.45 mm) to 625 mils (15.88 mm).

The specified pipe wall thickness for 12-inch (30-cm), 14-inch (35-cm), 16-inch (40-cm), and 20-inch (50-cm) pipe is 480 mils (12.19 mm), 510 mils (12.95 mm), 540 mils (13.72 mm), and 157 mils (14.48 mm), respectively. The specified casting tolerances are 60 mils (1.52 mm) for 12-inch (30-cm) pipe and 80 mils (2.03 mm) for 14-inch (35-cm), 16-inch (40-cm), and 20-inch (50-cm) pipe. The measured pipe wall thicknesses were greater than the original specifications in both cases. Since graphitization, pitting, and the internal condition of the pipe can influence the accuracy of such ultrasonic test measurements, the obtained thickness values were considered inconclusive in this evaluation. In both cases, obtaining pipe samples for wall thickness determinations was not practical because of the nature of the pipelines and the physical difficulty in removing pipe samples.

Reduction of pipe wall thickness caused by general corrosion can be estimated from the graphitization measurements. The pipe surface in Case A was graphitized to a depth of 50 mils (1.27 mm) and corresponds to an overall reduction of approximately 10 percent in 20-inch-diameter (50-cm) pipe wall thickness. The pipe surfaces in Case B were graphitized to a depth of 60 mils (1.52 mm) to 125 mils (3.18 mm), which corresponds to an overall reduction in pipe wall thickness of approximately 20 percent.

Pit Depth. Pit depths were measured using a pipe pit gauge. In Case A, the maximum depth of observed pitting was 130 mils (3.30 mm); the greatest pit depth in Case B was 290 mils (7.37 mm). Visual observation of both cases revealed that superficial pitting (roughening of the pipe surface) was widespread beneath the graphitized pipe surface.

Pipe Joints. In Case A, one pipeline joint was exposed by digging approximately 2 feet (60 cm) below the surface. The exposed joint was leaking slightly and hydrogen sulfide odor was present.

A pipeline joint was also excavated and inspected in Case B. The pipeline joint retainer ring was in poor condition. Significant metal loss of the retainer ring had occurred because of corrosion activity. Near one of the bolt holes, 1/4 inch (6.35 mm) to 3/8 inch (9.53 mm) of metal had been lost to corrosion. Both general corrosion and pitting corrosion were apparent on the retainer ring. The external surfaces of the ring bolts were in good condition.

Chemical Analyses and Electrical Testing

In both cases, soil samples were collected from each excavation at the depth of the pipe. The soil samples were analyzed for electrical resistivity, pH, and chloride and sulfate content. These soil properties affect the corrosion rates of buried metallic pipelines.

Electrical Resistivity. Corrosion is an electrochemical process in which a metal returns to its natural state as an ore. Soils with low electrical resistivity more readily conduct electrical current associated with corrosion and are therefore more corrosive than soils with high electrical resistivity. Soil resistivity measurements were used to evaluate the corrosiveness of the soils around the pipeline. Table 1 relates soil resistivity to the expected degree of soil corrosiveness.

Table 1
SOIL RESTIVITY COMPARED TO SOIL CORROSIVENESS

Corrosiveness	Soil Resistivity
Very Corrosive	0 to 3,000 ohm-cm
Moderately Corrosive	3,000 to 10,000 ohm-cm
Slightly Corrosive	10,000 ohm-cm

Case A had a soil electrical resistivity of 26,000 ohm-cm, which indicates a slightly corrosive soil. Case B had an average soil electrical resistivity of 550 ohm-cm, indicating very corrosive soil conditions.

pH. The pH value is a measure of acidity or alkalinity of the soil. A pH of less than 7 indicates acidic conditions, and a pH of greater than 7 indicates alkaline conditions. When all other factors are the same, an acidic soil is generally more corrosive to cast iron than an alkaline soil.

Case A had an average soil pH of 5.5, which indicates slightly acidic soil conditions. Case B had an average pH of 7.1, indicating nearly neutral soil conditions.

Chlorides and Sulfates. As cast iron corrodes, the corrosion products formed usually act as a protective film that inhibits further corrosion. Chlorides and sulfates present in the soil can interfere with the formation of the protective corrosion product layer and allow the cast iron to continue to corrode at the initial rate.

The chemical soil analyses in Case A indicated that the chloride and sulfate concentration ranged from 4 to 10 ppm (mg/l) and 2 to 24 ppm (mg/l), respectively. These low concentrations would not increase the corrosion rate of the pipeline. Chemical soil analyses in Case B indicated that the chloride and sulfate content of the soil ranged from 400 to 1,300 ppm (mg/l) and from 130 to 410 ppm (mg/l), respectively . The high concentration of these chemicals is conducive to a higher rate of cast iron corrosion.

STATISTICAL ANALYSIS OF PITTING

A statistical analysis of the pitting corrosion of both pipelines was performed using the information gathered during the limited field inspections. These statistical analyses were used to estimate the condition of the entire pipeline.

Soil-related corrosion of several metals has been the subject of an extensive study by the National Bureau of Standards (NBS) (Reference 1). The NBS study began in 1922 and included a field burial program in which different metals were exposed to a variety of soils. The metallic samples were excavated and examined after certain periods of time to determine the extent of corrosion. The test results showed that the depth of pitting is best related to time and the soil parameters by the following equation:

$$P = kT^{n} \qquad (1)$$

where T = time in years

k,n = constants that depend upon actual soil conditions

In 1968, John R. Rossum (Reference 2) used the electrochemical theory of underground corrosion to derive equations for depth of pitting of metals in poor, fair, and well-aerated soils. The derived equations were of the form $P = kT^{n}$, and the theoretical values of n were in agreement with the values of n empirically derived by the NBS. The constant k was expanded to include soil resistivity and pH, two parameters that influence corrosion in the electrochemical theory. The resulting general expression for depth of pitting is shown below.

$$P = K_{n} \; (10\text{-pH})^{n} \; p^{-n} t^{n} \qquad (2)$$

where P = depth of pit in mils

p = soil resistivity in ohm-centimeters

K_{n}, n = aeration constants

t = time in years

As the number of pits examined increases, the probability of finding a pit with a depth greater than another pit also increases. The probability of maximum depth of pitting is, therefore, a function of area. An empirical equation for the relationship between maximum depth of pitting and area was developed by Scott (Reference 3) and is shown below:

$$P = K_a A^a \tag{3}$$

where P = depth of pit

K_a = deepest pit in unit area

a = constant, depending on soil and metal

A = area

Combining equations (2) and (3), Rossum's general equation for predicting maximum depths of pits is:

$$P = K_n K_a \ (10\text{-pH})^n \ p^{-n} \ t^n \ A^a \tag{4}$$

where A = area in square feet

n = constant, depending on soil and metal

By measuring these soil parameters and using the empirical values derived from the NBS tests, it is possible to plot pit depth against time to establish a corrosion rate that is predictable, assuming the soil parameters remain unchanged.

If the pipe wall thickness is known, then the total number of pipe wall penetrations in a given time can be calculated by rearranging Rossum's equation shown below

$$L = A\left(\frac{KnKa}{z}\right)^{1/a} \left(\frac{t\ (10\text{-pH})}{P}\right)^{n/a} \tag{5}$$

Where z = pipe wall thickness

L = number of penetrations in a given time

The measured values of soil resistivity, pH, and estimated aeration were used in Equation A to calculate the theoretical maximum pit depth at each excavation. The theoretical values were compared to the actual maximum pit depths found in the field . A good correlation existed between theoretical and actual maximum pit depths.

The average soil resistivity and pH values found along the pipelines were used in Equation B to calculate the statistical number of pipe wall penetrations per mile of diameter pipe per year. The number of penetrations was calculated for soils of good or fair and poor aeration depending upon the soils condition.

In soils of fair or good aeration, a precipitate ferric corrosion product will form over the pit. When cathodic reaction causes the deposit to form, corrosion is slowed because the diffusion rate of oxygen is decreased as the deposit thickens.

In soils of poor aeration, the ferrous ions migrate away from the pit before they are oxidized and do not retard the corrosion rate.

The results of these calculations for Cases A and B are shown on Figures 3 and 4, respectively.

EVALUATION OF RESULTS

The overall reduction of pipe wall thickness in Cases A and B because of corrosion appears to be approximately 10 and 20 percent, respectively, as determined by observed external pipeline graphitization.

Localized corrosion in the form of pitting is moderate in Case A and substantial in Case B. Statistical analysis using actual soil characteristics measured in the field indicate that in Case A approximately 11 to 41 pipe wall penetrations per mile have occurred in 35 years and in Case B about 60 to 70 pipe wall penetrations per mile have occurred have occurred in 20 years. This corresponds to an average of one penetration per ten pipe lengths in Case A and one penetration per four pipe lengths in Case B.

Penetration of the pipe wall will eventually result in a leak. The cement mortar lining inside the pipeline may, at least for a time, prevent or retard leakage at many locations where the pipe has been penetrated by pitting. Since the condition of the cement mortar lining was unknown, its effectiveness in preventing leaks could not be estimated.

As the number and size of pipe wall penetrations increase, the number of leaks will also increase. This leakage will probably not be visually detectable because of soil conditions around the pipelines. However, with the increase in the number of leaks and an increase in graphitization, the mechanical strength of the pipe decreases and the probability of a pipeline break increases.

Corrosion products were also observed on the retainer rings at pipeline joints in Case B. Leaks may begin to occur at pipeline joints.

Based on the field investigation and statistical analysis, the estimated remaining useful life for the Case A pipeline is 10 to 15 years and 5 to 10 years for the Case B pipeline.

Cathodic protection was considered as an alternative to replacement of the pipelines. Cathodic protection is an electrical method of corrosion control for buried metallic pipelines. However, it will only prevent corrosion of the pipeline from the time it is installed and cannot replace lost metal or otherwise improve the structural integrity of the pipe. In addition, cathodic protection requires electrical continuity of the pipeline. Both cases have joints that are not electrically continuous. Electrical bonding for continuity would require excavation of every pipe joint. Also, operation of a

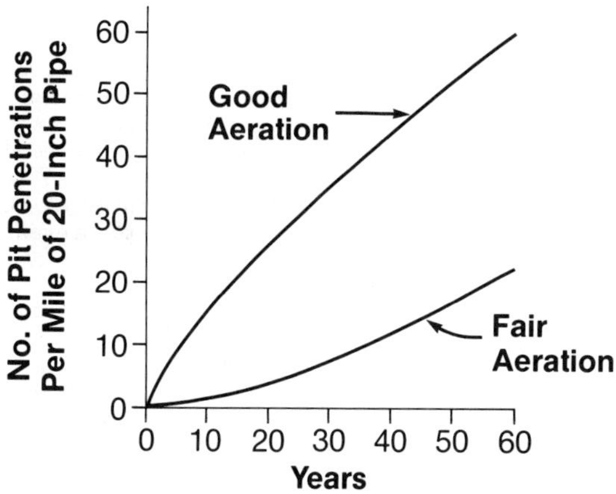

FIGURE 3
CASE A
Penetrations per Mile

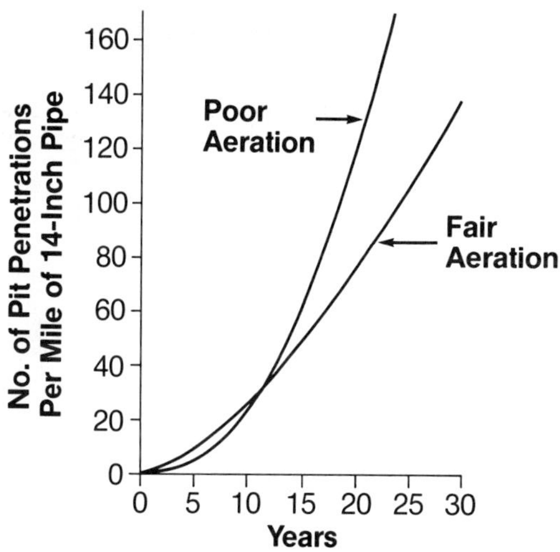

FIGURE 4
CASE B
Penetrations per Mile

cathodic protection system could accelerate the corrosion of adjacent buried pipelines because of stray electrical currents.

SUMMARY AND CONCLUSIONS

Field testing and statistical analyses indicate that the two pipelines are being corroded and that selective dissolution of the iron is occurring. Monitoring of both pipelines was recommended and replacement in the near future (5 to 10 years) was recommended. Cathodic protection for the replacement pipeline was also recommended. Cast and ductile iron is a still a suitable pipe material but the pipeline must be installed with corrosion control to prevent corrosion damage.

ACKNOWLEDGMENTS

Credits

This paper was presented at the second specialty Conference on Pipelines in Adverse Environments, San Diego, California (November 1983).

REFERENCES

1. Romanoff, Melvin. Underground Corrosion. National Bureau of Standards Circular 579. Washington, D.C.: U.S. Government Printing Office. 1957.

2. Rossum, John, "Prediction of Pitting Rates in Ferrous Metals from Soil Parameters," American Water Works Association Journal. 1968.

3. Scott, G. N. "Adjustment of Soil Corrosion Pit Depth Measurements for Size of Sample." Proceedings American Petroleum Institute Vol. 14, No. 204. 1934.

AN INSIDE LOOK AT THE PIPELINE

B. VerNooy & Scott Jordan*

ABSTRACT

Instrumented pipeline pigs are self-contained devices which travel through a pipeline propelled by the flow of fluid carried in the pipeline. They measure various parameters and record the data for later evaluation. The integrity of the pipeline can be judged from the information provided.

This paper reviews the history and current status of several types of instrument pigs including those that measure pipeline geometry, leakage, corrosion, and other flaws.

DEFINITION OF INSTRUMENT PIG

Before we can evaluate the instrument pig, we need to look at its background and arrive at a definition. If we define an instrument pig as a tool to detect and/or locate some pipeline fault not obvious to the aboveground observer, then the first instrument pig was the first pig that got stuck in the pipe. No matter what the original function of that pig might have been, and no matter how it was finally located and removed, the process of getting it out determined the location and nature of a fault. If the deficiency was with the pipeline structure or its fluid or solid contents, the pig performed a valid and valuable service in pinpointing the defect, even though this particular service may not have been appreciated at the time. If the failure lay with the pig itself, the pig was inadequate for its intended use, and the costs incurred would be classed as educational. In the early days of pipelining, many pigs were installed in launching traps with more hope than promise that they would be removed from a receiving trap.

Pigs which managed to negotiate the pipeline to the receiver, but arrived with physical damage above and beyond normal wear and tear, could often be analyzed to determine the nature, if not the location, of some obstruction. In the evaluation, one had to consider the possibility of damage resulting from operational error, of course. It was difficult to discern the difference between a pig that had been forced under the trap valve disc while it was opening, and one

*B. VerNooy, Vice President, Consulting Services
Scott Jordan, Senior Special Projects Engineer
T. D. Williamson, Inc., Box 3409, Tulsa, Oklahoma 74101

that had crunched its way through a flattened pipe; but, careful analysis of the scars made a surprisingly accurate conclusion possible, and thus a damaged pig was, in effect, a recording instrument. Fortunately, we have learned a lot about pipeline and pig operations, as well as about pipeline and pig design. Significant pig damage or pig loss is unusual today.

GAUGING PIGS

The first pigs expressly designed to detect and locate faults in a pipeline were gauging pigs. These pigs had an oversize steel disc on the leading face, and were run through new or reconditioned pipelines. The construction contract contained a clause requiring the

SMALL DIAMETER GAUGING PIG WITH GAUGING PLATE BETWEEN BUMPER AND FRONT CUP

passage of a pig of specific diameter completely through each line section before that section was accepted. This run proved a substantially round opening from one end to another, and these tools were often called prover pigs. In the days of gas welding, the gauging plate was stellited or hardened around the edge to remove the welding stalactites that protruded into the pipe. The diameter of the gauging plate was a matter of arbitration; it was ordinarily specified to be somewhere between 90 and 95 percent of the inside diameter of the pipeline. A few were furnished with larger diameters, and many had their diameters reduced in the field while the inspector was gone to lunch. Unbelievable as it sounds, one pipeline company specified an extra heavy plate with a rounded leading edge, so that it could be forced through obstructions to "round out" their 30-inch pipe. In the days of heavy-wall, mild steel pipe and relatively small diameters, the gauging pig was a valuable legal instrument for pipeliene acceptance; and, it certainly brought the lunchboxes, rocks, and dead rabbits out of the pipe.

STUCK PIGS

As steel improved and pipelines got larger and thinner, the owners were understandably concerned about the increasing possibility of damage to the pipe by a gauging flange. Some switched to aluminum gauging flanges or thin sheet-metal flanges, notched to bend easily and stay bent to indicate the greatest degree of diametral reduction. In the process, however, one of the advantages of the gauging plate was lost. In smaller lines, the gauging pig was propelled by air at relatively low pressure. A chain or noisemaker aided in following or leapfrogging the erratic pig travel; and, if the pig encountered a dent, it stopped. A continuing pressure would cause the cups to leak, and an ear to the ground could usually pinpoint the pig and the obstruction. It wasn't very sophisticated, but it was effective because it forced the contractor to cut and repair the pipe to get the pig out. The larger the pipe, the more erratic the pig ran. With more square inches to act on, a smaller pressure moved the pig. With a larger volume of air to compress, the pig sat longer, waiting for enough pressure to move it away from a valve or fitting or heavy wall pipe; and, when it did move, it did so rapidly and forcefully. Following the pig on foot became impractical, if not impossible, and pigs were equipped with radioactive tracers so they could be located without digging out and tapping the line. At higher velocities, the pig often passed the obstruction and would arrive with a distorted gauging plate to indicate the

LARGE-DIAMETER FOUR-CUP GAUGING PIG. GAUGING PLATE DISTORTED FROM PASSING PIPELINE DENT.

degree but not the location of the fault. The increased use of hydrostatic testing suggested that the gauging pig do double service by acting as the fill pig at the head of the column of test water. Now, the pig had the energy of tons of moving water behind it, and it couldn't quickly stop this mass, so the pig deformed and passed the obstruction. And, as a fill pig, it wasn't very effective. The standard gauging pig is sold on the basis of low cost and quick delivery. It is a short, heavy, pig with its weight concentrated at the front so that it quickly wears out in the construction environment of dirt and debris in the line. Neither the design nor the environment are conducive to a good, lasting seal.

Since a number of contractors "lost" pigs containing radioactive materials (after they were out of the pipeline), the use of isotope tracing has been subject to so much control that it is seldom used. Other means of pig location have been developed, such as noisemaker wheels, radio transmitters, and pingers; but, the development of pipeline instrumentation using radioactive materials has suffered. Some of the radiation principles used in oil well surveys could have pipeline applications, for example.

The greatest investment in a pipeline was the tonnage of steel in the pipe. To reduce this weight and cost, pipe materials and processing were improved. The pipe got stronger, but thinner, and made larger sizes or higher pressures more economical. Because the pipe was thinner and its yield point had been pushed up closer to its ultimate strength by cold expansion, the pipe became more sensitive to notches, gouges, and other deformations which reduced the wall thickness and caused stress concentrations. Failure was less straightforward and was more likely to occur after a period of operational cyclic stress reversals than in the initial hydrostatic test. The pipe had to be handled more carefully to avoid dents, grooves, and arc burns. The old gauging pig could now be a gouging pig with the ability to inflict an unknown level of damage.

KALIPER PIG®

One of the simplest and most effective of the several instrument pigs available today has found a market because changes in pipe and procedures made the gauging pig less effective and more hazardous. A pig carrying a continuously recording internal caliper was designed so that it could traverse a round or flattened pipeline without metal-to-metal contact with the pipe wall. It continuously records the minimum inside diameter of the pipe. It can be propelled by air or the test water or the pipeline fluid column displacing the test water, and it has conical cups which will seal in out-of-round pipe. The recording tape advance is controlled by pulses from a pair of odometer rollers so that the tape moves in direct proportion to the movement of the pig through the pipeline. Originally intended for new construction as an alternate to the gauging pig, the KALIPER pig is as easy to handle as the gauging pig and provides a permanent visual record. Without further processing, the tape can

®KALIPER is a trademark of T. D. Williamson, Inc. registered in the U.S. Patent Office and in other countries.

KALIPER PIG. PIPELINE DIAMETER
REDUCTIONS ARE MEASURED BY FINGERS
AND RECORDED ON INTERNAL INSTRUMENT

be analyzed to evaluate and to locate diametral reductions, both from their position on the tape, and in reference to welds and other known benchmarks normally recorded.

Of course, the original gauging pig would not have been used in an operating pipeline; and, of course, the diameter of an operating line can also be distorted, most often by the earth-moving machinery of a third party working in close proximity to the pipeline. Wrinkles, flattening, and dents can also result from settling, from thawing or frost heave, or from any change in support due to significant movement of the pipe and/or its supporting soil. River crossings, permafrost areas, earthquake and landslide zones, and offshore currents can cause failures which are preceded by detectable buckling or flattening. Runs of a KALIPER pig can be made without shutdown in almost any pipeline, and some pipelines run these pigs on a regularly scheduled basis as a safety check or in conjunction with other inspection pigs for a more detailed survey.

The caliper mechanism can be reversed to record the maximum instead of the minimum inside diameter of the pipe. This inside-out pig has been used to locate stretched pipe as indicated by continued input of water during hydrostatic testing without an accompanying increase in pressure. Usually, this is accidental because of testing errors or installation of a joint of substandard pipe. If we are really concerned with the most efficient use of our resources, we should take advantage of this procedure. Why not test every pipeline to the point where the first yield occurs, now that we can find and replace those joints? After replacement of the stretched joints and a successful test to a slightly lower pressure, most pipelines would find that they could safely increase their operating pressure and throughput. Of course, this proposal is not without problems. The physical problems can be resolved. Steel mills might oppose it as not being to their best interests.

The KALIPER pig, as it is marketed today, does not have the structure or function of the original internal pipe-measurement pig. The first onstream internal measurement tool was built by T. D. Williamson, Inc., as a study tool for Battelle and Texas Eastern research on pipe reinforcement. This tool simply assumed a pipe diameter was a circle, and a circle is determined by three points, and the relative position of one point to another pair of points will determine the diameter of the circle. The tool was very light, with a small, clockwork recorder and spidery arms to the pipe wall. It verified that thinwall pipe wrapped with piano wire could be used in higher pressure service, and brought up questions about other line diameters. For regular extended use, the tool was too elementary and too fragile. It was reproduced by Texas Eastern in much stronger form and a second record trace added to detect each weld. The tape was driven directly by the rollers, eliminating the time-based clockwork device.

LEAK DETECTION PIGS

The information the KALIPER pig provides is often valuable in finding points that will leak in the future. It is useless in the location of leaks which are not accompanied by distortion, but a number of instrument pigs have been made with the express purpose of locating leaks in pipelines. Leaks can be a nuisance, a hazard, a financial loss, and a civic responsibility. A lot of thought has been given to finding them from inside the pipeline. Patents have been issued for pigs that would find leaks by sensing the difference in pressure in front of, behind, and between the sealing elements of pigs. One patent suggests a heater in the center of the pig and sensing thermometers around it. With no leak, the flow is all straight down the pipe, and the pig's thermometers, all equidistant from the central heater, all read the same. But, at a leak, the flow would move toward the hole in one side of the pipe, and one thermometer would get hotter than the rest. Another pig was to pull an open-ended plastic tube behind it. The tube would slide along the inside of the pipe wall until it came to a leak. Then, the differential pressure would hold it to the pipe wall; the pig would sense the added load, and record the leak. Still another scanned the pipe, looking for the point where the fluid exit would lose pressure and lower the temperature of the pipe wall. These and many others were based on sound principles and could probably be made to work in some condition. Unfortunately, those conditions do not include gasoline in turbulent flow, or natural gas at twenty miles per hour.

A leak detection system developed by Conoco was approaching mechanical success when it expired from environmental considerations.

It involved the introduction of a small isolated batch of radioactive liquid into the pipeline. At a leak, this short, half-life material seeped to the outside of the pipeline just as the gasoline had before and after its passage. The pig followed a few hours later, and its scintillometer detected the residual radioactivity through the pipe wall. This triggered an internal camera to photograph a clock and meter. By comparing flow and time, the leak could be located. It took a lot of dedicated laboratory and field work to find a material that would adhere to the soil longer than it stuck to the inside of the pipe. Background levels had to be evaluated. Some very ordinary-looking cow-pastures suggested someone should file for uranium rights. Ultimately, the developers could find every artificial leak put on the line. But, they never found a leak they didn't know was there and the whole project folded when the legal department pointed out that there would be no defense if they ran a batch of radioactive material under a group of grand champion bulls which subsequently failed to produce.

ULTRASONIC LEAK DETECTION

Dr. Charles Feldstone, manager of the behavioral science section of Southwest Research Institute's bioengineering division, said his studies showed that dogs were more competent in finding gas leaks than all of the various machines in use. Unlike detector instruments, they can discriminate between pipeline gas and the scents of other types of hydrocarbons which confuse conventional devices. Is there perhaps an overlooked relationship between leak-detector dogs and leak-detector pigs? Of course, the dog detects from outside the pipe, and pipelines, particularly those under hydrostatic test, often use a leak detection service to externally pinpoint hard-to-find, very small leaks. The leaks are located by listening for them with sophisticated equipment which amplifies the noise in specific frequency ranges. If this will work from the outside, why not do the same thing from within?

Mr. H. J. EnDean of Gulf Research patented a tool to do this in 1959. The concept was too far in advance of the technology at the time; and, 15 years after his invention, Shell, Exxon and Texaco tried it again. They each developed their own electronic circuits, a variety of carriers and display methods, and their own interpretation of the ultrasonic noises generated by a high-pressure liquid leaking through a "typical" pipeline fault into the surrounding soil. Although their details were dissimilar, the ultrasonic method has come closer to success than any other, and probably holds the most promise for the future. One research group picked four specific frequencies and continuously recorded the amplitude of each of them. This produced a lot of information, some of which could be later interpreted as pump noise or construction work, and it also produced yards and yards of records. Another group picked arbitrary limits and printed the pipeline footage on a tape each time the pig heard the presumably significant noise. This produced only a few inches of tape and eliminated the subsequent analysis and argument.

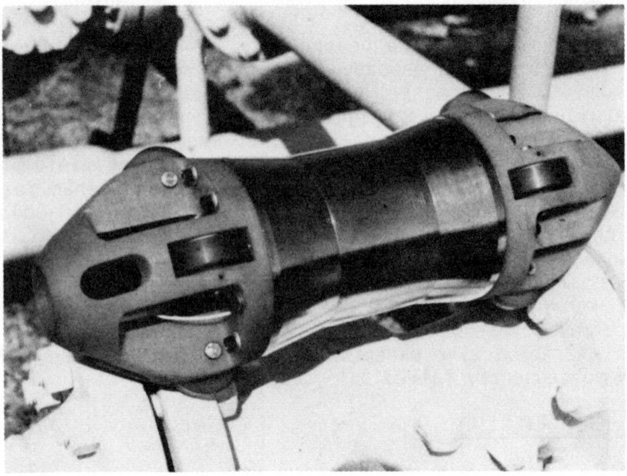

PROTOTYPE ULTRASONIC LEAK
DETECTION PIG USED ROLLERS
RATHER THAN CUPS

Finding leaks goes beyond the instrument into policies and politics. The only successful and continuing use of the ultrasonic leak-locator pigs is in Europe where the pig run was approved as an alternate to a witnessed shut-down pressure test of the pipeline every month. This pig passes one artificial leak as it enters the line and another just before it is removed. If it records both of these and nothing else, then it can be assumed there are no additional unknown leaks between the two artificial leaks. At least, none as big as or with the same characteristics. In America, however, people were not interested in proving there were no leaks in a large-diameter, new, clean oil line. They wanted to find out where the leaks were in small, rusty, dirty, old pipeline. And, therein lies the problem. Ultrasonic leak noises overlap the ultrasonic noises generated by pig cups sliding on pipe, or by rollers turning on axles, or by fluid flow around corners, or through rust, scale and dirt, or past pigs, or any unexpected combination thereof. In large-diameter, clean pipe, a pig can be moved and supported relatively quietly by a very small differential pressure. In dirty, out-of-round, six-inch pipe, no pig can sneak through that quietly. As far as can be determined, there is no successful leak detection pig being broadly marketed today. This does not mean that there are not a lot of people working on it. But, perhaps the answer is to find the fault before it leaks.

CORROSION DETECTION PIGS

Corrosion is the culprit behind many leaks, and corrosion costs are a large part of pipeline maintenance expense, either prevention of corrosion or repair of corrosion, preferably the former. It would be a real money-saver if corrosion control could be applied in doses proportional to the need instead of indiscriminately. It would also be helpful if the electrical currents that cause corrosion and those that control it could be more easily and directly monitored. A patent was filed in 1953 for a corrosion pig made in two parts. Once out of the trap, the front unit would tow the trailing unit on a long cable. Constant electrical contacts with the pipe wall could then be monitored over a significant span, and the current flow direction and level recorded. Flow away from any point would indicate stray current leaking onto the pipe. Flow toward a point would indicate a holiday in the pipe coating, or a point where current left, taking some of the iron pipe along with it. The only thing wrong with this concept is that it has been reevaluated every few years and the technology to make it work apparently hasn't arrived yet.

One way to prevent internal corrosion, and the leaks resulting therefrom, is to internally coat the pipe. Pigs for squeezing a batch of paint and spreading it evenly on the pipe wall may be quite specialized, but could not be classed as instrument pigs. But, how about the pig with sensors to travel through the line, locate unpainted bare spots, and spray paint on them? It doesn't collect information, but when it comes out with paint left in its tank, one could assume all the bare spots were painted.

PHOTO PIGS

Ray Watts of El Paso Natural wondered what kind of a coating job he was getting in a 12-inch line, and in 1969 he made a camera container to photograph it. The container was installed in a pig and took pictures about every forty feet. Since the visibility was good for about four feet, this gave him 10 percent inspection. The 16-millimeter film was single-frame advanced on a time basis and was good for about 20 to 30 miles of pipe. Later, the pig was run in a 30-inch line and worked well in coated pipe but couldn't see through the dust storm in the uncoated pipe. The pig was run onstream in natural gas.

Another photo pig was an offshoot of the pipeline caliper. The first model was simply a trailer behind a caliper pig and actuated to take a series of photos when the caliper sensed a dent. It could also be actuated on a time of footage basis, or by a signal from the ground above the pipeline. This pig took a number of successful photographs of the inside of a gas pipeline at over 500 psi and 12 miles per hour. The good news was that the pictures were surprisingly clear and showed 6 to 12 feet of 30-inch pipe well enough to identify the longitudinal and circumferential welds. Some of the pictures included what appeared to be 2-inch balls of water, perhaps bounced up by the pig, but never explained. The bad news was that the dent the pig was supposed to photograph caused the photo-trailer to overreact, and the result was a series of excellent close-ups of the opposite but undented pipe wall. An improved version was built on a single, unjointed, two-cup body which carried both the caliper mechanism and the camera. More lighting and larger film was provided and a counter was photographed at the same time so that each frame could be identified by pipeline footage. Laboratory testing showed that the use of color and the larger film added a great deal to the quality and anlysis value of the photos. This pig was never marketed because there was no apparent urgent customer demand at that time.

It was obvious that the justification for being able to have photos of the inside of the pipe was there. Obstructions, repeatedly pinpointed by internal measurement but lacking in any external evidence, need to be evaluated more positively. The diameter reduction under the middle of the Ohio River will cost a fortune to uncover; surely it would be better to take a picture of it first. So, the concept of taking pictures inside the operating pipeline has not died. The Peabody Testing Service (Magnaflux) developed a free-running photo pig. They had a service capability in the past for a tool to operate with a crawler in empty pipe of limited lengths. The photo pig was developed for use in operating pipelines. Rather than the general 360-degree longitudinal view of the pipeline, as in previously mentioned pigs, the Magnaflux tool aims its lenses directly at the pipe wall for a closeup view of wall conditions. One camera is aimed at the bottom where most corrosion may be expected, and a greater area of coverage can be obtained by adding more cameras. The pig is weighted to avoid rotation. Test runs were made in Tennessee Gas Company's offshore lines. Problems, such as coping with liquids in two-phase flow, still exist. The pig is now marketed by GEO Pipeline Services.

PRESSURE AND TEMPERATURE SENSING

A series of pigs were made in the 1950's, based on a family of recorders that could provide a continuous trace of the fluid temperature, the fluid pressure, and the differential pressure across the pig. Many runs were made in both gas and liquid lines, and a lot of information of considerable interest, but limited economic value, was gathered. For example, the fluid pressure could be plotted against mileage to verify the pressure drop per mile and, therefore, the relative efficiency of each segment of the pipeline. Or, the fluid pressure could be used to verify the static head or elevations. The fluid pressure and temperatures were used on one mountainous pipeline to select the optimum sites for adding compressor stations. The differential pressure recording showed an increase as the pig picked up natural gas liquids condensed in the lowest point of a river crossing. The temperature recording indicated the cooling effect on the pig as the load decreased when these same liquids evaported. The differential pressure recording predicted the pipeline caliper. It sensed the extra effort required to push the pig into heavy wall pipe or past a dent. One pipeline had repeated cleaning pig damage, but a check of all the gate valves showed them to be wide open. A differential pressure tape showed which one had the faulty indicator. A shortcoming of these pigs was that they recorded on a time rather than distance basis. This required continuously following the pig to check what time it passed specific key points. In a gas line, the expansion and compression of the gas with a varying pig resistance gave erratic pig speeds. These surges increased and, of course, so did the pig velocity as the pig traveled from the higher to lower pressure through the pipe. Today, one would move the record paper in proportion to the distance, rather than by time, and avoid the need to race the pig to the next road crossing.

FLAW DETECTION

If we could find the faults in a pipeline, we would not only be able to prevent small leaks, but disastrous blowouts as well. These faults can be in several major categories. One is the result of corrosion, internal or external, which either concentrates itself in one spot to penetrate the pipe wall, or spreads itself over a larger area until the pipe wall is weakened to the point of rupture by internal pressure. A second type of fault is a longitudinal crack, generally resulting from an incomplete longitudinal weld. The initiating lack of fusion may be on the inside or the outside, or both (circumferential welds almost never fail on an operating pipeline). The third mode of fault is more general and includes voids, laminations, or inclusions of foreign material in the base metal of the pipe wall. Most of this type of physical anomaly was present in the material before the pipe was rolled. In high-strength pipe, hard spots have been found to have been caused by transport and exercise of those points of the pipe supported on skids. Faults like this are invisible to the human or X-ray eye. Visible faults are generally gouges caused by handling the pipe, by arc burns during welding, or by subsequent third party damage. There are several instrument pigs that have the capability to locate and evaluate these faults.

ULTRASONIC FLAW DETECTION

There are two basic non-destructive testing methods adapted to inspect pipe for the kind of anomalies mentioned above. One method uses ultrasonics. In this method, as in the ultrasonic thickness gauge, a pulse of high-frequency sound is sent into the pipe wall, and a pickup transducer (sometimes incorporated into the transmitter) receives and measures the time of travel of the signal through the pipe wall and back. Any major change in density, such as at the outer surface of the pipe, acts as a reflector. By calculating the time from signal initiation to reflected wave reception against the known speed of sound in the pipe material, the pipe wall thickness can be accurately gauged. If the transmitter is angled, it can scan a greater volume of the pipe wall on its inward and outward paths. If there is an inclusion, a void, a lamination, or any interruption in the path, it will show up. If the wave path is directed circumferentially around the pipe and includes the longitudinal weld, it can find a lack of fusion. From a theoretical standpoint, this is the preferred system. It works so well on the bench that wall thicknesses can be measured within a few thousandths of an inch with hand-held instruments. At 15 or 20 miles an hour in a pipeline whose inner surface may not be clean, let alone smooth, the problems increase. Holding the transmitter and receiver in any known, consistent relationship to the pipe wall is one problem. Maintaining a consistent and effective sound path into and out of the pipe wall is another. Several ultrasonic instrument pigs have been designed; a few were built; one or two were temporarily marketed. None are enjoying widespread use today. Since this is one of the best theoretical tools for the job, surely a number of researchers are presently working to overcome the application problems.

An ultrasonic imaging system was publicized in 1977 by Holosonics, Inc. In this case, an acoustic imaging pig was proposed that would scan the pipe wall for thickness, faults, and even stress corrosion, and generate a magnetic tape recording that could be converted to three-dimensional pictures. The device was apparently quite bulky because of its mechanisms and power requirements, so it was suggested for 48-inch or larger pipe. The principle has been proven with an external girth-weld inspection tool that surrounded the pipe with a circumferential scanner, and was backed up with a trailer-load of auxiliary equipment and power supply. Apparently, the pig has not been built.

The American Gas Association, through its pipeline research committee, initiated a project with Rockwell Corporation to develop an ultrasonic inspection pig with the following objectives.

A. Locate stress corrosion cracking, corrosion pitting, wall thinning, gouges, and other flaws with a high degree of accuracy.

B. Improve the state of the art in characterization of defects.

C. Detect both ID and OD flaws.

D. Locate longitudinal cracks.

E. Operate automatically by recording data on-board and processing data at a later time with electronic data processing equipment.

F. Operate at high pipeline speeds.

G. Negotiate sharp bends.

Rockwell accomplished the early laboratory phases of the research and developed ultrasonic transducers which did not require the couplant normally associated with this method of inspection. Today, work continues on the prototype pig development and testing phase of the project.

MAGNETIC FLUX FLAW DETECTION

The other metal inspection means is the magnetic flux system. In this case, the pig is equipped with two separate circumferential magnetizing brushes or poles, one representing a north pole, the other a south pole. Arranged between the poles are transducer coils.

As the pig moves through the pipeline, the wall is magnetized between the poles. If the transducer coils pass a flaw in the pipe, they detect the magnetic flux leakage caused by the discontinuity. British Gas, AMF Tuboscope, and Vetco use variations of this magnetic flux analysis system. In addition to corrosion, inclusions, laminations, and even the magnetic anomalies of hard spots can sometimes be found. Because the path of the pig is longitudinal, it is more difficult and less positive in the detection of longitudinal faults, such as cracks or incomplete longitudinal welds. Even though the operating speed is limited, it is not unreasonable, and the magnetic flux instrument pig is presently marketed with enough successes to keep at least two or three companies in the business.

INSTRUMENT PIG LIMITATIONS

The problem with any pipe wall inspection system, whether it is ultrasonic or magnetic or some secret process under development, is the sheer volume of information it must generate. In a 48-inch pipeline, the pig must inspect an acre of pipe surface every 1,200 yards (one hectare every 2.7 kilometers). This requires a lot of power. That means many batteries and a lot of space and weight. The mounting of a large number of small, individual, sensing units requires a complex mechanism to position them in an appropriate relation to the pipe wall to accommodate the average pipeline with not only bends, wall thickness changes, side outlets, valves and fittings, but diametral reductions exceeding five percent. To date, this has meant a heavy complex pig with major jointed segments and minor jointed arms. As a result, one area of failure has been the inability of such a complex tool to safely negotiate a pipeline with unexpected internal surprises. With experience and mechanical and electrical improvements, this problem will be reduced. TransCanada Pipeline has built a more compact tool based on the magnetic flux principle, using permanent magnets. The pig has been used within the TransCanada system and to a limited degree in other pipelines. British Gas uses their pipeline survey tool regularly and reports excellent results with the magnetic flux system. Again, this is with a large, jointed, heavy pig.

Attempts have been made to reduce the battery load by supplementing the stored power with power generated by the travel of the pig or its bypass. Wheels rolling on the pipe wall with an absolute minimum torque load, consisting of only axle friction and a magnet passing a reed switch, must incorporate redundancy to measure the distance through the pipeline without slippage. A wheel driving a generator would have more resistance and could presumable work only with the addition of teeth or pins hard enough to mark the pipe. Use of the bypass through the pig to drive a turbine or hydraulic air motor has failed repeatedly because of the inherent dirt and wax in oil lines and the dust-storm the pig raises in a gas line. There are both U.S. and U.K. patents on a solid-piston pig with no bypass and a propeller on the front to generate electric signals as the pig passes through the oil. Both the inventor and the patent offices failed to recognize that the oil would be moving at the same speed as the pig and nothing would rotate the propeller.

The pipeline pig that inspects acres of pipe surface must also collect and store that information. Upon removal of the pig from the pipeline, the information must be converted to a more usable form, usually visually analyzed, multiple-channel tape. Analysis is time-consuming and requires skill and judgement, and perhaps a little artistic license. Advances in computer technology are allowing the analysis of recorded data to be performed more rapidly and accurately. But even with a computer, there are limitations of data volume, processing time, and software design. As with all computer applications, the system is only as good as those who design it.

Again, experience and physical improvements will improve the results. If one is aware of a fault and the tape is subsequently analyzed, the known fault can generally be located on the tape. This indicates that the improvement lies more with the analysis than with the data collection.

No discussion of pipe wall defects can avoid stress corrosion. Although it may have electrical, chemical, or thermal causes, a survey of those parameters will not predict it. Stress corrosion can certainly result in pipe failure. And, if it could be detected at its inception, it could be corrected. Many research groups have attempted to define the cause and effect parameters of stress corrosion so that a tool could be developed to find it. Much of the work is in progress and all of it proprietary. British Gas has specified one path of investigation. They are studying elastic wave techniques for the analysis of stress corrosion, as well as fatigue and other fault developers. British Gas, understandably, is reluctant to go into detail, but they certainly have made the largest research investment into the art and science of pipeline inspection by instrumented pigs. One day, the pipeline industry will benefit from this massive research and development project. Today, we can see that British Gas is proving their premise that no single technique will successfully locate the various kinds of pipeline defects.

THE FUTURE

The present status of pipeline inspection by pigs is in its infancy. We cannot forecast the range of instrumentation for the future because the demand and emphasis will change as it has in the past. If our regulations say a pipeline must be at least one meter below the surface and the operator must prove it regularly or be subject to penalties, then the economics dictate the answer. If a certifiable record cannot be obtained from the air, can it be recorded from inside the pipe? This is already important enough, offshore, that the Danish government refused to allow a gas line to operate through Danish waters until it could be proven that it was protected by cover. The Norwegian Institute for Atomic Energy is working on a pipeline instrument pig for this purpose.

Corrosion surveys, subject to weather, judgement, personnel problems, and increasing costs, may get to the point where the cost of development of a survey pig can be justified. Pipelines built under adverse climate, geography, and politics can encourage the expenditure of millions of dollars to develop a tool for a very narrow, specific purpose. The Alyeska curvature pig to detect very small pipe movement was an example.

It is interesting that dozens of pipeline survey pigs have been proposed, many developed to some incomplete stage, and a few put into operation. To date, less than half a dozen are proven by survival in the marketplace. The future will be better because current, limited usage gives general approval of the concept as a valid method of resolving pipeline problems. Don't expect to see many instrument pigs in the years to come - they'll be there, but out of sight.

This report on the background and status of pipeline instrument pigs was initiated with the intent that it be a short synopsis, mainly based on the author's experience, but as more and more questions arose, the bulk of information expanded and it was obvious that half the story would have been inadequate. The authors apologize to those who feel the article tells them more than they wanted to know about a relatively narrow technology and point out that the most difficult part of the job was to keep it from being longer. The authors also regret that some readers will find that they have omitted reference to their favorite device or have not examined in depth the intricacies of neutron/gamma ray surveys as applied to pipelines. The conclusions drawn herein are those of the authors and not necessarily in accord with those of their employer, T. D. Williamson, Inc., or of the sponsoring society, The American Society of Civil Engineers, both of whom the authors thank for giving them the opportunity to express their views. The authors take full responsibility for statements made and statements omitted, but acknowledge that this report would not have been possible without the generous and understanding help of a number of people and organizations who supplied previously unrecorded data, who refreshed their memory, and who pointed out where they were in error. Some, but certainly not all, of these include:

D. R. Anderson, Alyeska Pipeline Service Company
Gerald Clerehugh, British Gas Corporation
Jack Fulsom, AMF Tuboscope
Gordon Griffiths, CE VETCO SERVICES
A. E. Knowles, British Gas Corporation
Don Lorenzi, Peabody Testing Services (Magnaflux)
Coleman McDonough, Peabody Testing Services (Magnaflux)
L. D. Payne, T. D. Williamson, Inc.
Earl Sturgeon, Tennessee Gas Pipeline Co.
Hershel Vanzant, T. D. Williamson, Inc.
Cordell Wagner, Texas Pipeline Company
J. Clifton Williams, Texas Eastern
Ray Watts, El Paso Natural (now with Phillips)

BIBLIOGRAPHY

"Radioactive Markers in Go-Devils - Safety Precautions"
Prepared by Ministry of Labour and National Service in cooperation with Atomic Energy Research Establishment, Harwell, U.K., March 1959

"Electronic Pipe Wall Survey"
Pipeline Industry, March 1969

"Pipeline Inspection Using an Intelligent Pig"
W. T. Flamauk (Southern Gas Board)
(Paper presented to London & Southern Gas Association, April 6, 1973)
Published in I.G.E. Journal, September 1973

"Development of an Instrument Pig"
B. Vernooy (T. D. Williamson, Inc.)
ASCE Presentation, July 1973, Reprint No. 2045

"Ultrasonic Device Detects Pinhole Oil/Gas Leaks"
Pipeline Industry, November 1973

"Intelligent Pigs"
J. L. Cordell
P.I.G. Journal, Autumn 1975, Volume 53

"Fracture Propagation in Gas Pipelines"
G. D. Fearnehough (British Gas Corporation)
3R International, September 1976

"A Survey of Companies on Corrosion Control Techniques"
K. Krider (editor)
Gas Digest, March 1977

"Ultrasonic Imaging Shows Corrosion"
R. C. Peterson (Holosonics, Inc.)
Oil & Gas Jorunal, May 30, 1977

"Locating Froeign Contacts of a Cathodic Protection System"
G. E. Cox & H. E. Holland (Florida Gas)
Gas Industries, August 1977

"New Ultrasonic Tool Checks Offshore Pipeline, Welds"
H. Jackson (Mat. Eval. NDT Co., Ltd.)
Oil & Gas Journal, March 13, 1978

"On-Stream Inspection of Pipelines"
C. W. Osborne (presented September 7, 1978, Bristol, England)
P.I.G. Journal, Autumn 1978, Volume 65

"On-Line Inspection in British Gas"
(A presentation by G. Clerehugh to the Pipeline Industries Guild in London, December 12, 1978; questions and answers, only, reviewed in P.I.G. Journal, Summer 1979, Volume 68)

"Texaco Develops Improved Ultrasonic Leak Detector"
N. E. Fluornoy & W. W. Schroeder (Texaco)
Pipeline Industry, March 1979

"Curvature Monitoring Confirms Trans-Alaska Pipeline Stability"
D. R. Anderson (Alyeska Pipeline Service Co.)
Pipeline and Gas Journal, July 1979

"Niet-destructief onderzoek"
Polytechnisch Fijdschrift, November 1979
(Netherland Engineering Society publication)

"Stuck Pig Removal No Easy Chore"
Alaska Construction & Oil, November 1979

"British Gas Develops On-Line Pipeline Inspection"
G. Clerehugh & A. E. Knowles (British Gas Corporation)
Oil & Gas Journal, December 3, 1979

"AGTL Examines Pig's Ability to Detect Corrosion Damage"
L. O. Rowland, Editor
Pipeline & Gas Journal, August 1980

"Camera Pig Checks Weldments, Internal Coating Offshore"
Fred Clarke (GEO Pipeline Services)
Pipeline & Gas Journal, March 1981

"New Techniques Improve In-Situ Pipeline Cleaning and Coating"
Fred Clark (GEO Pipeline Services)
Pipe Line Industry, April 1982

"Linalog Inspection of In-Service Pipelines: It's Practical Applications"
Chris Polland (AMF Tuboscope, Inc.)
ASME Presentation, January 1983, Houston, Texas

"Here's Experience of British Gas with On-Line Inspection"
Gerald Clerehugh, et al (British Gas Corporation)
Oil & Gas Journal, January 31 and February 7, 1983

Annual A.G.A. Pipeline Research Reports
"NDT Inspection of Transmission Pipelines"
Project PR-73-57 (Ultrasonic Pig Development)

"Corrosion Detection & Evaluation"
C. C. Kammerer, F. H. Stuckenberg, F. E. Sugg
(A review of all known corrosion detection and evaluation practices
for NASA, Marshall Space Flight Center)

"Electronic Survey Helps Assure Integrity of Offshore Pipelines"
W. M. Rogers & H. N. Duckworth (AMF-Tuboscope)
Offshore Technology Conference, Paper Number OTC-1854

NACE - National Association of Corrosion Engineers
"RP-01-75 Control of Internal Corrosion in Steel Pipelines"

"Stress Corrosion Crack Detection in In-Place Pipelines Utilizing an Instrument Pig"
W T. Walters (AMF-Tuboscope)
Corrosion 73, Paper Number 98

THE NEED FOR GUIDELINES FOR THE JOINT OCCUPANCY OF UTILITY AND PIPELINE CORRIDORS

Walfred E. Hensala, M. ASCE[1]

Abstract: The joint occupancy of utility and pipeline corridors has become common as a direct result of the growth of transportation systems, the utilization of natural corridors, land use and regulatory requirements. Guidelines for joint occupancy would facilitate the route selection process as well as the construction, operation and maintenance of facilities in the corridor, and further the technology of pipeline practice.

Continuing growth and development, designation of special land management areas on public lands, and extensive utilization of natural corridors through the growth of transportation systems have brought about the joint occupancy of pipeline and utility corridors. When joined with land management fiat, based on regulatory requirements to limit the proliferation of rights of way and de facto land use practices, these factors have made the joint usage of corridors a cardinal parameter in the pipeline route selection process.

Pipeline route selection is a deductive process; it begins by projecting a straight line between the origin and destination, then finds the most logical route by evaluating the natural corridors and assessing the land uses that may affect the routing. It then addresses the regulatory requirements. Most pipeline projects will cross public lands making it necessary to comply with the requirements of land management agencies in the route selection process. In addition, many states have facility siting regulations which may determine requirements of linear projects and affect the route selection process. The environmental review required of all major projects and most smaller projects also influences, and in many cases may govern, the route selection for a project.

Environmental review and land management regulations are based on reasonable objectives to attain the greatest benefit to the public from the use of a common resource and to minimize environmental impacts. They have led to use of the corridor concept for linear systems in the regulatory process.

1 Senior Engineer, Transmission Systems, Northwest Pipeline Corporation, 295 Chipeta Way, Salt Lake City, Utah.

The perceived validity of the corridor concept led to the study[1] conducted by the Department of Interior, (DOI), Bureau of Land Management (BLM) (July, 1975) on the desirability and necessity of a system of transportation and utility corridors. As it was evident that the corridor concept must address lands other than Federal lands to assess the effects of corridor planning, all lands to be crossed by the corridors were considered. The need to include transportation and utility requirements in land use planning and an attempt to coordinate the decision making process in project approval, as well as land management and environmental review were addressed in the study. As the competition for the use of limited land resource was felt to reflect many concerns too involved for local agencies to address, the study was conducted at the national level.

Considered in the study were two corridor concepts, planning corridors and joint use corridors.

Planning corridors, based on coordinated land use planning, are broad areas with flexible borders. They were considered on broad policy issues including environmental and social impacts, conceptual engineering and design, topography, safety and security, and possible future uses.

The study, which included information and assistance from other Federal and state agencies as well as industry, concluded that a system of planning corridors is desirable, feasible and needed. The study also concluded that insufficient information is available to make rational decisions, and until further data are available and properly analyzed, decisions on corridor planning will continue on a case-by-case as origins and destinations of corridors are more readily assessed.

A joint use corridor is defined as "a narrow strip of land with restricted boundaries on which facilities of the same or different systems are placed adjacent to each other in as close proximity as practical and feasible[1]." In addition to the broad concepts included in planning corridors, joint use corridors include in their analysis the type and size of systems, physical aspects of the site, potential interference between systems, mitigation practices for interference, costs; system safety and maintenance, and the effect of the systems on resource and environmental values.

Compatibility or non-compatibility of different types of systems was reviewed, and system requirements of sensitivity to terrain features, grades and alignment were considered. Influence of soil types on different systems was briefly analyzed.

Among the disadvantages identified in the study are the requirements for restricted construction procedures and conditions, and adverse effects that are the result of interaction between different types of systems.

The study concluded that joint use corridors cannot be identified in advance of specific project proposals as each right of way for a specific type of system has special requirements for working space and construction.

The study also stresses the need for coordination by land management agencies in developing corridor planning. In February 1980, an interagency agreement between the BLM, and the Forest Service (FS), Department of Agriculture, for transportation and utility corridor planning was executed to coordinate planning. In July 1981, the Western Utility Group, an ad hoc committee of representatives from natural gas, electric and telephone utilities in eleven western states, noted in a report[2] compiled to identify governmental regulations that could be changed to increase benefits or decrease costs, that corridor planning by the BLM and the FS could result in discontinuity in corridor designations between the two jurisdictions. Planning in the two agencies is not consistent in scheduling and separate processes of wilderness area review, not included in corridor planning, were being conducted.

The study also indicated that coordination is required with potential users of the corridors. However, agencies are still inclined to retain technical expertise from other governmental agencies, not from the corridor user. Some potential users remain unconvinced that agencies have the expertise to assess priorities, engineering problems, and costs.

Structured corridor planning was also studied in Montana on a state basis,[3] and in Alberta, Canada in conjunction with the development of the Alberta tar sands.[4] The corridor concept is also applied by agencies in reviewing major project proposals for authorizations to proceed. At a local level, it is also less judiciously used by land management agencies in granting rights of way for linear projects. Whether enforced by agency practices, a result of existing land use practices, or imposed by the restrictions of natural corridors, the corridor concept is a fact of life in the route selection process for linear projects.

As the use of a common corridor is often a compromise from preferred routing, and the management of the corridor focuses on factors other than the operation and maintenance of the facilities with the corridor, some guidelines for the corridor concept appear advisable. The guidelines should incorporate the expertise of the user, who plans, constructs, operates and maintains the facilities in the corridor.

The guidelines should identify compatibility, as well as differences of the various facilities in the corridor, recognizing operating and maintenance requirements, as well as construction. The guidelines should address the need for protection of facilities during construction and maintenance of other facilities, the various codes and standards governing construction, maintenance and operation, and provide a mechanism for surveillance during construction and maintenance.

The control and administration of the corridor, the allocation of space within the corridor, and the establishing of necessary priorities is not normally vested in a user of the corridor. Yet each user plans, designs, constructs, operates and maintains the facilities independently.

There is an obvious need for close coordination with a feed-back process. Established guidelines could aid in providing such a process.

Construction activities in a corridor require special techniques, methods and precautions when in close proximity to existing systems to prevent damage to facilities and hazards to construction personnel and the public. Construction requirements may result in limitations on the kinds of facilities that can be placed in a joint usage corridor, or impose the use of techniques that result in excessive costs not anticipated in early planning.

Established corridors with existing facilities may also have limited access for construction as a result of the characteristics of the existing systems. Again, preliminary planning may not address the existing limitations adequately.

In addition to the constraints placed on the user of the corridor, by concentrating activities and facilities within the corridor the effects on adjacent lands are concentrated, lessening the value of the lands to the landowner and limiting the potential for development. The rights and uses in a corridor must be carefully defined, as what is perceived as a primary right by the landowner may be perceived as a secondary right by the user.

Joint use also creates possibilities for disruptions of service, reducing the reliability and security of systems. The vulnerability of the systems to natural catastrophies may be increased. Risk assessment and the planning for emergencies that result from disaster appear to be only superficially addressed in corridor planning and use.

There exists a possibility for some reduction in the costs of maintaining facilities in a corridor through cooperative procedures. In existing shared corridors such cooperative efforts often exist, but at a local level with little recognized structure. The question of assignment of liability and responsibility for cooperative procedures may well limit their applicability. Appropriate guidelines could aid the development of cooperative maintenance.

Many of the problem areas of joint corridor use have already been addressed through the previous experience of corridor users. By incorporating the results of past experiences, together with new concepts, perhaps realistic guidelines could be developed. In may be difficult to recommend specific practices, because they are based on site-specific conditions. However, a check list of items that should be considered could help to ascertain that even the obscure factors are properly addressed.

Coordinating all aspects of pipeline engineering and technology involving joint occupancy into a set of guidelines could result in the furthering the technology of pipeline practice.

APPENDIX - REFERENCES

1. U.S. Department of the Interior, Bureau of Land Management, "The Need for A National System of Transportation and Utility Corridors", Washington, D.C., July 1975.

2. Western Utility Group, "Federal Land Use Regulations Adversely Affecting the Western Utility Industry, A Report to the President's Task Force on Regulatory Relief", Published by Pacific Gas and Electric Company, San Francisco, California, June 1981, pp. III.1.1-III.2.7.

3. Breuninger, Roy H., "Utility - Transportation Corridor Study in Montana", presented at the September 24, 25, 1981 ASCE Conference on the Joint Usage of Utility and Transportation Corridors, held at Houston, Texas.

4. Weir, C.H., "Multi-purpose Transportation Corridors", presented at the September 24, 25, 1981 ASCE Conference on the Joint Usage of Utility and Transportation Corridors, held at Houston, Texas.

Utility Corridors for Pipeline Construction

Ralph H. Sandmeyer, Jr.*

For over a century now, buried pipelines transporting crude oil, natural gas and refined petroleum products have been operating safely and efficiently. With over one million miles of pipe installed, certainly the optimization of land for construction has been and will continue to be paramount. Although the cost of utility easements has skyrocketed, particularly in urban and developed areas, the cost of rights-of-way typically represents only 3-5% of total installed costs. However, the ability to acquire construction rights-of-way and use of the right of eminent domain, if necessary, is sometimes critical to project scheduling and project economics. The use of existing utility rights-of-way, namely pipeline, powerline and railroad easements can often be beneficial, economical and even the only resort to placement of a new facility. However, the random use of utility rights-of-way, without proper reconnaissance, can also lead to construction and operational problems that should be avoided. Consequently, this paper deals with the utilization of utility corridors for pipeline construction, by depicting a number of construction scenarios dealing with the proper (and also the improper) placement of new pipelines, giving due consideration to engineering, environmental, geologic, safety and cost considerations.

The first consideration for any pipeline is determining a constructable route from point A, the origin, to point B, the destination. In today's age, engineers must closely evaluate the archaic straight line routing approach of the past and allow for some leeway for cost factors relating to difficult sideslope construction, environmental and local social concerns, and ever increasing urban development. Just a small increase in pipe length can often avoid difficult terrain or right-of-way problems and eliminate other factors that make construction costs high and/or lengthen the project schedule.

To visually show how a slight increase in line length can easily allow a reroute around a problem area, take an 8-1/2 X 11 inch piece of grid paper and a 12-inch flexible plastic ruler. Holding one end at the paper's edge and deforming the ruler in an arch or S curve vividly displays

*Senior Project Manager, Willbros Butler Engineers, Inc., Post Office Box 45207, Tulsa, Oklahoma 74145

how an urban, sensitive, or area with difficult terrain can be avoided with only a slight increase in length. (Each 1/4-inch deflection equates to a 2.25 percent increase in line length.) An extreme case would be indicated by placing the 12-inch ruler as shown below. At a scale of 1 inch = 10 miles, this would graphically be equivalent to taking a 110-mile long west-to-east pipeline and increasing the length 10 miles (9 percent increase) in order to avoid the problem area by deflecting the route 20 miles to the north.

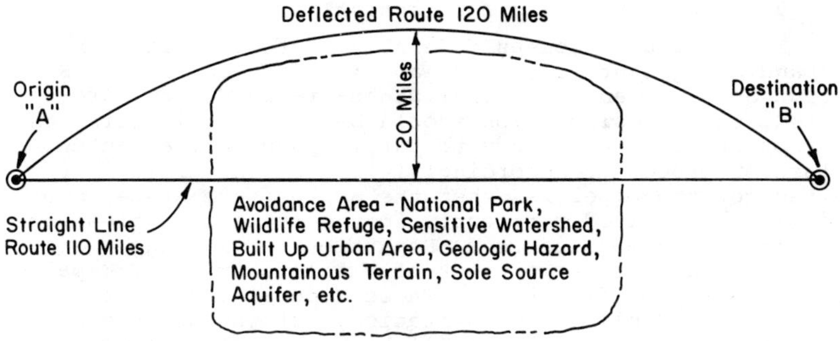

Difficult permitting requirements and high construction costs (sometimes 2-5 times normal) can often be avoided by a relatively small increase in total pipeline length and cost. These minor deviations in the route can often help obtain crucial permits, expedite the project completion, and avoid years of delay and the resultant escalation which is often devastating to project economics.

Route Selection Criteria

In all cases, the pipeline location studies should focus primarily on the search for a suitable pipeline right-of-way alignment. However, several permitting agencies recommend that wider (up to 1 mile) corridors be identified and studied before permits can be obtained. If this is the case, the route selection process must be expanded in scope, as described in the following paragraphs.

First, a pipeline route location engineer obtains all available maps of the general area. The most useful maps are the USGS topographic maps, NOAA navigation charts, county road maps, and U.S. Forest Service maps. From these and other maps, the terrain, population centers, rivers, lakes, roads, railroads, powerlines, and pipelines can be observed.

If the pipeline route crosses environmentally sensitive areas, a special multi-discipline route selection team should be formed to select primary and alternative routes. The team may consist of members with the following expertise:

1. Pipeline route location engineer
2. Pipeline construction specialist
3. Geologist (landslide, earthquake, etc.)
4. Hydrologist (water crossing studies)
5. Biologists (terrestrial, aquatic, botanist)
6. Archaeologist (cultural resource specialist)
7. Reclamation (soil scientist)

If the route crosses federally controlled land (Bureau of Land Management, U.S. Fish and Wildlife Service, U.S. Forest Service) and a federal Right-of-Way Grant is required, then consideration should be given to inviting the appropriate federal agency to participate in the route selection process. Accordingly, if a state permit is required for the pipeline, then the appropriate state agencies should be consulted very early in the route selection process (i.e., Water Resources, Department of Ecology, Department of Fisheries, Department of Wildlife or Game, Department of State Lands, etc.). These federal and state agencies may decline the invitation, but the gesture can be of utmost importance to establishing a beneficial working relationship and obtaining the necessary permits.

The "utilities corridor" concept should be used whenever possible; that is, routing the pipeline on or along existing rights-of-way of other pipelines, powerlines, and/or railroads whenever there is suitable terrain and adequate space for construction, the additional length is not excessive, and when construction and/or operations will not create hazards to the adjacent utilities or endanger public safety. The use of utility corridors is particularly important in environmentally sensitive areas and in congested areas where the pipeline must pass through population centers or industrial areas. The general procedures and criteria used in selecting the pipeline route are described below.

Following utility corridors, natural mountain passes and other geographical avenues, the route location engineer and multi-discipline route selection team plots several preliminary alternative routes, then evaluates each by individually weighing the construction variables and environmental concerns associated with each route. Route selection criteria of primary concern both from an economic and environmental standpoint are:

- length of route, which directly affects the investment and future costs;

- o terrain conditions, which affect construction costs and future maintenance costs;
- o most acceptable location for inland waterway, river and important stream crossings;
- o cost of rights-of-way and damage settlements;
- o accessibility from existing roads to facilitate construction and maintenance; and
- o avoidance of populated areas and environmentally sensitive areas (such as national forests, watersheds, aquifers, wild and scenic rivers, wildlife refuges, sites with historic or archaeological significance, scenic areas, or areas where future developments might present problems for the pipeline. This can require extensive coordination, numerous meetings and supplemental reports to document proposed mitigation (if avoidance is not practical).

These route selection criteria are based on standard industry practices and include guidelines established by Department of Transportation regulations, Part 195, "Transportation of Liquids by Pipe Line" and ANSI Standard B31.4, "Liquid Petroleum Transportation Piping Systems." Areas routinely avoided during pipeline route selection are:

- o military installations;
- o national, state, and local parks;
- o cemeteries;
- o airports;
- o interstate highway crossings;
- o landslide areas;
- o earthquake fault areas;
- o archaeological sites; and
- o environmentally sensitive areas, such as wildlife preserves

Several route alternatives may be plotted initially, but some may be eliminated at once because of excessive length, rugged terrain, or important environmental concerns. Following this process, the route selection team observes the remaining viable route alternatives by auto, helicopter, and even on foot in critical areas. Additional effort should be expended in evaluating mountainous routes; major river, inland water, highway and railroad crossings; and environmentally sensitive areas. The areas that must be avoided are noted and plotted on the maps.

When this air and ground reconnaissance has been completed, the route location engineer returns to the office with the team input and revises the route accordingly. In some instances, geotechnical, ecological, and pipeline construction experts may be required to perform further reconnaissance if critical or sensitive areas are crossed. Each viable pipeline route alternative is again evaluated for the route selection criteria previously discussed, and construction cost comparisons of route alternatives are developed as well as economic and timing comparisons of environmental and other issues. Since quantitative values are difficult to assign to environmental concerns, a subjective evaluation and decision must be made by the route selection team to select the preferred route.

Use of Existing Pipeline Rights-of-Way

Existing pipeline rights-of-way usually afford the best potential multi-use corridor for new pipeline construction. Many pipelines have multiple or assignable rights for additional pipelines. This not only saves time and money in acquiring new right-of-way but often eliminates the need for eminent domain proceedings and other costly and time-consuming legal situations.

Although terrain along an existing pipeline is normally suitable for a new pipeline, encroachment by existing landowners or other adjacent utility easements often precludes additional multi-use. Each route location exercise must have the multi-discipline team approach to weigh all variables. An example of multi-use to safely construct and operate a 42-inch diameter pipeline on a 90-foot construction right-of-way is depicted below:

Typical 42-Inch Pipeline Right-of-Way in Level Terrain

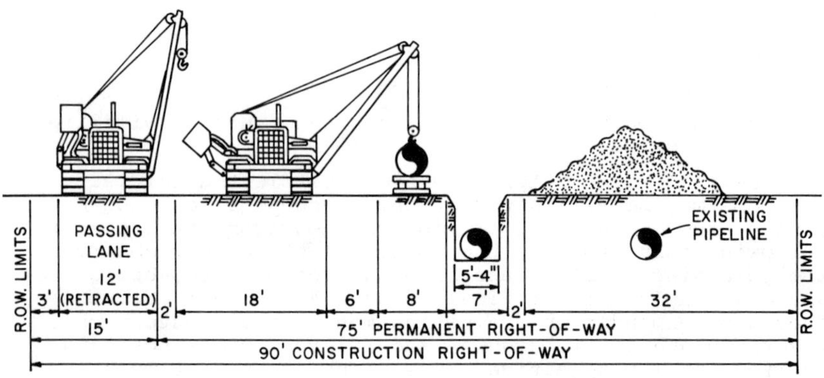

Each utility may have its own separation distance depending on the age of the existing facility, commodities transported, space available and other engineering and safety considerations. Although there are no universal standards for pipeline separation, normally a 10-foot or greater distance is specified, unless special circumstances exist.

Under a normal cross-country construction scenario, the following construction and operation space is required:

Typical Pipeline Easement Requirements

Pipeline Size (Diameter)	Temporary Construction Width (Level Terrain)	Permanent Width (Operations & Maint.)
6-14 inch	50 feet	30 feet
16-24 inch	70 feet	40-50 feet
26-36 inch	80 feet	50-60 feet
40-42 inch	90 feet	75 feet
> 42 inch	100 feet	75 feet

In terrain where extensive cutting is necessary, additional temporary spoil storage width is required.

Lesser widths can be accommodated with special consideration. Construction widths may be narrowed for short distances if necessary, but the above widths must be maintained over long distances for efficiency and safety reasons. Construction widths in difficult mountainous areas can vary considerably depending on the methods necessary to safely complete the construction. For example, the construction width can sometimes be narrowed by spreading the pipeline trench spoil on the equipment side of the right-of-way, working on top of it, and then backfilling the trench. Sometimes existing pipeline easements which are not adequate in width may be used for temporary spoil storage or a portion of the existing right-of-way may be used to build the pipeline, thus reducing the width of new right-of-way required.

However, if the existing right-of-way is used, the contractor must not allow heavy equipment to travel and impose heavy loads to the existing pipeline overburden. This is normally accomplished by locating the work side and access road for the new pipeline on the outside of the existing right-of-way and only allowing traffic to cross at existing roadways where deeper burial and casing pipe protect the pipeline.

Permanent easement widths can and do vary widely, and due to encroachment problems encountered in the past, industry is presently favoring wider permanent easements such as those shown above. In the past, permanent easements varied from the width of the pipe only (undefined easement) to 50 feet in width for a single pipeline, with 30, 40 or 50 feet being common widths used by many pipeline operators.

Use of Existing Powerline Corridors

The multi-use of electric transmission line right-of-way is generally advocated by both electric utilities and pipeline operators. The shared corridor often eliminates a lengthy search for suitable rights-of-way. In forested areas, multiple use requires less clearing, and usually provides many access roads for hauling men, construction equipment and pipe along the right-of-way. Cadastral surveys and land acquisition are greatly simplified and expedited. The use of existing high voltage electric powerline rights-of-way for pipeline construction requires special consideration because the facilities are dissimilar and their routing criteria vary substantially.

A powerline in level terrain follows a straight line where practical, much like a pipeline. However, in hilly or mountainous terrain, there are many differences:

	Pipelines	Powerlines
1. Use of existing valleys and natural mountain passes	Yes	Yes
2. Sideslope terrain	Avoid if possible	Can accommodate easily
3. Acceptable construction gradients	0 to 20% common 20 to 45% over short distances with assistance 45 to 100% by winching materials & equipment from the top or bottom (in very short distances)	Generally no limitations, towers often placed on edges of cliffs while conductors span the valley.
4. Clearing forested right-of-way	Must completely clear trees including stumps	Clear trees, stumps left in place

When pipelines and powerlines share the same right-of-way, the primary concern is safety during construction and operation. The pipeline trench must be located a sufficient distance from powerline towers to avoid undermining towers or guy wire supports. Typical clearances are as indicated in the sketch and tables shown on the following page:

Typical High Voltage Powerline Right-of-Way in Level Terrain

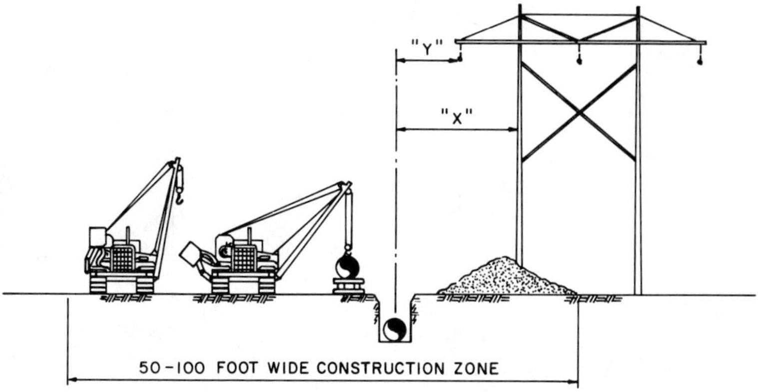

*Typical Mechanical Clearances (X)

1. Wood poles, guy anchors . . . . 25 feet
2. End of counterpoise (ground mats) . . 16 feet
3. Footings of steel towers . . . . . . 32 feet

*Minimum Clearances Between Equipment and Transmission Line Conductors (Y)

Voltage	Feet	Voltage	Feet
34.5 KVAC	15	500 KVAC	20
115 KVAC	15	± 400 KVDC	17
230 KVAC	15	± 500 KVDC	21
345 KVAC	16	± 600 KVDC	25

*Consult the appropriate utility for exact specifications.

Safety Considerations

To reduce the dangers of electrical shock due to induced voltages from the AC powerline before pipe is lowered in the ditch, grounding ties and portable steel mesh grounding mats with cable bonds must be used. During pipeline operations, certain voltages may still exist. The induced voltages may be reduced by maintaining electrical bonds, potential gradient control systems and insulated flanges. Paralleling DC transmission lines requires special design considerations due to potentially high ground currents.

The typical construction scenario can vary from level terrain to extreme sideslopes. As shown on the following page, paralleling a powerline along steep sideslopes can require massive excavations and a wide construction zone.

Typical Powerline Right-of-Way with 30% Sideslope

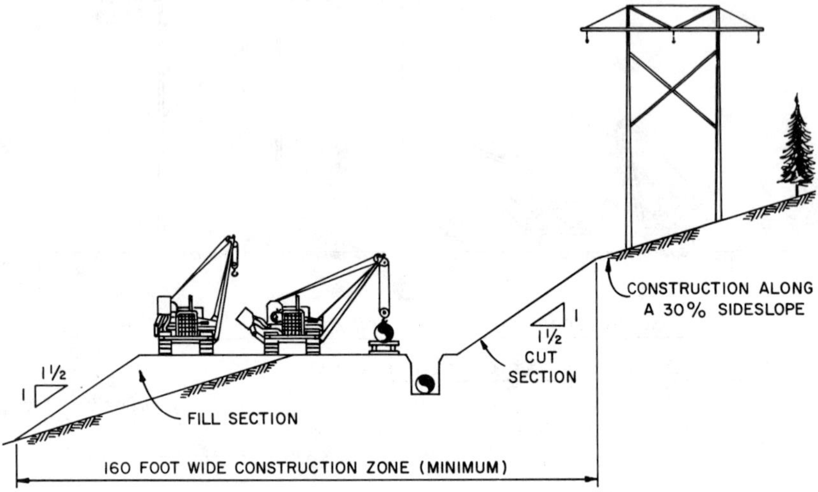

Note that the pipeline construction scenario shows pipeline equipment working on the fill section downhill from the powerline, with a balanced cut and fill to minimize excavation. Equipment is working on the far side of the right-of-way to maximize the distance from the high voltage conductors.

Use of Existing Railroad Rights-of-Way

For many years, railroad rights-of-way have been safely and successfully used for pipelines. Railroad right-of-way is often helpful and convenient for placement of pipelines, especially for urban pipelines where private right-of-way is simply not available. Due to the wide variance in routing criteria, careful study by engineering, geotechnical and environmental specialists is essential before deciding to use railroad right-of-way for any pipeline project.

Railroad right-of-way is best utilized for pipeline placement if the following criteria are met:

1. Level terrain (with few if any cuts, fills, trestles or switchbacks).
2. Wide right-of-way (100 ft. or greater) to allow adequate working space for construction.
3. Lack of parallel utility and communications systems (electrified systems have many wooden poles which block construction work space).

4. Unpopulated areas (it is often not desirable to use railroad right-of-way through metropolitan areas, if the populated area can be skirted on private right-of-way). Railroads traditionally pass through the heart of cities where no land for construction work space is available.

The typical level railroad right-of-way suitable for joint occupancy by a pipeline is shown below:

Typical Railroad Right-of-Way in Level Terrain

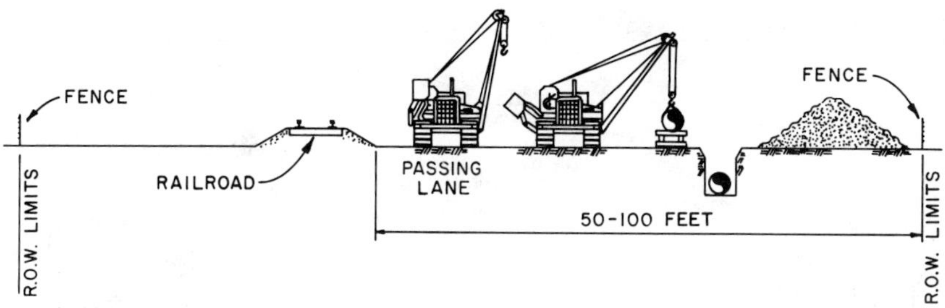

Pipelines are usually best suited for sharing railroad rights-of-way if they are relatively small in diameter (12 inches or less), require a 50-foot construction width and if the terrain (perpendicular to the right-of-way) is flat. Since railroads are built to gradual grades (usually 3 percent or less), a pipeline can easily be constructed along the gradient of the route.

The most difficult aspect of building a pipeline along a railroad is construction and stabilization through the cut and fill sections that accompany all railroads in hilly or mountainous terrain, and the work space that is needed for conventional pipeline construction. An example of two improper places to share railroad rights-of-way are as follows:

Cut Section-Not Useable for Efficient Pipeline Construction

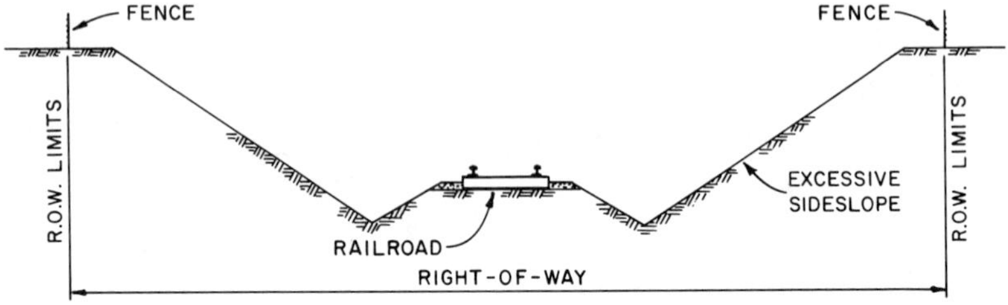

Fill Section-Not Useable for Pipeline Construction

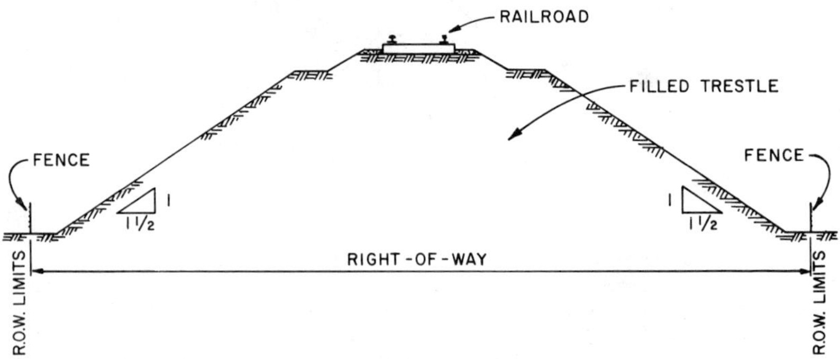

The use of railroad rights-of-way for pipeline placement, where numerous cut and fill sections exist, would require special pipeline construction techniques. This can result in construction costs 3 to 5 times higher than placing the pipeline on adjacent private rights-of-way where the route can follow natural terrain and use conventional construction procedures.

Summary

The utilization of existing utility corridors for pipeline construction must carefully be evaluated for each site-specific situation. Arbitrarily following existing corridors is not always a preferable solution. A multi-discipline team of specialists with hands-on experience is essential in weighing the engineering, construction, environmental, economic and safety aspects involved. The end result of such an exercise is the optimization of project economics and improved land utilization.

STIFF SOILS - AN ADVERSE ENVIRONMENT FOR LOW STIFFNESS PIPES

K. K. Kienow,[1] Member, R. C. Prevost,[2] Member ASCE

INTRODUCTION

A flexible pipe is generally defined as one which can deflect two percent or more without significant structural distress.

The load-carrying ability of a flexible pipe is related to both the stiffness of the pipe and the stiffness of the soil in which the pipe is embedded. Properly designed and installed, the pipe will deflect in an elliptical shape. The loads are redistributed over the long term, resulting in increased lateral pressures. The pipe tends to function as an arch structure. The final soil pressure distribution around the pipe ring is principally a function of the ratio of pipe stiffness to soil stiffness. "Flexible pipe" include: Corrugated metal pipe (CMP); ductile iron pipe (DIP); mortar-lined steel pipe (MLSP); and plastic pipe, which include the thermosetting plastics such as fiber-reinforced plastic (FRP); reinforced plastic mortar pipe (RPM); and thermoplastics such as acrylonitrile, butadiene, styrene (ABS), polyvinyl chloride (PVC), polyethylene (PE), and high-density polyethylene (HDPE). This paper will refer primarily to gravity flow plastic pipes for storm drainage, culvert, and sanitary sewer applications.

Definition of Pipe Stiffness: In the United States, the inherent strength of a flexible pipe is termed "pipe stiffness" (PS) and is defined as F/dy per unit of length:

$$PS = F/dy = EI/.149\ R^3 \text{ psi } (N/cm^2)$$

where F = deflecting force per unit length along pipe, pounds per inch (Newtons per cm)

dy = resulting vertical deflection inches (cm)

E = modulus of elasticity of the pipe wall, psi (N/cm^2)

I = moment of inertia per unit length of pipe, in^4/in (cm^4/cm)

R = mean radius of pipe, inches (cm)

(1) P.E., Corporate Engineer, Hydro Conduit Corporation, Corona, CA.
(2) Dr. Ir., Civil Engineer, Washington, D.C.

Pipe stiffness is determined by ASTM test method D2412, the standard for the parallel plate loading test for plastic pipe, and is measured at five percent deflection in the parallel plate test. "Pipe Stiffness Factor" is defined as PSF = EI/R^3 psi (N/cm^2) in the United States. In order to avoid confusion, "Pipe Stiffness" in psi will be used throughout this paper.

Two important factors affecting pipe stiffness determination are the duration of the loading time in the parallel plate test, and the amount of deflection at which pipe stiffness is measured.

EFFECT OF DURATION OF STIFFNESS TEST

Figure One illustrates the effect of the duration of loading time on pipe stiffness for PVC and HDPE pipe. The decrease in pipe stiffness with time is the result of creep under load. "Long-term" modulus and pipe stiffness is determined by projecting data for several years out to 50 years on a logarithmic plot.

In the United States, the short-term stiffness has sometimes been used for deflection calculations. An appropriate "lag factor" to account for the long-term soil consolidation is applied to the calculated deflection. The usual lag factor used is 1.5. The lag factor does not account for the decrease in pipe modulus which occurs with time. Moser (1977) recommends that the short-term stiffness be used for all calculations (including buckling) for PVC pipe. For HDPE pipe, Gulf (1980) recommends the use of long-term stiffness for buckling calculations. Taprogge (1981) recommends the German practice for plastic pipes, i.e., using short-term properties for initial deflection calculations, and long-term (50-year) properties for long-term deflection calculations and buckling analysis.

An approximate relationship for long-term (50-year) pipe stiffness (and modulus) for various materials is as follows:

HDPE long-term = 16% initial modulus
HDPE long-term = 33% 24 hour modulus
PVC long-term = 33% initial modulus
RPM long-term = 50% initial modulus

For HDPE, the creep rate is much more rapid than for other thermoplastics. From Figure One it can be seen that "pipe stiffness" decreases to 50 percent of the initial value in a 24-hour test.

The German Industrial Standard, DIN 16961, Part 2, "Pipes of Thermoplastic Materials with Profiled Wall, General Quality Requirements and Test Methods", applies to PVC, PE, and HDPE profile wall pipes. DIN 16961 defines six standard stiffness classes of pipe, and requires that the stiffness be determined at 24-hours and three percent deflection. The specified "proof load" for the class is applied to the pipe in a parallel plate type test and remains on the pipe for 24-hours. At the end of the 24-hour period, the change in vertical diameter is measured. The maximum allowable change in diameter (deflection) is three percent of the pipe diameter.

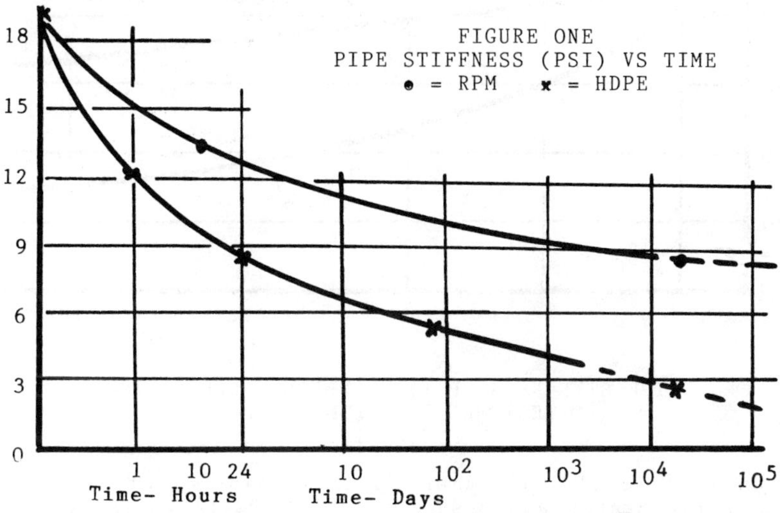

Pipe tested for "pipe stiffness" (Ring-steifigkeit) according to DIN 16961, will have half of the "pipe stiffness" measured according to ASTM D2412, if all pipe properties are the same. Conversely, a pipe with a PS of 25 psi as measured by DIN 16961, would have a PS of about 50 psi if measured according to ASTM D2412.

Pipe stiffness is defined by DIN 16961 as:

$$S_{R24} = .1548F/(dy)L$$

where F = proof load in Newtons

S_{R24} = pipe stiffness N/cm^2 (@ 24-hours)

dy = .03 times diameter (cm)

L = length of test pipe in cm

EFFECT OF REQUIRED DEFLECTION IN STIFFNESS TEST

Figure Two illustrates the effect of the deflection at which pipe stiffness is measured:

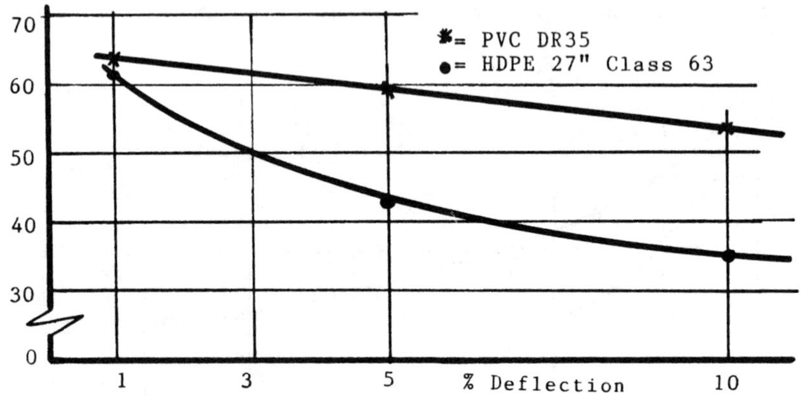

PIPE STIFFNESS (PSI) VS PERCENT DEFLECTION
TEST METHOD: ASTM D2412
FIGURE TWO

Until recently, all thermoplastic pipe manufactured in the United States was tested per ASTM D2412 which requires that pipe stiffness be determined at five percent deflection. As mentioned previously, the German specification for HDPE, PVC, and PE pipe requires that stiffness be determined at three percent. The United States manufacturer of HDPE profile wall pipe measures the "stiffness" of their product at one percent. The relative effects of five, three, and one percent stiffness determinations, in combination with the variation in test duration of several minutes (U.S.) to 24-hours (Germany) is illustrated in Table One.

TABLE ONE

METHOD	PVC	HDPE
USA INITIAL @ 1%	10.7	10.0
USA INITIAL @ 5%	10.0	6.9
24 HR. GERMAN @ 3%	8.8	4.1

VARIATION IN PIPE STIFFNESS F/dY WITH TEST DEFLECTION AND TEST METHOD. USA PER ASTM D2412, GERMAN PER DIN 16961 PART TWO

SIGNIFICANCE OF PIPE STIFFNESS

As pipe stiffness decreases for a given soil stiffness, there is a point where the pipe deforms in a rectangular, rather than elliptical, shape; the Modified Iowa Formula will underestimate vertical deflection by a factor of four to six or more; and haunching and adequate sidefill compaction becomes difficult, if not impossible. Schrock (1983) and Carlstrom (1982) have reported that strain levels in rectangularly distorted pipe are virtually impossible to calculate and may be as high as 10 or more times the strain calculated if an elliptical shape is assumed. That a lower limit on pipe stiffness exists has been suggested by the authors of ASCE Manual of Practice No. 37 (WPCF MOP No. 9), the United States Bureau of Reclamation, the German Sewer Technical Association (ATV), the ASCE Continuing Education Short Course Curriculum on "Plastic Pipe Design", and the Transportation Research Board of the National Research Council. The new (1982) ASCE Manual of Practice No. 60, "Design and Construction of Gravity Sanitary Sewers", includes the following warning paragraph concerning low pipe to soil stiffness ratios:

"For small deflections, the vertical deflection dy may be assumed to approximately equal the horizontal deflection dx in Eq. 9.17 (The Modified Iowa Formula). Much research is being conducted on the application of Eq. 9.17 to pipe having a low ratio of pipe to soil stiffness where the vertical deflection may not be assumed to approximately equal the horizontal deflection. The result of this research is beyond the scope of this chapter but should be reviewed by the designer as a part of the design process."

STIFFNESS OF PROFILE WALL PIPES

The introduction of several types of plastic "profile wall" pipes in the past several years represents an effort by the plastic pipe industry to produce pipe larger than 27-inch diameter, until recently the largest size solid wall direct burial gravity plastic pipe manufactured.

A PVC profile wall PS 46 pipe in sizes from 18-inch through 24-inch has been marketed for several years. A proposed ASTM specification is currently under consideration which will allow a PS 10 class of PVC profile wall pipe in sizes up to 48-inches.

A recent entry into the United States profile wall pipe market, high density polyethylene (HDPE) profile wall pipe does not utilize the traditional method for defining pipe stiffness, but has introduced a new term, "Ring Stiffness Constant" (RSC). For a given RSC value, pipe stiffness is an inverse function of diameter.The pipe stiffness (PS) of a

"Ring Stiffness Constant 40" pipe will vary from about 11 psi for a 27-inch diameter pipe to 3.0 psi for a 96-inch pipe. Figure Three illustrates the trend toward lower pipe stiffness values which has occurred in the past few years.

STIFFNESS FACTOR IN THE MODIFIED IOWA FORMULA

The formula most commonly used in the United States for predicting the deflection of a flexible pipe under a given soil loading is the Modified Iowa Formula:

$$dx = D_1WKR^3/(EI + .061E'R^3)$$

The denominator in the Iowa Formula includes two terms: the left-hand term representing the contribution of the pipe itself to deflection, and the right-hand term in the denominator representing the contribution of the soil to deflection.

The Modified Iowa Formula was originally developed for steel pipe (corrugated metal pipe). When the ASCE Manual of Practice No. 37, "Design of Gravity Sanitary and Storm Sewers" was published in 1969, it was recognized that one should not allow the soil stiffness term in the denominator of the Iowa formula to become excessively predominant. The following is quoted from the ASCE Manual No. 37, page 219:

"The first term in the denominator, EI, in Equation 15, reflects the influence of the inherent strength of the pipe on deflection; whereas the second term, $0.061\ E'R^3$, reflects the influence of the passive pressure on the sides of the pipe. The second term may be excessively predominant in the case of large diameter pipes, with the result that a very light-weight pipe may appear to be satisfactory. Since the pipe wall must have sufficient local strength in bending and thrust to develop and utilize the passive resistance pressure on the sides of the pipe, it is recommended as a practical measure that the value of EI should never be less than about 10 to 15 percent of the term $0.061\ E'R^3$."

The Modified Iowa Formula was developed from tests on steel pipe. The design of flexible pipe manufactured from a thermoplastic or thermoset material rather than steel involves consideration of both long and short term properties. While the modulus of elasticity "E" of steel is on the order of 30 million psi, and is relatively unchanged with time, the modulus of elasticity of plastic materials changes significantly with time. In the case of high density polyethylene, the long-term modulus (50 years) is on the order of 30,000 psi, 1/1000th that of steel pipe, while the short-term, or initial modulus, is on the order of 180,000 psi. PVC pipe initial modulus ranges from 400,000

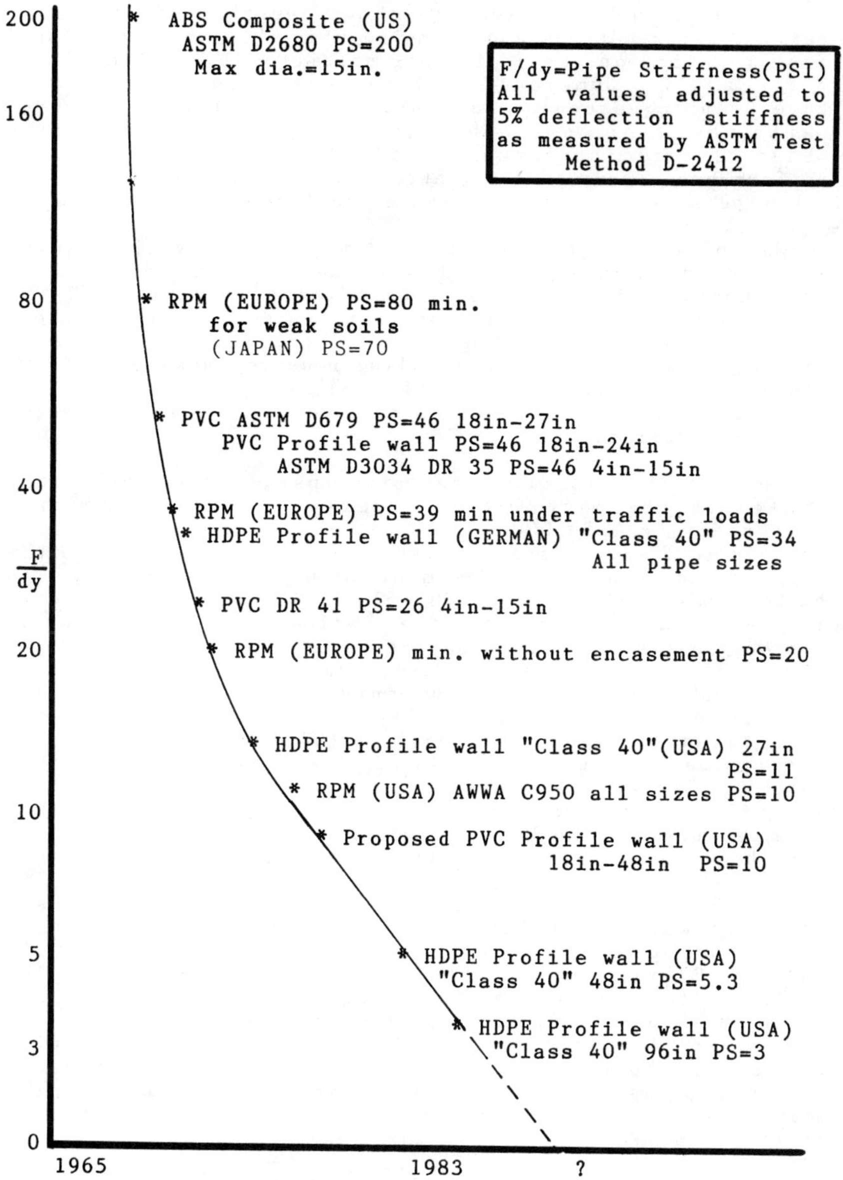

DECREASING PIPE STIFFNESS TREND

FIGURE 3

to 500,000, and decreases to about one-third in the long term. Design methods and formulae which have been developed from research on steel pipe are not directly applicable to plastic pipe. Numerous modifications to the Iowa formula have been suggested by Watkins, Molin, Carlstrom, Howard, Gaube, Mueller, and others, in attempts to reconcile observed field performance with Iowa formula predicted performance. Many of these attempts involve corrections for low pipe stiffness or low pipe to soil stiffness ratios.

One of the major differences is obviously the marked decrease in long-term modulus. The initial modulus is important,as the pipe must be sufficiently stiff to allow compaction of the bedding material beside the pipe without excessive vertical elongation. Long-term performance related to deflection and buckling require knowledge of the long-term characteristics of the plastic raw material. The reduced, or long-term modulus, must be used when calculating long-term performance.

Most of the plastic sewer pipe installed in the United States in the past several years has been SDR35 pipe conforming to ASTM D3034 or ASTM F679. F679 includes pipe from 18 to 27-inch diameter, while D3034 includes pipe from 4 to 15-inch diameter. The minimum pipe stiffness required by both D3034 SDR35 and ASTM F679 is 46 psi. This value is in the mid-range of the stiffness values used by Spangler in his original steel pipe experiments at Iowa State, in which pipe stiffness varied from about 25 psi to 61 psi. In the past, RPM pipe has been the only gravity pipe material marketed in the United States with a stiffness significantly below 46. ASTM D3262 for RPM sewer pipe requires a minimum stiffness of 10 for pipe 24-inches and larger. Current estimates are that there are some $60 to $80 million in lawsuits pending over RPM water and sewer pipe failures in the United States. Most of the failed pipe have been in the ground six to ten years.

J-M Pipe has introduced a PVC sewer pipe which is a spirally-wound, profile-wall PVC section with a pipe stiffness of 46, the same as smooth-wall ASTM F679 18 through 27-inch diameter plastic sewer pipe. The pipe stiffness for pipe manufactured according to ASTM F679 is a constant value of 46, regardless of pipe diameter. The RPM sewer pipe stiffness of 10 also applies to all pipe diameters above 24-inches. Stiffer reinforced plastic mortar pipe is manufactured in Europe by Hobas with a pipe stiffness of 38, and in Japan by Kabota with a minimum pipe stiffness of 70.

STIFFNESS EFFECTS ON INSTALLATION

As ASCE Manual No. 37 points out, one of the obvious reasons a zero-stiffness pipe is not practical is that in order to develop the side support upon which the pipe depends to support the vertical load over the top of the pipe, some degree of compaction must be achieved in the material adjacent to the pipe during installation. In order to achieve a given degree of compaction in the soil beside the pipe, a proportional amount of energy must generally be imparted to the soil by mechanical means. The degree of compaction achieved is directly related to the amount of energy imparted to the soil. The compactive effort, in addition to compacting the soil immediately adjacent to the pipe, will cause the soil to move laterally towards the pipe and trench wall, forcing the sides of the pipe wall in towards the center of the pipe. The stiffer the trench wall soil and the more flexible the pipe, the more vertical elongation of the pipe will take place. Howard (1981) discussed this phenomenon and established broad criteria for pipe stiffness to avoid vertical elongation of the pipe during installation. Jeyapalan (1982) also demonstrated this vertical elongation due to sidefill compaction using a finite element, soil-structure model. One of the significant errors commonly made in evaluating field deflections is to neglect the vertical elongation of the pipe before backfill is placed.

At some lower pipe stiffness value, it becomes a physical impossibility to compact the material beside the pipe itself. In a study of field performance of plastic pipe of varying stiffness, McGrath (1981) reported higher levels of compaction apparently could be achieved against stiffer pipe. McGrath suggests deflection values to be added to the calculated deflection ranging from .5 percent for stiff pipe (PS > 100) in 95 percent Proctor bedding to 10 percent for low stiffness (PS < 40) pipe in low density soil without proper haunching. On the basis of research conducted by the United States Bureau of Reclamation, Howard (1981) reported that at a pipe stiffness value of 34 psi, pipe bedded in a "moderately stiff" soil (compacted to 85 percent Proctor or more) will deform in a rectangular, rather than elliptical, shape. Rectangular deformation leads to high strain levels at the short radius "corners" of the rectangle. The "flattening" of the top of the pipe increases the potential for buckling failure of the top of the pipe. Schrock (1983) terms the strains developed in rectangularly deformed pipe "incalculable".

Schrock (1982), in the course syllabus for the American Society of Civil Engineers Continuing Education Shortcourse on Plastic Pipe Design, Chapter 7, includes an extensive discussion of the potential problem due to low-stiffness pipes in stiff soils, and refers to the German Technical

Sewer Association ("ATV") system for designing flexible pipe.

Chapter 7 includes Figure Four which relates vertical deflection, dy, to horizontal deflection, dx, for a given pipe to soil stiffness ratio, EI/R^3E'. From this relationship, it can be seen that there is a lower limit on pipe stiffness for a given soil stiffness, at which point the pipe deforms in a vertical direction while the horizontal dimension remains unchanged. This is represented by the vertical portion of the curve. Figure Six from Carlstrom (1982) illustrates this. Observed dy/dx relationships for laboratory tests conducted by Moser (1980) are plotted on the ASCE/ATV curve, Figure Four, and show greater distortions than predicted by ATV. The results of Spangler's original tests performed at Iowa State on corrugated metal pipe are also plotted on the ASCE design syllabus curve. The relationship of vertical to horizontal deflection for Spangler's original CMP tests is in excellent agreement with the curve developed by ATV.

Perhaps the most striking illustration of the effect of pipe stiffness on pipe performance is illustrated in Table Two and Figure Five, which compare average and maximum observed field deflections for PVC pipe on 27 projects in the Netherlands, KOMO (1982). These deflections were observed two to six years after installation. The relationship between pipe stiffness and both average and maximum deflection is quite obvious. At pipe stiffness values of 50, average deflections are four percent with maximums on the order of eight percent. At a pipe stiffness of 40 psi or less, maximum deflections exceed ten percent. At 10 psi pipe stiffness, maximum deflections are twenty-two percent. The difference between observed average and maximum deflections varied from two to ten percent for pipe stiffness values of 115 and 7 psi, respectively.

RECTANGULAR DEFORMATION

There are several reasons a rectangularly deflected pipe shape is to be avoided:

1. The formula for the calculation of strain levels in a deflected pipe <u>assumes an elliptical shape</u>. At the "corner area" or short-radius curvature portions of a rectangularly deformed pipe, stress levels are high and unpredictable. Molin (1971) determined that the error may be as high as a factor of two. Schrock (1983) states that it may be on the order of 5 to 10 or more. Figure Six is taken from field observations reported by Carlstrom (1982) and indicates an increasing strain rate with time in spite of a decreasing deflection rate in a low stiffness pipe bedded in a stiff soil. Rectangular deformation is evident in the pipe shape.

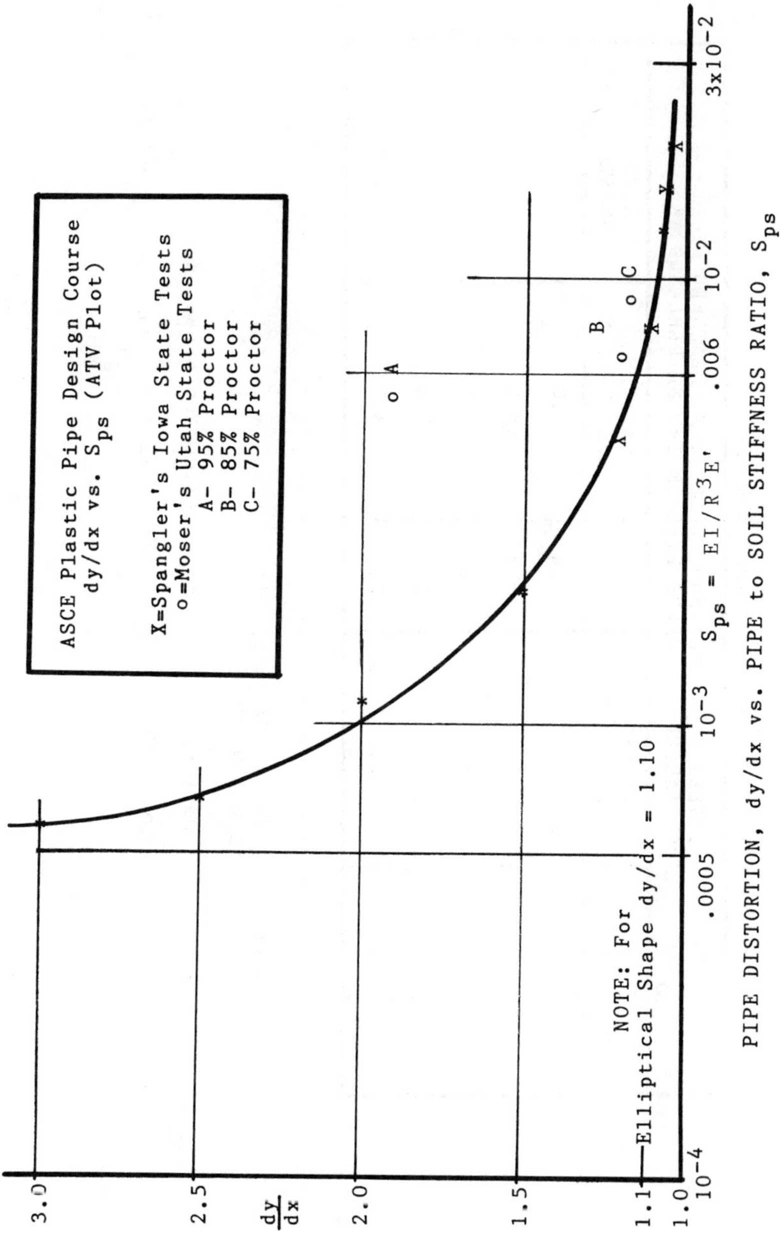
ASCE Plastic Pipe Design Course
dy/dx vs. S_{ps} (ATV Plot)
X=Spangler's Iowa State Tests
o=Moser's Utah State Tests
A- 95% Proctor
B- 85% Proctor
C- 75% Proctor
NOTE: For
Elliptical Shape dy/dx = 1.10
$\frac{dy}{dx}$
3.0
2.5
2.0
1.5
1.1
1.0
10^{-4}
.0005
10^{-3}
.006
10^{-2}
$3x10^{-2}$
A
B
C
$S_{ps} = EI/R^3E'$
PIPE DISTORTION, dy/dx vs. PIPE to SOIL STIFFNESS RATIO, S_{ps}

FIGURE 4

TABLE TWO

AVERAGE AND MAXIMUM DEFLECTIONS MEASURED IN INSTALLED PVC SEWER PIPE IN THE NETHERLANDS **					
DR	P.S.	AVERAGE DEFLECTION %	S %	MAXIMUM DEFLECTION %	NUMBER OF INSTALLATIONS
65	7	17.0	4.4	28.0	3
51	14	11.0	3.8	18.0	5
41	28	7.4	1.3	12.0	7
34	50	4.5	1.0	8.4	7
26	115	2.0	0.4	3.7	5

DR = Dimension Ratio, Diameter/Wall Thickness

PS = Pipe Stiffness, $F/dy = EI/.149R^3$ (psi) = $(4.47E)/(DR-1)^3$
E assumed = 400,000 psi

S = Standard Deviation for average deflection measured

** Data from KOMO (1982), group "C", diameter range 250 to 400 mm. (10 in. to 16 in.) Pipe in place for two to eight years

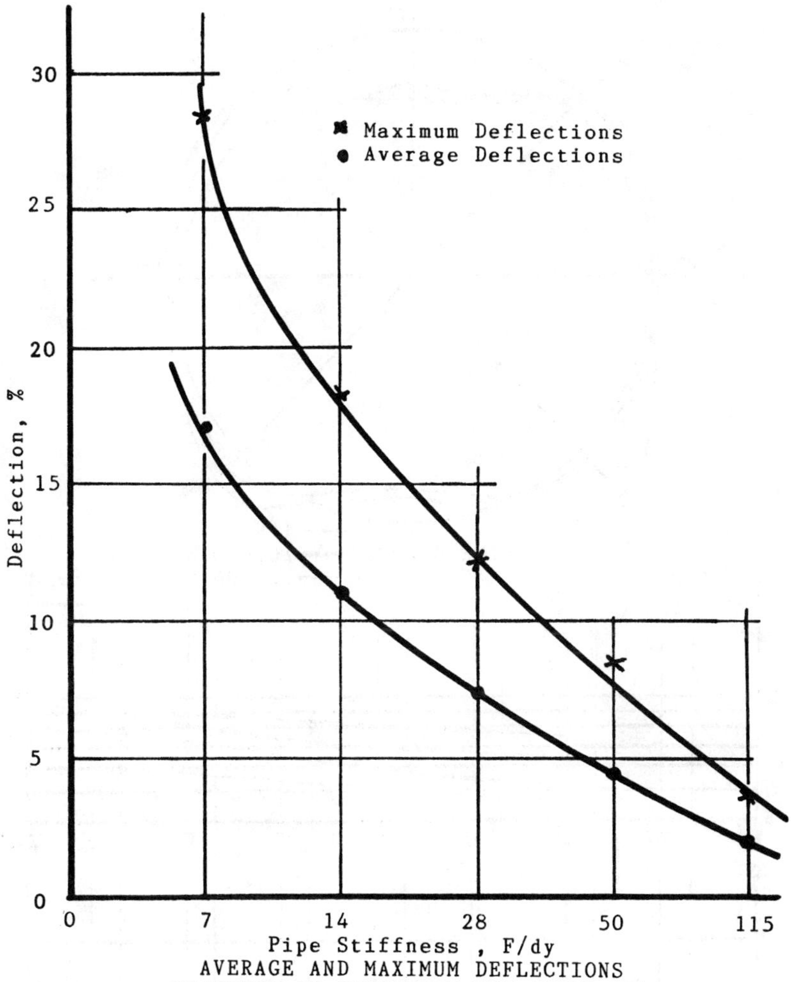

AVERAGE AND MAXIMUM DEFLECTIONS
MEASURED IN INSTALLED PVC SEWER PIPE
IN THE NETHERLANDS (KOMO STUDY)

FIGURE 5

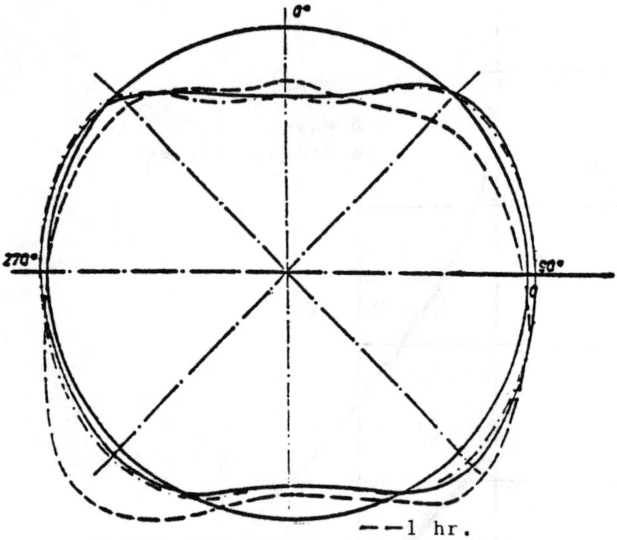

ACTUAL PIPE SHAPE

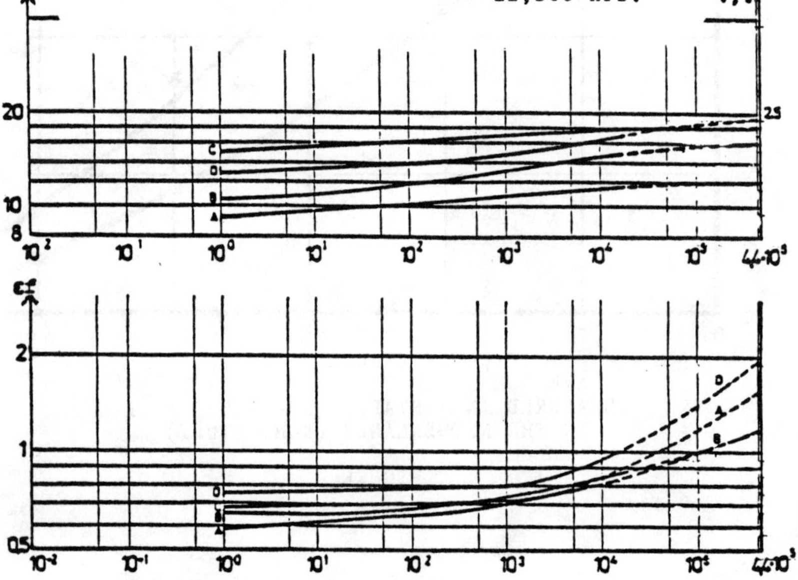

DEFLECTION AND STRAIN VS. TIME (HRS.)
From TUV Study- Schrock (1981), Carlstrom (1982)

FIGURE 6

2. The buckling calculation on which the long-term resistance to buckling failure of flexible pipe is based includes a correction factor for the amount of deflection. For pipe deflected five percent in an elliptical shape, the correction factor is 0.64, i.e., the buckling strength of the pipe is reduced by 36 percent. When the pipe deforms rectangularly, rather than elliptically, the factor by which buckling strength is reduced is unknown, since the radius of curvature of the crown section is increased significantly. A second stage, in fact, of the rectangular deformation phenomenon, may be reversal of curvature and buckling of the pipe wall. This was evident in tests conducted by Howard (1981) in which a 13 psi stiffness steel pipe deformed rectangularly and buckled at 18 percent deflection.

3. Some plastic materials are susceptible to strain corrosion in particular environments. High density polyethylene, for example, is subject to strain corrosion when exposed to common household detergents. The high strain levels caused by rectangular deflection may accelerate strain corrosion failure of plastic pipes.

4. In the case of RPM pipe, gel coat cracking may result from local deformation of the wall, followed by deterioration of the glass fibers and collapse. This can occur in spite of the fact that "deflection" measured as the change in vertical diameter from the nominal diameter may be only a few percent after installation and backfill. Recent extensive failures of PS = 10 RPM sewer pipe in Fairbanks, Alaska, after only six years in the ground, may be due to this phenomenon.

5. The recent introduction of "profile wall" PVC pipe in 18 through 48-inch diameters, and "profile wall" HDPE pipe up to 96-inch diameter, results in pipe with greater extreme fibre distances than solid wall pipe. For a given deflection, or radius of curvature in a rectangular-shaped deflected pipe, the strains at the outer edges of the "profile" section will be considerably higher than in a comparably shaped or distorted solid wall pipe, and may lead to long term local buckling of the profile section. Large shear stresses will exist at the junction of the "profile" section to the barrel, or pipe wall section.

6. Distortional strains, where the pipe to soil stiffness ratio is low, will be much more significant in the new "profile wall" pipes. Distortion may have an adverse effect on joints and on the spiral seam in the pipe wall. Moser (1980) conducted load-deflection tests on 24-inch profile wall PVC pipe, with a nominal pipe stiffness of 46 psi. Soil densities in the test included 75, 85, and 95 percent Proctor densities. For the pipe bedded in 95 percent Proctor density bedding (Test No. 12), spiral seam

failure occurred at a deflection of only one percent of the diameter. In the other tests, the pipe deflected to as much as 16 percent with no seam failures observed, when bedded in less dense soils. In Test No. 12 (95 percent Proctor), the calculated ratio of pipe stiffness factor to soil stiffness factor was eight percent, assuming an E' of 1500 for the soil. This is just below the recommended minimum of ten to fifteen percent in ASCE Manual No. 37. Observed vertical deflection was 1.9 times the horizontal deflection.

Installation, deflection, and related jointing problems in a recent 1000-foot-long, 60-inch "profile-wall" HDPE test section of sanitary sewer installed in Dallas, Texas, were so severe that subsequent re-excavating, re-rounding, and re-installing the pipe cost in excess of a half-million dollars. Pipe stiffness of the 60-inch pipe was 5.5 psi, measured at one percent deflection, equivalent to a DIN 16961 pipe stiffness of about 2.1 psi. This project probably represents the lowest stiffness large diameter gravity pipe ever installed under direct burial in the United States or Europe.

MINIMUM STIFFNESS CRITERIA FOR PLASTIC PIPE

1. Limiting Distortion Basis (ATV): From Figure Six it is apparent that the minimum and maximum radius of curvature of the deflected pipe changes significantly when dy/dx = 2.0. The buckling resistance of the pipe at the crown is reduced significantly, and the strains at the short radius (2 o'clock and 10 o'clock) portions of the pipe will be much greater than the strains which would exist in an elliptically deformed section.

Moser's (1980) tests with 95 percent compacted backfill on 24-inch profile wall pipe exhibited spiral seam failure apparently due to local distortions at a dy/dx value of 1.9. It would seem prudent to limit the distortion factor dy/dx by limiting the pipe to soil stiffness ratio, $S_{ps} = EI/R^3E'$. Limiting S_{ps} to .002, one can calculate the minimum long-term pipe stiffness required for a given soil E' value in order to limit distortion dy/dx, to a maximum of 1.5. The minimum long-term pipe stiffness factor (PSF) for a given E' can be calculated as:

$$PSF = EI/R^3 = .002E'$$

Since pipe stiffness is $(PS) = EI/.149R^3$

the minimum long-term pipe stiffness can be calculated as:

$$PS = .002E'/.149 = .0134E'$$

Long-term pipe stiffness (PS) and resulting initial PS values for HDPE, PVC and RPM can be calculated using the correction factors for variation in test method given in Table One. (Correcting for 1, 3, or 5 percent deflection stiffness, and the long-term modulus decrease values for each material given on Page Two.)

TABLE THREE

MINIMUM STIFFNESS VALUES ($EI/.149R^3$) for dy/dx = 1.5 MAXIMUM						
SOIL	TEST METHOD --- ASTM D2412			DIN 16961*		
E'	RPM @ 5%	PVC @ 5%	HDPE @ 1%	RPM	PVC	HDPE
1000	27	40	78	19	24	33
2000	54	80	155	38	48	65
3000	81	121	233	57	72	98

* 24 hr stiffness @ 3% deflection
NOTE: Above listed values are for initial pipe stiffness.

2. ASCE Manual No. 37 Basis: As discussed previously, ASCE Manual No. 37 recommends that the pipe stiffness term, EI, in the denominator should not be less than 10 to 15 percent of the soil stiffness factor, $0.061\ E'R^3$).

$$EI > .10 \text{ to } .15(.061E'R^3)$$

Since $F/dy = EI/.149R^3$, we can write:
PS = .10 to .15 x (.061E'/.149) = .10 to .15 x (.409E')
Then, minimum initial pipe stiffness values (correcting for variation in stiffness test methods) become:

TABLE FOUR

MINIMUM PIPE STIFFNESS PER ASCE MANUAL NO. 37				
SOIL	TEST METHOD ASTM D2412			DIN 16961*
E'	RPM @ 5%	PVC @ 5%	HDPE @ 1%	24 hr.@ 3%
1000	41 - 60	41 - 60	58 - 84	31 - 45
2000	82 -120	82 -120	116 -168	62 - 90
3000	123 -180	123 -180	174 -252	93 -135

* Applies to all thermoplastic pipe tested per DIN 16961.

3. Other Stiffness Criteria:

a. Carlstrom (1982) suggests the following minimum initial stiffness values for GRP pipe:

TABLE FIVE

MINIMUM INITIAL PIPE STIFFNESS FROM CARLSTROM (1982)		
INSTALLATION CONDITION	EI/D^3 N/m^2	PS psi
Minimum without concrete encasement	1250	10
"Normal" without traffic loads	2500	20
Pipe under traffic loads	5000	39
Pipe Installed in weak native soils	10,000	78

b. Chambers (1980), in Highway Research Program Report No. 225, suggests for PVC and PE pipe:

"Proposed specification SGH PVC-TD requires a pipe stiffness of 50 psi, but a higher stiffness may eventually be required to improve resistance to installation deflections" and for polyethylene: "The adequacy of the proposed pipe stiffness of 50 psi has not been fully verified under conditions of shallow burial and heavy, sustained traffic."

c. Howard (1981) suggests that rectangular deformation will "probably" occur when pipe with a pipe stiffness f/dy less than 34 psi is installed in bedding which is moderately compacted, i.e., 85 to 95 percent Proctor density. Assuming an E' of 1000, this corresponds to an S_{ps} value, EI/R^3E' of .005, very close to the limit of .006 which Carlstrom (1982) defines as the point where Spangler's strain factor (D_f) equation is no longer applicable.

d. ASTM D2321, the installation specification for thermoplastic pipe, includes this note: "Pipe having a stiffness of 25 psi or greater has been found to perform satisfactorily when installed in accordance with this specification." (Appendix X.1)

e. Watkins (1978) stated his research found an "optimum stiffness" to exist at about 42 psi: "...the Federal Highway Administration Report (Figure 2.7) shows a relative optimum at SDR 38--(PS = 42 psi)--just as the USU tests did. True, these minimums are not as pronounced as the USU tests show; however, as stated earlier, the test conditions were not the same and only three tests were conducted--hardly adequate to establish a meaningful trend. Since the original USU report on PVC pipe, several years of additional testing have been completed. Similar test results have been found."

4. Dynamic Buckling Failure

Nearly all of the existing design methods consider the pipe-soil system as a static rather than dynamic system. Chambers (1980) points out that the adequacy of a pipe stiffness of 50 psi has not been fully verified under conditions of heavy sustained traffic. At covers of six feet (two meters) or less, the pipe and soil envelope are not at rest, but rather a constantly changing dynamic system. As heavy wheel loads pass over the pipe, the pipe is alternately loaded on one side, pushed away from the load into the bedding on the opposite side, and the process is reversed as the load passes over to the other side of the pipe. The result is a constant repetitive series of small movements of the pipe and soil over long periods of time.

The pipe can be considered as a ring subjected to unconstrained slow buckling forces under earth and live load, as pipe and soil deform over the long term. In the analysis, the pipe is considered a ring with the horizontal dimensions fixed, as predicted by Spangler. Only the upper half is considered, as an arch that is fixed at its ends. The tangent to the arch at springline is assumed to be vertical. (Carlstrom and others have reported that, contrary to theory, the upper half of the pipe is the most highly stressed, and most failures have been observed in the upper half of the pipe.)

The buckling formula from Roark, "Formulas for Stress and Strain" is:

$$P_a = 8EI/R_a^3 \qquad \text{(Equation 1)}$$

where P_a is the buckling pressure and R_a is taken as the radius of curvature at the crown of the deformed ring. Since strains and curvature vary together, and the strain factor D_f is on the order of 3.5 (assuming no rectangularization takes place) we can write:

$$R_a = (1 + 3.5\ dx/D)R \qquad \text{(Equation 2)}$$

Excessive deflections and strains would be prevented if the vertical pressure P_v met the following condition derived from Equations one and two:

$$(P_v/8)\ (1 + 3.5dx/D)^3 = EI/R^3 \qquad \text{(Equation 3)}$$

Since $PS = F/dy = EI/.149R^3$ the minimum pipe stiffness required for a given deflection dx/D, is:

$$PS = F/dy = (P_v/1.19)\ (1 + 3.5\ dx/D)^3 \qquad \text{(Equation 4)}$$

Equation Four establishes the relationship between the long-term pipe stiffness, the loading, P_v, and the allowable deflection dy/D, which effectively places limits on bedding, compaction, and effectiveness of haunching.

For a pipe under an H20 wheel load of 16,000 pounds, and a cover of 3-1/4 feet (one meter), the load P_v is about 9.25 psi (58,000 N/m^2). The following minimum pipe stiffness values are calculated for 5, 7½, and 10 percent deflections:

TABLE SIX

MINIMUM INITIAL STIFFNESS BASED ON DYNAMIC BUCKLING FAILURE UNDER LIVE LOADS

PERCENT DEFLECTION	TEST METHOD---ASTM D2412			DIN 16961*		
	RPM @ 5%	PVC @ 5%	HDPE @ 1%	RPM	PVC	HDPE
5	25	39	74	17	23	29
7½	31	47	89	21	28	35
10	38	57	108	26	35	43

* 24-hour stiffness @ 3% deflection

The pipe stiffness values above may be adjusted based on expected frequency of occurrence and load intensity. Infrequent occurrence of H20 live loads may permit the stiffness requirement to be relaxed due to the decreased risk of deformation. Improved installation techniques and inspection during construction reduce deflections, and the required pipe stiffness.

High water table, poor soil, poor installation and inspection would raise the requirement. Where bedding soils consist of fine-grained materials and high ground water exists, siesmic liquefaction is a possibility and the full unconstrained hydrostatic loading of saturated soil should be used for the buckling analysis.

If pipe to soil stiffness ratio, S_{ps}, discussed previously is less than .002, rectangular deformation and flattening of the crown portion of the pipe may result in buckling failures at pipe stiffness values higher than those listed in the above analysis.

The results of the various minimum pipe stiffness criteria are summarized in Figure Seven.

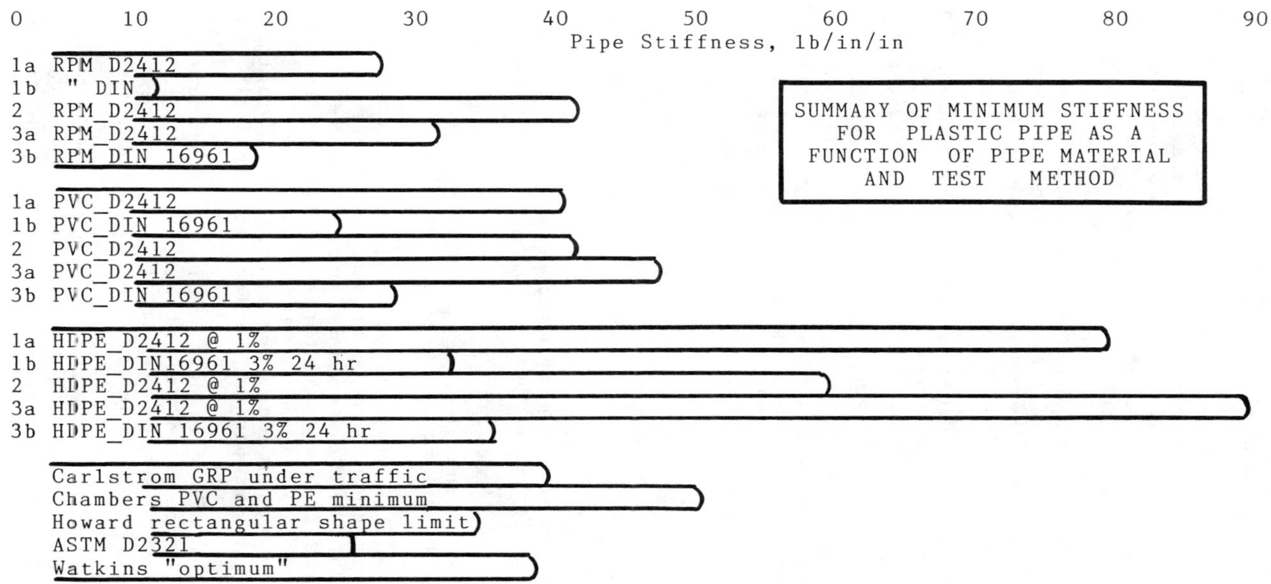

MINIMUM STIFFNESS CRITERIA:

1a ASCE/ATV Limiting distortion, dy/dx, to 1.5 maximum. ASTM D2412 @ 5% * (E'= 1000)**
1b Same as above, except tested in accordance with DIN 16961, 24 hr stiffness @ 3% dy
2 ASCE Manual 37 criteria, EI> 10% of $.061E'R^3$ ASTM D2412 @5% * (E'= 1000)**
3a Dynamic buckling limit @ dy= 7½%, ASTM D2412 @ 5% deflection.*
3b Same as above, except tested in accordance with DIN 16961, 24 hr stiffness @ 3% dy

* 1% for HDPE as recommended by US manufacturer.

** For higher E' values, minimum stiffness requirement will increase.

FIGURE 7

SUMMARY

While the Modified Iowa Formula will indicate that a ZERO STIFFNESS pipe will deflect only five percent under 16-feet of fill when a soil E' value of 1000 is used, no one has yet advocated zero stiffness pipe. From Figure Three, zero stiffness pipe does appear to be on the horizon. Much research and field experience has indicated that minimum pipe stiffness levels do indeed exist, and must be applied in design if we are to control deflection within acceptable ranges and limit extreme moments and strains in the pipe wall resulting from pipe distortion.

Variations in both the definition of pipe stiffness and stiffness test methods have served to confuse the issue. From the analysis of various stiffness criteria, some conclusions are evident:

1. **RPM** pipe in the traditional United States stiffness of 10 psi is probably inadequate under urban traffic conditions. An initial stiffness of 30 appears to be adequate when E' is limited to 1000 and long term deflections are limited to five percent.

2. **PVC** solid wall pipe as generally specified in the United States, with a stiffness of 46, is adequate, but not overly conservative. Profile wall pipes with a stiffness of 46 psi will probably perform satisfactorily if E' and deflection is limited so that extreme distortions do not destroy the integrity of the pipe wall seam and/or induce local buckling of the "profile" portion of the pipe wall.

Profile wall PVC pipe with a stiffness of 10 is grossly inadequate for all but shallow, temporary installations, where they will not be subject to traffic.

3. **HDPE** - Successful HDPE profile wall pipe experience in Germany has often been used to promote it's use in the United States. If the German test method were used, (stiffness measured @ 24 hours and three percent deflection) a stiffness of 30 to 35 would be adequate for an E' of 1000 or less. For the United States test method, i.e., initial stiffness measured at one percent deflection, initial pipe stiffness minimums of 80 to 90 psi are required. This seemingly high requirement is due to the large difference between the DIN 24 hour and United States initial test (factor of 2.0) compounded by the DIN three percent deflection in the stiffness test and the manufacturer's one percent requirement (factor of 1.24). Thus, for equivalent pipe performance, the United States pipe must be $2\frac{1}{2}$ times the German HDPE stiffness. The HDPE stiffness question is further confused by the introduction of "Ring Stiffness

Constant", rather than use of the conventional pipe stiffness definition. Nesbeitt (1980) has calculated that HDPE profile wall pipe stiffness $EI/.149R^3$ at five percent can be approximated from the "Ring Stiffness Constant" or "Class" as:

$$PS = RSC/.120\ Dm$$

where: Dm is the mean pipe diameter in inches, and RSC is the "Ring Stiffness Constant" or "Class".

Adjusting for the difference in stiffness measured at five percent to a one percent stiffness test, from Figure Two,* $PS_{5\%} = 0.70PS_{1\%}$ and the equation becomes:

$$PS(5\%) = RSC/.17\ Dm$$

As an approximation, the 5% initial pipe stiffness in psi can be estimated as six times the "Class" divided by the diameter in inches. For example, a "Class 40" 60 inch diameter HDPE pipe will have a 5% initial pipe stiffness of (6x40)/60 = 240/60 = 4 psi.

CONVERTING TO EI/D^3

To convert pipe stiffness $F/dy=EI/.149R^3$ (psi) to EI/D^3 (N/m^2) multiply F/dy by 128. For example, 10 psi F/dy is equal to 1280 N/m^2.

* Note: PVC data from Uni-Bell (1979) Handbook, Figure 13, page 138; HDPE data from Law Engineering Testing (1981) Report on 27-inch "Class 63" Spirolite HDPE profile wall pipe.

Figure One RPM data from Carlstrom (1982), HDPE data from South German Plastics Center (1982), tests on German HDPE profile wall pipe.

REFERENCES

ASCE-Manuals and Reports on Engineering Practice, No. 37, "Design and Construction of Sanitary and Storm Sewers", American Society of Civil Engineers, New York, New York, 1969. (WPCF-Manual of Practice-No. 9)

ASCE-Manual and Reports on Engineering Practice, No. 60, "Gravity Sanitary Sewer Design and Construction", American Society of Civil Engineers, New York, New York, 1982. (WPCF-Manual of Practice-No. FD-5)

ATV Guidelines "Richtlinien fur die Statische Berechnung von EntwasserungsKanalen und - leitungen", ATV, Augustin, BRD, January 1980. (Guidelines for the Static Calculation of Drainage Pipes and Lines, by the German Wastewater Engineering Society.)

Carlstrom, B. (1982) "Calculation of Circumferential Deflections and Flexural Strains in Underground GRP Pipes Used for Non-Pressure Applications", Europipe '82 Conference Papers, January 1982, Basel, Switzerland.

Chambers, R. E. (1980) "Plastic Pipe for Subsurface Drainage of Transportation Facilities", National Cooperative Highway Research Program Report No. 225, Transportation Research Board, National Research Council, Washington, D. C., October 1981 (with T. J. McGrath and F. J. Heger).

DIN 16961, Part 2, "Rohre aus thermoplastischen Kunststoffen mit profilierter Wan-dung und glatter Rohr-Innen Flache", Deutsche Normen, September 1977 (DIN Specification for Pipes of Thermoplastic Materials with Profiled Wall and Smooth Inside, General Quality Requirements and Test Methods.)

Gulf (1980) "Spirolite Pipe Product Data", Spiral Engineered Systems, Division of Gulf Fabricated Plastics Company, Norcross, Georgia, 1980.

Howard, A. K. (1981) "Diametral Elongation of Buried Flexible Pipe", Proceedings, International Conference on Underground Plastic Pipe, ASCE, New Orleans, Louisiana, April 1981.

Jeyapalan, J. K. (1981) "Soil Structure Interaction Analyses of Plastic Pipes", ASCE Convention and Exhibit, New Orleans, Louisiana, October 1982 (with F. Osequeda and W. J. Horn).

KOMO "Vervorming smet ingen aan operationele PVC - Straatrioleringen" Stichting KOMO, Rijswijk, Netherlands, June 1982.

Law Engineering Testing Company, Report of ASTM D2412 Parallel Plate Test on HDPE Pipe, January 18, 1981, Atlanta, Georgia.

McGrath, T. J. (1981) "Field Performance of Buried Plastic Pipe", Proceedings, International Conference on Underground Plastic Pipe, ASCE, New Orleans, Louisiana, April 1981 (with R. E. Chambers).

Molin, J (1971) "Principles of Calculations for Underground Plastic Pipes--Loads, Deflection, Strain", Owens-Corning Fiberglas Technical Report, Brussels, 1971.

Moser, A. P. (1977) "Design and Performance of PVC Pipes Subjected to External Soil Pressure", Buried Structures Laboratory, Utah State University, Logan, Utah, February 1977 (with R. K. Watkins and O. K. Shupe).

Moser, A. P. (1980) "Structural Performance of 24-inch Diameter Ribbed- PVC Pipe Subjected to Soil Loadings", Buried Structures Laboratory, Utah State University, Logan, Utah, August 28, 1980, (with O. K. Shupe).

Nesbeitt, W. (1980) Uni-Bell Plastic Pipe Association, Evaluation of Spirolite literature, June 2, 1980, Dallas, Texas.

Schrock, B. J. (1981) ASCE Continuing Education Services, "Plastic Pipe Design", New Orleans, Louisiana, April 1981, (with J. W. Pierce).

Schrock, B. J. (1983) "Plastic Pipe Overview--1983", Europipe '83 Conference Papers, May 1983, Basel, Switzerland.

South German Plastics Center, "E Modulus from Flexural Test per DIN 53547", July 13, 1982, Wurzburg, BRD.

Taprogge, R. H. (1981) "Large Diameter Polyethylene Profile Wall Pipes in Sewer Application", ASCE, New Orleans, Louisiana, April 1981.

Uni-Bell Plastic Pipe Association, "Handbook of PVC Pipe Design and Construction", Dallas, Texas, 1979.

Watkins, R. K. (1978) "Discussion: Predicted vs. Actual Deflections in PVC Sewer Pipe", with A. P. Moser and O. K. Shupe, Utah State University, Logan, Utah, June, 1978

WATER SUPPLY TO NAMAQUALAND

D STEPHENSON
MASCE

N D GRAHAM
MASCE

SYNOPSIS

The supply of potable water over a distance of 83 kilometres and with a static head drop of 990 metres posed unusual problems. Owing to the rugged nature of the terrain and the corrosive nature of the surface soils, the pipeline was laid above ground. Temperature variations are extreme and required careful attention to be paid to pipe jointing. Economical design of the pipeline was achieved by providing break pressure tanks and valves at three points. Synchronisation of the control valves was required to prevent surge. The use of air valves was minimised by careful design. The rapid construction programme was facilitated by the use of computers for plotting sections and analyzing the main. The supervision and coordination of the project were also intensive for the rapidly constructed pipeline.

1. INTRODUCTION

The town of Annex Kleinzee lies is an arid region, Namaqualand, on the West Coast of South Africa and serves several diamond mines in the region. Although sea water is available for industrial purposes the domestic water supply was a problem.

A water study revealed that the most economical supply was a connection from an existing supply line from the Orange River to Nababeep (see Figure 1). At Nababeep, 83 km inland, a reservoir 990 metres in elevation above Kleinzee could supply sufficient water to meet the demand.

The pipeline route falls steeply initially, followed by 65 km of fairly flat terrain (see Figure 2 for a profile). The pipe, which supplies 38 litres per second, comprises 33 km of 200 mm diameter and 50 km of 250 mm diameter steel pipe. A branch pipe to the town of Port Nolloth required a certain minimum pressure in the line and this is indicated on the section.

*Professor of Hydraulic Engineering, University of the Witwatersrand, 1 Jan Smuts Avenue, Johannesburg, 2001, South Africa.
**Senior Divisional Engineer, Anglo American Corp of S.A Ltd. P O Box 1587, Marshalltown, 2107, South Africa.

On account of the highly saline soil the corrosion protection system for a buried pipe would have been elaborate and costly. As there was little inconvenience in laying the pipe above ground it was constructed on concrete support blocks at each joint. The temperature variations in Namaqualand can be severe, from -10°C to 50°C in a year. Severe temperature expansion and contraction of the steel pipe was therefore anticipated. Special precautions had to be taken with the joints to ensure they did not deteriorate. To prevent damage to the lining and overcome temperature stress, flexible joints were provided.

2. DESIGN AND INSTALLATION OF THE PIPELINE

Pipe Support and Jointing System

The construction of the 83 kilometre length of 250 mm and 200 mm bore steel pipe above ground required a number of structural and laying improvisations. Supports were constructed of precast concrete plinths embedded in concrete bases. Two supports were provided per 9,1 metre pipe length. The jointing system used was Viking-Johnson flexible couplings. On one side of each coupling the pipe was clamped to the support and on the other side the pipe was free to slide in order to accommodate temperature movements. It was expected that temperature movements of up to 9 mm in each pipe length would be experienced.

It was acknowledged that, in this application, the use of the Viking-Johnson coupling was novel. Tests were done on the couplings to check their reliability under extreme conditions. A test joint was set up in a hydraulic test rig in a laboratory. The pipe and joint were pressurised and temperature movements in the couplings was simulated at various deflection angles. Deflection angles up to 7° at a pressure of 30 bars were accommodated without leaks. The only possible problem due to expansion and contraction appeared to be a shuffling of the collar towards one side of the pipe joint which could then expose the other end. Central registers were provided on the couplings to prevent this.

The possibility of wind blown sand getting into the coupling between the rubber ring and the pipe surface was considered and such possibility was simulated in the laboratory. It was found that the tight fit between the rubber ring and pipe excluded any sand. A few leaks were however experienced on the pipeline after installation.

A compromise between the height of concrete plinth and the permissible angles of bending in the vertical plane was necessary. Owing to the undulating ground and in order to keep the pipeline straight in some sections excessively high plinths would be required. This would require adequate lateral bracing and structural rigidity. A computer curve-fitting programme and profile-plotting procedure was developed to optimise the height of the plinths.

Vertical Profile Design

The testing programme established that the joints could tolerate 7° deflection without distress. Since, however, experience on other pipelines constructed with Viking-Johnson couplings had demonstrated

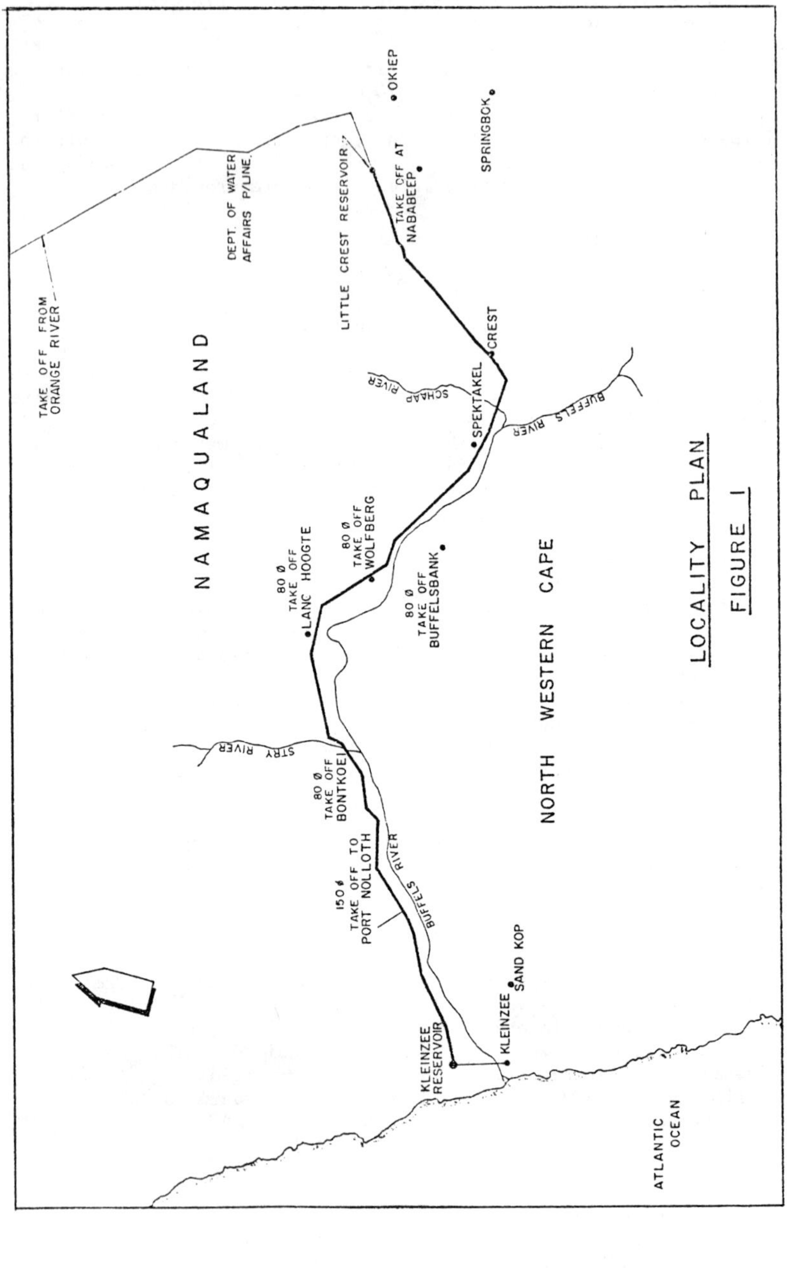

LOCALITY PLAN

FIGURE I

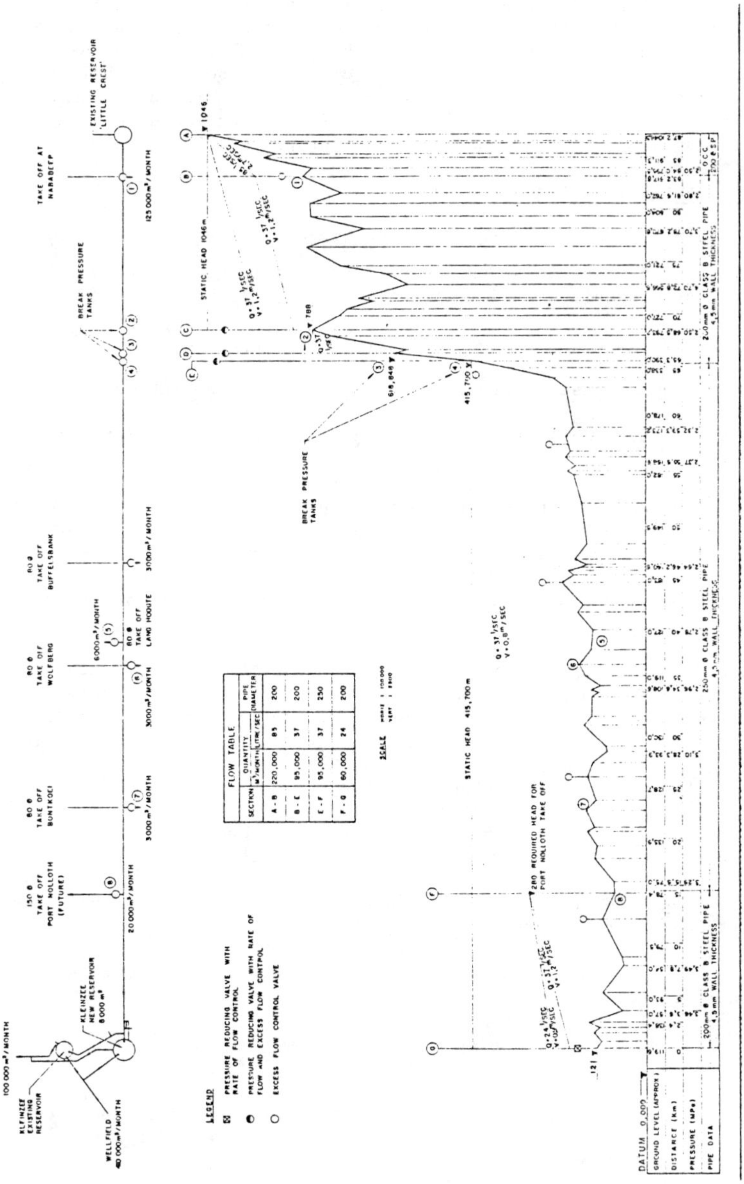

FLOW TABLE			
SECTION	QUANTITY		PIPE DIAMETER
	m³/MONTH	LITRE/SEC	
A - B	220,000	85	200
B - E	95,000	37	200
E - F	95,000	37	250
F - G	60,000	24	200

FIGURE 2

that angular joint deflecting in excess of 5° created occasional problems, a decision was taken to limit the vertical or horizontal deflection to a maximum of 4° at any joint. The vertical alignment was therefore designed to fit the terrain as closely as possible while at the same time minimising the construction costs. The calculation of the vertical profile to optimise the cost of the line was done using a computer programme written especially for the purpose [1].

Outline of the Calculation Technique

Design parameters for the alignment were as follows:

(i) Angular deflection at the joints was limited to 4° without thrust blocks.

(ii) Standard pipe lengths of 9,1 metres were employed.

(iii) Standard precast plinth sizes of 0,500 metres height were used.

(iv) Any vertical deflection greater than 4° required fabrication of a bend.

(v) The thickness of blinding under the plinths was limited to 50 mm. Any additional level adjustment was made up by excavation or in-situ concrete work.

A typical geometric longitudinal section of the pipeline is as shown in Figure 3.

The costs for blinding, excavation, in-situ concrete and fabricated bends being known, the problem became one of selecting a configuration that fitted the ground profile, according to the above parameters, for the least cost.

In outline the analysis technique consisted of:

(i) Fitting a suitable curve to the surveyed ground levels to approximate the terrain as closely as possible.

(ii) Fitting cord segments to the curve, each segment being one standard pipe length.

(iii) Calculating the cost implications of this particular curve selection.

(iv) Modifying the curve selected until the minimum cost was achieved.

A cubic spline curve fitting subroutine was written which fitted a cubic function to the ground levels. The computed curve was then lifted off the ground to accommodate the plinths. The next step in the calculation was to fit straight segments to the curve, each segment representing the standard pipe length. For every pipe joint a cubic function and equation for a circle was solved simultaneously. (See Figure 4).

Cost Calculation

Once the mathematical curve fit was obtained it was relatively simple to calculate the cost implications for the curve. The Computer program examined all joint deflections. When vertical bends were found in excess of 4° an adjustment was made to flatten the curve at that point (by weighting the ground levels in this area in the direction necessary to flatten the curve). The process was repeated until an optimum cost was achieved.

The computer programme also included a number of other options for modifying the curve, namely; manipulation of the end conditions to the cubic spline, modification of the number of data points used in calculating the cubic function and adjustment of the function by hand after visual inspection of the computer plot of the ground and pipeline profiles.

Corrosion Protection

A potentially serious problem of external corrosion with a buried pipeline was identified. In addition 28 km of the line was laid adjacent to 32 kV electrical transmission lines. The decision taken to install the pipeline above ground presented other problems, however, namely:

(i) Corrosion and mechanical damage resulting from wind blown sands during the severe sandstorms which are prevalent in this region.

(ii) Heat which affected pipe and water temperatures.

The problems were countered in the selection of the external coating system, namely:

(i) An ethyl based inorganic zinc silicate primer used to resist the effects of sand abrasion; and

(ii) Alkyd aluminium employed on top of a buffering zinc phosphate coating to reflect the radiation of the sun.

Internal corrosion was not considered a hazard but scaling was, due to the nature of the transported water. It was therefore decided to line the pipeline to maintain the flow characteristics throughout its life. It was considered that the extreme temperature ranges mentioned previously precluded the use of bitumen linings, and an epoxy was employed for the internal coating of the pipe.

3. AIR VALVES

Customary procedure for the installation of air release and air vent valves is to provide such valves at ½ to 2 kilometre intervals. The cost of such valves and the structures to protect them would have been out of proportion to the cost of the rest of the pipeline if this procedure had been adopted. Therefore research was instigated into the effects of omitting many of the air valves. Although it is normal procedure to install small-orifice air release valves on rising

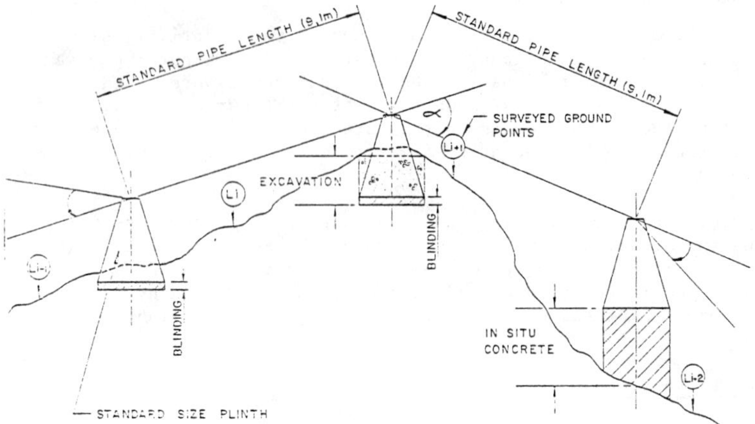

FIGURE 3

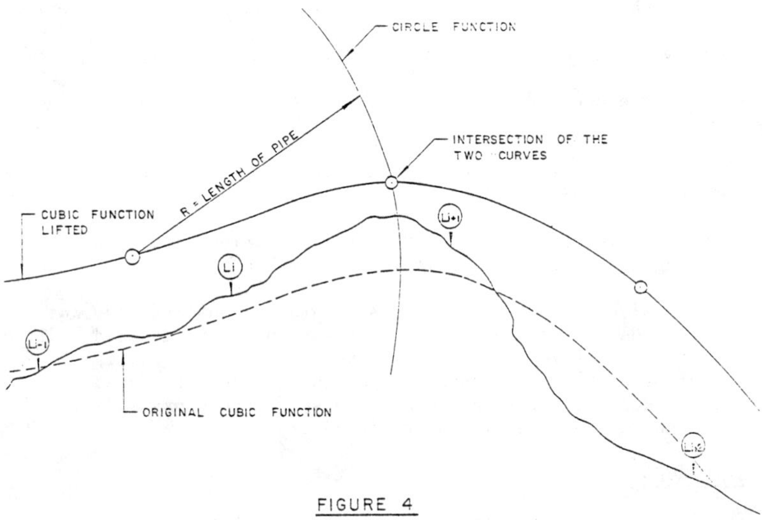

FIGURE 4

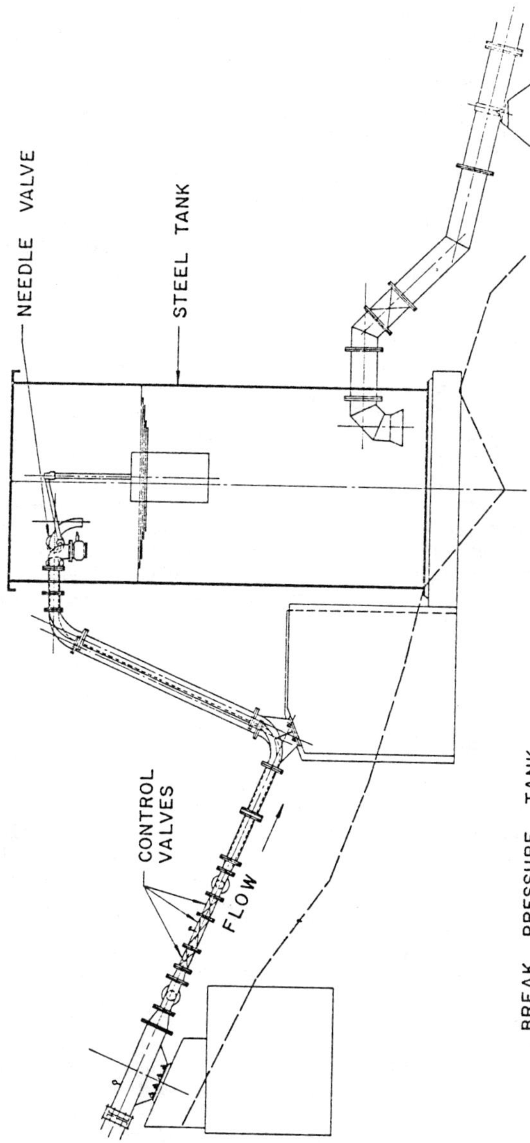

BREAK PRESSURE TANK

FIGURE 5

sections of pipeline there were few such sections on this particular line. A rise in pipeline profile with respect to the hydraulic gradient can reduce pressure which, in turn, would release air from solution and this air would have to escape via air release valves. In this case the hydraulic grade line was effectively parallel to the pipeline profile or else pressure increased downstream.

The installation of air vent valves i.e. large orifice type, also was minimised. The pipeline is structurally strong enough to resist vacuum pressures but the possibility of restrictions in flow rate during scouring required some large orifice air valves. For filling the pipe, the large orifice air valves were also of use, as air pockets could restrict the capacity of the pipeline.

Research indicated that if an air valve was not installed at a rise in the pipeline, flow would be restricted to a small cross-sectional area of the pipe so that the velocity under the air pocket would be high. Therefore there would be a hydraulic jump at the downstream end of the air pocket which would entrain air. The rate of air entrainment was in all cases greater than the anticipated rate of introduction of air from upstream. Therefore it could be expected that such air pockets would wash out or else become small enough to have negligible effects on the head loss of the pipeline. Where the pipeline profile is relatively flat the air pocket would wash along bodily and eventually be released when it reached an air valve or alternatively be entrained in the flowing water to be released at the downstream pressure reducing tank or at the reservoir. (Stephenson[2]).

4. BREAK PRESSURE ARRANGEMENTS

The pipeline was constructed of 4,5 mm thick Grade B steel. This was the minimum thickness considered desirable from the point of view of corrosion and handling and the steel was the highest grade which could be economically used in this situation. The head in the pipeline therefore had to be maintained at less than a certain maximum to avoid overstressing. The large static head was broken by installing break pressure arrangements at three points. The pipeline profile indicated in Figure 2 is relatively uniform for most of the downstream length, i.e. the lower 65 kilometres, so that the break pressure arrangements were installed on the steep escarpment. The entire downstream section could therefore not experience higher static heads than those in the lowest break pressure tank. This tank was located in a position suitable to provide the necessary hydraulic gradient. Similarly, other tanks were located to provide the necessary hydraulic gradient but to avoid overstressing the pipe. A number of alternative break pressure systems were considered. Initially in-line pressure reducing valves were considered without the use of break pressure tanks. However, a number of unfortunate previous experiences with such valves were of concern. The possibility of such valves failing could not be ruled out and the consequences of pressurising the entire pipeline would be severe. The water is hard so that scaling of the pilot tubes of in-line control valves and consequent malfunction was a distinct possibility. A fail-safe system therefore had to be installed. Break pressure tanks were considered to be the most fail-safe method of breaking pressures.

In order to avoid high velocities in the pipe, especially in the short length of rapidly falling pipe to the last break pressure tank, in-line control valves were also installed. The last break into each tank and the downstream reservoir was done with a needle type float controlled valve. These valves were considered to be corrosion and cavitation resistent which was necessarily required in the last stage of the pressure reduction to atmospheric pressure. They were also sized to deliver the design flow at the pressure to which an upstream in-line pressure reducing valve was adjusted.

Approximately 10 metres upstream of the needle float valve in-line pressure valves were installed. These valves are known to cavitate with extreme pressure reduction ratios so they could not be used in isolation. The maximum practical pressure reduction ratio was considered to be 3 : 1. The accompanying Figure 5 illustrates the pipework and break pressure tank arrangement typical for each of the three upstream break pressure locations.

The upstream break pressure tank posed a problem due to the wide variation in head on the tank. The possibility of a take-off upstream of the tank had to be considered. With the draw off at a high rate the pressure on the break pressure tank could be as low as 30 metres whereas without any draw-off the head could be as high as 140 metres. The maximum head is anticipated at a flow of 37 litres per second. The static pressure on the other hand would rise to 260 m. The valves, especially the needle valves, had to be sized on the lower pressure whereas the breaking of pressure by the upstream valve was required for the higher pressure. The pressure breaking ratio in the extreme case of no upstream draw-off therefore exceeded the desired ratio for no cavitation. As this occurrence was not likely to be regular this was accepted. A further problem arose due to commercially available valve sizes. In order to standardise, 50 mm needle valves were used at each tank which required a head of 30 metres to discharge 38 litres per second. The upstream pressure control valves were 80 mm nominal bore and these required a head of at least 20 metres to pass the discharge required. A slight reduction in the maximum inflow to the top tank was therefore to be expected when the Nababeep draw-off was at its maximum.

In order to avoid hunting of the valves the following procedure was adopted. Each successively lower length of pipe was adjusted to operate at a slightly lower design flow. This means that when the downstream pipe was operating the inflow to each tank would be slightly throttled, thereby avoiding hunting, i.e. successive opening and closing. When the downstream reservoir was filled, successively higher pipe lengths would shut down.

The pressure control valves also functioned as flow control valves. A break in the pipeline could go undetected for many hours owing to the long length of pipeline. The only way of detecting a break appeared to be by observing excessive flow velocities. Flow measuring orifices were therefore installed downstream of each break pressure tank and when the head loss across this orifice exceeded a specified limit the upstream flow and pressure control valve would automatically shut. This prevented each tank and pipeline running dry. Thus after the break was remedied there would be sufficient water in the pipeline system to get water to the Kleinzee reservoir rapidly again. Excess flow control valves were also installed along the last 65 km length of pipeline to Kleinzee in case of breaks along that length of line.

Problems were experienced with filling of the line as the flow rate through the orifices was often excessive before the line was filled and this automatically shut the flow control valves thus sealing off inflow to the break pressure tank and drawing in further air. The operating procedures laid down therefore required that the flow control system be isolated while the line was being filled and commissioned. The only way of adjusting the pressure controls and the flow controls was by site adjustment.

It was therefore necessary to adopt a conservative attitude to avoid overstressing the line and pressures were adjusted by initially fully throttling the pressure which reduced flow and adjusting the control valve gradually until the flow rate measured into the break pressure tank was sufficient.

Water Hammer

The flow control valves could shut over a few seconds and water hammer had therefore to be considered. Although the water hammer pressures would not be abnormally high compared with operating pressures they were nevertheless avoided in order to prevent jolting of the couplings, support plinths and pipework. The two short lengths of pipeline between the top and bottom break pressure tanks were not protected with surge relief valves. This is because the needle float valve would shut over an acceptable length of time compared with the wave reflection time for these short lengths of pipe. The needle control valve shut at a rate depending on the rate of water rise in the break pressure tank. As these tanks were only 2 metres in diameter this could occur over 3 minutes. This was reasonable for the shorter pipe lengths but in the case of the longer lengths, slower closing was required. The upstream tank was 4 m in diameter which limited the rate of rise in water level. In the case of the reservoir at Kleinzee, although the water surface area of the reservoir would rise slowly the size of the inlet pipe to the siphon tank had nevertheless to be throttled. The wave travel time of this lower length was 65 seconds up to the next break pressure point and a valve closure of at least 20 minutes was considered desirable. This was accomplished with a throttled siphon filled float chamber. Nevertheless in order to accommodate malfunctioning of the valves, surge relieve valves were installed at this reservoir and at the upstream break pressure tank. These surge relief valves were adjusted to open when the pressure exceeded 50 metres above static. The latter valves were entirely hydraulic as were the pressure and flow control valves.

5. CONCLUSIONS

The pipeline was designed as a gravity main to take advantage of the natural fall in terrain. To cater for the problems which the harsh environment presented special precautions were taken.

The 83 kilometre line was laid and installed within 12 months and was commissioned at the beginning of 1981. After initial teething troubles and adjustments it is now discharging the design flow within the acceptable pressure requirements. Although there were a few collar leaks, these are now all sealed.

REFERENCES

1. Rogers, D F and Adams, A J. : 1976 Mathematical Elements for Computer Graphics. McGraw-Hill New York.

2. Stephenson, D., 1981 : Pipeline Design for Water Engineers, 2nd Edn. Elsevier, Amsterdam.

PIPELINES IN ADVERSE ENVIRONMENTS

John C. Seeley, P.E.*

This paper deals with water intake lines in large bodies of water such as the Great Lakes and the connecting waterways. Specifically, the discussion will describe problems encountered with two intakes that were constructed in the early 1900's and the methods used to rehabilitate the pipelines for continued use.

Inland water supplies are generally from either groundwater or reservoirs. However, the cities located on the shorelines of the Great Lakes have access to the largest supply of fresh water on this planet. Most of those communities take their water supply from this vast resource, and the intakes for these systems are, in many ways, as unique as the Lakes themselves.

Typically, a water intake line into one of the Great Lakes consists of a pipeline laid in a trench under the bottom of the lake. The intake lines are normally several thousand feet long. This is due to the fact that 30-40 feet of water depth is needed to get the best water quality, and to prevent damage from shipping and ice formation. At the desired water depth the pipe turns up and into a "crib" constructed of timbers, about five to six feet above the lake bottom.

The intake pipeline must be designed at sufficient depth so that it is protected from the effect of wave action on the lake bottom, as well as wind currents and natural currents in rivers. Problems are often encountered in the excavation to achieve the required depth, because many times the lake or river bottom is of rock, or sand and sometimes muck.

A pumping station is generally constructed at the lakeshore. Water flows by gravity from the crib, through the intake line, and into a shorewell. It is then pumped to a water treatment plant. Treated water is then pumped into the distribution system.

Both the pipelines described in this paper were constructed before 1910. However, the inherent problems with construction in the Great Lakes, such as control of backfill and the danger of sudden, violent storms warranted the consideration of cost-effective measures to rehabilitate the old pipelines. While the measures were successful, there were problems associated with the projects.

LOCKPORT, NEW YORK

The City of Lockport, New York is a community of 25,000 population located on the Erie Barge Canal at a point where there is a double set of locks. The principal industry is the Harrison Radiator Division of General Motors. Water use in the City is approximately 11 MGD, of which about 75% goes to industrial customers.

*Life Member-American Society of Civil Engineers, McNamee, Porter & Seeley, 3131 South State Street, Ann Arbor, MI 48104.

The Niagara Escarpment runs through the middle of Lockport, and this results in a difference in elevation of 100 feet or so from one side of town to the other. Prior to 1903, the water supply for the City came from the Erie Canal. However, in 1905 it was determined that the east branch of the Niagara River would provide a better water supply. Subsequently, a pumping station was constructed and a 48" diameter, riveted steel intake line with flanged joints was constructed into the river. In addition, 17 miles of pipeline from Lockport to North Tonawanda was also constructed.

In 1928, a water filtration plant was constructed to purify the Niagara River water. These facilities are still in use today.

In the 1970's the City engaged McNamee, Porter and Seeley to determine the cause of taste and odor problems in the municipal water supply. As a part of this study, samples of Niagara River water were taken from:

1. The point of intake on the west side of the river.
2. The low service pumping station on the east river bank.
3. The finished water from the treatment plant.

No problems were detected in the water from the intake. Yet both problems were present in the samples from the other two locations. This indicated there was a leak in the intake lines. To verify the existence of a leak, Great Lakes Diving Consultants of Traverse City, Michigan were engaged to inspect the lines. The distance from the pumping station to the intake point is 2,584 feet. A diver was sent through 1,000 feet of the pipeline from each of the access points. About 620 feet from the pumping station, the diver found that one of the old, flanged joints had failed and was permitting some type of industrial waste to enter the pipe. This waste material was dissolving the tar coating on the inside of the pipe. The combination of industrial waste and dissolving tar produced the bad taste and odor in the water.

Considering the age of the pipe, and the fact other joints might fail in the future, it was decided that it would be most cost-effective to reline the entire length of the intake line. The City Council authorized McNamee, Porter and Seeley to prepare plans and specifications for the project. Herbert F. Darling Company of Williamsville, New York was awarded the construction contract to install a polyethylene liner with an outside diameter of 44 inches and walls about 1½ inches thick. The annular space between the liner and the intake pipe was grouted with cement. The polyethylene liner was purchased in 30 foot lengths and joined by butt fusion to make pieces 800 feet long which were strung out beside River Road.

The configuration of the intake line at the pumping station presented a real problem. The end of the pipeline was 25 feet below grade and was at a 90° angle to River Road, the main thoroughfare through town. This would not permit the contractor to align the liner directly with the intake. Consequently, it was necessary to pull the liner around the 90° bend through an opening in the concrete wall and then into the intake line. The contractor provided a curved, sheeted trench for pulling the pipe around the 90° bend. The trench also provided a uniform slope to the 25' drop from the ground surface to the intake line.

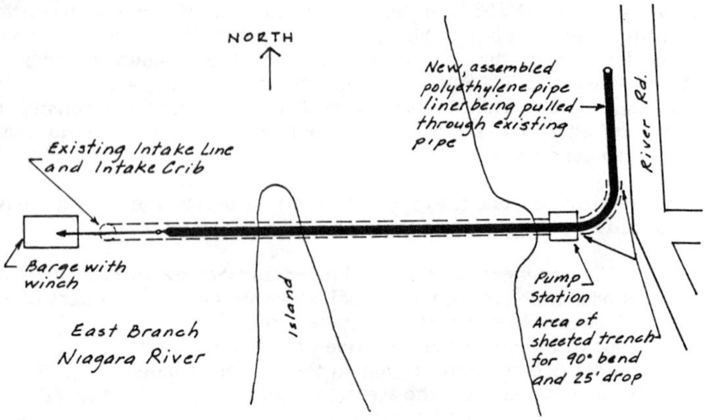

Plan view. The liner was inserted into the intake line at the pumping station and pulled through the pipe by a winch located on a barge. The 90° bend presented a problem in installation.

The river current and water temperature also required special consideration. The Niagara River has a current of about 10 miles an hour and the water temperature varies with the season from 32° - 70°. The contractor anchored a barge equipped with a winch out in the river over the intake point. A heavy cable was then strung from the barge through the intake pipe to the pumping station to provide a means for pulling the liner through the original intake pipe. A special harness was provided for attaching the cable to the liner.

Harness was attached to pull polyethylene liner through the original intake pipe.

The contractor allotted two days for pulling the liner into place. Work was begun in the afternoon of the first day. The liner was placed in the trench and pulled around the 90° bend just into the mouth of the intake line before the day was over. The next day when work resumed, the liner would not budge. All efforts were to no avail. In fact, the pulling assembly separated from the pipe, so the pipe was removed.

One afternoon a few days later, efforts were resumed and the same thing happened. The next morning, the pipe would not go around the bend. Leaving the pipe overnight at the cool temperature inside the old pipeline caused the polyethylene to loose its flexibility.

It was decided then that an early morning start would be made with the pipe on top in a straight mode. The entire assembly would be pulled into place in one day. The pulling operation was halted only twice, just long enough to allow for butt fusing the 800 foot lengths together. To everyone's relief, no trouble was experienced with the pipe taking a "set" in the 90° bend. It was apparent that the reason for the trouble had been that the liner in the curved section of the trench had taken a set over night, and this prevented its movement the next morning.

The polyethylene liner has a coefficient of expansion of $8x10^{-5}$in/in/^{o}F. According to the supplier, this would be offset by the grouting between the liner and the steel pipe. However, there is a sluice gate at the pumping station and its operation could be affected by even a slight movement of the liner. To preclude the possibility of movement, the end of the pipe at the pumping station was securely anchored by bolting the pipe to the structure.

After the liner was pulled into place and the grouting completed, the taste and odor problems at the Lockport water works were solved. The new liner has been in service for six years and frequent checks with City personnel indicate the intake line is operating satisfactorily.

ALPENA, MICHIGAN

Alpena, Michigan provides water service to approximately 17,000 people. The City is located 250 miles north of Detroit on the shore of Thunder Bay off Lake Huron at the mouth of the Thunder River. The principal industries are Huron Portland Cement Company, Fletcher Paper Company, Abitibi Paper Company and Besser Manufacturing Company.

The water use is approximately 4 MGD. About 50% goes to industry. In the early 1900's the first municipal water supply was provided by constructing a 42-inch, riveted steel intake line 2000 feet into Thunder Bay. This line was unique because it had two intake points -- one 1000 feet from shore in about 12 feet of water and another 2000 feet from shore in about 14 feet of water. At the outer crib a tee and a short extension of pipe had been provided for a possible future extension into deeper water.

The intake pipe was laid in a shallow trench, since the lake bottom is composed of 2 or 3 feet of sand over ledge rock. The cost of excavating the ledge rock limited the depth of trench.

At the shore, water flows by gravity from the intake line into a shore well which is 30 feet in diameter and equipped with screening facilities. Until the early 1920's the water was not treated. However, since 1920 the water has been filtered and chlorinated.

For a number of years City personnel had experienced trouble with sand accumulating in the intake line. McNamee, Porter and Seeley was retained to determine how the sand was getting into the line, and how it could best be removed. Great Lakes Diving Consultants of Traverse City, Michigan were engaged to inspect this intake line as they had in Lockport. The divers found that many of the flanged joints were partially open which permitted sand to seep into the pipe.

It appeared that the sand had stabilized. That being the case, if it were cleaned out, there was a good chance the line would stay clean for several years. A contractor was engaged to clean the line using a "pig". After the cleaning, the line was inspected and found to be free from sand. However, another inspection conducted about a year later, showed that a considerable amount of sand had accumulated in the pipeline. This indicated that more extensive measures would be required.

McNamee, Porter and Seeley proposed that the 42-inch pipe be relined with a 33-inch OD polyethylene pipe which would have sufficient capacity to meet the water supply needs of the City. The City authorized the firm to prepare drawings and specifications for relining the intake.

To maintain an adequate water supply, the contractor was required to provide a temporary intake and diesel pumping equipment during the construction period. The plans and specifications also required the contractor to remove the blank flange from the intake extension at the outer crib and to pull the liner into the intake line from that point. The construction contract was awarded to Lakeshore Engineers of Muskegon, Michigan.

The contractor started work in May 1978, constructed a temporary intake and set up temporary pumping facilities for supplying water to the treatment plant. Arrangements were made with the City for the use of a City Park adjacent to the plant to assemble the polyethylene liner pipe after butt fusing. The pipe for this project was also purchased in 30 foot lengths. However, in this instance, the contractor assembled the full 2000 feet for the liner is a continuous tube by butt fusing the joints. While the liner pipe was being assembled, divers checked and cleaned the 42" steel intake line to make sure nothing would interfere with the pulling of the liner once it was started.

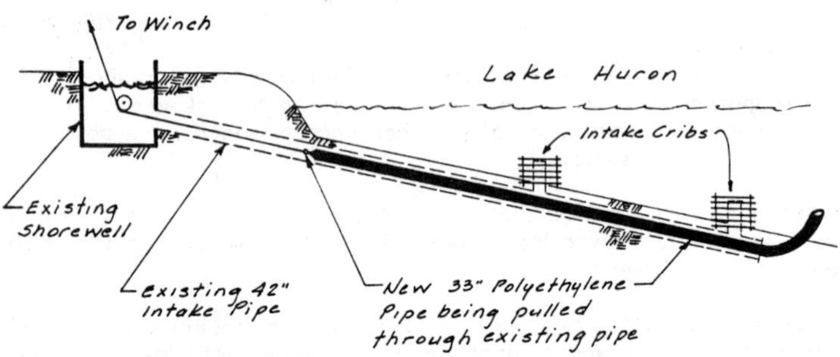

Cross-section view. Polyethylene liner was inserted into original intake line from barge anchored offshore. Winch installed in pumping station on shore pulled liner through pipe.

In the Lockport project, the pipe was pulled from the shore into the intake line. However, in this project the Contractor set up a large winch in the pumping station at the shore end of the intake and ran a cable out to the end of the pipeline. In accordance with with specs, the blank flange at the outer end of the intake was located and removed to provide access to the pipe. Then on a calm day in August, the 2000 foot liner was pulled out into the lake by a tug, threaded into the 42" pipe and then pulled through the intake line by means of the cable and winch.

The pulling operation was accomplished without complication. However, due to the high coefficient of expansion in the polyethylene, the liner had to be securely anchored at the pumping station, the intermediate intake point and the outer intake. The installation was further complicated by the need to cut an opening in the liner at the intermediate intake.

The old steel pipe from the shore well to the pumping station, a distance of 300 feet, was found to be badly corroded. Instead of relining that pipe, it was decided to remove the section and encase the liner in concrete to a minimum thickness of 6".

Construction was completed in September 1978 and the City started using the lined intake. The intervening years have demonstrated that relining the intake line solved the problems of sand infiltration.

SUMMARY

The rehabilitation of the intake lines described in this paper have proven to be a worthwhile investment for the Cities of Lockport and Alpena. Frequent checks with operating personnel since the completion of the projects have confirmed that the intakes are operating satisfactorily with no sign of the original problems.

Experience with installing the liners led to some conclusions that may prove useful to other communities. Polyethylene pipe has three characteristics which require careful attention:

1. Due to the coefficient expansion of $8x10^{-5}$in/in/^{o}F, if it is important to keep the ends or any other part of the tube in a fixed position, the liner has to be securely anchored at that point.
2. The liner is also subject to collapse when the space between the polyethylene and the original pipe is grouted. This can be prevented by taking care to maintain the pipe full of water with no air pockets during the grouting process.
3. In pulling the liner in place, if any deflection is required, it is essential to keep the polyethylene pipe moving in order to prevent the liner from taking a "set" that will prevent further movement. The problems can be prevented by limiting stoppage to brief periods such as for butt fusing.

FINITE ELEMENT TECHNIQUES FOR PIPELINES SUBJECTED TO EXTREME LOADS

Dennis Row,[1] Graham H. Powell,[2] Mary W. Goodson[3]

INTRODUCTION

The design of major gas and oil pipelines in arctic environments requires consideration of severe loading conditions that are not encountered in regular pipelines. Furthermore, conventional code pipe stress requirements, which are based on elastic behavior, are too restrictive for extreme loading conditions. To overcome this problem, and at the same time produce a safe and economical design, limit state criteria can be used in conjunction with inelastic finite-element analyses.

This paper considers techniques which can be used to perform inelastic analyses of pipelines and to establish corresponding design criteria. In particular, the paper addresses the following topics:

1. Procedures for constructing a suitable finite element model, including inelastic behavior of the pipe, nonlinear and time-dependent behavior of the soil, and the effects of frost heave.
2. Computational procedures for performing nonlinear analyses of the finite element model, considering temperature, pressure, soil settlement, frost heave and other effects.
3. Procedures for developing compressive stress limits, based on local wrinkling of the pipe.

Modeling procedures for the pipe and soil are considered first, followed by the nonlinear computation procedure. Some example analyses obtained using a specially developed computer program are then described. Finally, a technique for establishing compressive strain limits is described. Because of space limitations, many details must be omitted.

MATHEMATICAL MODEL OF PIPELINE

General—To be practical, the finite element model for a pipeline must be simple, yet must model the essential features of pipe and soil behavior. It is particularly important that inelastic behavior of the pipe steel be modeled realistically, that pressure and thermal expansion effects be treated rationally, and that nonlinear effects in the surrounding soil be considered. In theory, exact modeling of the pipe and soil is possible, using shell elements for the pipe and solid elements for the soil. In practice, however, a number of simplifying assumptions must be made. Several assumptions which are believed to be reasonable are as follows;

---

[1]President, SSD, Inc., Berkeley, California
[2]Professor of Civil Engineering, University of California, Berkeley, California; Consultant, SSD, Inc.
[3]Senior Engineer, SSD, Inc.

1. The problem is two-dimensional. This assumption is not essential, but it greatly simplifies the analysis and covers the majority of practical cases.
2. The pipe can be modeled as a series of inelastic beam-type finite elements.
3. The pipe cross section remains circular. This is a questionable assumption, because it implies that cross section ovalling is insignificant. Elements based on this assumption are not suitable for station piping, where the pipe bends may be of short radius and significant ovalling may occur.
4. Inelastic behavior can be modeled using the Mroz theory [1]. This is a flexible and computationally convenient theory, which considers yield under biaxial stress (hoop and longitudinal), nonlinear strain hardening, and arbitrary cycling loading.
5. The soil can be modeled as a nonlinear Winkler foundation (i.e. a series of uncoupled nonlinear springs). Transverse and longitudinal soil supports should both be considered.

For analysis, it is convenient to divide any length of pipeline into straight and curved segments, and to divide each segment into a number of pipe elements (Fig. 1). Longitudinal and transverse soil supports can be specified at the nodes within segments, and anchors may be specified at the joints between segments. The theories for the pipe and soil elements using this type of model are summarized in the following sections.

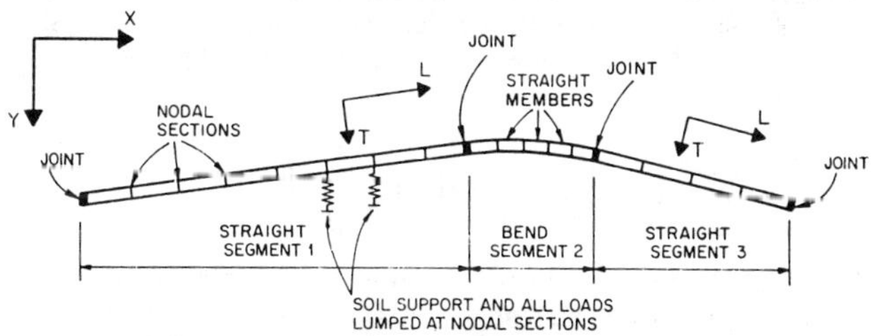

Fig. 1 Idealization for Analysis

SUMMARY OF PIPE ELEMENT THEORY

A suitable pipe element is illustrated in Fig. 2. A "slice" is considered at each end, the cross section at each slice is divided into a number of subelements and biaxial stress-strain behavior is considered in each subelement. Hence, detailed stress and strain results can be calculated at close intervals along the pipe length.

To obtain the element stiffness, the steel is assigned a multilinear stress-strain relationship. For such a relationship the Mroz theory permits a tangent stress-strain relationship to be established in the form

$$\begin{Bmatrix} d\sigma_a \\ d\sigma_h \end{Bmatrix} = [D] \begin{Bmatrix} d\varepsilon_a \\ d\varepsilon_h \end{Bmatrix} \tag{1}$$

in which σ_a, ε_a = longitudinal stress and strain; σ_h, ε_h = hoop stress and strain; and [D] = constitutive matrix. In a pipe, the hoop stress depends on the internal pressure. Hence

$$d\sigma_h = \frac{R_i}{t} dP \tag{2}$$

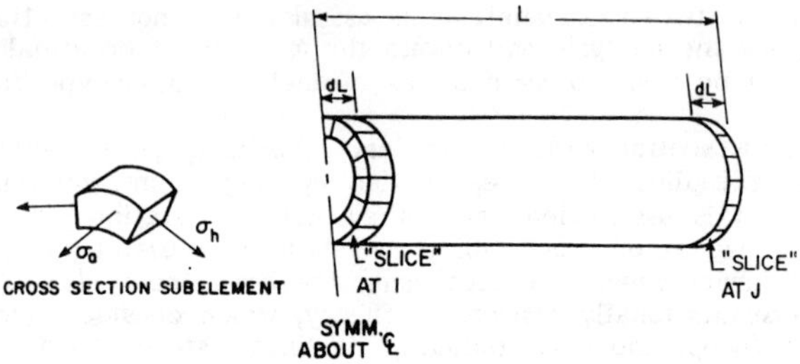

Fig. 2 Pipe Element

in which dP is the increment of internal pressure; R_i = inside pipe radius; and t = pipe wall thickness. Using Eq. 2, Eq. 1 can be modified to given an effective longitudinal stress-strain relationship in the form

$$d\sigma_a = D' \, d\varepsilon_a \tag{3}$$

The actual value of D' depends on the current and past states of stress and strain. The value follows from the Mroz theory.

Once Eq. 3 has been established for each subelement in a slice, the stiffness for the slice can be constructed (essentially by addition), in the form

$$\begin{Bmatrix} dP \\ dM \end{Bmatrix} = [k_s] \begin{Bmatrix} d\varepsilon \\ d\psi \end{Bmatrix} \tag{4}$$

in which P = axial force; M = bending moment; ε = strain at pipe axis; ψ = curvature; and $[k_s]$ = slice stiffness matrix. Eq. 4 is evaluated for the slices at each end of an element. The slice stiffness can then be assumed to vary linearly over the element length. An element stiffness matrix can thus be constructed. The stiffness matrix for the complete pipe follows, using conventional direct stiffness procedures. Nodal loads for the element assemblage, including thermal expansion effects, are constructed using similar principles.

SOIL MODEL

The deformations of a pipeline depend greatly on the transverse strength and stiffness of the supporting soil. A convenient procedure is to model the soil as a number of uncoupled "springs," each consisting of four components as shown in Fig. 3a. The most important of these components is the inelastic spring, which models the basic nonlinear behavior. The remaining three components (a spring and two dashpots) can be used to model creep effects.

The inelastic spring component must be assigned properties which account for nonlinear behavior, including initial loading, inelastic unloading, development of gaps between the soil and pipe under reversed loading, and seasonal variation of uplift resistance. This last feature is important for cold region pipelines because it allows the uplift resistance to be varied from summer to winter, recognizing that thawed summer soil will be much weaker than frozen winter soil. The basic nonlinear properties might be defined by multilinear curves, with options for the forms shown in Fig. 3b. These basic curves would then be modified to account for unloading, gap opening and seasonal strength variation.

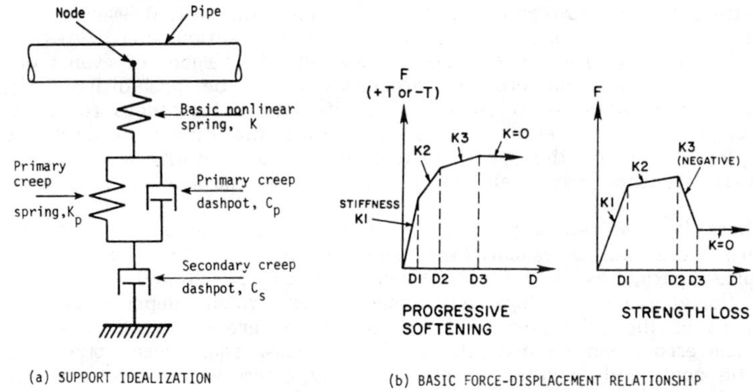

Fig. 3 Transverse Soil Support

The three creep components allow nonlinear primary and/or secondary creep to be modeled. The spring and dashpot properties can be specified to vary with both temperature and load. Creep analysis can then be carried out step-by-step through time, the properties being varied for each node at each step, depending on the current bearing pressure at the node and the current soil temperature.

The deformations of a pipeline may also depend substantially on the longitudinal cohesive resistance between the pipe and the soil. As for the transverse soil support, the longitudinal resistance can be modeled using discrete nonlinear springs.

COMPUTATIONAL PROCEDURE

Pipelines can be subjected to complex load sequences, including cold bending of field bends; test pressure; operating pressure and temperature; soil movement (including heave and/or creep); pressure and temperature reduction during shutdown; etc. When nonlinear behavior of the pipe and soil are considered; it is not possible to perform a separate analysis for each loading and then superimpose the results, simply because superposition no longer applies. Hence, it is necessary to select design loading sequences and to confirm, by analysis, that the design criteria will not be exceeded during the life of the pipeline. The choice of appropriate design sequences requires careful consideration by the designer.

Nonlinear analysis strategies are notoriously unreliable, especially when complex loadings sequences are considered, and typically require a great deal of skill and experience on the part of the analyst. Such lack of reliability is not acceptable, however, in a computer program intended for design use. Experience has shown that the most reliable strategy is of the "event-to-event" type, in which the nonlinear behavior is monitored as the analysis proceeds, and the system properties are changed each time a significant nonlinear "event" occurs. This closely simulates what happens in nature, which is why the strategy is reliable. In nature, however, the number of events is infinite, because the properties change continuously. In a practical analysis the number must be finite, and for efficiency must be kept reasonably small. A reasonable compromise is to use an event-to-event strategy with "overshoot" permitted beyond the actual event. For example, yield at a subelement in a pipe cross section constitutes an event. If each new yield is treated as an event, the number of events can be very large, and the cost of computation can be prohibitive. If, however, a yield overshoot tolerance is specified (plus tolerances for other types of events), the event occurs only when the stress exceeds the nominal yield value by the specified tolerance. As a result, several cross section subelements may yield at once, and the number of events is reduced.

There is, however, a penalty to be paid when overshoot is allowed. If there were no overshoot, the solution would follow exactly the "equilibrium path," as in nature. When overshoot is allowed, equilibrium errors (unbalances) develop, with magnitudes which depend on the magnitudes of the tolerances. If the tolerances are set small, then the equilibrium errors can be controlled. Nevertheless, equilibrium corrections should be continually applied to eliminate any errors which develop. This can be done by calculating the equilibrium error after each event, and applying corrective loads to eliminate the error during subsequent analysis steps. It is sound policy to print out the unbalanced forces at frequent intervals, to confirm that no significant errors have developed.

SAMPLE APPLICATIONS

Design of oil and gas pipelines in arctic regions requires detailed nonlinear pipe-soil interaction analyses because of the extreme nature of the loadings.

Particular examples of severe loading conditions that may be encountered in design are as follows.

1. Frost Heave: When chilled gas is pumped through heave susceptible soils, large heave displacements can occur due to the creation of a frost bulb. When a heave susceptible zone is surrounded by nonheaving zones, large differential vertical displacements can occur in the pipeline, causing high shear and moment in the pipeline combined with large pipe wall bearing forces.
2. Thaw Settlement: When warm oil or gas is transported through permafrost regions, the ground will tend to thaw and settle. If thaw susceptible regions are adjacent to soils that will not settle significantly, large differential settlements can occur over time.
3. Fault Movement: If pipelines are buried in seismically active regions and the route crosses active faults, it may be necessary to design for differential transverse and longitudinal displacements. These differential movements can occur over a short section of the pipeline, producing a guillotine effect.

In the three cases described above the loading can be classified as secondary, in the sense that it is an imposed displacement phenomenon. Typically in all design applications for frost heave, thaw settlement and faulting, the design movements are sufficiently large that some level of inelastic straining must be accepted in the pipe wall. To evaluate the state of pipe stress and strain under a given extreme loading situation requires the design engineer to perform an inelastic pipe-soil interaction analysis. In general it is necessary to consider the full sequence of loading that the pipeline will potentially experience throughout its design lifetime. For example, this may include: hydrostatic testing, operating conditions and either frost heave movements or thaw settlement, possibly in conjunction with seismic traveling waves and faulting.

The scope of this paper is not sufficient to address all applications in detail. To give some insight into the complexity of the analysis problem, and illustrate practical solutions, a frost heave analysis example is discussed in the following section.

FROST HEAVE ANALYSIS

General—Frost heave is a complex phenomenon which depends on the soil properties, the rate at which heat is extracted from the soil, the rate of growth of the frost bulb, the pipe bearing force on the ditch bottom, and other factors. A convenient computational procedure is to introduce ditch bottom displacements as initial deformations in the transverse support springs (equivalent, in effect, to thermal expansion or contraction of these springs). For ditch bottom movements of known magnitude, initial deformations of these magnitudes can be introduced into the springs. For deformations due to frost heave, the initial deformations can be calculated at each step of an analysis, using an appropriate heave theory and considering the current state and past history of the system.

Heave Model—Fig. 4 shows a frost bulb cross section with growth ΔD in some time period Δt. One particular model for frost heave growth is to assume that the amount of heave, ΔH, (or ditch bottom movement) in time Δt is given by

$$\Delta H = \alpha \Delta D \tag{5}$$

where α is an expansion factor which may be specified to depend on the bulb base bearing pressure, P, and the frost bulb depth, D. In this particular model P is calculated as follows.

$$P = \gamma (D_o + D) + \frac{S}{B_S} + \frac{F}{B_F} \tag{6}$$

where γ = effective density of ice-soil mixture; D_o = burial depth; s = effective shear force acting along bulb side; F = pipe-soil interaction force; B_S = effective bearing width for shear force; B_F = effective bearing width for pipe interaction force.

At any given time and any given position along the pipe, the values of D, S, B_S, B_F are known from user-defined input time histories, and F is equal to the current transverse soil reaction. The bulb base bearing pressure, P, is calculated using Eq. 6 and α is then obtained by linear interpolation from the input D, P relationships. Having determined α and knowing ΔD, the increment of ditch bottom displacement in the subsequent time step is determined from Eq. 5.

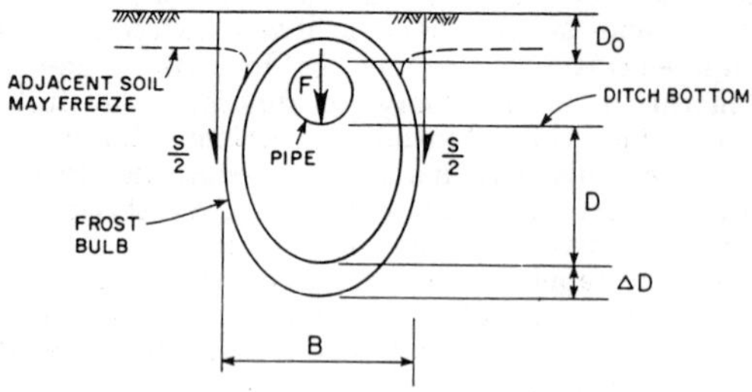

Fig. 4 Frost Bulb Model

Another very important consideration for frost heave analysis is the seasonal variation of soil uplift resistance and the method in which it is modeled. In the nonheaving zones directly adjacent to the heaving zone, the pipe tends to be lifted out of the trench. The pipe is restrained by the uplift resistance of the frozen backfill. In the winter months the soil increases in stiffness and strength, which increases the uplift resistance and subsequently increases the transverse bearing reaction and pipe bending moments under the continuing heave displacements. In the spring and summer the uplift resistance decreases, leading to seasonal relaxation in pipe deformation and bearing pressure. Two alternatives are considered here for modeling and analysis of the seasonally varying uplift resistance, as described in the following section.

Soil Uplift Resistance—Available data for frozen soils suggests that the principal behavior controlling the uplift resistance is similar to secondary creep response. That is, for a given level of load there appears to be a corresponding strain or displacement rate. The proportionality constant, C_s, is the secondary creep dashpot in the transverse soil model shown in Fig. 3a. In the model used in this study, the secondary creep constants, C_s, are defined over a range of transverse bearing reactions, R, for a given temperature, T. By inputting tables of C_s versus R and different temperatures, the model can account for the seasonally varying uplift resistance. For the purpose of analysis C_s-R tables are generated corresponding to winter and summer temperatures. Values during the spring and fall are then obtained by simple linear interpolation between these limits. For the example illustrated here the uplift soil creep response was given in the form

$$\dot{\Delta} = A\ R^n \qquad (7)$$

where $\dot{\Delta}$ = uplift soil deformation rate; and A,n are soil-specific constants. For the particular soil considered in this example, the summer and winter values are as follows;

		n	log A
Summer Temperature	(T_s)	3.68	- 6.71
Winter Temperature	(T_w)	7.74	-21.18

($[R]$ = K/ft; $[\dot{\Delta}]$ = ft/yr; 1 K/ft = 14.59 kN/m, 1 ft/yr = .305 m/yr)

Using the relationship of Eq. 7, the values of C_s can be determined for a given R as follows,

$$C_s = R/\dot{\Delta} \tag{8}$$

These values are then tabulated for a complete range of R for both winter and summer and then input into the model. In addition, it was assumed that the static soil resistance spring (see Fig. 3a) had a value of K = 854 K/ft (12460 kN/m).

Under differential frost heave, both bulb heave and soil creep are continuously occurring. In the model used here, the analysis is carried out by alternately performing heave and creep analysis steps. In each heave step a static ditch bottom displacement profile movement is calculated and in each subsequent creep step the soil deforms under creep relaxation. The sequence of the analysis of a single year, starting in summer, is illustrated schematically in Fig. 5. In any given analysis it may be necessary to subdivide each season into a number of sequences. This division within a season is sometimes necessary to allow a different time step size over numerically sensitive periods. For a given load sequence, the following data is specified; starting and ending soil temperatures, T_{js} and T_{je}, respectively; the number of analysis time steps, n_j; and the time step size, Δt_j.

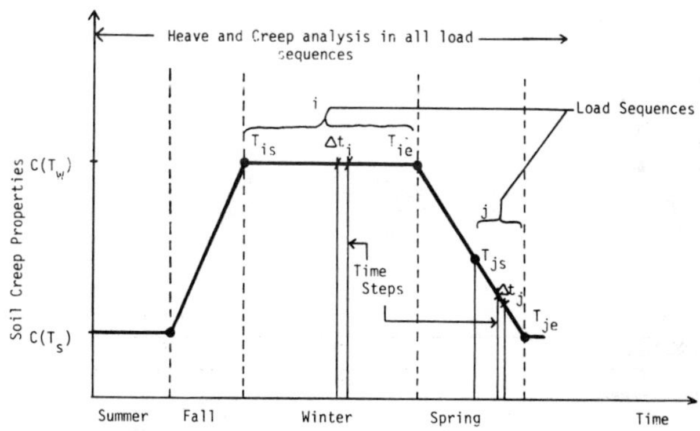

Fig. 5 Analysis Sequence

An alternative procedure is to represent the soil uplift resistance in terms of a simple static uplift resistance relationship whose maximum resistance is assumed to vary seasonally. The motivation to represent the soil resistance in this manner is that the frost heave analysis can then be performed pseudostatically, with less numerical sensitivity and decreased computation time.

One approach is to construct isochronous force-displacement relations. If a secondary creep model is loaded at a specified value of bearing, R_i, for a fixed time period, t, the total deformation is

$$\Delta_i^c = R_i t/C_{si} \tag{9}$$

where C_{si} corresponds to the uplift reaction, R_i. If a complete range of uplift reactions are considered, a plot of R_i versus Δ_i can be produced for a particular soil type in a given season.

Isochronous relationships were developed for the 5-month winter season and the 2-month summer season. The winter isochronous curve was approximated by a trilinear relationship, and this was used as the basic transverse static soil resistance relationship. The resulting static force-deformation curve for the soil is shown in Fig. 6. The total soil deformation at a bearing load level, R_i, is the sum of Δ_i^c (Eq. 9) plus the elastic spring deformation $\Delta_i^e = R_i/K$. The seasonal variation in soil uplift resistance is controlled by specifying the uplift strength limits as a function of temperature. These limits were taken as the bearing force at 2 inches displacement on the summer and winter isochronous curves. The uplift strength at intermediate temperatures (or times) is determined by linear interpolation.

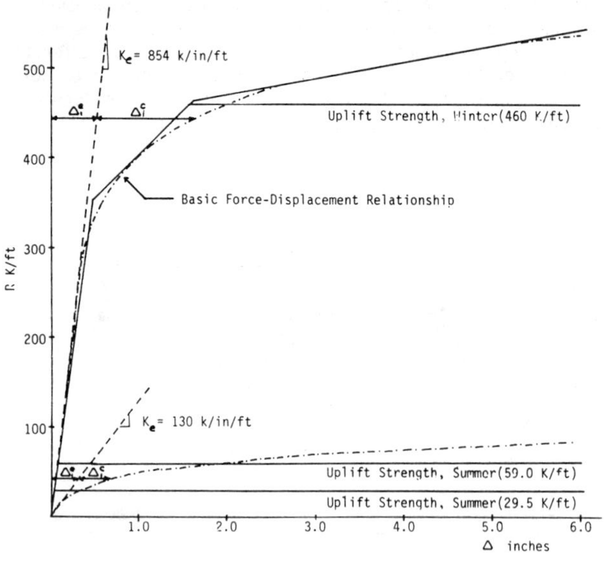

Fig. 6 Transverse Soil Model for Uplift Analysis

The frost heave analysis sequence using static soil uplift is similar to that described above for creep (Fig. 5). The analysis is performed by interweaving heave and uplift analysis steps. In each heave step a static ditch bottom displacement profile increment is calculated and in each subsequent uplift step an adjustment is made for the incremental change in soil uplift strength. The uplift procedure is simpler, in that uplift analysis is only performed in the periods when the strength is varying (i.e. fall and spring). Furthermore, the uplift procedure can produce reliable results with larger time steps and hence reduced computation time. This is an important consideration when parametric heave studies may have to be performed considering the design life of the pipeline.

Results—The idealized pipeline-soil model shown in Fig. 7 was subject to frost heave and analyzed using both creep and static soil resistance models using the PIPLIN [4] computer program. A heave span of 40 ft (12.19 m) was assumed with an abrupt transition to the nonheaving zones. The analysis uses symmetry to reduce the model size in half. The heave model data used is listed in Table 1. A soil density of 100 lb/ft^3 (15.71 kN/m^3) was assumed with a pipe burial depth of 30 in (0.762 m). The results of the five year heave analyses are shown in Fig. 8. The variations in the maximum axial compressive strain and the maximum uplift bearing force are shown over the five year period. As the analysis enters each spring there is reduced resistance to upward pipe movement in the nonheaving zones, with a subsequent relaxation in the maximum axial compressive strain. As time progresses the maximum compressive strain levels off even though heave is continuing to occur. This phenomenon occurs because the bulb base bearing pressures have increased to such a level that the incremental ditch bottom movements are small for increments in bulb growth (Eq. 5), combined with the seasonal relaxation effect from uplift resistance. The creep and static uplift methods give maximum compressive axial strain values within 10%. Using the seasonal static strengths from Fig. 6 has resulted in overestimates of the uplift bearing forces. A second heave analysis using a summer uplift strength of 29.5 K/ft (430.4 kN/m) was performed. The results were in good agreement with the creep analysis. Further study is required to determine an appropriate method of estimating the equivalent static uplift strengths from the basic creep data.

TABLE 1 HEAVE DATA

Frost Bulb Properties Versus Time

Time (Months)	Bulb Depth Below Pipe, D_o (in)	Bearing Width, B_F (in)
0.0	24.0	48.0
2.0	44.0	136.0
4.0	56.0	160.0
6.0	64.0	176.0
12.0	84.0	216.0
18.0	100.0	248.0
24.0	114.0	276.0
36.0	132.0	312.0
60.0	160.0	368.0

Expansion Ratio Versus Bulb Base Pressure

Pressure (lb/ft^3)	Expansion Ratio
200	.10
300	.07
700	.05
1300	.02
2000	.01

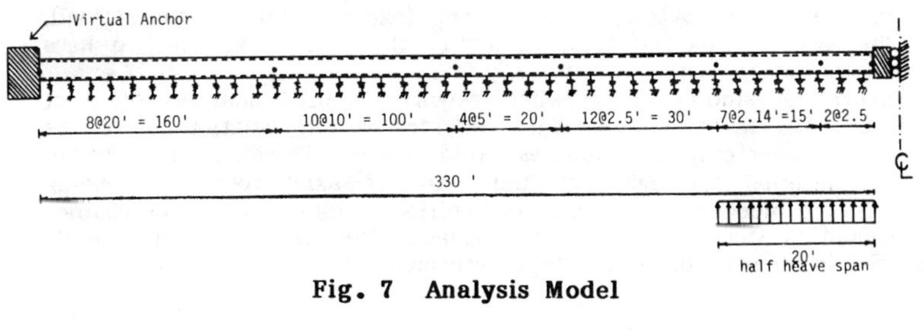

Fig. 7 Analysis Model

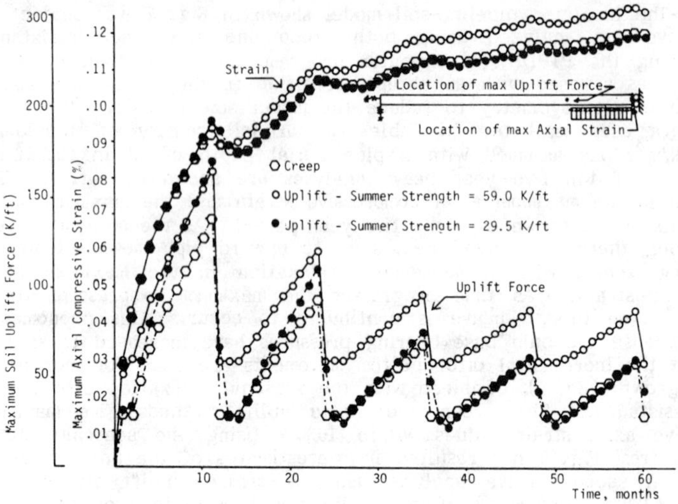

Fig. 8 Heave Analysis Results

DESIGN LIMITS

The type of analysis described above calculates information on the inelastic behavior of the pipeline (subject, of course, to qualifications which result from the assumptions made in developing the mathematical model). The criteria used to check the structural adequacy of the line must similarly be based on inelastic considerations.

For the pipe, criteria must be developed to guard against both rupture and excessive plastic deformation. Criteria for rupture are typically based on fracture mechanics principles, and are beyond the scope of this paper. The most important criterion for plastic deformation is one which will ensure that a local wrinkle will not develop in the pipe wall. If such a wrinkle develops, the pipe cross section at the wrinkle loses much of its strength in bending. If the soil is soft, substantial deformation of the pipe can occur. Repair of the pipe will usually be necessary, and in extreme cases a leak may develop at the wrinkle.

A convenient means of defining the design limit state is in terms of the longitudinal strain in the pipe. An analysis of the type described above will not consider wrinkling directly. Hence, a wrinkling strain limit must be determined by separate studies (experimental or theoretical). When this limit is reached in an analysis, the pipe is assumed to have exceeded its design limit.

Theoretical studies of wrinkling require detailed analyses of short lengths of pipe, taking into account the stress-strain relationship of the pipe steel, imperfections (such as weld high-low) which can trigger wrinkling, internal pressure; and the ratio of axial force to bending moment on the pipe cross section. In addition, ovalling due to overburden loads should be taken into account, because this effect is not directly considered in an analysis of the type described.

Wrinkling analyses are most appropriately carried out using the finite element method. The pipe is modeled as a mesh of shell elements, considering imperfections, inelastic behavior, large displacements and nonproportional loading. Sophisticated analytical techniques are needed because the onset of wrinkling is characterized by loss of strength. If the analysis technique is not able to trace out the descending load-displacement curve in the postwrinkling state, the critical strain may not be determined accurately. Nonlinear solution strategies based on finding solutions for specified loads are unsuitable, because no solutions exist for loads above the critical load. Strategies based on finding solutions for specified displacements [2,3] are thus necessary. Although wrinkling analyses are complex, they can be performed with existing techniques. These techniques have been used in extensive studies of parameters affecting the local buckling of pipelines and the rational development of compressive strain design limits [5].

CONCLUSION

Modeling and analysis techniques for investigating the inelastic behavior of pipelines under extreme loading have been briefly reviewed. These techniques are complex, but have been developed into practical tools which can solve a wide variety of practical pipeline problems, and provide a designer with a rational methodology to quantify pipeline behavior under extreme load.

ACKNOWLEDGMENTS

The authors wish to acknowledge the Northwest Alaskan Pipeline Co. who funded the studies discussed in this paper; and Gulf Interstate Engineering Company/Michael Baker Jr. - Joint Venture for their advice and assistance.

APPENDIX I - REFERENCES

1. Mroz, B., "On the Description of Anisotropic Work Hardening," Journal of Mechanical Physics and Solids, Vol. 15, pp. 163-175 (1967).
2. Riks, E., "An Incremental Approach to the Solution of Snapping and Buckling Problems," Int. J. Solids and Structures, Vol. 15, pp. 529-551 (1979).
3. Powell, G. and Simons, J., "Improved Iteration Strategy for Nonlinear Structures," Int. J. Num. Meth. in Engrg., Vol. 17, pp. 1455-1467 (1981).
4. "PIPLIN-III, Computer Program for Stress and Deformation Analysis of Pipelines," SSD, Berkeley, California.
5. Row, D.G., Powell, G.H. and Morris, G.R., "Local Buckling Analysis of Pipelines," Second International Offshore Mechanics and Arctic Engineering Symposium, Houston, Feb. 1983.

MODIFIED REEL METHOD FOR SUBSEA ARCTIC PIPELINES

Peter W. Bryce, P.Eng., Member ASCE [1]
Jeffrey M. Key, P.E., Associate Member ASCE [2]

SUMMARY

With the anticipated discovery of commercial offshore oil fields in the Canadian Beaufort Sea and in United States North Slope Alaskan waters, it is imperative that appropriate designs and methods for installation of submarine pipelines be developed. These methods must be compatible with the very short open-water season in these regions, especially where ice conditions do not permit through-ice installation techniques. Consideration must also be given to permafrost shore crossings, speed of assembly, construction of the relatively shallow shore approaches and minimizing downtime.

A number of pipeline installation methods, which involve the use of reel pipelaying vessels, are compatible with these requirements and appear to be technically feasible for use in Arctic waters. One installation concept uses an existing reelship which is reloaded at the site with pipe from auxiliary reels mounted on supply barges. In this way, a caravan of barges can transport many miles of pipe to the site for installation in one work season. Another concept uses these same reel-equipped supply barges without the reelship. Instead, a special installation vessel temporarily attaches to each barge separately as the pipe is laid from the reels on that barge. This paper describes these concepts, discusses the potential advantages, and provides a schedule for installation of a typical subsea pipeline in the Arctic.

The use of reel pipelaying vessels for installing pipelines in Arctic waters may result in significant time and cost savings.

INTRODUCTION

The search for new sources of hydrocarbon energy has carried the petroleum industry into increasingly hostile offshore environments. Recent exploration has confirmed the presence of major hydrocarbon accumulations on the Continental Shelf of the Alaskan and Canadian Beaufort Sea, which extends some 75 miles (120 km) from shore to water depths of approximately 590 feet (180 m). This area is shown in Figure 1.

---

[1] Gulf Canada Resources, Inc., Frontier Development Division, 401 - 9th Avenue S.W., Calgary, Alberta T2P 2H7 CANADA

[2] SF/Braun and Santa Fe Offshore Construction Company, 505 S. Main Street, Orange, California 92668-0401

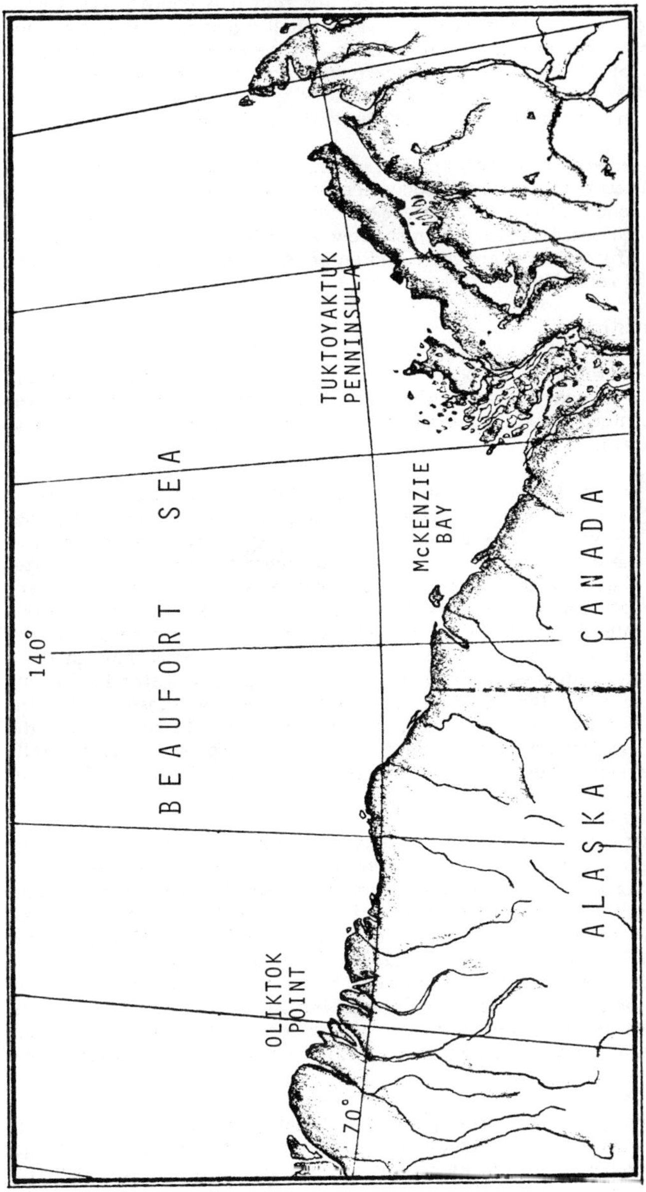

APPROXIMATE SCALE: 1"= 75 MILES

FIGURE 1

As a southern extension of the Arctic Ocean, this region is characterized by a seasonal ice cover which presents formidable challenges to the design and operation of exploration and production systems. Since ice covers the sea for at least nine months of the year, most offshore construction activity, with a few exceptions, has taken place in the short open-water season; in particular, drilling operations have been performed from islands in winter and summer.

Plans for the next several years call for an escalation of exploration activity, and a concurrent detailed evaluation of production structures and systems to bring Beaufort Sea oil economically to market. The exploration effort will involve the building of additional artificial islands in shallow water (up to 30 feet (10 m)) and the increasing use of innovative concrete and steel structures in the intermediate water depths 30-130 feet (10-40 m).(1)

Environmental Conditions

The Beaufort Sea environment has been described as one of the most hostile in the world with operational constraints ranging from the extremely low temperatures and darkness of winter to subsea permafrost (2). However, sea ice is the most significant constraint in this area.

The southeastern Beaufort Sea is usually ice covered nine months of the year. With the onset of winter in early October, freeze-up commences and progresses seaward with the growth of landfast ice. This ice may be subdivided into two zones, a smooth area extending 16 miles (26 km) offshore to about the 9 m water depth and beyond that a rough pressure-ridged area extending to the 60-75 feet (18-23 m) water depth range where the landfast ice stabilizes in early January. The undisturbed winter ice within the landfast zone reaches a thickness of 6 feet (2 m) by May. Multi-year ice floes and small glacial ice islands may, in some years, be incorporated into the landfast ice. A recurring open-water lead exists along the outer edge of the landfast ice which opens and closes under the influence of offshore and onshore winds, respectively. At its outer edge, the landfast ice is characterized by a dense pressure ridging which is partially grounded on the seabed. These have an average thickness on the order of 25 feet (8 m).

Beyond the landfast ice is a transition or seasonal pack ice zone. Here, the ice canopy is in sporadic motion which creates changes in ice type, floe size, and ice thickness distributions. This area is also characterized by first-year pressure ridges which are produced by the motion and resulting deformation of this sea ice. Most pressure ridges in this area have sails in the 3-6 feet (1-2 m) range; however, occasionally they may exceed 25 feet (8 m) above sea level and have keels exceeding 100 feet (30 m). Multi-year ice floes, 12-25 feet (4-8 m) in thickness, multi-year pressure ridges, and glacial ice islands up to 100 feet (30 m) thick may also be incorporated into the ice cover. These features which sometimes drift into the zone of operations pose the main hazards to structures and pipelines.

Beyond the seasonal pack ice zone exists the polar pack which is comprised primarily of old multi-year ice. The overall movement of the ice in the seasonal and polar pack ice zones is in a clockwise rotation in the Beaufort Sea. Yearly ice movements on the outer edge of this gyre average under 2 miles (3 km) per day, but in spring can reach 15 miles (24 km) per day.

Breakup occurs in late June or early July. Summer ice conditions are variable depending on the direction of the prevailing winds. Figure 2 shows the number of days of open water by years at a typical drillsite in the Beaufort Sea in 100 feet (30 m) of water. Some years no open water exists, while in other years there have been 120 days. This graph does not include those periods of thin ice during which vessels can also operate with icebreaker support.

Waves are relatively minor because of reduced fetches and the presence of sea ice. Significant waves can reach 16 feet (5 m) with currents generally less than 1.6 foot per second (50 cm/sec). Total darkness in winter, extended periods of extreme cold, drifting snow and fog hamper operations in the Beaufort Sea.

Design Considerations

Design criteria are based on a statistical evaluation of available data (3). For construction purposes, a two-year return period is usually specified for meteorological and oceanographic design criteria. For all other aspects of design, a 100-year return period is specified except for ice scour, which is discussed below.

The design criteria for the safe installation and operation of subsea pipelines will consider the following:

o Oceanography and Meteorology; specifically:

  Wind
  Precipitation and temperature extremes
  Waves
  Currents
  Sea ice
  Bathymetry
  Tidal variations

o Geotechnical; specifically:

  Soil properties
  Permafrost distribution
  Seismicity

o Ice Scour

  Ice scour is a phenomenon peculiar to this region and is caused by the grounding of large ice features such as pressure ridges and ice islands (4). In water depths between 50-100 feet (15-30 m), the scour intensity is quite high with mean frequencies of 10-20 scours per mile (6-12 scours per km) and mean depths of 3-6 feet (1-2 m). Scours of up to 12-16 feet (4-5 m) have been encountered in areas with thick soft sediments. In area where scouring occurs pipelines must be installed below the original sea floor to minimize the risk of ice contact. A return period of 1,250 years is used as a design basis for ice scour.

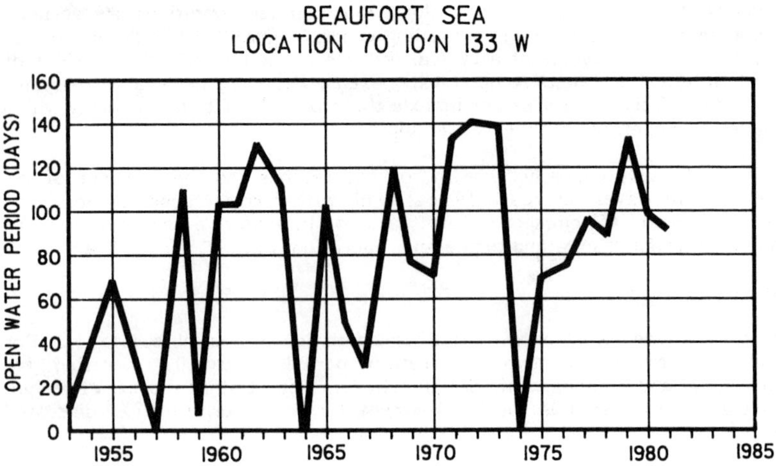

VARIATION IN OPEN WATER SEASON

FIGURE 2

Construction Methods

There are several pipeline installation methods which can be considered feasible for use in Arctic areas. Although there are numerous variations of each method, the four primary techniques may be described as:

(a) Installation by conventional laybarge
(b) Towing or pulling pipe strings into location
(c) Ice-based installation
(d) Installation by reel vessel (Reel Method)

This paper will discuss some aspects of the potential use of the Reel Method in this environment.

MODIFIED REEL METHOD

The Reel Method has been proven to be a rapid, cost-effective approach to subsea pipelaying when a suitable re-spooling base can be located within 30-60 miles (50-100 km) of the pipeline right-of-way. This method has not been previously considered applicable for Arctic areas because of the remote location and expense of mobilizing and maintaining a shore-based re-spooling yard. Two alternative methods are proposed which avoid the necessity for such a shore-based yard. The first method considers transporting the total pipeline length on reels mounted on supply barges, from which a pipelay vessel would re-spool after exhausting its initial supply of pipe(5). Briefly, the sequence would be as follows: the reel vessel would be loaded to capacity at a fabrication yard in the south. At the same time, depending on the total amount of pipe to be laid, the supply barges, on which perhaps one to four large reels are mounted, would also be loaded. All vessels would travel to the installation site. The lay vessel would initially lay pipe until its supply is exhausted. Then it would back up to the side of a supply barge where it would spool up with pipe from one of the supply reels, as illustrated in Figure 3. (Vertical supply reels are shown in the figure. However, both horizontal and vertical supply reels appear to be feasible and the choice would depend on the specific project.) This loading operation would take a total of approximately one to two days. After reloading, the lay vessel would continue to lay pipe. In this manner, the reel vessel would be supplied with enough pipe to keep it busy for the entire work season. (This procedure is explained in more detail below.)

The second method similarly considers transporting the complete pipeline on auxiliary reels mounted on supply barges. In this instance, the conventional reel vessel is replaced by a specially-constructed, dynamically positioned "installation barge." This vessel would support only a straightener, and ramp/stinger assembly - it would not contain its own reel. The barge would temporarily attach to each supply barge, positioned so that it is fed from one supply reel at a time. This is illustrated in Figure 4. Thus, each supply barge would be temporarily transformed into a reel pipelaying barge when connected up to the installation barge. The installation barge would tension and control the descent of the pipe to the sea bottom. Use of a single portable installation barge avoids the cost of providing each reel with the necessary equipment. A tugboat would tow the supply/installation barge pair along the pipeline right-of-way, steered by the DP thrusters mounted on the installation barge.

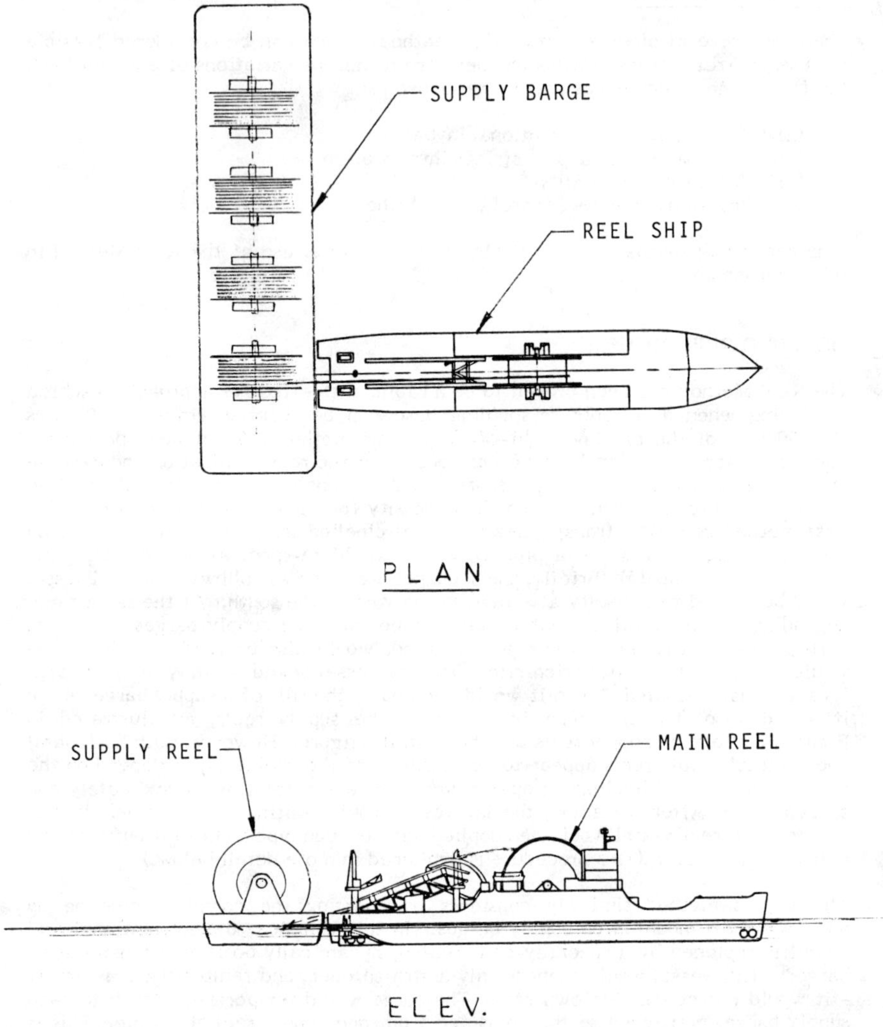

FIGURE 3

There are advantages to this installation barge concept:

(1) The time-consuming respooling operation is eliminated since pipe is laid directly from the supply reels, one at a time. Also, changeovers from one empty reel to the next full one could be done faster than if a separate reel vessel were involved. This results in having many more days available for actual pipelaying, making better use of the short construction season.

(2) Since there are a limited number of reel vessels in the world, it may be necessary to construct a new pipelaying vessel. It would be less expensive to construct an installation barge than a new reel barge simply because no main reel is required. (Supply barges must be constructed in either case.) The new barges could be custom built for shallow-water pipelaying.

It should be noted that potential cost savings may be realized since these alternative methods avoid the cost of setting up a fabrication yard in the Arctic and minimize the use of expensive Arctic labor.

Draft Considerations

When operating near shore areas in the Beaufort Sea, special consideration must be given to limitations imposed by the shallow water depth. A safe distance must be maintained between the bottom of the reel vessel hull and the sea bottom. When pipelines are laid in very shallow (less than 20 feet (6 m)) waters, the choice of pipelaying equipment may be limited to a flat-bottom barge with horizontal reels. Depending on the particular project, it may be economical to construct the special shallow-draft equipment referred to in the preceding paragraphs.

Supply Barges

The two concepts for the supply barge mentioned above are illustrated in Figures 3 and 4, where a 100-foot x 400-foot (30 m x 122 m) flat barge is equipped with three or four reels. The outer diameter of each reel might be approximately 80 feet (24 m), and each would hold as much as 10 miles (6 km) of 14-inch (.4 m) pipe. The reels would be equipped with a motor drive and brake for control of the pipe during loading and unloading. The number of barges required would be dictated by the amount and size of pipe to be installed in a single work season.

The reloading operation would begin after the pipelaying vessel docks up to the supply barge. A wire rope is brought from the pipelayer through the supply barge straightener assembly, and is attached to the end of the pipe on one of the supply reels. The supply reel brake is used to apply back-tension as the wire rope pulls pipe from the reel through the straightener and onto the pipelay vessel. As shown in Figure 4, there is a certain amount of residual curvature in the pipe as it comes off the supply reel before it reaches the straightener. Therefore, it may be necessary to provide a pipe support assembly off the side of the supply barge. This will depend on pipe size and amount of back-tension applied to the pipe.

Supply barges might be moored near the pipeline right-of-way in a protected area, or towed along as the lay vessel moves forward along the pipeline route. The costs and risks of the latter must be weighed against the increased time for the lay vessel to travel over to a moored supply barge for reloading. Another option

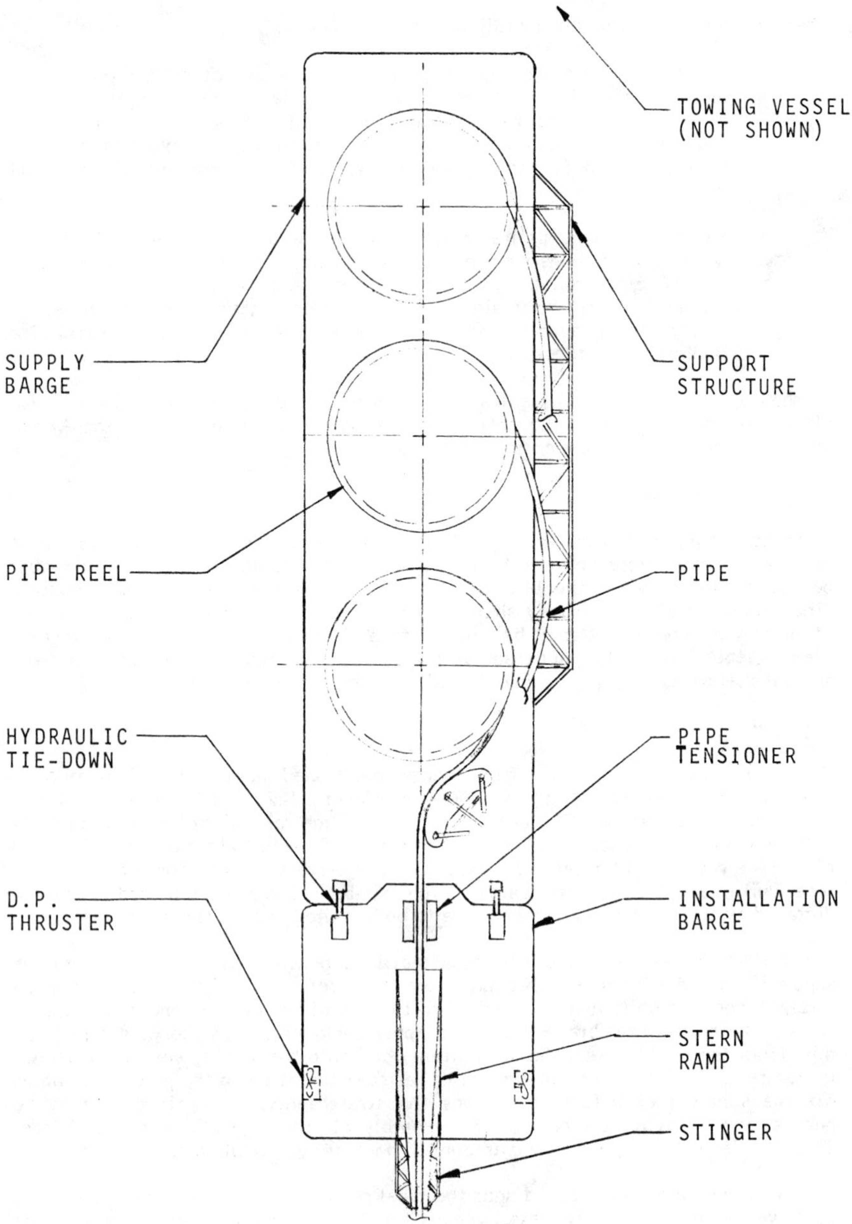
TOWING VESSEL (NOT SHOWN)
SUPPLY BARGE
SUPPORT STRUCTURE
PIPE REEL
PIPE
HYDRAULIC TIE-DOWN
PIPE TENSIONER
D.P. THRUSTER
INSTALLATION BARGE
STERN RAMP
STINGER

FIGURE 4

would be to locate and moor supply barges at various protected locations along the route, so that whenever the pipelay vessel needs to reload there is a supply barge nearby.

SPECIFIC APPLICATION OF REEL METHOD

To better understand the potential benefits of using the Reel Method for installing Arctic pipelines, the construction sequence for a specific project is outlined. It is assumed for this example that a conventional dynamically positioned reel vessel will lay the pipeline, and reload from supply reels as explained previously (5).

Project Data:

Number of pipelines	2
Length (each)	50 miles (80 km)
Diameter	14 inches (.4 m)
Pipe coating	1-inch thermal insulation (25 mm)
Water depth	65 feet (20 m) to shore (majority 30-65 feet (10-20 m)
Open-water season	60 days
Project location	Canadian Beaufort Sea

The main reel on the reel vessel will hold approximately 8 miles (13 km) of 14-inch (.4 m) pipe. This leaves 92 miles (150 km) of 14-inch (.4 m) pipe to be stored on the supply reels. Preliminary fabrication work is assumed to take place in a yard on the West Coast of North America.

This project will require the fabrication of nine auxiliary reels; each contains slightly more than 10 miles (16 km) of pipe. These reels will be mounted horizontally (reel rotates about a vertical axis) on supply barges. Each barge will support three reels; consequently, three supply barges are required. (For this example, the reel vessel shown in Figure 3 is used with supply barges similar to the one in Figure 4. Other combinations could also have been chosen.) The loaded barges will be towed up to the Beaufort Sea where they will be used to re-spool the reel vessel at the offshore site. The reel vessel will reload eleven times (total of twelve loads) from the barge-mounted reels to complete the project within one work season.

Summary of Installation Sequence

(1) Fabricate auxiliary reels, install on modified barges.

(2) Purchase 100 miles (161 km) of thin-film epoxy-coated 14-inch (.4 m) by .625-inch (16 mm) pipe, with delivery to West Coast.

(3) Weld pipe together into 1500-ft. (457 m) long strings, X-ray, apply weld joint coating (shrink sleeves, etc.), and spool onto reels.

(4) Sail reel vessel and tow three cargo barges, each loaded with three reels of pipe, from West Coast to Point Barrow.

(5) Transit from Point Barrow to jobsite. Note: Estimate excludes any allowance for delays at Point Barrow awaiting ice clearance.

(6) Install two shore approaches by pulling pipe from reel vessel; assume shore pull length requires no additional buoys for buoyancy. A small winching station must be set up on shore prior to shore pull.

(7) Lay two 50-mile (80 km) pipelines. Note: Estimate excludes any time for pipeline tie-ins, buckles, or hydrostatic testing; but includes connection of reel loads.

(8) Return barges and reel vessel to West Coast and remove pipe reels and other equipment.

Time Estimate

Total reel vessel Utilization	110 days
Total on-site time including transit to/from Point Barrow	55 days
Total auxiliary reel supply barge utilization (includes modification work)	150 days each

The length of the work season is assumed to be 60 days; consequently, a 5-day reserve period is provided for making the subsea tie-in, and as a buffer against unforeseen delays. Both pipelines are installed in a single work season, an accomplishment that other installation techniques are not likely to achieve.

Major Assumptions and Observations

(1) The reel vessel is working near the West Coast prior to mobilization. The supply barges are available in the West Coast area as well.

(2) The pipe is to be welded together in a yard on the Coast. Additionally, the West Coast is a good area to fabricate and install the pipe reels onto the supply barges.

(3) Weather downtime at the jobsite is assumed to be 10 percent during the pipelay operations. This is reflected in the time estimate. The reel supply barges will be moored in protective waters.

(4) It is assumed that the shore approach, including the necessary ditch, is available on arrival of the reel vessel at the jobsite.

(5) The reel vessel's layrate is assumed to be 1,200 feet (366 m) per hour less 10 percent weather downtime. This results in a layrate of approximately 5 miles (8 km) per day, not including diving, inspecting, equipment maintenance, etc.

ADVANTAGES OF MODIFIED REEL METHOD

If the development of the production facilities requires the installation of multiple small-diameter pipelines, the reel method could provide several significant technical and economic advantages:

- A reel vessel could lay up to 25 miles (40 km) of pipe per day, depending on pipe size, anode spacing, and other factors. Hence, it is possible to lay a very large quantity of pipe in a short work season. Relatively short periods of good weather may be used effectively when such high laying rates are obtainable. This could reduce the risk of non-completion and potentially accelerate an overall construction schedule.

- The use of a dynamically positioned (DP) vessel eliminates anchor-handling. Not only is this a cost-saving feature, it also reduces the impact on the environment; there are no anchors to rupture existing lines or disturb fragile bottom conditions. A towed or self-propelled vessel is generally faster and more maneuverable than one that must use an anchor mooring system, resulting in more rapid positioning of the reel vessel.

- Almost all welding, inspection, and testing of the pipe may be performed in a warm climate, under controlled shop conditions.

- No shore facility is required in the Arctic region (unless a small winch pad is necessary for a shore pull) because pipe fabrication and assembly is done elsewhere. This results in major environmental and economic benefits and significantly reduces the size of the permanent work force.

AREAS FOR ADDITIONAL STUDY

(1) Pipelaying Equipment - The selection of the appropriate pipelaying equipment for a particular project should be made only after a careful detailed investigation of existing reel pipelay vessels and preliminary designs of the supply and installation barges described previously. There are many factors which become significant regarding equipment selection, and this task will be a major portion of the total engineering effort.

(2) Insulated Pipelines - It may be necessary in certain circumstances to insulate subsea pipelines. This would be required primarily to avoid degradation of shallow permafrost which could cause unacceptable settlement. Insulated pipelines may be installed by the reel method if the insulation is resilient enough to pass through the straightener and tensioner without damage, and if it is flexible enough to be spooled and unspooled without cracking. It is suggested that a high-quality, neoprene coating may be considered for such an application. The insulation properties will often be adequate to meet design requirements. Each project must be examined with respect to what is needed for thermal insulation before selecting an appropriate pipeline installation method.

(3) Bundled Pipelines - Typically, pipeline bundles have been installed by spooling pipe onto one or more smaller deck-mounted reels in addition to the main reel on the reel vessel. The lines are unspooled from all of the reels simultaneously and strapped together just before passing over the stern. The same procedure could be used in Arctic construction. However, in certain situations it will be desirable to encase all of the bundled pipelines in a larger steel pipe jacket. Reasons for this would be to protect the pipelines during installation or to permit the application of thermal insulation around the bundle on the inside of the jacket. Preliminary engineering studies indicate that laying of a single-conductor jacket bundle with a foam-filled

annulus from a reel vessel is feasible. Although field tests have been successfully performed, it has not yet been demonstrated conclusively that jacketed bundles may be spooled onto a reel and subsequently installed. Thus additional study will be necessary to determine whether the Reel Method is an appropriate installation technique for jacketed pipe bundles.

(4) Material Selection - The choice of construction materials, particularly steel pipe and thermal insulation is extremely important and worthy of further study. The pipe must remain ductile, and there should be no degradation of insulation properties even in sub-freezing water.

CONCLUSIONS

By constructing special supply barges which support vertical or horizontal reels, it is possible that large quantities of pipe may be installed in a single short construction season using the Reel Method. The supply barges may be used to reload a conventional reel vessel, winterized as necessary, or may be connected up to a specially-built installation vessel which would lay the pipe directly from the supply barge. Although equipment costs would be considerable with either concept, in many instances they would be more than offset by technical advantages and economic savings resulting from single-season installation (versus two or more seasons for other construction methods). Hopefully, equipment costs could be amortized over several projects. The method may be used to lay insulated pipelines, in many cases, as well as uninsulated lines.

There are certain limitations on the use of the Reel Method in Arctic areas, and there are aspects that require additional study. Overall, however, the method seems to possess considerable potential for use in the Arctic as well as in other remote parts of the world where no shore-based facilities exist. It would appear that the necessary technology and installation procedures are available to apply this method where suitable conditions exist to warrant its consideration.

REFERENCES

1. Maxwell, S. V., McGonigal, D. and Graham, C., 1983, "Artificial Islands and Steel Structures in the Beaufort Sea." Proceedings from Arctic Technology and Policy Conference, March 1983 at M.I.T., Cambridge, MA.

2. Hnatiuk, J., 1976, "Results of an Environmental Research Program in the Canadian Beaufort Sea," Eighth Annual Offshore Technology Conference, Houston, Texas.

3. R. J. Brown & Associates, 1981, Environmental Impact Statement: Pipeline Installation, Protection and Repair Feasibility Study, an Engineering report submitted to Dome Petroleum, Ltd., August 1981.

4. Hydrocarbon Development in the Beaufort Sea and the McKenzie Delta Region, 1982, Environmental Impact Statement Volume II - Development Systems, Dome Petroleum, Ltd., Esso Resources Canada, Ltd. and Gulf Canada Resources, Inc.

5. SF/Braun, 1982, an Engineering Feasibility Study on the Use of the Reel Method for Installation of Pipelines in the Canadian Beaufort Sea, submitted to Gulf Canada Resources, Inc.

PIPELINE CONSTRUCTION IN ARCTIC & SUB-ARCTIC REGIONS

MICHAEL J. WAGNER, P.E. *
ASCE MEMBER

With few exceptions, pipeline transport is the most efficient method of transporting hydrocarbon energy supplies from arctic regions.

Relative to lines constructed in more moderate climates, construction costs are higher due to lower labor productivity, higher equipment capital and operating costs, as well as higher costs for all elements of project support. Both the environment in which the construction takes place and special line design features contribute to higher costs.

This paper discusses many of the unique labor, equipment and support service requirements of a major arctic mainline project.

This paper discusses many of the unique equipment, labor and support service requirements of a major arctic mainline project.

Advances in technology are affecting the nature of the pipeline construction industry through the utilization of complex system designs including very large diameter, high pressure lines built of special materials suitable for demanding service conditions and adaptable to environmental extremes. Such designs and quality of materials demand revised field procedures and stricter assessment of the contractor's quality of workmanship than every experienced before.

It has been said that the politics of energy have clearly become more influential than the economics of energy. This is certainly true with respect to the approvals required of a multitude of government agencies before projects may proceed to construction and throughout the actual construction phase. The cost element of construction operations too is becoming more significant to the contractor as each new project

*President, O. J. Pipelines Inc., 4796 Business Park Blvd., Suite #5
Anchorage, Alaska 99503

begins. When assessing the announced major projects anywhere in the world, contractors are focusing on the possible financial impacts of shortages of skilled labor, experienced field supervision and inspection staff the substantial investment requirements for increasingly specialized construction equipment and the cost impact of new or modified construction techniques. More onerous quality control standards and measurement methods employed by client, third-party, and government inspection agencies are cause for further concern when assessing potential financial impacts.

The outlook in the mid-70's for the start-up of major projects throughout North America and particularly through arctic regions was very bright. In anticipation of the massive requirements of skilled labor and specialized equipment for these mega-projects, contractors geared up. However, in the early 80's the picture changed dramatically. Mega-projects were put on hold as economic downturns stalled financing while environmental and native land claim disputes curtailed regulatory approvals.

Today's pipeliner - both designer and constructor - must provide new technology and methodology to build facilities in remote areas which are environmentally and geotechnically hostile. Fortunately, many innovations in the pipeline industry have now been thoroughly field tested and the contractor can face northern and arctic projects with a variety of proven time and money-saving techniques and equipment adaptations.

Following is a step-by-step summarization of the major components of a mainline pipeline construction project. Specific areas of expertise required for arctic and sub-arctic projects are set out as extensions of the conventional pipelaying techniques.

RIGHT-OF-WAY

In arctic and sub-arctic areas clearing and grading crews utilize relatively lightweight tracked units to prepare the right-of-way driving the frost into the soil a sufficient distance to support heavier units of equipment. In the high arctic where delicate permanently frozen soils are encountered, alternate methods of right-of-way preparation have been developed. Work pads of granular material or snow construction are used. Such methods are extremely important in permafrost areas due to restrictions placed on grading operations resulting from recognition of the unique geotechnical properties of permafrost soils. The challenge is to preserve or improve the insulating layers overlaying the permafrost. Where tundra is encountered, this delicate insulating layer must be stripped from the ditch line and replaced following backfill. This procedure is similar to the method used in the south for protection of fertile topsoil layers. Extensive testing has been done on various adaptations of equipment for stripping frozen topsoil. In addition, both permanent and temporary access roads must be built to accommodate construction activities and maintenance operations. Temporary ice roads are preferred for winter construction as they require no reclamation and cause minimal damage to the tundra. Permanent access roads can be constructed using conventional methods except in unstable terrain where roads will be constructed from a combination of insulating material and gravel layers.

DITCH

A great deal of brute force is required to excavate ditch through frozen terrain. Massive hydraulically operated rippers mounted on D-9 and D-8 class tractors are used to break up frozen soils. Wheel-type ditchers are often used to dig a shallow pilot ditch immediately in front of larger wheel ditchers.

During winter construction backhoe ditch is kept to a minimum as it results in a spoil bank of large chunks of frozen earth that cannot be backfilled directly onto the pipe without risking structural damage to the line and coating.

Development of a new class of arctic pipeline ditcher has advanced tremendously in recent years. Ditchers weighing in excess of 150,000 pounds capable of digging the most difficult of permafrost soils have been extensively tested and re-modified. The result is a number of state-of-the-art machines available to the contractor.

Conventional ditch blasting techniques have not proven successful for excavating seasonally frozen or permafrost soils. It has been found, however, that cutting a slot along one or both sides of the ditch limit offers an effective fracture face with the effect of limiting the explosive action to only the ditch line.

Traditionally, the pipeline contractor excavates the ditch ahead of the bending and welding crews in order that the pipe can be made to accurately conform to the ditch bottom. Utilizing winter construction techniques, the pipe is bent and welded ahead of the ditch, allowing the pipe gang to proceed at their best pace without being hampered by the erratic progress of the ditch. The big advantage to this simple reversal of form is that the pipe is ready for lowering into the ditch immediately behind the ditching machine. By leaving the ditch open for a relatively short period of time, the ditch bottom and spoil bank do not freeze thus reducing the risk of damage to the pipe and coating. After backfill, a crown of frozen material is placed over the ditchline. When the frost goes out of the ground in the spring, this crown of earth settles into the ditchline, reducing the possibility of sunken ditch.

STRINGING

As the right-of-way is prepared, pipe is strung ahead of the pipe laying crews. Pipe is normally moved to the work site by truck direct from the pipe mill or from nearby railroad sidings. In remote northern regions, however, pipe as well as other critical materials such as valves and fittings are placed in stockpile sites selected and prepared well in advance of actual construction. This advance stockpiling is required to accommodate the seasonal limitations of barge systems in the north which often provide the only viable mode of transport for large amounts of equipment and supplies.

Pipe transported to the stockpiles by barge, rail or truck is typically 40 to 60 feet (12.2 to 18.3 meters) in length. Depending on the contour of the terrain to be followed by the pipeline route, pipe lengths may be joined together at the storage sites to form double-jointed sections of up to 80 feet (24.4 meters) long. Movement of these longer joints of line pipe require specially constructed steering trailers to navigate the temporary access roads leading to the right-of-way.

In the case of a large diameter arctic oil pipeline where the route traverses sensitive permafrost areas, the line is elevated above grade on pipe supports. In many cases pipe bents are supported on piles frozen in place many of which may be installed to depths exceeding 60 feet (18.3 meters).

BENDING

The pipe must be bent to conform to both alignment and contour as the line snakes across the terrain following a right-of-way secured by easement agreements. Bends are achieved through a series of short pulls by a hydraulic bending machine to the exact degree marked by the engineer. On larger diameter pipe, an internal bending mandrel prevents "egging" during the pulls.

The bending operation is typically limited, in the southern latitudes, to a minimum ambient temperature of -20°F (-28.9°C). Such a restriction to a large extent is related to the ductile characteristics of pipe manufactured for these more temperate locations.

In the arctic, bending must often be performed at temperatures below -35°F (-37.2°C). The options available which will ensure that this operation can continue at such low temperatures are to heat the pipe prior to bending and to develop pipe specifications which will assure ductile behavior at the low ambient temperature.

In addition to temperature constraints on the bending operation, size of pipe in terms of both diameter and wall thickness can dictate significant revisions to standard bending procedures and equipment.

Also, the increasing use of pre-coated and insulated pipe presents new challenges to the bending operation as protection of the coating during the bending pull becomes a major concern.

WELDING

Welding operations in arctic conditions require numerous adaptations due to severe cold conditions. Welders must be protected and the sensitive welding operations housed in portable heated shelters, pipe ends must be preheated and completed welds wrapped with heat retaining blankets to control the rate of cooling.

Due to the tremendous magnitude of the welding operation for large diameter pipelines from the arctic, the demand for highly skilled pipeline welders is a critical concern.

The use of a mechanized welding system on such lines serves to minimize the manpower and quality control difficulties experienced in the past on arctic projects. Extensive use of such systems on sub-arctic and southern projects has proven them to be significantly more efficient in terms of welding production rates, repair rates and overall quality control. In addition, the demands of exacting welding codes and specifications appropriate for new pipe compositions are more easily met on a consistent basis using the mechanized system.

With regard to manpower requirements, while the use of a mechanized system has not eliminated the need for skilled pipeline welders, it has significantly reduced the absolute level of individual expertise and experience levels required of the welders. Use of mechanized systems has proven to be less physically demanding on the tradesmen with resulting higher production rates - a major plus in extreme northern conditions where worker productivity is significantly reduced. Experience with mechanized welding on southern mainline projects has shown a full mainline complement of experienced downhand welders can be upgraded and qualified on the equipment in a matter of a few days at either a job-site or southern based training facility.

Wider acceptance of field use of the gas metal arc welding process has had an effect on non-destructive testing procedures. Systems employing an innovative ultrasonic girth weld scan are starting to be used. Compact scanning devices travel the weld circumference on a band with nearby recording equipment providing a continuous digital readout as well as a permanent printout of possible weld defects.

COATING

Line travel coating techniques with their requirements for heated materials shelters and special heat application to the pipe prior to coating have been virtually eliminated. New methods for yard coating of the pipe and superior grades of yard coating materials have proven more cost effective with a more consistent quality. With field coating required only for welded joints and tie-in sections, a very costly, time consuming aspect of the field work has been reduced. Technological developments in field applied coatings, such as a type of spray-applied fusion bonded epoxy used extensively on southern mainline projects, have dramatically affected that operation as well. With field tested specialized equipment such as skid-mounted sand blasting rigs for coating preparation and mobile spray applicator rigs, this operation has been effectively streamlined.

Large diameter pipelines carrying heated crude oil in the arctic will generally be designed for installation on above-ground support structures as previously described. Even though the line will not be in contact with the soil, insulation is required to prevent heat transfer by radiation to the underlying soils. The supply and application of this insulation represents major logistics and construction efforts not normally associated with pipelines designed for southern latitudes.

LOWER-IN

The pipe is normally welded up into long sections 2,000 to 4,000 feet (915 to 1,220 meters) in length which are then lowered into the ditch. In the case of above ground construction, the pipe is welded up directly on pipe supports with gaps left periodically to be tied in at a later time. In certain cases such above ground piping may be welded up into three or four joint sections at a pipe assembly yard, insulated then transported to the field for a series of frequent tie-in welds on the vertical support members. The objective in this case is to move more of the welding and coating operations from the field to a sheltered environment.

Protecting the pipe coating through rocky terrain is a challenge. Just ahead of the lower-in operation, soft soil or sand is placed in the ditch bottom to form a protective padding. Innovations in the area of pipe protection include field tested application of a type of urethane foam "pillow". These pillows are set along the ditch bottom to cushion the pipe when it is lowered in place and then sand or other padding materials are placed around and over the pipe for further protection during the backfill operation. Padding material characteristics are usually specified in the client's construction specifications.

As the line crosses muskeg or unstable soils, concrete weights are placed on the line to provide negative buoyancy. For a recent northern oil pipeline project a new saddle type of weighting device was specified. This weight, known as a Geotextile weight, is made of a durable stitched fabric which is filled with competent native spoil then placed over the pipe in saddle bag fashion.

A major challenge faced by the contractor with respect to concrete weight or continuous coating manufacture in northern regions is the lack of suitable aggregate supply. Such restrictions plus the added inconvenience of cold temperature conditions greatly adds to construction costs.

RIVER CROSSINGS

The installation of major river crossings is carefully programmed into the construction schedule. Often special crews are assigned to installation of these crossings while the mainline construction spread continues its overland progress.

A trench is excavated across the river to design depths that will ensure adequate cover over the installed pipe. The client's engineers will have prepared a special design for each major crossing. The crossing design will require the installation of bolt-on concrete weights or continuous concrete coating thus providing sufficient negative buoyancy to ensure that the line will remain stable when in place.

The pipe is made up in long sections on one shore. If continuous concrete coating has not been specified, then weights are bolted on. A pulling head is welded on to the end of one section, a cable strung to a winch on the opposite shore is connected and the section is then pulled into the previously excavated ditch. The next section is then welded to the end of the first and the second pull is made. This process is repeated until all of the sections have been installed and the pipe emerges on the opposite shore.

A number of major river crossings will be encountered on any of the proposed pipelines leading out of the arctic. The ecological impact of crossing major water courses has brought about extensive environmental controls which necessitate exact timing and methodology for river crossing construction. Sensitive fish spawning areas must be crossed outside the spawning season with the least disruption of natural habitats. Extreme care must be taken in excavating river banks comprised of sensitive permafrost. Major water crossings completed during the summer months must not inhibit the passing of barges up and down the river as these water courses represent the chief transportation system for the north.

Directional drilling techniques currently being used in the south could be of tremendous advantage especially on these arctic projects. This technique involves boring from one bank completely under the river bed to the opposite bank with no disturbance to the river itself. Major difficulties with this technique include the inability to bore through boulder infested soils inherent in many regions of the north.

TIE-INS

Specialized tie-in crews follow the main pipelaying operations joining long sections of previously lowered-in pipe, highway and railway crossings as well as stream and river crossings.

Careful control of the tie-in operation must be maintained in arctic regions. The demand for such control is related to the detailed stress analysis of the line in which all forces imposed on the pipe must be considered.

To minimize the longitudinal stress problem, it is necessary to raise the temperature of the lowered-in sections of pipe to a predetermined temperature prior to joining the long sections together. This may be accomplished by either delaying the tie-in operation until such time that ambient temperatures warm the pipe sufficiently or by the application of heat directly. Since one of the prime objectives of arctic pipelining is to minimize damage to thermally sensitive soils through application of winter construction techniques, the second alternative becomes the most likely choice.

BACKFILL & CLEANUP

Soil excavated from the trench or select imported materials are carefully replaced over the pipe to avoid damage to the protective coating. Topsoil or the vegetative permafrost insulating layer is then carefully replaced and debris removed. All land owner requirements and regulatory agency stipulations with respect to the final condition of the right-of-way must be met.

Three erosion control measures are typically employed in the construction of conventional pipeline systems when the route follows extreme grades. The concern is that rains and melting snow may easily wash away padding and backfill material if special precautions are not taken. First, sacks filled with sand are stacked in the ditch and around the pipe to form breakers which divert water that would otherwise run along the pipe eroding padding and backfill. An innovation in this area is the application of urethane foam breakers which are formed directly into the ditch around the pipe. This system has been field tested and proven to be significantly more efficient than traditional sand bag breakers. The installation of cross drainage from one side of the right-of-way to the other is a second method often employed with the objective of re-establishing surface drainage patterns. Revegetation is the third principal erosion control measure. Seeding is usually accomplished immediately following completion of construction.

In thermally sensitive areas these procedures must be supplemented by additional steps to maintain stability. The delicate insulating layer of tundra previously stripped during the clearing operation must be replaced or an insulating layer of wood chips installed following backfill. In areas where right-of-way cuts are made, surface vegetation is stripped from the entire area prior to grading. Following restoration of these cuts, the right-of-way is seeded in most cases. However, in sensitive slope areas wood chips may be required as is the case with the ditch line.

All unstable soil excavated from the ditch is removed and replaced with select granular material. Upon completion of the backfill operation the entire area is seeded with mixtures of fast-growing native grasses . Erosion control must be maintained during the first summer while more permanent vegetation becomes re-established during succeeding growing seasons.

None of these steps involve totally new technology but the logistics effort required for execution on the scale of a major arctic pipeline project located in a remote area poses a tremendous challenge to those performing the work. Final cleanup operations present scheduling difficulties as the short growing season may require re-vegetation throughout successive seasons.

TESTING

Each section of the pipeline must be tested to meet rigid requirements established by the regulatory authorities before a pipeline can be put into operation. Test pressures applied to the completed line are far in excess of those allowed during operation of the system. Water is the most common test medium used. Techniques have been perfected which permit the hydrostatic testing of lines to be completed during the harsh northern winter season.

Open ends of the line are enclosed under canvas to prevent water in exposed sections of the pipe from freezing during the test. Water is heated and circulated through the test section until it reaches a specified temperature just above freezing. The section is then shut-in and high pressure pumps put on the "squeeze". Pipeline codes generally require the test pressure to be maintained for a twenty-four hour period with critical sections such as river crossings requiring up to two shorter duration "pre-tests". A dead weight tester is used to measure the exact pressure at predetermined intervals during the test with chart recorders providing permanent test records. Upon successful completion of the test, the pressure is reduced and the line dewatered. Precautions are taken to insure that the water used during the test is returned to natural water courses free of contaminants which may have harmful effects on local ecosystems.

An alternate to the heated water technique for pressure testing of larger diameter lines is the use of water-methanol mixes for smaller line sizes. Proper disposal of the water-methanol mix at the conclusion of the test represents an important consideration for the contractor.

In certain circumstances an owners specifications may permit a high pressure air test in lieu of a hydro test.

EQUIPMENT

In addition to the specialized construction techniques required for winter construction in arctic and sub-arctic regions, the impact on equipment design and operation is also a major factor to be considered by the contractors.

For the most part, equipment available for use by contractors is standard regardless of where the contractor is going to work - whether it be the deserts of the Middle East, the mountains of South America or the muskegs of northern Alaska and Canada.

It is a well known fact that many steels exhibit poor ductile properties as temperatures drop. Subsequently, as the temperature drops, the frequency of equipment breakdowns increases, especially for older equipment more subject to parts fatigue than newer units. The costs related to equipment breakdowns are high indeed considering the costs of labor standby, repairs and production losses. Overall economy and productive efficiency demands maximum utilization of equipment suitably modified to perform in severe arctic winter conditions. Such parts as undercarriages, dozer arms, swing frames, yokes, blades and booms must be manufactured from or reinforced with alloys resistant to low temperature brittleness. Most tracked equipment requires heated cabs and ice lugs. All units require suitable block and internal heaters, and automotive equipment operated away from the main work areas must be equipped with emergency heaters, sleeping bags, water and food rations. With few exceptions, units are required to run twenty-four hours a day to prevent icing.

Standardization of such basic items as fuel is an important factor when planning major projects in the remote arctic. Diesel fuel, because of its relatively low cost and its high utilization in heavy equipment and large transportation units, has become the norm. A special winter grade diesel is required as the standard grade would jell at between -20° and $-30^{\circ}F$ (-28.9° and $-34.4^{\circ}C$). Similarly, special winter grade lubricants are designed for the extremely cold temperatures. Conventional greases become so hard at sub-zero temperatures that they are virtually ineffective as lubricants.

Pneumatic control systems have likewise proven relatively ineffective in the arctic. The slightest trace of moisture causes valves and orifices to freeze off, even in the simplest of systems. Air brakes on trucks, for example, require extreme precautions to avoid failure. Compressor air intakes are filtered and discharge is run through a scrubber to remove any traces of moisture before entering the breaking system. Even then troubles occur. Electrical control systems are considered the most reliable in the arctic followed by hydraulic and mechanical with pneumatic control systems ranked a distant fourth.

TRANSPORTATION & SUPPORT FUNCTIONS

The procurement, transportation and maintenance of material, equipment, facilities and personnel for the construction of major arctic and sub-arctic pipelines are a truly monumental undertaking. Given the quantities of materials and equipment to be purchased and transported, the distances to be traversed, the limited work site accessibility, the number of personnel to be accommodated and the on-going support required, it would be an enormous task even if all-weather highway, railway and communication services were available throughout the north. Unfortunately, they are not. The movement of millions of ton-miles of material, equipment and personnel with the relative inaccessibility of any pipeline route north of the 60th Parallel and under severe seasonal limitations is a major problem.

A large portion of materials and supplies for the western Canadian arctic presently move by barge down the Mackenzie River from the Hay River railhead. Seasonal limitations severely restrict traffic on the river to about three months. As the summer transportation period is well in advance of winter construction schedules, extensive and precise planning must go into the determination of supplies and equipment quantities required throughout the scheduled construction period.

Another major consideration is the method of transporting goods from the existing highway and wharf sites to the pipeline right-of-way. The preparation of conventional access roads for this purpose is precluded in many areas by the environmental difficulties associated with grading. A whole new assortment of transport vehicles has been developed in recognition of this problem. Tracked vehicles and low pressure rubber tired vehicles are proving effective in traversing all but the most rugged terrain. These low ground pressure machines are able to travel at acceptable speeds with little or no damage to the environment if properly operated.

The most obvious transportation means for northern services and possibly the most economic is the common sled. The sled is one of man's earliest machines. Cat trains made up of successive units of sleds pulled by tandem axle trucks have been extensively tested and modified for use in the north. Major research in this area has resulted in the evolution of cat train designs to a workable, relatively fast means of transporting large amounts of equipment in remote areas.

Air transportation, of course, plays a major role in logistics planning for northern projects. Aircraft such as the Hercules and heavy lift capacity helicopters have proven very successful for arctic support operations.

CAMP FACILITIES

Remote construction sites necessitate the use of camps to provide accommodation for construction workers. Camps for 400 personnel are not uncommon and accommodation for upwards of 1,000 is foreseeable in the near future on arctic projects.

Modern camps provide comfortable living quarters, washrooms, laundry, kitchen, dining room and recreational facilities, communications center, office facilities, supply warehouse and repair shops.

Good morale is of paramount importance on remote northern camp jobs where personnel are required to work under the extreme conditions of cold temperatures and virtual darkness during much of the winter construction season. Top consideration is always given to the comfort, convenience and welfare of workers. Special attention is focused on high quality food, a quality water supply, modern sewage disposal facilities, laundry service, canteen, recreational facilities and fire protection.

The implications of fire in such remote locations under the extreme weather conditions encountered are of prime concern to those planning such projects. From past experience, electrically heated buildings have proven appropriate in this regard. Electrical generating capacity to service such camps constitutes another major logistic consideration as does the fuel required just to operate the generators on a round-the-clock basis. Environmental considerations associated with water supply and sewage treatment add further to the list of logistics priorities.

COMMUNICATIONS

A key element in the successful completion of remote northern pipeline construction projects is the requirement for a reliable communications network. Effective coordination and control efforts demand that continuous communication be maintained between the northern construction camps and southern points of manpower and materials supply.

Likewise, a proper mobile communications system must be available to each field construction unit to maintain control of the production rate of each highly integrated pipeline spread.

Although the costs are significant, advances in satellite communications systems have resulted in availability of a high quality, reliable method for voice and data transmission connecting client and contractor centres. Satellite systems are also available to provide more sophisticated entertainment features for workers on site.

With advances in computer technology, project control from southern centers presents few problems. Sophisticated computer systems at southern centers controlling payroll, engineering and accounting functions can be readily linked to on-site remote equipment systems.

****

I hope that it is clear from this general summary of approaches to construction for northern pipeline projects, that success of this type of construction effort relies heavily on detailed advance planning and thorough preparation. Pipelines are still the most viable long-term answer to the dilemma of efficient and economic transportation of northern hydrocarbon reserves. The pipeline construction industry, in my view, is meeting the northern challenge through implementation of new or modified construction techniques and by approaching its work in the north with special consideration to its unique environmental and socio-economic conditions.

Cleaning and burning operation

Snow road construction

Gravel work pad construction

Drilling operation on rock right-of-way

Right-of-way preparation

Topsoil stripping of frozen materials

Large Arctic wheel ditcher

Roc saw excavating

Pipe stockpile site

Self-steering stringing trailer

Stringing pipe on the right-of-way

Bending operation

Preheating pipe in preparation for welding

Portable welding shelter

Mechanized welding system internal welder

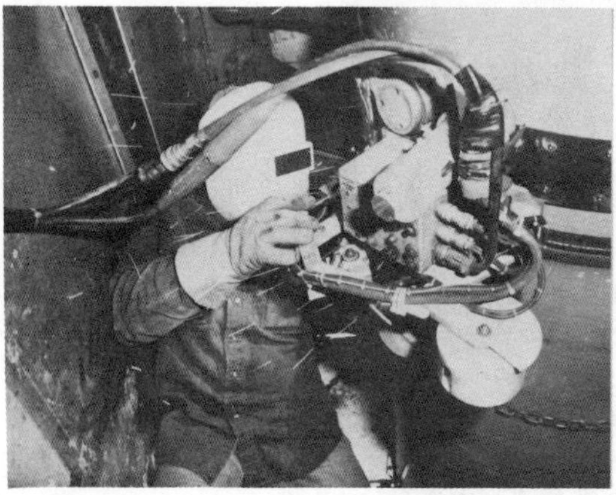

Mechanized welding system external welder

Asbestos covered blanket used to retard cooling rates

Self contained portable joint coating rig

Field applied insulating materials

Yard coated insulated pipe

Completed vertical support members

Pipe installed on the vertical support members awaiting insulation

Foam pillows installed in the rock ditch

Sand padding operation

Special concrete protection device for an above ground installation

Note insulation pad on which the pipe is placed

Select material for backfill being placed

Mormonboard placement of backfill

Dragline excavating for a river crossing from a barge and mats

Sideboom tractors carrying on a river section of continuous concrete coated pipe

Note the clam operating from a spud barge cleaning the ditch just prior to the pull of the section

Hydrostatic test heads

Mobile housing for the pressure testing recording equipment

Final cleanup results

Low ground pressure fuel truck

Half-track utility vehicle

Heavy load capacity transporter

Low ground pressure pipe transporter

Arctic sleigh

Hercules Aircraft

Typical cross country
pipeline camp

DESIGN OF SINGLE PILE SUPPORTS FOR ARCTIC PIPELINES

H. W. K. Bird[1] and K. J. Nyman[2], A.M. ASCE

ABSTRACT

Procedures for designing single pile supports to meet the intent of governing codes must consider the buckling as well as the bending and axial behavior of the pile. Analysis of the buckling of a cantilever pile partially embedded in frozen soil is facilitated if an equivalent buckling length can be assessed.

In this paper, a general procedure is developed to determine the equivalent buckling length of cantilever piles partially embedded in frozen soils.

INTRODUCTION

Pipelines in the Arctic are typically constructed above the ground surface wherever the pipeline route must traverse thaw unstable permafrost. In the past, pipeline support systems have consisted of a two pile arrangement with the pipelines resting on a common cross beam. Recently, however, support systems have, where possible, been optimized to a single pile arrangement. (See Fig. 1)

The design methodology for pipeline supports considers vertical and lateral stability of both the pile and the permafrost. This methodology is presented in Fig. 2 in the form of a design flow chart. Generally, frost jacking forces within the active layer will control the depth of pile embedment in permafrost required to resist vertical loads. The section size of the pile and the allowable "free height" above permafrost are governed by the lateral and buckling loads applied to the pipe support. This paper is concerned with the lateral stability analysis and in particular, with the buckling behavior of the partially embedded single pile supports.

PROBLEM STATEMENT AND LITERATURE REVIEW

The direct analysis of partially embedded piles has been addressed by many researchers in the past (1, 2, 3, 4. 5, 6, 7, 8, 10). Numerical techniques are also available to study the buckling behavior of

---

1. Associate Engineer, SF/Braun, 505 South Main Street, Orange, CA 92668.

2. Senior Geotechnical Engineer, ARCO Pipeline Co., Alaskan Division, 515 South Flower Street, Los Angeles, CA 90071.

Figure 1 — SINGLE PILE PIPELINE SUPPORT

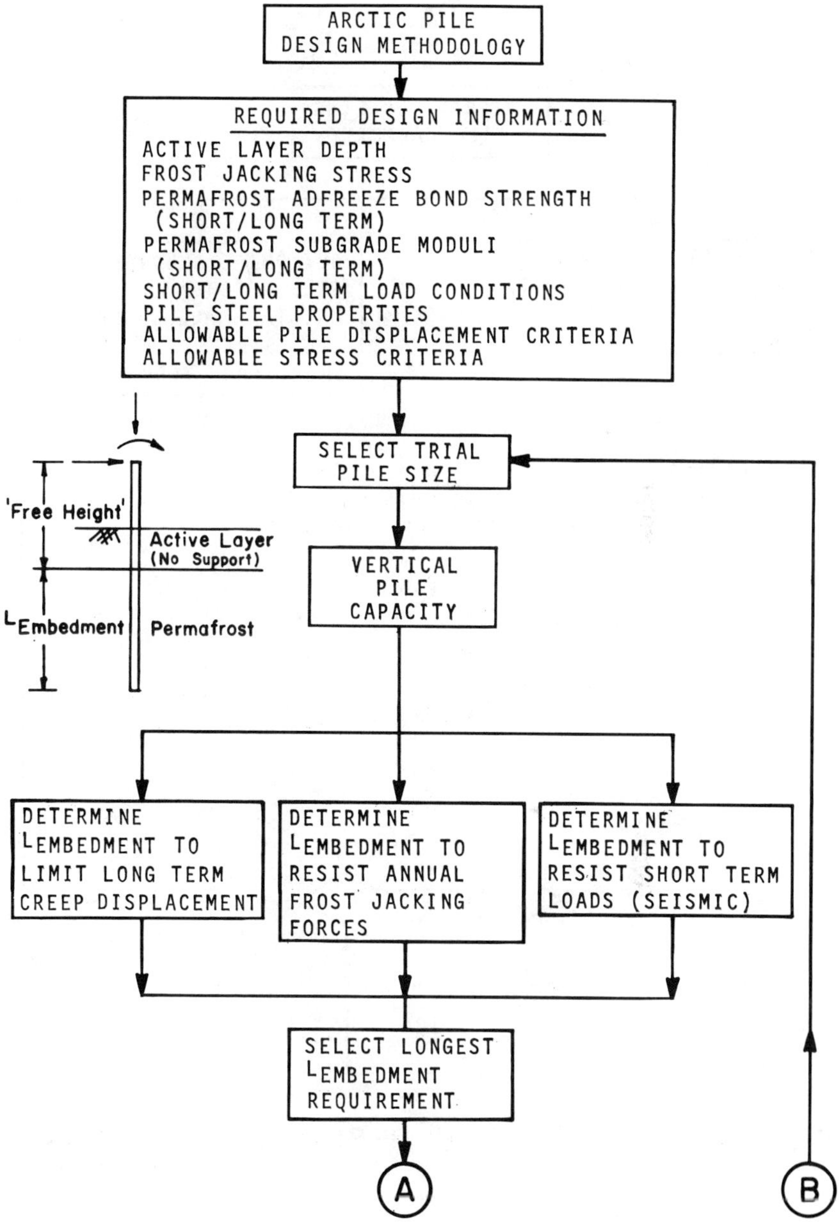

Figure 2a – ARCTIC PILE DESIGN METHODOLOGY

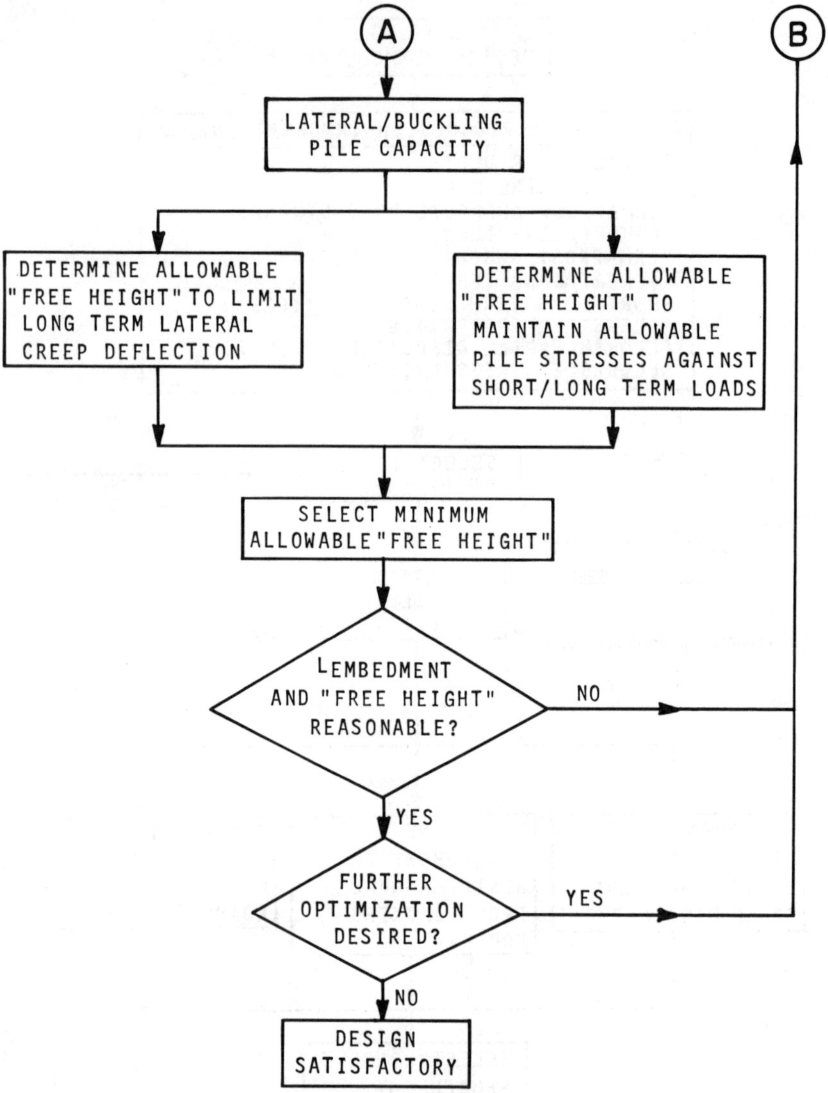

Figure 2b — ARCTIC PILE DESIGN METHODOLOGY

partially embedded piles. Matlock and Ingram (1963) propose an incremental load method to determine the load at which a "Winkler foundation" pile model experiences either adverse displacements or numerical instabilities (buckling). An appropriate design can be achieved with this approach. Factors of safety and allowable loads, however, are left to the discretion of the designer.

From the viewpoint of the designer, it is more desirable to model the partially embedded pile as a fixed end cantilever beam, such that the buckling analysis can be performed in accordance with established codes and criteria. The partially embedded pile is assumed to behave as a free standing cantilever rigidly fixed at some depth below the ground surface.

Davisson and Robinson (1965) point out that the most common assumptions regarding the depth to the point of fixity do not address the actual requirements of the soil-structure interaction problem. Instead, Davisson and Robinson suggest that a reasonable fixed end cantilever beam approximation can be accomplished by finding a point of fixity such that the deflection and rotation of the pile head and the buckling load are the same as for the partially embedded pile. Lee (1968) reported favorable results when using this approach to predict the buckling capacity for axially loaded model piles tested in sand.

In the case of single pile supports, the eccentric moment and lateral loads at the pile head are significant and should, therefore, be included in the buckling analysis of the pile. In this paper, the approach of Davisson and Robinson is extended to consider partially embedded piles subjected simultaneously to moment, lateral, and axial loads. The numerical and analytical techniques used to analyze a pile partially embedded in a "Winkler" medium and used to obtain an equivalent fixed end cantilever system are presented. The results of this analysis are put into a general design procedure which may be used to evaluate the combined stress interaction equation and, hence, design single pile supports for lateral and live loads in accordance with standard design codes.

NUMERICAL ANALYSIS OF A PARTIALLY EMBEDDED PILE

Analysis of a partially embedded pile as shown in Fig. 3 is based on numerical solution of the governing fourth order differential equation proposed by Hetenyi (1946) which is expressed as

$$EI\,(d^4y/dx^4) + P\,(d^2y/dx^2) + k_x y = 0 \quad \ldots\ldots\ldots\ldots (1)$$

where EI is the flexural stiffness of the pile, P is the axial load, x and y are the coordinates as shown in Fig. 3, and k_x is the subgrade modulus. The term $k_x y$ represents the soil reaction per unit length of the pile. The soil response is modelled in the form of a "Winkler" foundation as described below.

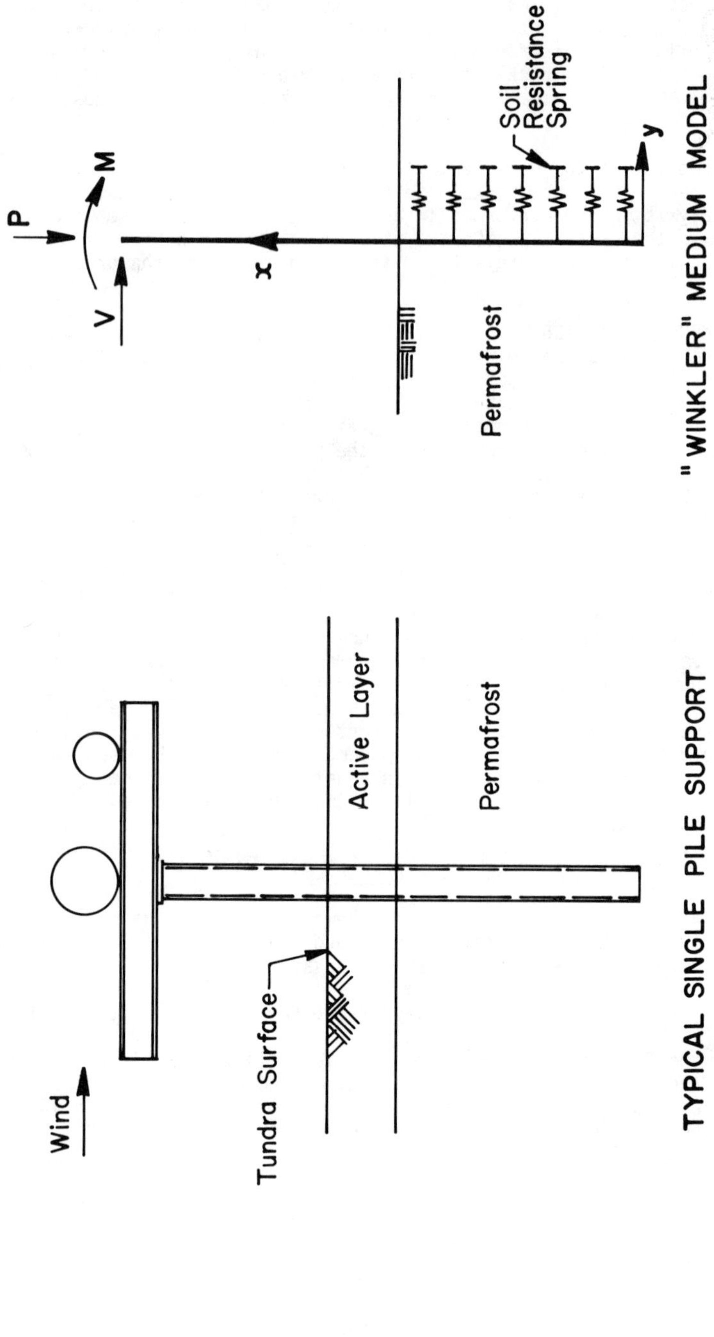

Figure 3 – " WINKLER " MEDIUM REPRESENTATION OF A SINGLE PILE SUPPORT

"Winkler" Foundation Model

One of the bases for the numerical approach to solve the laterally-axially loaded pile problem is the assumption developed by Winkler (1867). The single pile subjected to combined lateral-axial loading is modelled as a beam partially embedded in an elastic-plastic foundation as shown in Fig. 3.

The subgrade material is represented as a set of discrete springs which are attached to the pile and act independently. Each spring has a load-displacement response (P-Y curve) which approximates the actual subgrade reaction against pile movement.

The subgrade "spring" modulus mathematically represents the development of soil force or pressure as a function of displacement (9). In its most basic form, the load-displacement relationship is idealized as an elastic-plastic approximation of the non-linear soil response. Thus, the soil restraint function may be defined by a unique point having the coordinates of ultimate soil restraint P_u, and corresponding yield displacement, Y_u, as presented in Fig. 4.

Permafrost subgrades are composed of a two phase matrix consisting of ice and solid soil particles. Thus, permafrost soils exhibit time dependent load-displacement characteristics. Under short term lateral loads resulting from either seismic acceleration, impact, or wind, permafrost soils exhibit relatively high subgrade stiffnesses. On the other hand, permafrost soils will "creep" or produce cummulative displacements under constant long term loads, depending upon the load magnitude, permafrost composition, temperature, and salinity of the ice-soil matrix. A general comparison of P-Y curves for short term and long term load conditions is presented in Fig. 5.

Difference Equation Solution

Numerical methods are used along with the "Winkler" foundation model to approximate the solution to the fourth order differential equation, Eq. 1, which governs the behavior of the partially embedded pile.

Eq. 1 is, by nature of the soil model, a non-linear differential equation and is solved iteratively with the aid of a computer by using the finite difference method. The pile behavior is modelled as a beam-column, the pile itself being divided into a number of segments of equal length. The iterations will converge to a solution of equilibrium between the soil and all pile elements provided that the pile does not buckle.

From the above numerical analysis, the deflection, slope, soil pressure, shear, and bending moment may be obtained for different points along the pile. Fig. 6 shows a typical deflected shape, slope, bending moment, and shear diagram and soil reaction for a partially embedded pile.

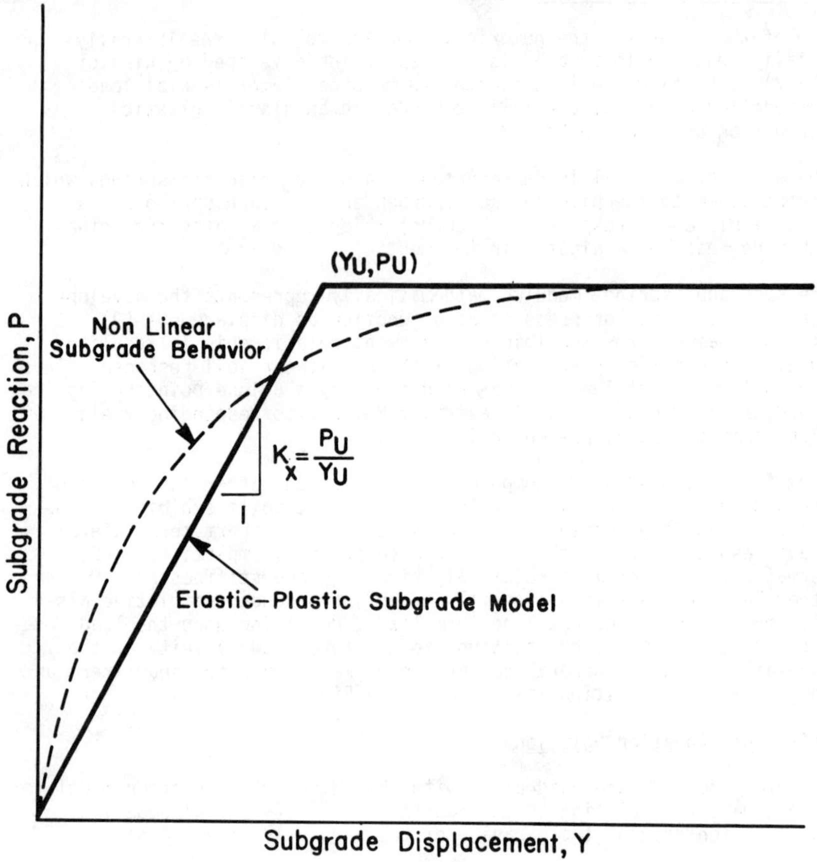

Figure 4– BILINEAR REPRESENTATION OF SUBGRADE RESPONSE

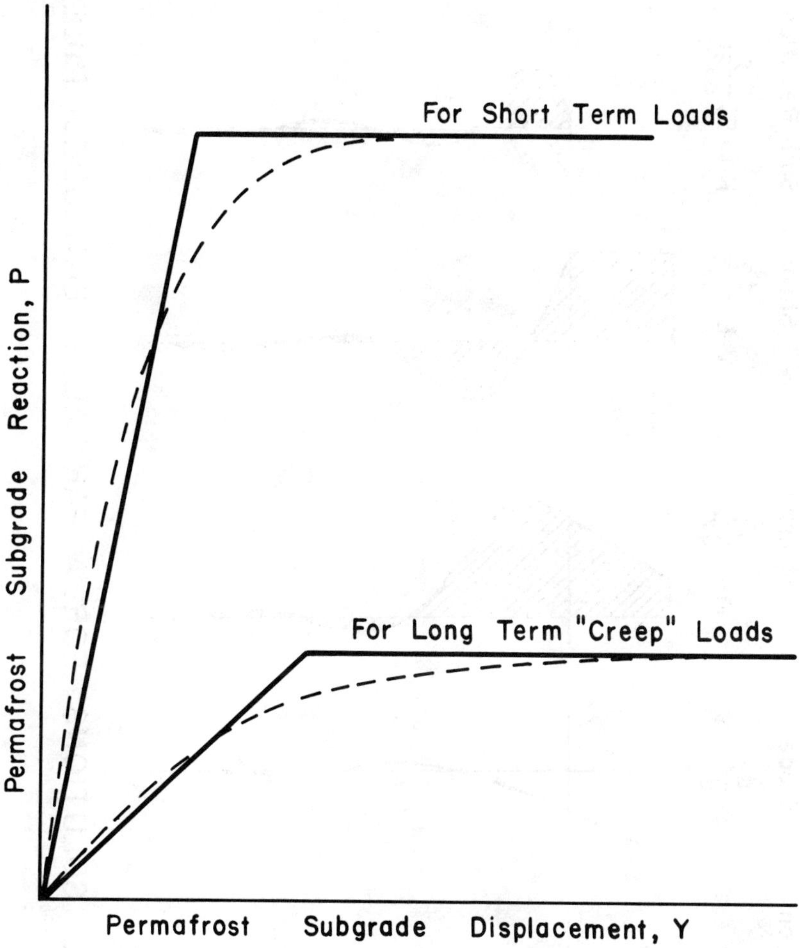

Figure 5 - TIME DEPENDENT TRENDS FOR PERMAFROST SUBGRADE REACTION

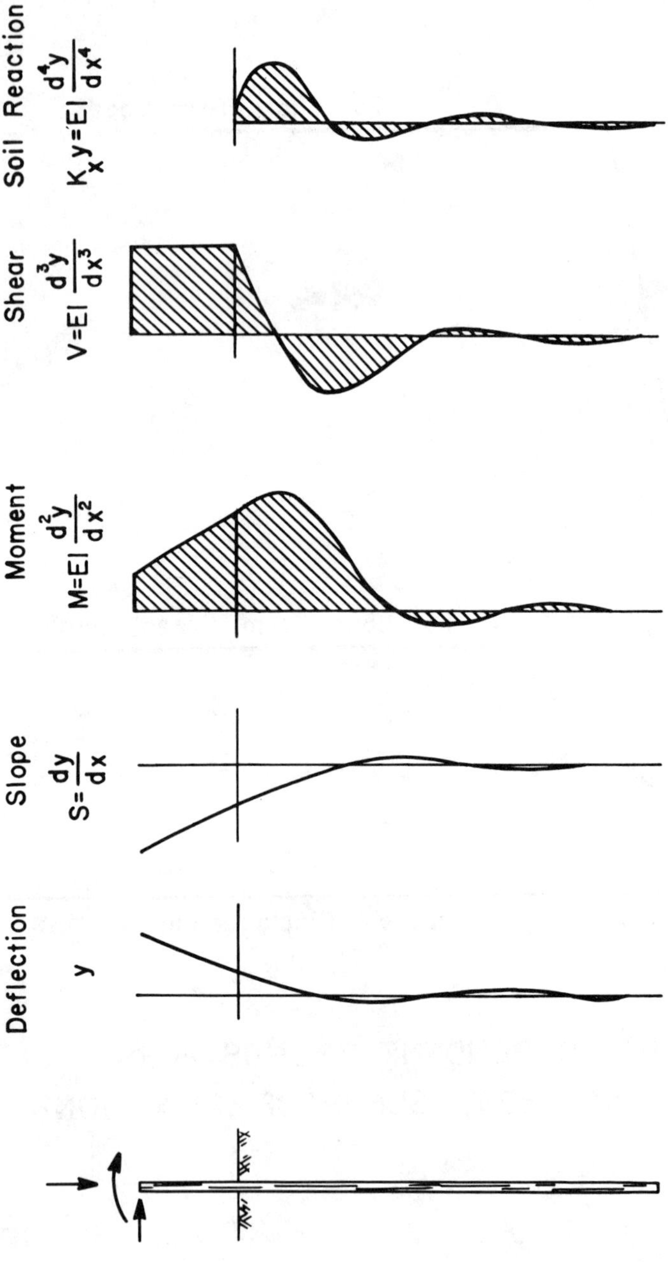

Figure 6 – TYPICAL SOLUTIONS FOR A PARTIALLY EMBEDDED PILE

EQUIVALENT FIXED-END CANTILEVER

The partially embedded pile considered above may be simplified to an equivalent cantilever fixed at some depth below the ground surface. This equivalent system is shown in Fig. 7. In order to obtain the equivalent fixed-end cantilever, the pile head displacement of the similarly loaded cantilever is set equal to that of the partially embedded pile.

The length of this cantilever is found by solving the governing second order differential equation for a cantilever system which is of the form

$$EI \frac{d^2y}{dx^2} + M_x = 0 \quad \ldots\ldots\ldots\ldots \quad (2)$$

where M_x is the bending moment at a distance x along the pile. Referring to Fig. 7, Eq. 2 may be written as

$$EI \frac{d^2y}{dx^2} = V(L_e - x) + M + P(\Delta - y) \quad \ldots\ldots\ldots\ldots \quad (3)$$

Using the following notation

$$j = \sqrt{\frac{EI}{P}} \text{ and } U = \frac{L_e}{j} \quad \ldots\ldots\ldots\ldots \quad (4)$$

A solution for the deflection y in Eq. 3 is written as

$$y = \frac{V}{P}[j\sin(\frac{x}{j}) + j\tan(U) - j\tan(U)\cos(\frac{x}{j}) - x] +$$

$$+ \frac{M}{P}[1-\cos(\frac{x}{J})]/\cos(U) \quad \ldots\ldots\ldots\ldots \quad (5)$$

The equivalent length L_e of the cantilever is then found by setting the displacement y, at $x = L_e$, in Eq. 5 equal to Δ the pile head displacement of the partially embedded pile obtained from the analysis described in the previous section. A Newton-Raphson iteration sequence is used to converge on the correct value for L_e. The length L_e defines the depth to the point of fixity.

Fig. 8 shows a plot of the depth of fixity versus the pile stiffness. Two trends are noted. Firstly, as the pile stiffness increases, the depth to the point of fixity increases. Secondly, with stiffer soil springs (short term load P-Y curves), the depth to the point of fixity decreases. If a relative soil-pile stiffness is defined, as the ratio of the pile stiffness to the subgrade modulus, it may be stated that as the relative soil-pile stiffness increases the depth of fixity of the equivalent cantilever increases.

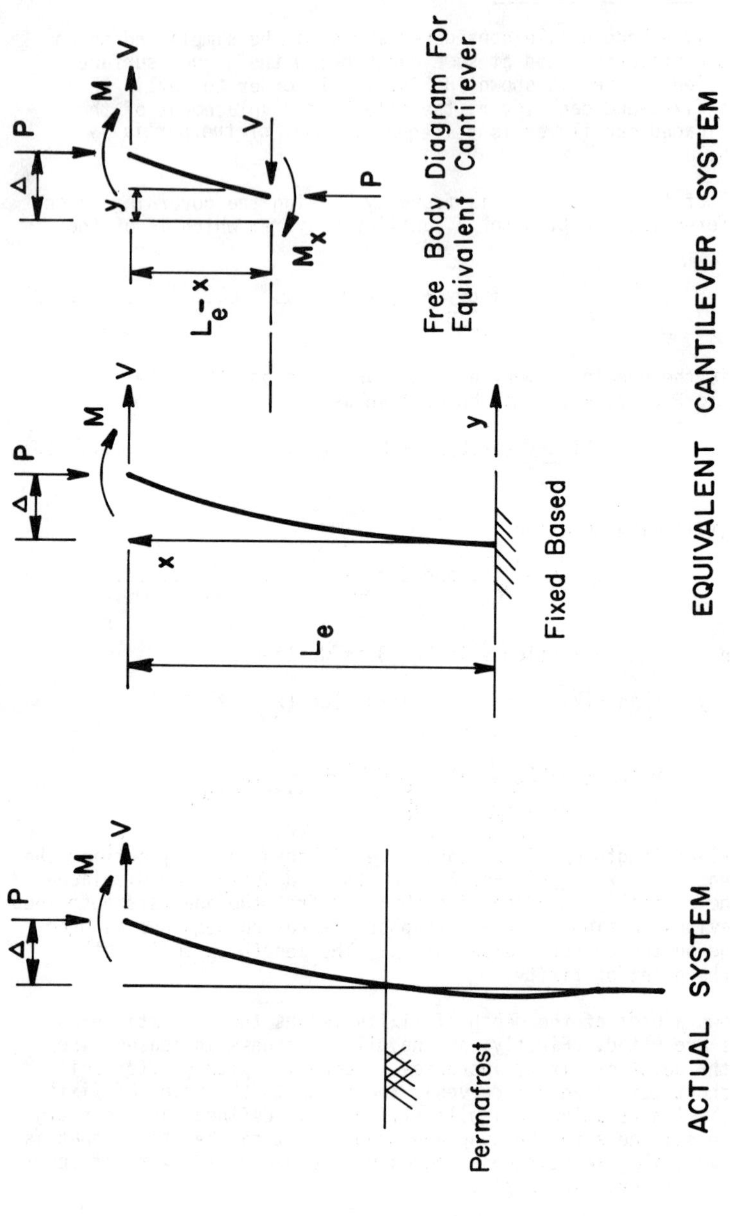

Figure 7 - EQUIVALENT FIXED END CANTILEVER

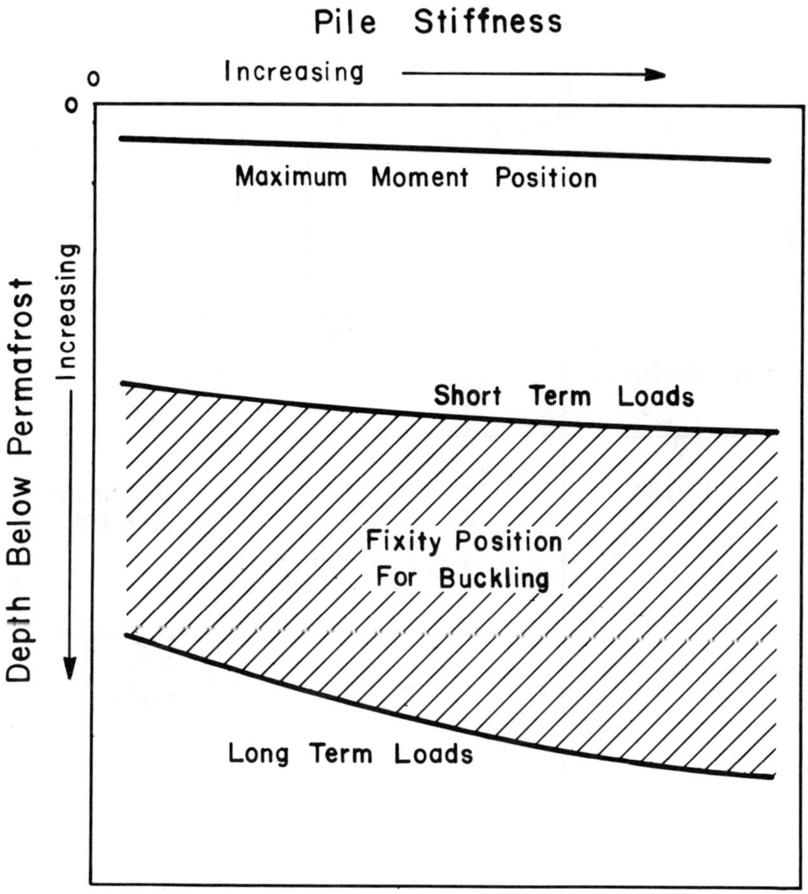

Figure 8-DEPTH OF FIXITY VERSUS PILE STIFFNESS

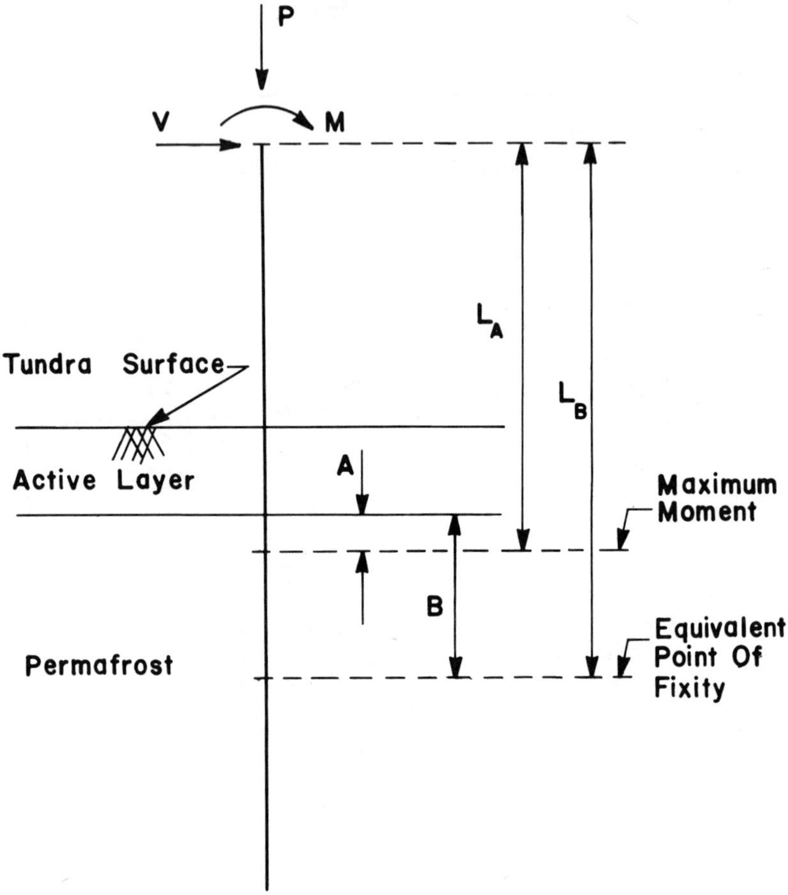

Figure 9 - EFFECTIVE LENGTHS FOR INTERACTION EQUATION

EVALUATION OF THE DESIGN INTERACTION EQUATION

Design of piles to meet the intent of governing codes normally necessitates the satisfaction of a combined axial and bending stress interaction equation. For the purpose of this paper, the interaction equation is used in its most simple form as follows

$$\frac{f_a}{F_a} + \frac{f_b}{F_b} \leq 1.33 \text{ (for seismic or wind loads combined with live loads)} \qquad (6)$$

where f_a = computed axial stress

f_b = computed compressive bending stress

F_a = allowable axial compressive stress

F_b = allowable compressing bending stress

The assessment of f_a and F_b is straightforward. To evaluate f_b and F_a, however, requires a more detailed analysis when considering the case of a partially embedded pile.

The compressive bending stress f_b, resulting from lateral loads, is computed using the lever arm L_A, which is shown in Fig. 9. The dimension A, the depth below the permafrost to the point of maximum moment, may be found from the bending moment diagram derived from the numerical analysis of the partially embedded pile. The eccentric moment resulting from vertical pipe loads should also be included in the computation of f_b.

The assessment of F_a, the allowable axial compressive stress, requires computation of the ratio

$$\frac{KL_e}{r} \qquad (7)$$

where K = effective length factor

L_e = equivalent unsupported buckling length

r = radius of gyration

Based upon the magnitude of this ratio, F_a is determined from equations which are associated with the Euler buckling solution for a column subjected to axial load.

Design codes do not recommend values for either the effective length factor K or the equivalent buckling length L_e to be used with partially embedded piles. Codes do, however, recommend values for a fixed end cantilever column subjected to axial load. Hence, the evaluation of F_a for the partially embedded single pile support is carried out using an equivalent fixed end cantilever approximation. The equivalent unsupported buckling length derived from this approach is shown in Fig. 9 as L_B; the dimension B being defined as the equivalent unsupported embedment length.

CONCLUSIONS

A general procedure for the analysis and design of single pile supports with regard to lateral stability is presented. The partially embedded pile is modelled as an equivalent fixed end cantilever in order to facilitate analysis in accordance with the existing design codes used in industry.

REFERENCES

1. Banerjee, P. K. and Davies, T. G., "The Behavior of Axially and Laterally Loaded Single Piles Embedded in Non-Homogeneous Soils", Geotechnique, Vol. 28, No. 3, pp. 309-326.

2. Davisson, M. T., "Estimating Buckling Loads for Piles", 2nd Pan American Conference on Soil Mechanics and Foundation Engineering, Brazil, Vol. 1, Jul. 1963, pp. 351-371.

3. Davisson, M. T. and Robinson, K. E., "Bending and Buckling of Partially Embedded Piles", Proc. 6th Int. Conf. Soil Mechanics and Foundation Engineering, Vol. 2, 1965, pp. 243-246.

4. Granholm, H., "On Elastic Stability of Piles Surrounded by a Supporting Medium", Ingeniors Vetenskaps Akad, Hand 89, Svenska Bokhendelscentralen, Stockholm, Sweden, 1929.

5. Hetenyi, M., "Beams on Elastic Foundation", University of Michigan Press, Ann Arbor, Michigan, 1946.

6. Lee, K. L., "Buckling of Partially Embedded Piles in Sand", J. of Soil Mechanics and Foundations Division, ASCE, Vol. 94, 1968, pp. 255-270.

7. Matlock, H. and Ingram, W. B., "Bending and Buckling of Soil Supported Structural Elements", 2nd Pan American Conference on Soil Mechanics and Foundation Engineering, Brazil, Vol. 1, Jul. 1963, pp. 373-388.

8. Reddy, A. S. and Valsangkar, A. J., "Buckling of Fully and Partially Embedded Piles", J. of Soil Mechanics and Foundations Division, ASCE, Vol. 96, 1970, pp. 1951-1965.

9. Terzaghi, K., "Evaluation of Coefficient of Subgrade Reaction", Geotechnique, Vol. 5, 1955, pp. 297-326.

10. Toakley, A. R., "Buckling Loads of Elastically Supported Struts", J. of Engineering Mechanics Division, ASCE, Vol. 91, No. EM3 Proc. Paper 4379, June 1965, pp. 205-231.

11. Winkler, E., "Die Lehre von Elastizitat und Testigkeit", (on Elasticity and Fixity), Prague, 1867.

THE SAN MATEO COASTSIDE REGIONAL OUTFALL

Michael A. Maggi,* M. ASCE

Abstract

California's Clean Water Grant Program has had a significant impact on most wastewater treatment and disposal projects undertaken in the State in recent years. Program requirements have lengthened the time required from inception to completion of many projects, frequently have influenced standards imposed by local regulatory agencies, and often had profound effects on design and construction. This paper traces the 11-year evolution of an ocean outfall project which is part of a grant-funded regional wastewater system being constructed by three California coastside communities.

Introduction

In 1972, three California communities located adjacent to Half Moon Bay on the San Mateo County Coastline began planning for a regional wastewater treatment and ocean disposal system. The City of Half Moon Bay, the Montara Sanitary District and the Granada Sanitary District formed a regional wastewater management agency which became the Sewer Authority Mid-Coastside (SAM). As indicated on Figure 1, the resulting project is located approximately 20 miles (32.2 km) south of the entrance to San Francisco Bay.

The Montara Sanitary District, located immediately north of Half Moon Bay, shares a common boundary to the south with the Granada Sanitary District. The Granada Sanitary District and the City of Half Moon Bay also have contiguous service areas and are located adjacent to the northerly and southerly ends (respectively) of Half Moon Bay. In 1972, the estimated population of the combined service area was 10,200 people (1) with each agency maintaining separate treatment and disposal systems.

Existing Systems

The Montara Sanitary District operated a secondary treatment system with a 12-inch (30.5 cm) diameter outfall

*Associate, Barrett, Harris and Associates, 3000 Alpine Road, Menlo Park, CA 94025

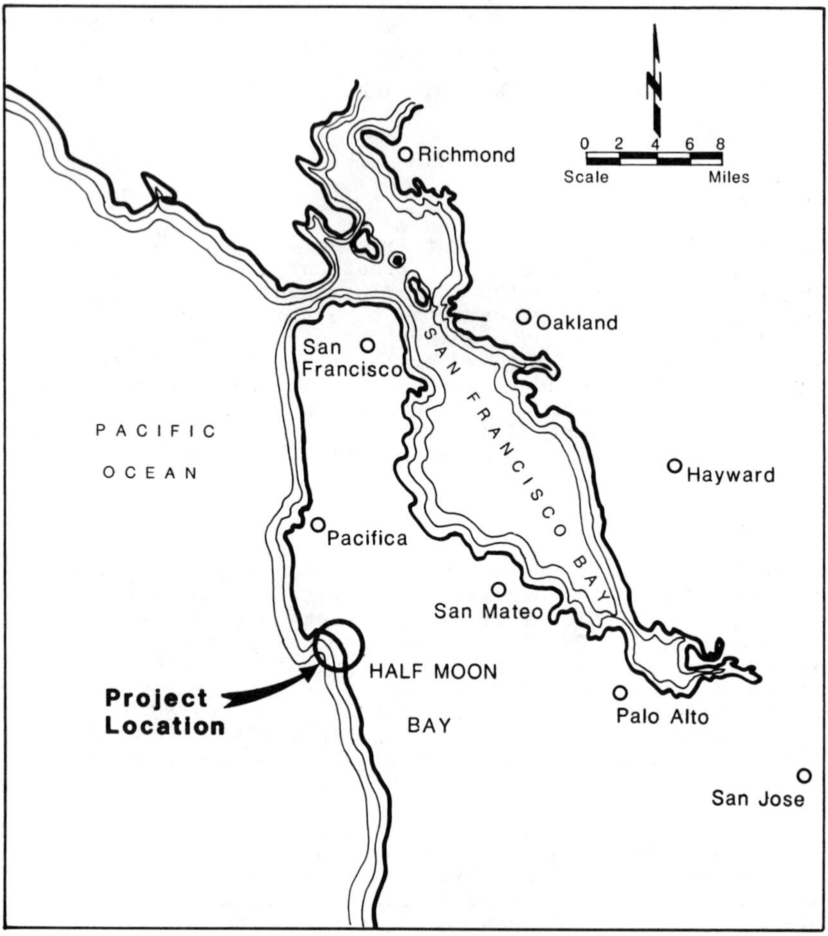

FIG. 1 -Location Map

extending 30 feet (9.14 m) offshore at Pillar Point and discharging about 7 feet (2.13 m) below mean sea level. The Granada Sanitary District operated a primary treatment facility discharging about 300 feet offshore through a 12-inch (30.5 cm) diameter outfall at a depth of approximately 5 feet (1.52 m) below mean sea level. Both systems discharged into shallow rocky areas which became part of the James V. Fitzgerald Marine Reserve.

In 1973 the City began construction of interim facilities; upgrading their primary treatment facility to provide secondary treatment, extending their outfall and adding a new multiport diffuser to provide an initial dilution of 100:1 at the point of discharge. The City's original outfall was installed in the late 1950's and had a history of breaking in the surf zone attributed to wave forces and shallow burial. On several occasions prior to 1976 the city had to make costly emergency repairs to the outfall (2).

The City's original 12-inch (30.5 cm) diameter welded steel outfall was extended with ductile iron pipe to a distance of about 1,600 feet (0.487 km) into Half Moon Bay. Also added was a 187 foot (57.7 m) ductile iron diffuser section with 35 ports, 1.5-inches (3.81 cm) in diameter, discharging at a depth of 30 feet (9.14 m) below mean sea level.

Regional Planning

The regional agency retained the engineering consortium of Barrett, Harris & Associates, Black & Beatch, and Resources Engineering & Management to provide planning, design and engineering services during construction. An extensive oceanographic study was done by Oceanographic Services Incorporated (OSI) to provide a basis for selection of an ocean outfall location.

The project report (4) completed in 1975 identified a number of alternatives for regional wastewater collection, treatment and disposal. The selected alternative called for consolidation of the regional treatment facilities at the site of the existing Half Moon Bay Treatment Plant; supported by an intertie system consisting of interceptor sewers and pumping stations, and an upgrading of the City's existing outfall system.

Project Setting

Half Moon Bay is a crescent-shaped open coastal area protected from severe sea conditions by Pillar Point Promontary to the north and to some extent by a shallow submerged reef system. (Refer to Figure 2). Pillar Point Harbor, located at the north end of the Bay, serves as mooring facility and launching area. As the only improved harbor along the Coast between Bodega Bay and Santa Cruz, Half Moon Bay's harbor provides an important refuge for mariners during inclement weather.

The Bay, which is approximately one mile (1.61 km) wide at its center, is bounded on the east by sandy beaches and

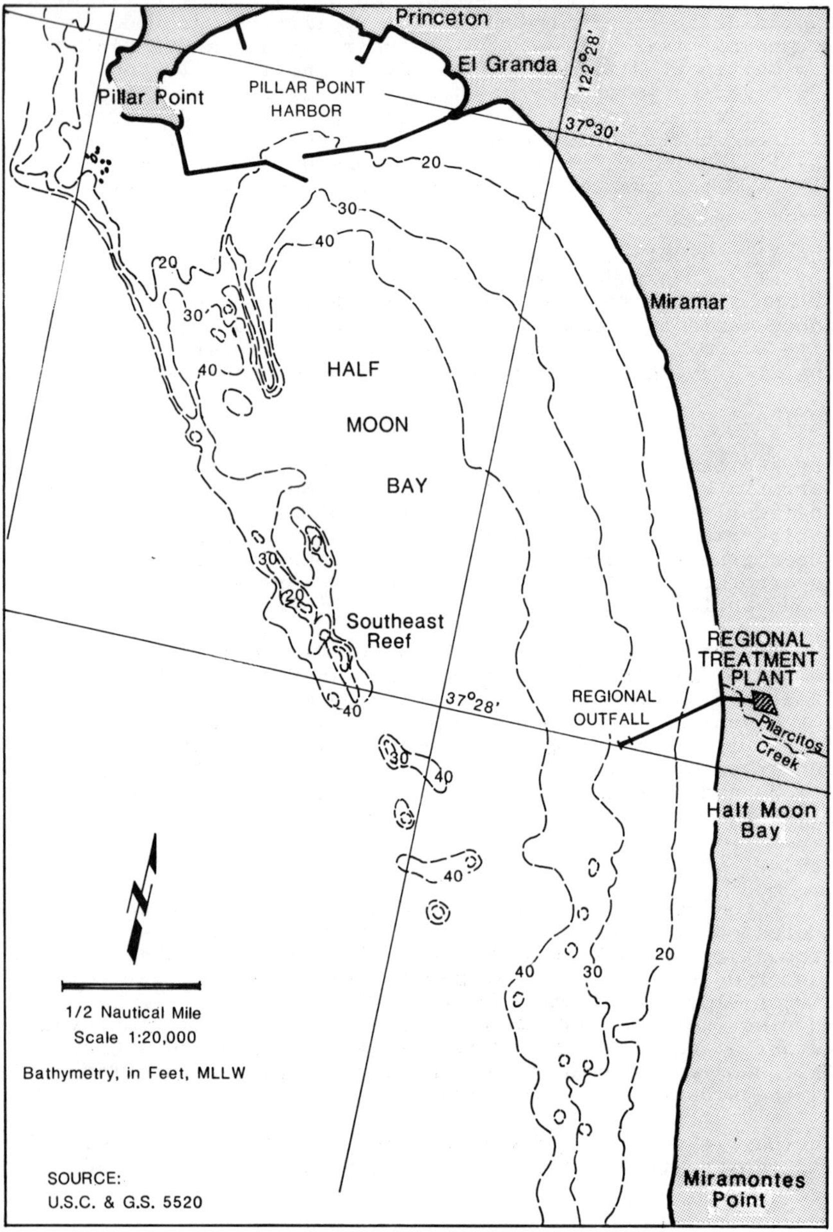

FIG. 2 -Vicinity Map

isolated from deep offshore waters to the west by the reef. The reef extends south parallel to the shoreline from Pillar Point towards Miramontes Point with an opening at the south end of the Bay. The reef is shallow but submerged along most of its length and is attributed to uplift on the western side of the Seal Cove Fault.

The Bay slopes gently to the west to a depth of about 65 feet (19.8 m) adjacent to the reef. The OSI report (3) indicates that sediments on the floor of the Bay are fine sand or shell fragments on the sloping areas with predominately hard mudstone (compacted silt) over the reef area. The southern most quarter of the Bay, near Miramontes Point, is accentuated by numerous rock out-croppings with variations in bottom elevation of from 4 to 10 feet (1.22 to 3.05 m). Reportedly no thermocline is present within the Bay.

Design Criteria

The "Water Quality Control Plan for the Ocean Waters of California", adopted by the State Water Resources Control Board (SWRCB) in 1972, restricted ocean discharges within the 30-foot (9.14 m) depth contour or 1,000 feet (0.308 km) of the shoreline, whichever is the greater distance from shoreline. The Ocean Plan further provided that bacterio-logical objectives, as determined by coliform concentra-tions, must be maintained throughout the water column in zones used for body-contact sports.

Actual discharge requirements issued by the Regional Water Quality Control Board include secondary treatment, a median receiving water coliform concentration not to exceed an MPN (most probable number) of 240 organisms per 100 milliliters, zero chlorine residual and a minimum 100:1 initial dilution at the point of discharge.

Because of the extensive recreational usage of Half Moon Bay Beaches, final design criteria was established in cooperation with the SWRCB Design Section after lengthy analysis of projected bacterial concentrations resulting from physical dilution only (ie: initial and transport dilution) without consideration of bacterial decay or disinfection. The agreed upon criteria included construc-tion of an outfall extending 1400 feet (0.427 km) from shoreline and sufficient backup chlorination equipment to insure continuous compliance with disinfection standards.

Outfall Design

Design of the regional outfall system was completed in 1976. The system was to include a new 18-inch (45.7 cm) diameter outfall, to operate in conjunction with the City's existing outfall, and new effluent pumping facilities. The new outfall was to extend 1460 feet (0.445 km) from the maximum reach of shoreline at extreme lower low water to maintain a 30 foot (9.14 m) water depth for discharge. The pipeline was to terminate in a 161 foot (49.1 m) long diffuser section with 24 ports, 2-inches (5.08 cm) in diameter, and spaced on 7 foot (2.13 m) centers.

The new effluent pumping station was to be located adjacent to the regional treatment facilities inside a flood protection dike surrounding the existing treatment plant. The design called for the outfall to extend under the dike and Pilarcitos Creek to the new surge chamber located in the sand dune area. From the surge chamber, the outfall extends across the beach and into the bay. A bore and jacked casing was used to extend the outfall under the dike and anchors, precast on the pipe sections, were used for the crossing underneath Pilarcitos Creek to prevent flotation and movement.

The pumping facilities were designed for use under peak flow conditions only, thus when pumping is required the outfall will convert from gravity to pressure flow. Air trapped by the conversion to pressure flow will escape through an air and vacuum valve mounted on top of the surge chamber.

Outfall Redesign

As a result of a capacity dispute between the regional authority and State regulatory agencies, the regional project was "shelved" in 1977.

In 1978, the City of Half Moon Bay's outfall pipeline suffered a failure in the surf zone rendering the pipeline useless without extensive and costly repairs. Failure of the City's outfall ultimately led to a revival of the regional project. The new outfall pipeline was redesigned to convey the entire capacity of the regional treatment facilities.

The regional outfall is centrally located in the Bay extending across an exposed beach with an active surf zone. The beach experiences a seasonal inshore-offshore sand movement common to open coast beaches. Storm waves erode the beach sand veneer during the winter months transporting the sand offshore and depositing it as sand bars. During the summer months the reverse phenomenon occurs reshaping the beach profiles. Evidence of a 5 to 8 foot (1.52 to 2.44 m) change in Half Moon Bay beach profile during a single year has been reported (5).

The original design for the regional outfall system called for pipe burial in sand at a depth believed to be just below the anticipated seasonal movement of beach veneer. Based on observations by local residents of a clay or "mud" underlying the sand veneer in the surf zone, further soil investigations during 1978 resulted in an increase in the pipe burial depth to extend into the non-eroded "clay" strata. Other related design changes included a graded granular bedding and backfill material with a 30-inch (76.2 cm) thickness of "Armor" rock, a concrete weight coat to provide a negative buoyance of 20 pounds per linear foot (29.8 kg/m) and the specification of 600 feet (0.182 km) of temporary sheet pile cofferdam through the surf zone for protection during pipe installation.

As a result of the capacity increase to provide for a peak design flow of 5.4 million gallons per day (0.237 m /s); the pipeline diameter was increased to 20-inch (50.8 cm) with 237 feet (72.2 m) of diffuser section extending approximately 2000 feet (0.610 km) offshore and into 30 feet (9.14 m) of water (MSL).

Initial Construction Project

Because of litigation brought by the State seeking to enjoin the member agencies of SAM to complete the regional project on a stringent schedule, SAM elected to bid the construction contract specifying a six-month completion.

A public opening of construction bids was held in May of 1979. However, one of the unsuccessful bidders filed a formal protest delaying award to the low bidder until the end of July 1979.

The low bidder, who proposed the use of an unconventional construction method, was not an established marine contractor but had previous outfall construction experience. The contractor intended to fabricate a mobile working platform (referred to by the contractor as a "tower") intended for use on all surf zone and offshore construction activities.

The tower was a welded frame structure fabricated from large diameter steel pipe which supported a platform elevated approximately 65 feet (19.8 m) above the bottom of the structure. The platform and frame were mounted on two large (8 feet x 9 feet x 110 feet/2.44 m x 2.74 m x 361 m) pontoons. The pontoons were designed to be filled with air by compressors (mounted on the top platform) to increase buoyancy when the tower was to be moved. When flooded, the pontoons would rest on the ocean floor providing stable operating conditions for the 100 ton (90.2 Mg) crane mounted on the top platform.

The tower was designed to advance only forward or backward in a straight line. The contractor set sea anchors offshore of the diffuser tied off to two wire ropes, 2-inches (5.08 cm) in diameter, that extended through the tower back to the beach. The tower was advanced along the wire ropes by wire clamps attached to hydraulic rams.

The contractor apparently intended to fabricate and assemble the tower offsite, however, shortly after award of the contract the contractor revised his plan and began construction on-site. The on-site construction necessitated moving the tower, when assembled, from the staging area across the beach to the surf.

The contractor apparently underestimated the time and manpower required to construct the tower resulting in the loss of the 1979 construction season. Upon completion of the tower construction in late Spring of 1980, the contractor began advancing the tower across the beach installing sheetpiling for the pipe trench shoring. Unfortunately, the force necessary to move the tower across the sand was also underestimated; causing further delay which was compounded by the Summer advancement of the beach.

In late September of 1980, the tower advanced sufficiently into the surf zone to utilize the buoyance of the pontoons, greatly increasing its mobility. However, storm waves in October of 1980 immobilized the tower in the surf zone requiring the contractor to construct a trestle for access and to modify the tower design. Shortly after modifications were completed the tower was disabled following a December 1980 storm requiring further extension of the contractors trestle. The contractor was unable to make repairs until June of 1981.

The contractor elected to use restrained joint ductile iron pipe (class 56) which had been specified as an alternate to welded steel pipe. The contractor constructed special forms to pour the 2-inch (5.08 cm) concrete pipe coating on-site.

During the Spring and Summer of 1981 the contractor began installing pipe in the surf zone from the trestle. Shoring problems were encountered in the surf zone and in September of 1981 SAM was notified that the contractor was unable to continue. The contractor constructed the pumping station, the surge chamber, and installed 1025 feet (0.312 km) of pipeline extending about 175 feet (53.3 m) into the surf.

Offshore Construction

A second construction contract for completion of the outfall was awarded to the Healy Tibbitts Construction Company in June of 1982. The contract required testing of existing work, installation of 1930 feet (0.588 km) of submerged pipeline and diffuser, and removal of existing temporary structures remaining from the previous construction. The contract specified a one-year construction period.

FIG. 3 -Derrick Barge Placing Armor Rock over an Offshore Pipe Section

FIG. 4 -The Spider II Shown Adjacent to the Trestle with the Tower Located Offshore

Healy Tibbitts elected to use cranes working from the existing 700 foot (213 m) long trestle to install sheeting and pipeline through the surf zone. A derrick barge (refer to Figure 3) and hydraulic pipe horse are being utilized for pipe installation offshore of the trestle. Additionally, Healy Tibbitts has used the Spider II (a mobile working platform - refer to Figure 4) to assist with construction work in the surf zone.

The contractor elected to lay pipe in both directions from the offshore end of the trestle, necessitating a special (closure) coupling on the inshore connection to previously installed pipe. Offshore, the pipe horse was used to place two 18-foot (5.49 m) pipe sections at the same time.

Placement of diffuser sections was accomplished in two steps because the horse could not accomodate the 9 foot (2.74 m) riser spools (for the discharge ports) when mounted on the diffuser sections. To overcome this problem, an intermediate set of flanges was fitted at midpoint of the spools roughly corresponding to the top of trench (i.e.; the ocean floor) so that the top half of the spool could be installed separately. The location of the intermediate joint also provided the opportunity to make the riser section extending above the ocean floor a "breakaway" section to limit potential diffuser damage. This was accomplished by securing the flanged joint with polyvinyl chloride (pvc) bolts.

Healy Tibbitts installed 1550 feet (472 m) of outfall pipeline between mid-August 1982 and mid-January 1983. During the Fall of 1982, construction was slowed by the unseasonal frequency of storm activity and the discovery of a mudstone outcropping extending into the pipe trench along the last 700 feet (213 m) of offshore pipe excavation. The contractor used the impact generated by dropping a heavy pointed H-beam (referred to as a chisel by the contractor) to break up the mudstone sufficiently to allow excavation with a clam bucket.

In January of 1983, weather conditions caused the contractor to discontinue operations for the duration of the stormy season. The current schedule calls for startup of operations in mid-April and final completion by the end of August 1983.

Appendix I. - References

1. "City of Half Moon Bay, Project Report for Water Pollution Control Facilities-1972", Barrett & Assoc., Yoder-Trotter-Orlob & Assoc., California, 1972, pp. 3-9.
2. "Preliminary Design Report, Sewer Authority Mid-Coastside Ocean Outfall",Mid-Coastside Area Consultants, California, August 1977, pp. II-3.

3. "Predesign Oceanographic Monitoring in Half Moon Bay, California, October 1975-October 1976", Oceanographic Services Inc., OSI#0480, October 1977.
4. "San Mateo County Mid-Coastside Project Report", Barrett & Assoc., Trotter-Yoder & Assoc., Resources Enginerring & Management, California, July 1975.
5. Scott, D.A., Miller, R.C. and Baird, C.W., "Oceanographic Survey of Half Moon Bay", Oceanographic Services, Inc., OSI#249-2, Santa Barbara, California, Dec. 1971, pp. 7-1.

DESIGN FEATURES AND CONSTRUCTION TECHNIQUES FOR OCEAN OUTFALL SCARBOROUGH, MAINE

Richard A. Jubinville, P.E., M, ASCE*
Robert D. Arsenault, P.E., AM, ASCE**

INTRODUCTION

Construction is proceeding on a 20-inch diameter ocean outfall that will discharge secondary effluent from the Scarborough Sanitary District's 1.8 MGD wastewater treatment facility. The 1,485 foot pipeline has been designed to withstand the effects of the severe northeast storms that batter this exposed shoreline off Prout's Neck on the southern Maine Coast. See Figures 1 and 2. Stringent environmental restrictions as well as the predominance of bedrock throughout the outfall's route, necessitated unique design features and construction techniques.

Regulatory agency restrictions do not allow construction activity during the summer months, when storm activity and wave heights are normally less severe than during the allowable construction period of November through May. Severe storms and the normally rough wave action at the site during the winter months have severly complicated construction efforts, and have resulted in two construction barges being driven onto the rocky shore after having broken loose from their anchor lines.

This paper will describe some of the design features incorporated into the project as well as the construction techniques utilized by the Contractor.

PRELIMINARY DESIGN STUDIES

Preliminary Design Studies were conducted in 1973 and 1980 for the purpose of determining the best location for the marine outfall, and to develop its final design criteria. The initial field studies were conducted in late 1973 - early 1974 by Whitman & Howard, Inc. together with EG&G Environmental Engineering

*Senior Associate, Whitman & Howard, Inc., Wellesley, MA.
**Superintendent, Scarborough Sanitary District, Scarborough, ME.

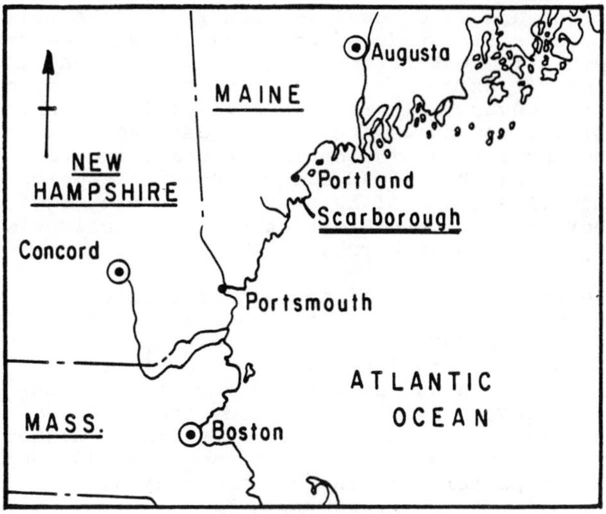

FIGURE. I: SCARBOROUGH, ME.
LOCUS PLAN

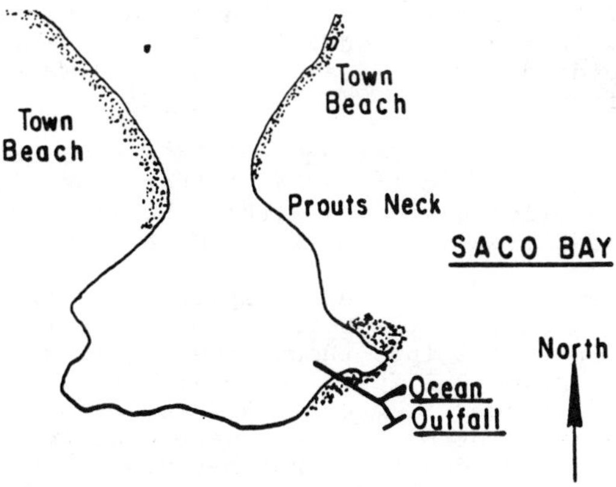

FIGURE 2: OUTFALL LOCATION
OFF PROUTS NECK

Services. This work involved: (a) the collection of ocean current and water quality data in Saco Bay; (2) a seismic-bathymetric study of proposed areas to determine water depth and ocean bottom geological data; and (3) calculation of maximum pollutant levels which must be expected for several alternative outfall locations. The purpose of this study was to collect data to be used in evaluating the alternative outfall routes, and to evaluate factors that would affect the final design of the pipeline. Results of this study were presented to the Scarborough Sanitary District and all applicable regulatory agencies in March, 1974.

In response to significant public concern over the perceived adverse effects of the outfall, a second study was conducted in 1980 by Whitman & Howard, Inc. and Normandeau Associates, Inc. In addition to recalculating the dilution and dispersion factors for the secondary effluent, this second study provided additional data in the areas of: (1) bathymetry; (2) bottom and surface current speed and direction; and (3) surface current drift patterns in the discharge area.

The bathymetric survey produced a contour map of the ocean floor along the proposed outfall route. The survey indicated that there was a 15-20 foot high ridge on the ocean bottom about 750 feet offshore. Any extension of the outfall beyond the ridge would considerably increase the outfall's construction cost without providing a significant increase in water depth over the diffuser.

Seismic studies performed in 1974 and visual observations in 1980 confirmed that there was virtually no sediment overlying bedrock within the study area, only small scattered pockets of gravel and cobbles.

Ocean current monitoring equipment used in both studies indicated that the near-bottom current velocities ranged up to 1.2 fps, though the majority of readings were below 0.1 fps. The near-surface mean current speeds ranged from 0.2 - 0.4 fps. The direction of the prevailing bottom currents were east-west, while surface currents tended to be north-south. The outfall diffuser was oriented in the east-west direction so that; (1) it would follow the natural bottom contours, thereby minimizing construction costs; and (2) onshore currents would be moving at right angles to the line of diffuser ports, providing for maximum dispersion of effluent.

A drogue study was performed in response to the fears of area residents that the effluent would end up on nearby public beaches. Drogue buoy tracking was conducted for both flood and ebb tides at the Southeastern and Southwestern ends of Prout's Neck. Results were difficult to interpret due to the inconsistent patterns of drift of the "window shade" type drogues. However even though the direction of drift was inconsistent, the drift speeds (which ranged from 0.1 - 1.0 fps) provided ample time for significant dilution of the secondary effluent prior to reaching land.

Effluent dilution calculations [1] at various distances from the diffuser indicated that at a depth of 40 feet a diffuser section 360 feet long with 36-3" diameter ports would provide dilution factors allowing the effluent to easily comply with the State of Maine's Class SA Marine Water Classification.

FINAL DESIGN DETAILS

The outfall pipeline consisted of three major sections; (1) a 300 foot land portion of 20-inch, diameter pipe; (2) an 825 foot underwater section of 20-inch diameter pipe; and (3) a 360 foot diffuser section of 16-inch diameter pipe. See Figure 3.

Ductile iron pipe was specified for the 300 foot long "land" portion of the outfall. The remainder of the outfall was specified as ultra-high density polyethylene pipe. This pipe material was attractive for many reasons: (1) it has a very high "C" factor resulting in maximum flow capacities; (2) it is virtually corrosion proof; and (3) after being butt-fused into long sections, it is relatively easy to install underwater.

One of its major disadvantages is its buoyancy, thus anchoring it becomes a major concern. After investigating several types of anchoring systems, it was decided that two piece concrete anchors, held together around the pipe by stainless steel bolts, would provide both adequate anchoring capacity and relative ease of installation. Buoyancy calculations showed that 1700 pound concrete anchors spaced at ten-foot intervals would be adequate provided the pipeline remained full of water. The weight of any backfill material and riprap was neglected as a safety factor.

The entire pipeline was specified to be installed within an excavated trench which averaged about 7 feet deep. This depth of trench placed the top of pipe about 4-foot below the existing ocean bottom. The entire trench was to be filled with crushed stone, and topped with riprap up to the ocean floor.

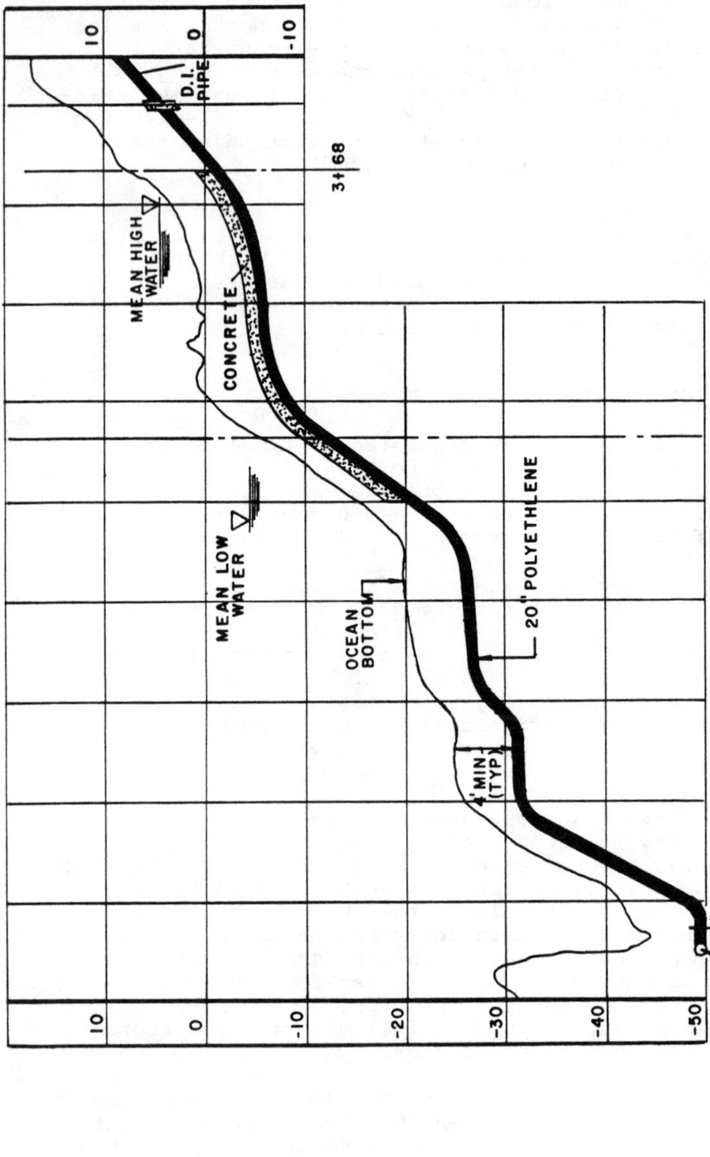

FIGURE 3 PROFILE OF OUTFALL & OCEAN BOTTOM

The outfall is "buried" in a trench for three major reasons: (1) being below the ocean floor, the pipe would not be subject to the lateral forces exerted by the severe wave activity at the project site; (2) the high level of fishing activity in the area (principally for lobster) required that the pipeline not be exposed for fear of becoming entangled in underwater fishing gear; and (3) the extremely sensitive nature of the project from a public relations viewpoint required that an ultra conservative design be adopted, with costs not being the principal restriction. A typical trench cross-section is shown in Figure 4.

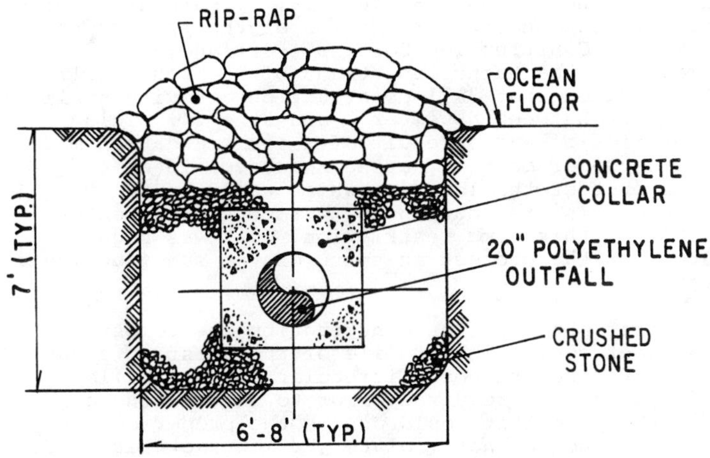

FIGURE 4 : TYPICAL TRENCH SECTION

Concrete Cap

Additional protection measures were taken in the area between mean high and mean low water elevations. A 3 foot thick concrete "cap" was poured over the pipe in this critical area in order to protect the pipe from the forces of the breaking waves. The most critical section was that between stations 6+00 and 7+00, where the steep rock face of the shoreline absorbs the full force of plunging waves at low tide. The cap terminated with an "end block" which completely surrounded the pipeline at station 7+10, where the individual concrete anchors were installed.

Riprap

Riprap was specified to be placed over the entire trench. Conservative estimates of maximum wave heights and maximum lateral bottom forces were used in calculating the minimum sized stone acceptable for riprap. Irregularly shaped rocks of at least 1,000 pounds would be adequate. It was found later that the blasted ledge from the trench provided most of the riprap necessary for the project.

CONSTRUCTION TECHNIQUES

The contract to construct the ocean outfall was awarded in February, 1982 to H. E. Sargent, Inc. of Stillwater, Maine for $881,878.00. H. E. Sargent, Inc. subcontracted the underwater portion of the project to General Marine Construction Corp. of Portland, Maine. The permit from the Maine Department of Environmental Protection stated that "all construction activity shall be carried out between November 1 and June 1". This permit restriction was a result of a recommendation made by the Maine Department of Marine Resources in an attempt to protect the lobsters which would normally migrate into this area during the warmer summer months. As a result of this time restriction, work was begun immediately on the project in spite of the sub-freezing temperatures.

The first phase of the construction was to establish an exact centerline profile of the existing ocean floor. Due to the extremely irregular ledge profile, this information was required prior to determining a grade for the pipe that would assure the proper cover over the pipe and yet not violate the manufacturer's standard for the allowable changes in slopes.

The critical points of the outfall and the two ends of the diffuser were laid out by triangulation from the shoreline. Hard hat divers were used to drill rock anchors at these points. A nylon cord, coded in ten foot stations, was attached between the rock anchors forming the proposed centerline location of the outfall.

After completing the layout of the centerline, the ledge profile was found by taking pneumofathometer readings at ten foot stations. These readings were adjusted for the local tide height at the exact time of the reading thus, producing an elevation relative to mean sea level.

Once the centerline profile of the outfall was established, the drilling and blasting of the ledge was started in the surf area. The force exerted by the breaking waves limited the hours of working time in this area to two hours on each side of low tide. An underwater track rig, controlled by a hard hat diver, was utilized to drill the ledge on the ocean floor. The most difficult section to drill was the area at the low tide interface at station 6+35. The ocean bottom in this area was on a 16% slope, thus limiting the mobility of the drill rig in the area that offered the least amount of allowable working time. Once the drilling was completed on this section, the divers were able to work beyond station 6+35 during high tide with little force being exerted on them from the normal tidal action. Once they reached a depth of about twenty feet, they were no longer influenced by the wave action and tide cycles and were able to work a standard working day.

The offshore portion of the trench was excavated by a crane on a 130 foot barge operating off anchor cables. The winch and anchor system gave the barge the necessary mobility to travel along the proposed outfall line. The excavated rock was stockpiled to the side of the trench to be utilized as armored rock backfill. The pneumo-fathometer was again used to verify the elevation of the trench bottom prior to placing the crushed stone bedding.

The pipe used on the project was a high density polyethylene pipe manufactured by Nipak. The pipe was delivered to the site in sixty foot lengths and butt fused into five larger sections. The sections of pipe were then joined together underwater with flanged connections. The joint between the polyethylene pipe and the ductile iron pipe on shore was also joined together with flanges and encased in concrete.

The first section of polyethylene pipe was four hundred and ten feet in length and extended through the surf zone. Once the trench was excavated in April and May of 1982 the pipe was pulled into position by utilizing the offshore crane in conjunction with a crane onshore. See Figure 5. Crushed stone was placed to the pipe's spring line and concrete anchors were placed to hold the pipe in position during placement of the concrete cap. Concrete was pumped into position to form the three foot thick cap. At the low tide interface, between station 6+00 and 6+50 additional concrete was utilized to embed large boulders to form a seven foot thick armored cap. See Figure 6. During the installation of the concrete cap, a sudden storm hit the

FIG. 5. Installation of 410 ft. section of 20" polyethylene outfall pipe through surf zone -Station 3+00 - 7+10.

coast. The concrete pump truck was sitting within the surf zone on a working level to the side of the trench. The storm arrived so abruptly that workers were unable to move the equipment before the building waves were breaking over the cab of the pump truck. The installation of the first section of polyethylene pipe was completed in May, 1982.

FIG. 6. Concrete pumping of protective cap over trench area.

Construction began on the second section of underwater pipe, which extended from station 7+10 to the angle point at station 10+21 in November, 1982. After the trench had been excavated to grade, the concrete anchors were installed on the pipe and a blind flange installed over both ends to provide a positive buoyancy in the water. The pipe was floated along side of the barge and lowered into the trench by introducing water into the pipe through valves that were installed in the blind flanges. Before this second section of pipe could be backfilled, a storm on December 16, 1982 drove the construction barge and crane onto the rocky shoreline. Subsequent wave action completely destroyed the 130 foot barge and pushed the crane into the ocean. General Marine did not re-establish a second barge on the site until mid March, 1983, when the backfill operation for the second section of pipeline was completed. This 310 foot section of pipeline was left exposed in its excavated trench for approximately three months. There was no damage to the pipe in spite of the severe winter storms that occurred during this period. This was undoubtedly due to two key factors: (1) the pipe was anchored in a 6-7 foot deep rock trench; and (2) the average 20 foot water depth helped protect the pipeline from wave forces, which would have been more significant at shallower depths.

The pipeline out to station 10+21 was completely backfilled and secured by April, 1983. During excavation operations beyond station 10+21 in early April, 1983, severe wave action broke the anchor lines on General Marine's second barge and drove it onto the shoreline. Salvage operations are currently underway to save both the crane and the barge.

Summary

At the time of publication of this paper in May, 1983 construction operations were at a standstill while salvage operations continued on the second wrecked barge. A third barge was being readied for arrival on-site in late May. The Contractor hopes to be able to get permission to work through June. Otherwise construction on the remaining 464 feet of pipe will not resume until November, 1983.

REFERENCES

(1) Calculations developed from the work by Rawn, Bowerman, and Brooks: Rawn, A.M., F.R. Bowerman, and N.H. Brooks, "Diffusers for Disposal of Sewage in Sea Water", Proceedings ASCE, Vol. 86 SA, 1960.

OCEAN OUTFALL DESIGN

TREATED WASTEWATER DISCHARGES

Peter W. Failla,[1] P.E.,M.ASCE
Charles M. Norkis,[2] P.E.
Francis Pandullo,[3] P.E., F.ASCE
Philip F. Healy,[4] P.E.

ABSTRACT

Ocean discharges of treated wastewater effluents into the Atlantic Ocean off Cape May County in New Jersey require an address of a number of environmental, oceanographic and hydraulic parameters. Satisfaction of the multiplicity of considerations imposed unique design criteria to achieve maintenance and protection of aquatic ecological systems and contact recreation resources, stabilization of the pipeline structures under varying bathymetric conditions and imposed wave and current forces, and maintenance of diffusor port velocities within an acceptable range under widely varying flow rates attendant to the substantial influx of tourists during the summer in relation to the permanent population. Alternate designs were utilized to foster competition amongst contractors and pipe manufacturers for the highly specialized work. Design and cost criteria were successfully attained as well as facilities operations.

INTRODUCTION

The development of major regional water pollution control facilities for Cape May County, New Jersey necessitated maintenance of local sewage collection systems, abandonment of antiquated treatment facilities and construction of regional collection, treatment and disposal facilities. Consideration of the unusual environmental amenities in Cape May County and the major

1. Vice-President, Project Manager, PQA Engineering Company
2. Chief Engineer, Cape May County Municipal Utilities Authority, Cape May, New Jersey
3. President, Project Director, PQA Engineering Company
4. Senior Project Engineer, PQA Engineering Company, Riverdale, New Jersey

tourist industry resulted in the adoption of ocean disposal as the most cost-effective solution for returning treated effluent to the environment.

Cape May County's environmental setting includes nearly 25 miles of barrier island beaches separated from the mainland areas by extensive wetland areas and back bay waters. The wetlands feature prime spartina alterniflora (salt-marsh) vegetation; the back bays support shellfish beds which in more recent times have been condemned due to pollution. Pollution has resulted from increased flow rates as increases in population continue to be evidenced, causing reduced efficiencies in existing wastewater treatment facilities which are incapable of meeting effluent quality requirements under current State and Federal standards. Effluents entering the back bay area, and back bay waters entering the ocean and moving along the coast have had high fecal coliform levels sufficient to effect beach closings in some areas.

Protection of the surface waters was mandated to maintain and enhance the tourist and shellfish industries and the long-term food chain productivity which abounds in this marine setting.

BACKGROUND

As a result of the water pollution problems identified in the surface waters in and about Cape May County, the Board of Freeholders of the County formed the Cape May County Municipal Utilities Authority (CMCMUA) as the implementing agency for the County-wide water pollution control program. PQA was retained in March 1973 as engineering consultant. At the same time the Federal Water Pollution Control Act Amendments of 1972 were in the early stages of implementation under the auspices of the United States Environmental Protection Agency (EPA).

PQA in conjunction with and under the direction of the CMCMUA set out to develop a County-wide Facilities Plan in compliance with the provisos of the Construction Grants Program. After extensive studies, interfacing with EPA and representatives of the New Jersey Department of Environmental Protection (DEP) and numerous public meetings and hearings, a plan was adopted designating four (4) regional service areas with separate secondary treatment facilities and the use of ocean outfalls as the most cost-effective means of processing and ultimate disposal.

To facilitate project implementation, including funding, the project schedule was established to provide a staging or phasing of design and construction activities from region to region. The initial work was undertaken for the Ocean City Region which will be the subject of the discussion herein.

DESIGN OBJECTIVES

The basic objective of the design was to disperse the treated effluent into the ocean environment to meet the more stringent criteria of the DEP and EPA at the CW-1/CW-2 interface. Coastal Water (CW) zone separation is established at 1500 feet (457 m) off the beach, or at 15 feet (4.6 m) of water depth, whichever is closest. The CW-1 zone extends from the shore to this limit.

Major design considerations were to address the hydraulic performance of the main barrel and diffuser under a wide range of flows, the structural integrity of the outfall and appurtenances and the dispersion of the effluent to comply with environmental criteria.

Three (3) criterion were the basic parameters addressed in the facilities planning phase to establish the size, location and configuration of the outfall facilities. Fecal coliform counts outside the mixing zone were to be below a most probable number (MPN) of 14 per 100 milliliters (ml). Dispersion of the effluent from the diffuser barrel was to achieve a 50 to 1 dilution in the plume at the water surface during maximum design flow rates. Fecal coliform at the CW-1/CW-2 interface was to be 50 MPN/100 ml.

The discharge criteria were prioritized based upon shellfish and contact recreation requirements for coliform. A cost/benefit analysis was used in lieu of strict adherence to the dilution ratio guideline to establish the length of the main barrel in conjunction with the length of the diffuser barrel. Increasing the diffuser barrel length and reducing the main barrel length resulted in substantial cost savings while maintaining the finally accepted dilution ratio of 45 to 1 and achieving compliance with the higher priority criteria involving bacterial levels.

To maintain structural stability and to avoid changes in littoral dynamics, a basic criterion established for the outfall was to construct the pipeline below the ocean floor along its entire length.

ENVIRONMENTAL CONSIDERATIONS

The ocean outfall is the last component of a regional water pollution control system that returns the waste stream to the environment in compliance with stand ards regarding Surface Water Quality and Shellfish Harvesting. Parameters applicable to the discharge of a secondary treated effluent which were investigated during a comprehensive ocean studies program conducted for this project in 1975 and 1976 include fecal coliform, dissolved oxygen, biostimulants, phytoplankton and relative toxicity. The study work was divided into specific areas of physical oceanography and of biological and chemical oceanography.

Physical oceanographic surveys were conducted to investigate the ocean current regime, mixing characteristics and density stratification of the coastal waters. These parameters influence the design of an ocean outfall.

Ocean current measurements demonstrated the existence of two characteristic conditions of water mass movements in the study area during the critical summer period. One is represented by negligible net movement over a period of several days. This period of stagnation was observed to be followed by a period of net ocean current moving the water mass along the shore approximately in the southwest direction. Both these conditions were reflected in projections of the outfall impacts on water quality.

The environmental effects of the ocean outfall operation on the coastal water quality were evaluated based on the compiled water quality data.

The evaluation indicated that fecal coliform bacteria concentrations due to the discharge of the treated and disinfected effluent are expected to be below all standards applicable to the area. Outside the initial mixing zone, the median concentration of fecal coliforms were projected to be approximately 2.3 MPN/100 ml, which is significantly less than the standard of 14.0 MPN/100 ml prescribed for shellfish harvesting areas by EPA. At the CW-1/CW-2 boundary concentrations less than 1.0 MPN/100 ml are expected 90% of the time. Comparing this with the standard of 50 MPN/100 ml set for the CW-1 primary contact recreation zone by DEP, it was concluded that public health will be more than adequately protected.

In addition to the physical oceanography study, specific biological and chemical oceanography covering fish and macroinvertebrates, icthyoplankton, phytoplankton, zooplankton, benthos, sediment chemistry and water quality were produced. Much of these data profile ambient environmental conditions prior to system operation and form the baseline for comparison of environmental quality modifications due to system operation and other influences that may be inappropriately related to the ocean outfall discharge.

FOUNDATION CONDITIONS

Stability of the pipeline in the ocean was a major factor to consider in the design of the facilities. Long-term variations in ocean floor elevations are critical in establishing a cost-effective profile for the pipeline to mitigate potential erosion which could result in a foundation failure. Also, the characteristics of the sub-bottom soils are critical regarding their potential to undergo liquefaction and to induce structural failure of the pipeline as the load carrying capacity of the soils are substantially reduced.

Two areas of investigation were undertaken to address the foundation conditions. A bathymetric survey was performed to determine existing bottom elevations in conjunction with historical data. A comprehensive soils investigation program was performed to determine the type of soils and their potential to liquefy and their behavior under the superimposed loads from the pipeline.

A historical bathymetric study was undertaken accessing profile data from as far back as 1893. Some data were available during the intervening years. The current bathymetry was obtained during a seismic survey along the route of the pipeline. Variations in ocean bottom elevations were predictably greater in the near-shore area than the off-shore area. Near-shore elevations varied nearly eight (8) feet; off-shore results approximated three (3) feet. The pipeline profile was established using the lowest elevations of the bathymetric envelope as the critical erosion profile.

The soils investigation program was conducted in three (3) phases - seismic, vibracore and boring phases. Seimsmic information was obtained on a broad basis to essentially establish the feasibility of utilizing a particular route and diffuser location to satisfy cost and performance criteria. Vibracore sampling allowed for the evaluation of the potential for liquefaction as well as to establish the soil profile. Soil borings were taken in the near-shore area to establish the state of consolidation and the load carrying capacity of the soils.

Soils mainly consist of compact to medium sands with some dense organic silt material found in the near-shore area that was over consolidated. Sufficient load carrying capacity was the resulting determination; however, the soils were deemed to have a potential to undergo liquefaction under induced stresses in the ocean regime. As a result, select sub-bottom and backfill materials were required to assure that the soils surrounding the pipeline were free-draining and, therefore, not susceptible to liquefaction.

DESIGN CRITERIA

To achieve compliance under current and future conditions attendant to this facility, a comprehensive list of criteria were developed which involved some unique features to facilitate the operational integrity and long-term viability of the components of the ocean outfall system. In addition, a contingency plan was developed to allow for maintenance and repair of the pipeline and appurtenances. This feature was particularly necessary because the outfall leading from the Ocean City Wastewater Treatment Facilities would be pressurized under most operating conditions. With the land based portion of the outfall transversing the barrier island in a west to east direction and the hydraulic grade being substantially above existing elevations on the island, it was necessary to install facilities to control flows to preclude the occurrence of major infrastructure damaged if the pipeline were impaired.

Considering that the cost for subaqueous construction in the Atlantic Ocean would be considerable and that the pipe material was less than 10% of the overall construction cost, the design life of the pipeline was established at 100 years. Similarly, the wave and current forces attendant to a design storm with a recurrence period of 100 years was established for structural and stability requirements. Detailed soils investigations indicated a potential for foundation sands to be subjected to liquefaction. Depth of excavation and the use of select bedding material and select backfill material to achieve the requisite ballast load on the pipeline were parameters addressed to provide the necessary integrity for the outfall pipe and diffuser system under the potential superimposed loads in the ocean regime.

The influx of tourists during the summer months created unusually wide ranges of flows projected to be handled by the regional facilities. A modular approach to the treatment facilities provided the flexibility of operational modes to maximize plant efficiency. Normally, the range of flows would not impact the outfall since the solids transport and scouring considerations in a raw wastewater line do not apply totally to an effluent line. Low velocities can be maintained in an outfall since the flow being carried is essentially devoid of solids. However, velocity variations in the discharge through the diffuser ports are of critical concern. High velocities incur scour or erosion of the ports; low velocities may allow the intrusion of the heavier ocean waters into the diffuser section and main barrel. Ocean water intrusion into the system can impact pumping operations due to flow stratification and system clogging associated with aquatic growth stimulation by wastewater effluent enrichment.

Design populations projected over the life of the project vary from a minimum of 12,500 in the winter under current conditions to over 136,700 as a maximum during the summer at ultimate growth in the service region. Summer to winter population ratios approximate 10 to 1 and are projected to approximate 8 to 1 in the future as more permanent residence inhabitation occurs.

Flows projected for initial operations were an average of 1.34 MGD (5.07 MLD) in the winter and 5.43 MGD (20.55 MLD). Projections fo the year 2000 when ultimate conditions are to occur were 2.50 MGD (9.46 MLD) in the winter and 8.24 MGD (31.19 MLD) in the summer. Using daily flow variation estimates from ASCE Manual No. 37[1] in conjunction with design requirements of DEP, the widest range of flows was established from 0.47 MGD (1.78 MLD) to 20.6 MGD (78.0 MLD). This flow range is too great to accommodate the diffuser port velocity operating range. Accordingly, facilities were incorporated at the treatment plant to store flow during the higher flow periods in a day for release during low flow periods to supplement normal flow rates and maintain a minimum discharge port velocity of 1.0 feet per second (0.30 m/sec) at all times. Maximum velocity through the port was held below 20 fps.

DESIGN FEATURES

The ocean outfall consists of a land section and a marine section of pipeline. Running in a west to east direction, approximately 1,000 feet (305 m) of 36" (91 cm) diameter pipe traverses the barrier island of Ocean City connecting the Ocean City Wastewater Treatment Plant, adjacent to wetlands and back bay areas, to a transition chamber on the east side of the island, just west of a bulkhead along the beach. The marine portion of the outfall extends about 300 feet (91 m) easterly from the transition chamber to the ocean and another 5,730 feet (1,748 m) into the ocean, terminating at a 620 feet (189 m) long, 24 inch (61 cm) diameter, diffuser section.

A contingency plan was developed to protect the land section and the existing infrastructure in the event of a pipe failure. Since the pipeline would be pressurized and the hydraulic grade line was well above existing grade under design conditions, a failure could involve substantial damage. Also, if a failure occurred, how would a repair be expeditiously made? To address this potential problem, a diversion chamber was installed immediately downstream of the effluent pumps at the treatment facilities. In the event of a failure, the plant effluent could be diverted and discharged to the back bay area.

The transition chamber located near the beach was designed to facilitate three (3) design considerations. In the first case, due to the wide variation in ocean bottom elevations in the near-shore area, the outfall pipe extending from the transition chamber was required to be substantially deeper than the pipeline leading across the island from the treatment plant, although both are well below sea level elevations. Secondly, to achieve an expeditious repair of the land portion of the outfall should a failure occur, it is necessary to isolate the pipeline from a backflow of ocean water that would hamper repair activities. Finally, in the event the ocean portion of the outfall required cleaning, a pig launcher would be required. The transition chamber was designed to facilitate the change in vertical alignment of the influent and effluent lines with the vertical differential being utilized to install the pig launcher and isolating valves. A pig retriever fitting was designed at the terminus of the main barrel of the outfall. The pig retriever fitting was installed, terminating in a blind flange on the main barrel slightly above the highest elevation of the ocean floor based on the bathymetric survey data.

A number of conditions needed to be addressed in the design of the diffuser section. The diffuser barrel connects to the main barrel just upstream from the pig retriever fitting. At the end of the 24 inch (61 cm) diameter diffuser a low level outlet was installed to allow purging of the diffuser if sea water gained entrance into the system. Since the diffuser barrel was to be buried, 12 inch (30 cm) diameter riser stacks were installed atop the pipe, 10 feet (3.1 m) on centers with port openings on both sides that would discharge at 28 feet (8.5 m) below sea level. The riser stacks were fitted with flanged connections that allowed for adjusting the elevations of the ports. To protect the diffuser barrel from damage from ocean fishing trawlers, the flange on the riser was designed as the weak link in the system. Termed a "breakaway" flange, this fitting would fail before a substantial load could be transferred to the diffuser barrel to effect a displacement.

The ports were designed with 2½ inch (6.4 cm) openings and 2½ inch (6.4 cm) by 2 inch (5.1 cm) adaptors which were initially installed to accommodate low flows during the early years of the design life. Concern that the 2 inch (5.1 cm)ports could clog resulted in the design flexability to utilize a 2½ inch (6.4 cm) opening in the later years. Each stack has a diffuser port on each side so that the flow direction could be changed by 180 degrees. Spectacle flanges were installed to allow for this change in flow direction.

Hydrodynamic wave analyses were performed to determine the loads on the riser stacks due to ocean current and wave forces. Oscillation frequency imposed on the stack was computed and found to be below its natural frequency. The loads on the riser stacks impart forces that tend to cause rotational displacement at the pipe joints. Joints and pipe cover were designed to withstand the superimposed loads.

Since ocean outfall construction is highly specialized, the design approach was to assure that competitive bids would be received. To accomplish this goal and in recognition of the unique loading conditions and resulting structural requirements, it was deemed necessary to perform two (2) designs. Concrete-coated steel pipe and prestressed concrete cylinder pipe were selected for this application. The steel pipe would be flanged and could resist rotational joint displacment; the concrete pipe had to be designed to include bolts to resist the thrust loads and sufficient bury in select graded stone material to resist pipe rotation at the diffuser barrel.

A buoy system was designed to identify the location of the diffuser in the ocean. Also, different cathodic protection systems were designed for the two (2) alternates.

CONCLUSIONS:

The design of the Ocean City Outfall involved many unique design considerations in meeting the needs of the project. Key elements in the design are summarized as follows:

1. Buried outfall and diffuser
2. Liquefaction control
3. Historical bathymetry for pipe elevations
4. Flow equalization for velocity control
5. Resistance to rotational displacement
6. Contingency plan for flow diversion
7. Main barrel cleaning appurtenances
8. Diffuser barrel purging appurtenance

9. Adjustable diffuser risers
10. Breakaway riser flanges
11. Spectacle flanges on risers
12. Adjustable port diameters

Alternate designs for steel and concrete resulted in very competitive bidding. Five (5) contractors submitted bids with the lowest being $8,837,000 for the concrete alternate. The second low bidder was at $9,098,604 or within 3%. The Engineer's estimate was at $9,267,054. Costs include the land and marine portions of the outfall.

The construction of the project was performed by Spencer White and Prentis, Inc. and met the schedule and cost objectives established. The facilities are presently operating successfully.

An Engineering Excellence Award from the Consulting Engineers Council of New Jersey was received in 1980 for the design of this project.

Appendix: References

1. Manual No. 37, Design and Construction of Sanitary and Storm Sewers, 1969, American Society of Civil Engineers, pp. 32-36.

Kenai, Alaska, Outfall Crosses Tidal Flat

Loren D. Leman, Associate Member,* and
Vince W. Rybel, Member ASCE**

The design of pipelines in tidal flat areas requires special attention if an economical, constructible project is to be achieved. The tidal environment, such as that off Kenai, Alaska, is unusual in that it seldom has sufficient depth for marine equipment nor sufficiently low tides for conventional dry-land techniques. The best design is usually one that enables the contractor to work as rapidly as possible by minimizing any time-consuming work required at the trench on the tidal flat.

The following case record describes how an outfall was economically designed and constructed for the City of Kenai, Alaska, in 1981.

INTRODUCTION

The Kenai, Alaska, secondary sewage treatment plant was expanded in 1980-81 from an average daily flow of 0.5 mgd (1.90 ML/d) to 1.3 mgd (4.93 ML/d). The expansion required a larger, 18-inch- (457-millimeter-) diameter outfall to accommodate the increased flow and to replace the existing damaged one.

Kenai is a city of 5,200 people situated at the mouth of Kenai River on Cook Inlet. The treatment plant is located 1,200 feet (366 meters) north of the Kenai River Cook-Inlet confluence.

The Kenai River has built a considerable delta at its mouth, which extends 3 miles (4.8 km) into Cook Inlet, 2-1/2 miles (4 km) northward, and 5 miles (8 km) southward. The Kenai River channel is 4 feet (1.22 m) deep at its mouth. The delta of the Kenai River is a broad tidal, flat whose general surface elevation is ±1 foot (.305-meter) above mean lower low water (MLLW). The project location is shown in Figure 1.

*Registered professional civil engineer, CH2M HILL, Denali Towers North, 2550 Denali Street, Anchorage, Alaska 99503

**Registered professional civil engineer, CH2M HILL, P.O. Box 428, Corvallis, Oregon 97339

The design of the Kenai outfall was subject to both administrative and environmental constraints. Administrative constraints consisted of directives from the Alaska Department of Environmental Conservation (ADEC) to prevent bacterial contamination of the publicly used beaches at the mouth of the Kenai River and to maintain a minimum discharge depth of 3 feet (.92 m) below the lowest water levels. (The latter directive was later waived.) The environmental constraints were more restrictive. The Kenai region is subjected to an average 20-foot (6.1-m) tide range; the maximum tide range is approximately 32 feet (9.76 meters). An extremely broad tidal plain is exposed twice daily. During the winter and spring, the Kenai River discharges large ice blocks into Cook Inlet; these are often washed ashore nearby. Between spring and fall, fishing boat traffic through the shallow river mouth is particularly heavy. In addition, earthquakes often occur in this part of Alaska.

The design of the Kenai outfall included route selection for the outfall pipeline, pipe material selection, determination of the outfall length, and design of the diffuser within the administrative and environmental constraints.

ROUTE SELECTION

Initially, two routes that appeared to conform to the environmental and administrative criteria were selected. One route followed the shoreline southward from the treatment plant to a point near the middle of the Kenai River channel at its mouth. This route paralleled the existing outfall to the Kenai River. The second route proceeded to the edge of the tidal flat, directly offshore from the treatment plant.

The first route was eventually eliminated from further consideration because of the ice and boat damage that the existing outfall had sustained. Also, dye studies had indicated effluent dispersion would be limited at the river discharge site. By elimination, the route across the tidal flat toward Cook Inlet was selected.

This second alternate is shown on Figure 2.

In order to reach a minimum 3-foot (.92-m) submergence under all flow conditions, a 3-mile- (4.83-km-) long outfall would be required. Field studies were undertaken to determine if the outfall could be shortened. These consisted of drogue studies to determine the potential for bacterial contamination at the shoreline and test pits to determine the nature of the soil materials in the tidal flat.

The design of the Kenai outfall was subject to both administrative and environmental constraints. Administrative constraints consisted of directives from the Alaska Department of Environmental Conservation (ADEC) to prevent bacterial contamination of the publicly used beaches at the mouth of the Kenai River and to maintain a minimum discharge depth of 3 feet (.92 m) below the lowest water levels. (The latter directive was later waived.) The environmental constraints were more restrictive. The Kenai region is subjected to an average 20-foot (6.1-m) tide range; the maximum tide range is approximately 32 feet (9.76 meters). An extremely broad tidal plain is exposed twice daily. During the winter and spring, the Kenai River discharges large ice blocks into Cook Inlet; these are often washed ashore nearby. Between spring and fall, fishing boat traffic through the shallow river mouth is particularly heavy. In addition, earthquakes often occur in this part of Alaska.

The design of the Kenai outfall included route selection for the outfall pipeline, pipe material selection, determination of the outfall length, and design of the diffuser within the administrative and environmental constraints.

ROUTE SELECTION

Initially, two routes that appeared to conform to the environmental and administrative criteria were selected. One route followed the shoreline southward from the treatment plant to a point near the middle of the Kenai River channel at its mouth. This route paralleled the existing outfall to the Kenai River. The second route proceeded to the edge of the tidal flat, directly offshore from the treatment plant.

The first route was eventually eliminated from further consideration because of the ice and boat damage that the existing outfall had sustained. Also, dye studies had indicated effluent dispersion would be limited at the river discharge site. By elimination, the route across the tidal flat toward Cook Inlet was selected.

This second alternate is shown on Figure 2.

FIELD STUDIES

In order to reach a minimum 3-foot (.92-m) submergence under all flow conditions, a 3-mile- (4.83-km-) long outfall would be required. Field studies were undertaken to determine if the outfall could be shortened. These consisted of drogue studies to determine the potential for bacterial contamination at the shoreline and test pits to

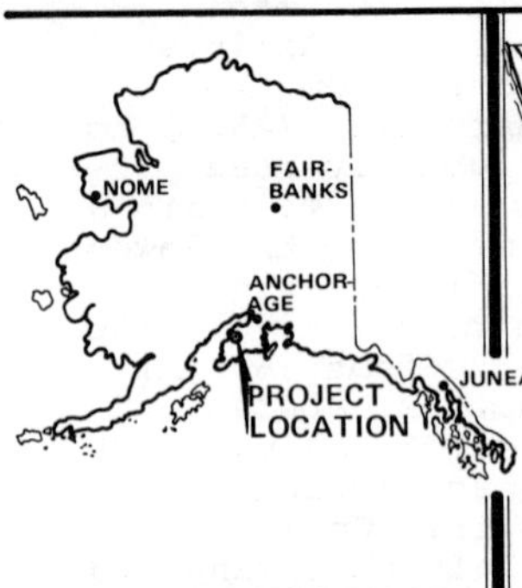

FIGURE 1 – VICINITY MAP

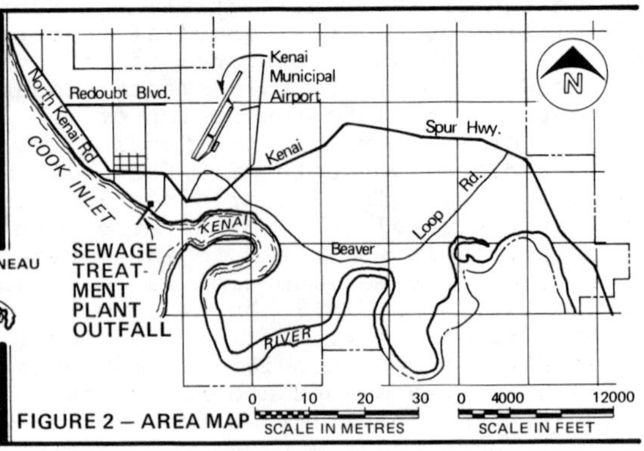

FIGURE 2 – AREA MAP

determine the nature of the soil materials in the tidal flat.

Surface drifter studies helped determine the currents, dispersion, and nearshore circulation patterns at the mouth of the Kenai River and along the tidal flat. The studies showed that the effluent circulation and dispersion would be strongly controlled by tidal currents, with modifications by wind-induced surface currents. The local topography of the Kenai River mouth and the tidal flat would further complicate circulation, causing accretion behind sandbars, sand spits, and tidal flat depressions.

Test pits on the tidal flat seaward of the sewage treatment plant were dug in the summer of 1980. The test pits indicated the beach consisted of clean sand and gravel. The inshore 600 feet (183 m) of tidal flat consisted of soft to firm cohesive sediments that readily supported the test trench side slopes. Offshore of this point, the fine-grained, cohesive soils were displaced by sandier materials, which caved readily when excavated.

DESIGN PROCESS

The cost of extending the outfall offshore until a 3-foot (.92-meter) minimum submergence was attained appeared excessive for such a small community. Accordingly, a variance was sought and obtained which allowed discharge near El. -5.0 feet (-1.53 m), MLLW. This elevation was chosen because it represented the seaward extent of the cohesive (noncaving) subsoils, and would hence simplify construction.

In order to provide an economical, constructible design, the project engineering had to be sensitive to the site constraints. The range in tides over the broad tidal flat were regarded as particularly important: over most of the outfall length, the water would be deep enough to support floating marine equipment only for a few hours a day.

It appeared that the outfall could be extended across the tide flat by using a conventional, dry land "trench and fill" method. With a limited time between low tide levels, any design details that caused even minor delays could have increased construction costs rapidly. The following design features simplified construction during the low tide period:

- Ductile iron pipe was used, mainly because its strength would not require imported bedding material. This would have been difficult to deliver on the soft tidal flat.

- The depth of cover was kept to 5 feet (1.53-meters) to minimize excavation and disturbance to the existing ground. This depth was felt to be sufficient to protect against the ice blocks that drift to the beach.

- The pipe was protected by anodes at 100-foot (30.5-meter) intervals. The anodes were attached before the pipe was brought onto the tidal flat to minimize work in the tidal zone.

- Easily assembled, push-on joints were used to speed the assembly. Each joint was bonded for continuity. The bonding detail allowed most of the work to be done on shore, out of the trench.

- The pipe was given a coal tar epoxy coating in the plant. No coatings or wraps were required at trenchside.

- The diffuser was a simple vertical riser encased in concrete to protect it from ice blocks. A 10-foot- (3.05-meter-) diameter corregated metal pipe was used as a rapidly erected form (Figure 3).

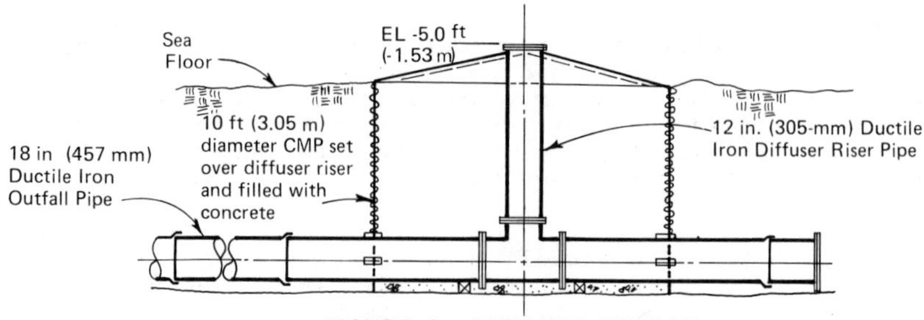

FIGURE 3 – DIFFUSER SECTION

CONSTRUCTION AND OPERATION

The outfall was constructed in the summer of 1981 at a cost of $120,000. The surface soils supported the wide-tracked construction equipment at first. A geotextile-supported access road was eventually needed to reach the diffuser site at low tide.

Most of the design features intended to ease construction proved effective. The outfall has been in successful operation since the fall of 1981. The pipe and diffuser have not been damaged by ice blocks or boat traffic and no contamination of the beach has been reported.

/CVR50/012

Cold Weather Effects on Underground Pipeline Failures

G. Hossein Bahmanyar, M. ASCE(1)
and
Tuncer B. Edil, M. ASCE(2)

Abstract

Buried pipes break more frequently during winter. A study of the influence of cold climate upon buried-pipe failures was undertaken utilizing an analysis of the past water-main break records and meteorological data for Madison, Wisconsin. The analysis was supplemented with the field data obtained from an instrumented section of a water-main in the same locality and the available data from a related laboratory model study of the problem. The analysis indicated that, during the winter months, a larger number of buried pipe breaks occur following a sudden decrease in the air temperature (a cold snap). The field study indicated, as well as did the laboratory study, that a cold snap is accompanied by increased pipe bending stresses in buried pipes. Most failures are circumferential and observed in smaller pipes [smaller than 6 in. (150 mm) in diameter] indicating that the increased bending stresses are responsible for the greater frequency of winter breaks. Changes in the depth and rate of ground freezing and, also, in the availability of groundwater are significant factors influencing the increased bending stresses in pipes even if they are buried at some distance below the frost depth.

Introduction

As the buried cast-iron pipes were increasingly installed in yesteryears, their breaks are becoming more numerous each year(5). It is important to note that the cost of buried pipe breaks has been estimated in excess of $600,000,000 per annum(14). Usually a combination of the causes is responsible, and the final one determines the timing of the break.

During the severe winter of 1976-77, the City of Madison, Wisconsin experienced an unusual increase in the number of water-main breaks. Figure 1 shows that every year a major portion of the water-main breaks occur during the winter months. The unversality of the greater number of breaks during winter is demonstrated by the data from Portland, Maine(10), New York(6), and Czechoslovakia(17). It is

---

(1) Senior Engineer, Milwaukee Testing, Division of Warzyn Engineering Inc., Milwaukee, WI 53227.
(2) Professor, Department of Civil and Environmental Engineering, University of Wisconsin, Madison, WI 53706.

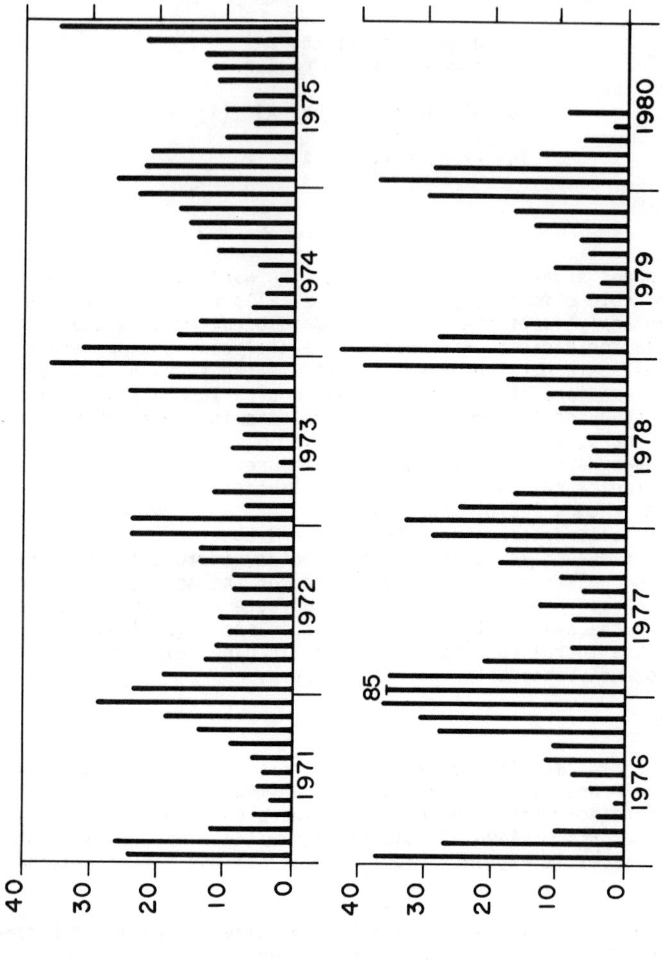

Figure 1: Histogram of the monthly water-main break (Madison, Wisconsin, 1970-1980)

recognizable that in the arctic and subarctic regions, freezing water in the mains could be a major contributor of the winter water-main breaks. But in the seasonally cold regions, the freezing of water-mains is not the likely cause of winter breaks since pipes are generally buried below the frost penetration depth. It can be inferred that the seasonal change in the climatic factors is the probable influential cause of the increased pipe breaks during winter. The main objective of the present paper is to report on an investigation of the relationship of the climatic changes and the water-main breaks.

Analysis of Pipe Breaks

The City of Madison, Wisconsin experienced one of the severest winters during 1976-77 as indicated by the freezing indexes given in Table 1 for a ten-year period. The below-freezing mean-air temperature during the freezing period for the 1976-77 winter is almost twice that for the normal period. It is noted that the largest number of water-main breaks occurred during 1976-77. Furthermore, a general semilogarithmic correlation is apparent between the below-freezing mean-air temperature and the total number of breaks in a given year, as shown in Figure 2. From the daily and monthly water-main break reports of Madison, Wisconsin and the monthly summary of the local climatological data, the following analyses are made.

Temperature-Break Correlation

The histogram of the monthly water-main breaks for the past ten years is shown in Figure 1. The peaks are positioned about January, which is the coldest month for the city. The annual summary of the local climatological data(11) states that the soil in Madison, Wisconsin is usually frozen from the first of December through most of March, with an average frost penetration of 26 to 30 in. (0.65 to 0.75 m). In the State of Wisconsin, it is the general practice to bury water-mains at a depth greater than 6.0 ft (1.8 m). Furthermore, the frost penetration is impeded by the heat dissipated from the water-main. Therefore, under the above-mentioned conditions, the increased failure of the water-mains during the winter months is not likely to be due to the freezing of the surrounding soils and subsequent freezing of the water in the pipes.

It is well established that the surface temperature is the most influential climatic factor of the heat conduction into and out of the ground(7). The surface temperature can be estimated from the air temperature, and there is an established relationship between the pertinent meteorological factors and surface temperature(12). To eliminate the effect of the long-term annual cycle on the short-term daily cycles, one must work only with the departure from the normal air temperature(13).

Based on these considerations, the daily water-main break data and the departure from normal air temperature for December, 1976 and January, 1977 (the severest period) were compared. Figure 3 shows the results which indicate that the number of water-main breaks increased a few days after a decrease in the air temperature was sustained for several days. Analysis of extensive data covering the

Table 1

Water-Main Breaks and Freezing Indexes
Madison, Wisconsin
1970-71 to 1979-80

Year (7/1 - 6/30)	Freezing Index °C Days	Below Freezing Mean Air Temp[1] °C	Total Number of Water-Main Breaks
1970-71	827	8.1	153
1971-72	771	6.9	168
1972-73	538	5.0	138
1973-74	611	7.2	175
1974-75	547	4.8	180
1975-76	500	6.7	183
1976-77	1040	11.6	287
1977-78	1049	9.5	188
1978-79	1023	8.3	194
1979-80	666	6.1	179
Average[2]	671	5.7	185[3]

1 Freezing Index Divided by Number of Days in Freezing Period.

2 From Normal Temperature Data for 1941-1970.

3 Average of this Column.

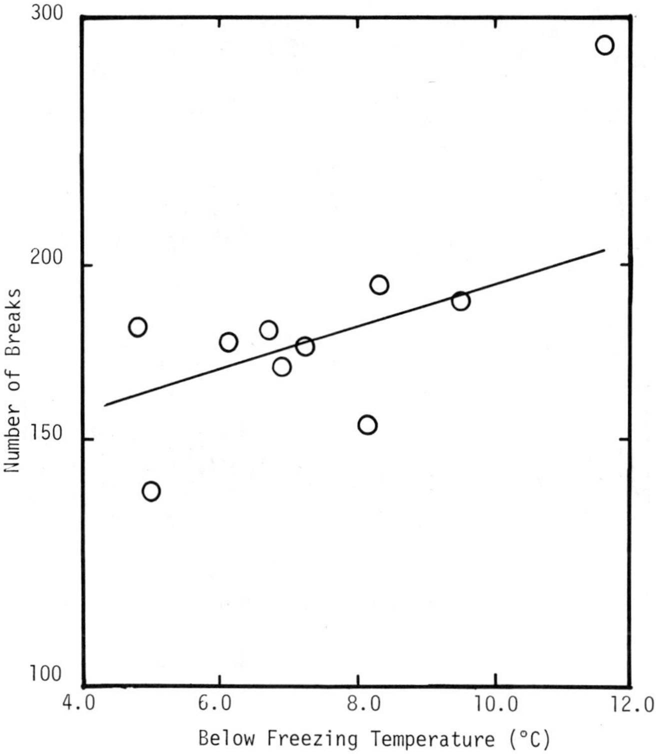

Figure 2: Number of water main breaks vs. below-freezing mean air temperature.

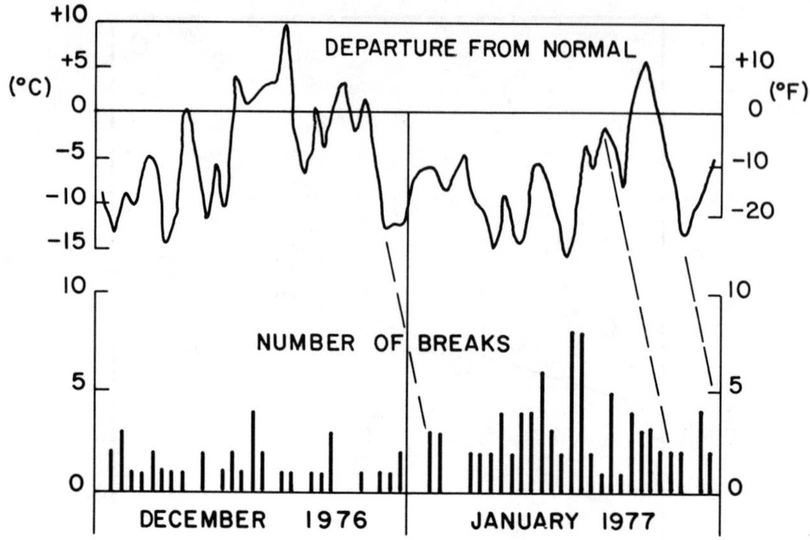

Figure 3: Water-main breaks and air temperature departure from normal, Winter 1976-77.

period 1969-1978 in a similar manner confirmed this basic observation(1). It should be mentioned that the water-main break reports indicated only the date of water appearance at the surface. The water from a fractured water-main may take more than a few days to seep through the pores, soil layers, frozen ground, and pavement to appear at the street surface. It also takes a while to notice a water-main leakage on the street surface. Nevertheless, a semi-quantitative relationship is observed between a period of sustained cold temperature (termed "cold snap") and an increase in the number of water-main breaks.

Pipe Size and Frequency of Breaks

Table 2 indicates that the frequency of circumferential breaks in pipes with a diameter of 6 in. (150 mm) and smaller is considerably higher than that in the larger diameter pipes. A circumferential break is caused by either a beam or shear loading. A beam loading occurs when nonuniform support, soil movement or surcharge exists. Figure 4 shows that the flexural modulus of pipes decreases sharply for pipes having a diameter smaller than 6 in. (150 mm). Thus, under similar loading conditions, the smaller pipes undergo larger bending stress. Also, the cross-sectional area of the pipes is proportional to their sizes; therefore, the smaller pipes are less resistant to the shear forces. Furthermore, the smaller pipes carry lesser amounts of water and, thus, being a smaller heat source, permit a deeper frost penetration in the soil above during freezing months. Deeper frost penetration results in higher stresses (as will be shown later) both in bending and in shear. Thus, the smaller pipes have the tendency to break more frequency, as is confirmed by the existing data.

The analysis of the water-main break data presented above demonstrates the relationship between air temperature changes in winter months and the breaks. Preponderance of circumferential failures and failures in the smaller pipes imply an increase in loading of the pipes by surrounding soil media during winter months.

Field Investigation

An opportunity for a field instrumentation and verification of the observations outlined in the previous sections became available. The Madison Water Utility provided access to a 4 in. (100 mm) cast-iron water-main at the University of Wisconsin-Madison campus. A 3- by 5-ft (0.9- by 1.5-m) pit was carefully excavated, minimizing any disturbance to the water-main. The field instrumentation included strain gages mounted on the water-main and thermistors positioned at various depths in the soil as shown in Figure 5.

During the period February to May, 1979, a set of data was obtained from the instrumented water-main section. Figure 6 shows the variation of temperatures and strains measured during this period. Only at the points under the pavement the soil temperature showed a damped-amplitude and lagged-phase relationship with the air temperature. The relationship observed can be described adequately by the heat transfer theory. The temperature of the soil surrounding the pipe never dropped below the freezing point, although the soil

Table 2

Water-Main Break Frequency
Number of Breaks per 100 km*/year
Madison, Wisconsin
Eleven-year Period 1970-1980

Mode of Break	38 mm (1.5 in)	50 mm (2 in)	75 mm (3 in)	100 mm (4 in)	150 mm (6 in)	200 mm (8 in)	250 mm (10 in)	300 mm (12 in)	400 mm (16 in)	500 mm (20 in)	750 mm (30 in)	Total†
Circumfer-ential			84.9	10.2	22.6	6.5	4.2	0.2				15.3
Longitu-dinal				1.3	0.8	1.3	0.8	0.5	0.6			0.9
Circum. & Longit.				0.2	0.3	0.3	0.4	0.5				0.3
Blowout		4.0		4.0	1.4	3.2	1.0	2.1				1.8
Joint	33.1		7.7	1.9	1.7	2.7	1.6	2.3		1.5	18.0	1.9
Split at Corporation				0.5	0.5	0.3	0.3					0.4
Sleeve					0.1							0.03
Others				0.2	0.3	0.4	0.4	0.2				0.3
Total	33.1	4.0	92.7	18.3	27.7	14.6	8.7	5.7	0.6	1.5	18.0	
												20.9

* Length of the mains in service by the end of 1975.

† Including all sizes.

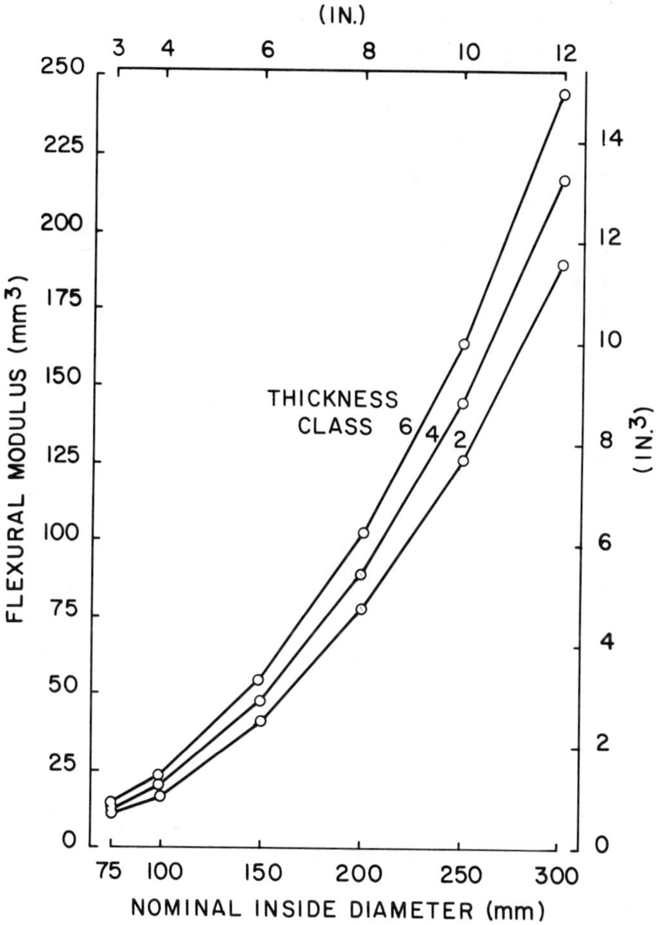

Figure 4: Flexural moduli (I/c) of cast-iron pipes.

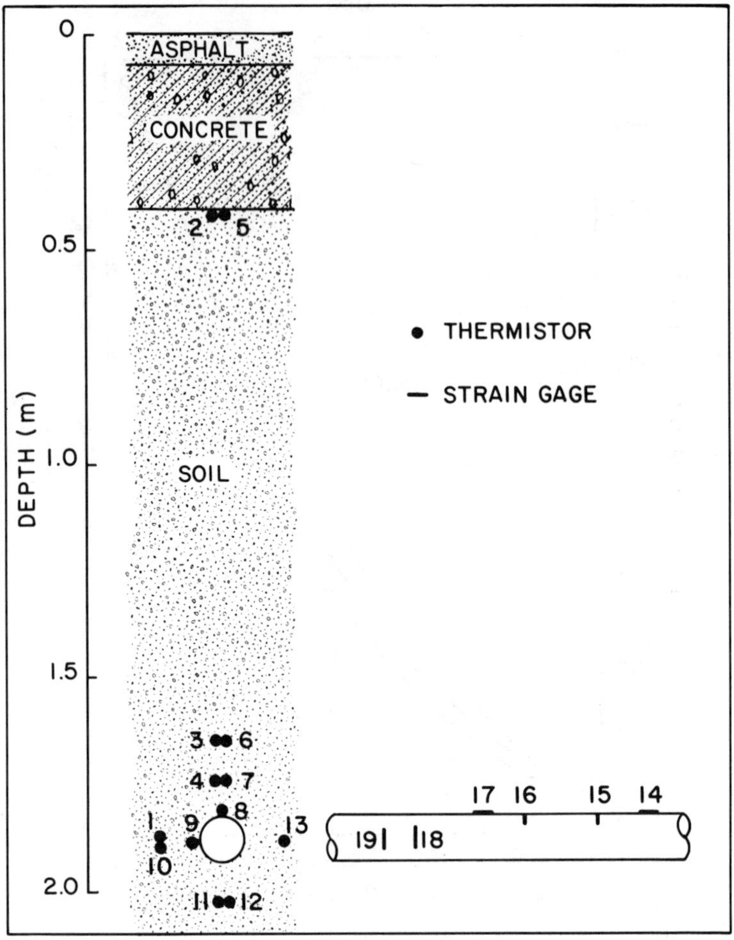

Figure 5: Position of strain gages and thermistors in the field.

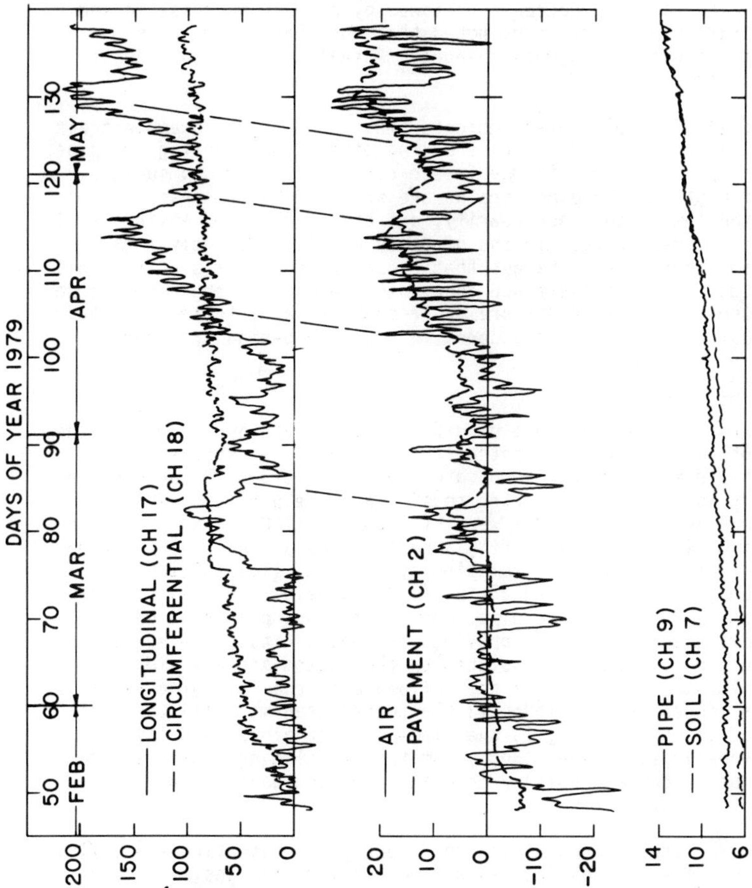

Figure 6: Pipe strains and air, soil, and pipe temperatures from field instrumentation.

layer close to the ground surface was frozen, reflecting the dominant influence of the temperature of the water in the pipe. The strain gage outputs, particularly the longitudinal strain, show noticeable build-up after a sharp decrease followed by a sharp increase in the air temperature. It should be noted that the measured strains are only indicative of the stress changes. Actual stresses were unobtainable from the recorded data.

The field data presented indicate that the changes in ambient air temperature may not lead to the freezing of soil surrounding the pipe; however, a part of the soil may freeze close to the ground surface. It is evident that the pipe strains are affected by temperature (not of the pipe temperature necessarily, but of the soil and environment). Therefore, stress changes in the pipe may result indirectly from significant temperature changes that occur in other parts of the soil medium such as the near-surface zone. In present day practice, these temperature-related loadings are not recognized and, therefore, they are not taken directly into account in the pipe design procedure.

Laboratory Study

An experimental study was undertaken to isolate and determine the effects of certain factors controlling the increased stresses in water-mains due to the air temperature variations(3). A large-scale laboratory model was constructed to simulate the problem. The principal components of the model were: (i) 36-ft^3 (1.0-m^3) soil container filled with silty fine sand, (ii) a heat source at the soil container base, (iii) a groundwater simulation system, and (iv) a 2.125-in. (54-mm) copper tubing buried 28 in. (0.72 m) below the soil surface. Strain gages mounted on the pipe, and thermistors placed at numerous places in the soil container provided the basic data. The model was placed in a cold room and insulated on the bottom and all sides. During this experimental work the influence of the following variables on pipe stresses was studied: (i) the thermal regime, (ii) the availability of groundwater, (iii) the pipe-end restraints, and (iv) the rate of freezing(2). The other parameters (including those related to the pipe, water and soil) were maintained constant during the experiments.

The model study, in addition to directly demonstrating the influence of soil freezing on increased pipe bending stresses, indicated that the maximum frost depth is the most significant thermal parameter that correlates with the bending stresses. A similar relationship between the frost penetration and the load on a buried pipe has been observed by other investigators(16). The model experiments also indicated that the rapid rate of freezing and the availability of groundwater significantly influences the increase of the pipe bending stresses while the pipe end restraints slightly affect (decrease) them during the freezing period.

The experimental results can be interpretted in terms of the water expulsion from the freezing front, as a result of a rapid soil freezing. With impeded drainage, this reduces the effective stresses in the soil and, in turn, reduces soil stiffness resulting in increased loading on the buried pipe. Fundamental basis for this type of behavior has been substantially developed in literature (8, 9 and 18).

Summary and Conclusions

Based on the analysis of the pipe break data and the air temperature records for a city located in a seasonally cold region and the field data from a buried pipe section in the same city presented here and supplemented by the available results of a laboratory experiment modeling the buried pipe problem, the following observations and conclusions are advanced:

1. There is a great frequency of buried pipe breaks during the winter months.

2. In seasonally cold regions, the number of breaks increases as a sudden decrease in the air temperature (a cold snap) is sustained.

3. A cold snap produces increased bending stresses in pipes although they are buried at some distance below the maximum frost depth.

4. Most pipe failures recorded are circumferential, probably due to beam bending.

5. Smaller pipes [smaller than 6 in. (150 mm) in diameter] are more susceptible to failure due to their relative weakness in beam bending.

6. The maximum frost depth is the most significant thermal parameter that correlates with the bending stresses.

7. Rapid rate of freezing and availability of groundwater influence pipe bending stresses significantly.

Acknowledgements

The authors wish to thank Madison Water Utility Department, its past director, Mr. L.W. Russell, and its staff, particularly Messers J.R. Blatz and D.M. Cawley for their assistance during the research program. Financial assistance was provided by the University-Industry Research Program of the University of Wisconsin-Madison, and the Madison Water Utility Department. Most of the drawings were prepared by Mr. J. Bejsovec and the manuscript was typed by Ms. D.K. Pintar.

References

1. Bahmanyar, G.H., "A Study of Buried Pipe Failures in Cold Climates," thesis presented to the University of Wisconsin at Madison, Wisconsin, in 1981, in partial fulfillment of the requirement for the Degree of Doctor of Philosophy.

2. Bahmanyar, G.H., "Measurement of Frost Loads on a Conduit Buried in Sand," Proceedings of 29th International Instrumentation Symposium, Instrument Society of America, 1983.

3. Carson, E., "Soil Temperature and Weather Conditions," ANL-6470, Argonne National Laboratory, Argonne, IL, 1961.

4. Edil. T.B. and Bahmanyar, G.H., "Stresses in Buried Conduits Caused by Freezing Front," Cold Regions Science and Technology, 1983, in press.

5. Hudson, W.D., "Reduction of Unaccounted-for Water," Journal of American Water Works Association, Vol. 56, 1964, p. 212.

6. Huguet, G.A. and O'Brian, C.V., "New York City Water Supply Infrastructure Study," Unpublished Technical Memorandum, U.S. Army Corps of Engineers, New York, NY, 1980.

7. Janson, L.E., "Water Supply Systems in Frozen Ground," Proceedings of First International Conference on Permafrost, NRC Publ. 1287, National Academy of Science, Washington, D.C., 1963, pp. 430-433.

8. Lachenbruch, A.H., "Some Estimates of the Thermal Effects of a Heated Pipeline in Permafrost," Circular No. 632, U.S. Geological Survey, 1970.

9. McRoberts, E.W. and Morgensterm, N.R., "Pore Water Expulsion During Freezing," Canadian Geotechnical Journal, Vol. 12, 1975, pp. 130-141.

10. Monie, W.D. and Clark, C.M., "Loads on Underground Pipe Due to Frost Penetration," Journal of American Water Works Association, Vol. 66, 1974, pp. 353-358.

11. National Oceanic and Atmospheric Administration, "Local Climatalogical Data, Monthly Summary, Madison, WI," U.S. Department of Commerce, Asheville, NC, 1982.

12. Noss, P.M., "The Relationship Between Meteorological Factors and Pavement Temperature," Proceedings of the Symposium on Frost Action on Roads, held at the Norwegian Road Research Laboratory in Oslo, Oct. 1-3, 1973, by Organization for Economic Cooperation and Development, Paris, France, 1973, pp. 21-25.

13. Panofsky, H.A. and Brier, G.W., Some Application of Statistics in Meteorology, Pennsylvania State University, University Park, PA, 1968.

14. Reedy, D.R., "Corrosion in the Water Works Industry," Materials Protection, Vol. 5, 1968, pp. 55-59.

15. Skaven-Haug, Sv., "Control of Frost Penetration in Norway," Proceedings of First International Conference on Permafrost, NRC Publ. 1287, National Academy of Sciences, Washington, D.C., 1963, pp. 268-272.

16. Smith, W.H., "Frost Loading on Underground Pipe," Journal of American Water Works Association, Vol. 68, 1976, pp. 673-674.

17. Sticha, Vaclav, "Optimum Depth for Laying Water Mains with Regard to Freezing," (Optimalini Hloubky Ukladany Vodovdniho Portrubi se Zretelem na Zamrzani), Vjzkumny Ustav Vodohospodarsk, Prague, Prace A Studie, No. 123, 1968, pp. 1-105 (in Czeck with English Summary).

18. Williams. P.J., "Pore Pressures at a Penetrating Frost Line and Their Prediction," Geotechnique, Vol. 16, 1966, pp. 187-208.

BOSTON MAIN DRAINAGE SYSTEM 100 YEARS LATER

Antoni J. Zelechowski, P.E., Member, ASCE*
and
Cathy C. Greene, Associate Member, ASCE*

Major portions of Boston's 100-year-old Main Drainage System must be replaced to address capacity problems and structural deterioration. Design which was already difficult due to large pipe sizes, poor soil conditions, and invert constraints became much more so in the crowded urban setting of Boston. In order to accommodate major transportation projects and utility interferences, experience, skill and flexibility have been required of the Boston Water and Sewer Commission's staff.

****

Boston's wastewater collection system is tributary to two regional treatment plants in three drainage systems. The oldest of these, the Boston Main Drainage System (BMDS), was constructed between 1877 and 1884 in order to abate a serious rise in the City's death rate.[3] As illustrated in Figure 1, the BMDS consists of a main interceptor, five branch interceptors, a pumping station, tunnel and ocean outfall. The interceptors receive dry weather wastewater flows and a small portion of storm flows. The majority of wet weather flows are diverted to combined sewer overflows which discharge to the nearest open body of water.

Some modifications to the system have been made during the past 100 years in order to divert flows for treatment at the Metropolitan District Commission's (MDC's) Nut and Deer Island primary plants, as Table 1 indicates. Despite these changes, the original interceptors and the Calf Pasture Pumping Station are still in use, with the latter now functioning as a storm and emergency relief.

Studies of the Main Drainage System

The first primary treatment plant for the Boston Area was built at Nut Island in 1951, followed by the Deer Island Plant in 1968. Although construction of these facilities eliminated the major points where raw sewage discharged into Boston Harbor, combined sewer overflows (CSO's) were still a significant source of pollution. The City of Boston received a Housing and Urban Development loan in 1965 to study the condition of the BMDS and recommend solutions to the CSO

*Project Manager and Engineer Administrator,
Boston Water and Sewer Commission
10 Post Office Square, Boston, MA 02109

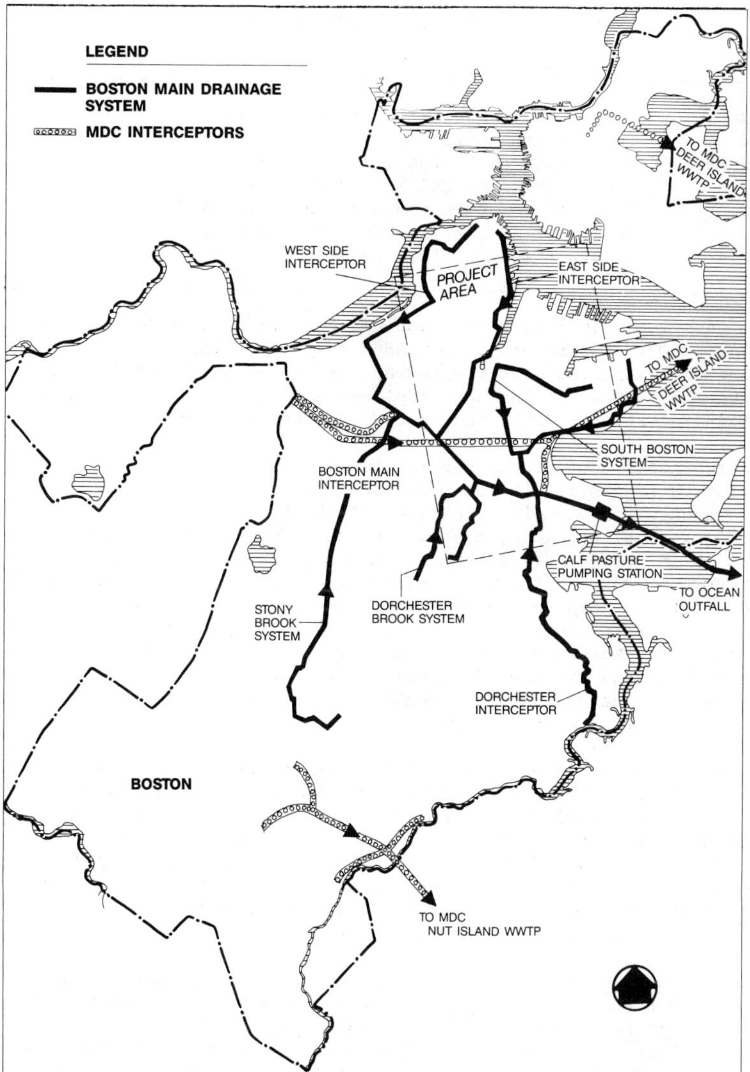

Figure 1

EXISTING BOSTON MAIN DRAINAGE SYSTEM AND MAJOR CONNECTION TO DISTRICT TREATMENT PLANTS

Table 1
DEVELOPMENT OF BOSTON's SEWER SYSTEM

Date	Activity
1833	City allows sewage discharge to drains
1884	BMDS begins operation
1895 to 1904	BMDS relieved by Nut Island Discharge
1951	Construction of MDC Nut Island WWTP
1968	Diversion of BMDS flows to Deer Island WWTP
1980	Design of new interceptors for BMDS begins
1982	Construction of first contract begins

problem. Completed in 1967 by the consulting firm of Camp Dresser and McKee, Inc. (CDM), the report recommended deep rock storage tunnels for collecting and disposing of CSO's and an initial construction program to improve the BMDS which included replacement of major sections of the Boston Main Interceptor (BMI) and East Side Interceptor (ESI).[1]

The tremendous cost of the storage tunnels stalled progress towards a CSO solution. However, it was clear that certain improvements to the BMDS were needed immediately. Portions of the BMI had collapsed in the early 1960's, requiring long, costly repairs. The ESI was also suspected to be in poor condition and was known to have insufficient capacity for the current flows, resulting in continuous CSO discharges to Boston Harbor. Despite these obvious problems and the possibility of federal and state grants of up to 80 percent (50 percent federal), The Boston Public Works Department was unable to obtain an appropriation to replace the BMI and ESI until 1971. Shortly thereafter, rumors that bills before Congress would increase federal grants to 75 percent resulted in further delays.

When adopted, the Federal Water Pollution Control Act of 1972 did increase the federal share. However, the regulations resulting from the Act required a supplement to the 1967 report. Prepared for the Public Works Department by CDM in 1975, the supplement recommended construction of a 36 to 42 in. (.9 to 1.1 m) diameter sewer to reverse the flow direction at the eastern end of the BMI, a new 96 in. (2.4 m) Boston Main Interceptor, and a new East Side Interceptor along the alignments shown in Figure 2, "Original Route."[2]

While the Supplement to the 1967 Report was being prepared, the MDC was completing a Wastewater Engineering and Management Plan for Boston Harbor - Eastern Massachusetts Metropolitan Area, the EMMA study. Published in 1975, the study recommended nine CSO treatment facilities in Boston, fed by collection conduits.[5] One of these conduits was proposed to run along the western edge of Fort Point Channel, the route also recommended for a portion of the New East Side Interceptor.

In order to receive state and federal grants, the City was required in September 1976 to incorporate the design of the CSO conduit into the interceptor replacement project. Having agreed to this condition, it seemed, for a short time, as if design might finally begin.

A New Agency Takes Over

It became evident in the 1970's that the City's sewers and storm drains were not being adequately maintained or upgraded because the Boston Public Works Department did not have sufficient funds. In an effort to correct this problem, the Boston Water and Sewer Commission (BWSC) was established on July 18, 1977. The Commission owns and operates Boston's water and sewer system and is financed through water and sewer rates and bond revenues. Although Commissioners are nominated by the Mayor and approved by the City Council, the BWSC is not a part of city government.

The Commission absorbed many of the Public Works Department's employees, but needed to hire complete administrative and financial staffs, fill its upper level engineering positions and establish a financial status. Creation of a strong engineering department was necessary before BWSC could take responsibility for administering the interceptor design and construction project.

Criteria Used for Design

In December 1979 the BWSC received $2 million in state and federal grants and, using the services of CDM, began design of the new sewers recommended in the 1975 Supplement Report. The interceptors were designed to carry peak dry weather flows, as calculated for the year 2030, plus an allowance for infiltration. An exception to this was the portion of the New East Side Interceptor above the proposed MDC CSO facility which also included an allowance for the one-year six-hour storm wet weather flows. Complicating the design was the need to provide for flow conditions from the South End area of the City - before and after major sewer and drainage construction projects were completed.

The new interceptor project had to receive connections from existing pipes not only at the upstream end but also from a number of pipes along its route. The downstream elevation was controlled by the invert of the bar rack chamber at the MDC headworks. The above constraints plus the need to cross under two subway tunnels with inverted siphons restricted the hydraulic grade line.

The soil and groundwater conditions along the interceptor route significantly affected the design and restricted construction methods. Because the entire project runs on the water side of the original 1630 Boston coastline, which was filled over many years (1806 to 1967), the fill material varied considerably from loose cinders and ash to sands, clays and gravel. In addition, obstructions such as granite walls, wood piles and sea walls were expected.

The general soil profile in the project area consists of 5 to 35 ft. (1.5 to 10.7 m) of fill material underlain with 0 to 30 ft. (9.1m) of organic soils. Below this lies 28 to 107 ft. (8.5 to 32.6 m) of marine deposits, 3 to 8 ft. (.9 to 2.4 m) of compact glacial till and finally a fractured siltstone bedrock.[4] As these ranges indicate, changes in soil conditions could occur over short distances.

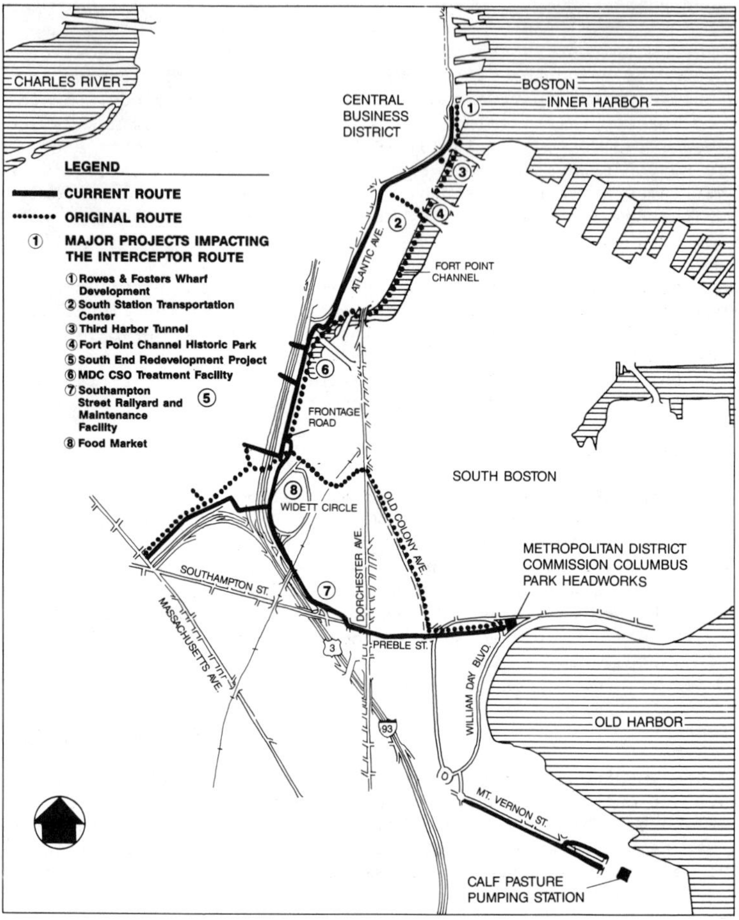

Figure 2

ROUTES OF NEW INTERCEPTORS

Most of the interceptor route would pass through one or more such layers. Where the pipe would be underlain by organic soils at depths too great to be over-excavated, pile support was called for in design, to be driven an average of 90 feet (27 m) to clay or bedrock.

The pipe invert would be below groundwater levels throughout its length, typically by as much as 20 feet (6 m). Therefore dewatering, predrainage, and lateral support requirements became critical parts of the design and, as seen in the three contracts bid to date, generated great concern from the bidding contractors.

Changes in Pipe Routes

When design began, it was assumed that the interceptor alignments would vary only slightly from those in the 1975 Report. Four major contracts were recommended in this report.

The first was for a new sewer at the eastern end of the BMI, called the Mount Vernon Street Sewer, which would direct sanitary wastewater flows from the surrounding area to the MDC Deer Island Treatment Plant. The old BMI would remain as the local storm drain and as a relief sewer when the Calf Pasture Pumping Station was operated.

The replacement for the section of ESI serving the South End would be the second contract. The New East Side Interceptor - South Branch (NESI-SB) would intercept the BMI at Massachusetts Avenue and run north-easterly between an expressway ramp and a 204 x 120 in. (5.2 x 3.0 m) combined sewer overflow conduit. (See Figure 2, "Original Route".) After jacking or tunneling under the expressway, the South Branch would join the New Boston Main Interceptor (NBMI).

The third contract would replace the upper end of the ESI. The NESI-North Branch would begin in downtown Boston at Atlantic Avenue, cross to the waterfront at Rowes and Fosters Wharves and run along the west bank of Fort Point Channel to the proposed MDC CSO treatment facility. Up to this point, the NESI-NB would consist of twin 84 in. (2.1 m) reinforced concrete pipes carrying dry and wet weather flows. Wet weather flows would be diverted to the CSO facility, while dry weather flows and tailings from the treatment process continued south in a single 72 in. (1.8 m) pipe to the NBMI.

The NBMI's (the fourth contract) role was to pick up flows from the North and South Branches. It was originally proposed to cross railroad property going east to the intersection of Dorchester Avenue and Old Colony Boulevard. Traveling down Old Colony Boulevard, it would reach Columbus Park and cross the park to the MDC Tunnel Headworks leading to Deer Island.

These alignments had been layed out five years earlier, as the shortest and least difficult alternatives. However, completion of a list of proposed or ongoing projects by other agencies or developers made it clear that design along the crosscountry and channel routes would not be simple. The railroad property and Fort Point Channel were proposed for revitalization with construction being planned by the

Federal Railroad Administration (FRA), Massachusetts Bay Transportation Authority (MBTA), Massachusetts Turnpike Authority (MTA), and Massachusetts Department of Public Works (MDPW), as well as private developers.

Two major pipe route changes have been made to avoid conflicts with above-ground projects (Figure 2). The first was to move the NBMI south, avoiding conflict with the FRA's Northeast Corridor Improvements Project. In the original alignment, the NBMI would cross under a major switching track on its way to Old Colony Boulevard. After discussing alternatives with the FRA, the BWSC agreed, in October 1980, to move the NBMI route down to Southampton Street, avoiding the track work.

This was a fairly easy concession, since BWSC staff had also questioned the route selected for the NBMI. Old Colony Boulevard was the shortest route to the MDC structure. However, it is a narrow commercial street surrounded by residential neighborhoods and a major traffic artery through South Boston. By moving the interceptor to Southampton Street, a wider road in a warehouse area, disruption to businesses and rerouting of traffic could be avoided.

The change in alignment of the NESI-NB took much more time and energy to finalize. Fort Point Channel had been selected in the 1967 and 1975 reports primarily to avoid Atlantic Avenue and its associated traffic and utility problems. But as design proceeded, the Channel also presented problems. In opposition were industries which use channel water for cooling, the Boston Aquarium, and abutters with marinas for whom the interceptors would restrict boat access. The MDPW had considered the same side of the Fort Point Channel for a third harbor tunnel for several years, but it was not until mid-1981 that this project began to look like a serious possibility. Since the interceptor and tunnel were both restricted by conflicting elevation, neither had much latitude for compromise. Coordination of the two projects was not feasible due to opposite slopes and different design and construction schedules.

Also located in the channel was an early 20th century non-reinforced concrete subway tunnel. Passing over this structure with the North Branch caused great concern. Piles would need to be driven on both sides of the tunnel and the interceptor bridged over.

In addition, the NESI-NB would pass beneath five low bridges with an average clearance of 15 feet (4.6 m). Since these were also in areas requiring pile support, dismantling of sections of the bridge decks might be necessary.

In desperation, BWSC halted design on the NESI-NB and asked their consultants to reconsider alternate routes. Three were proposed, but each had problems.

A single shallow tube tunnel under Atlantic Avenue was first evaluated. However, it was considered complicated because it would pass through a mixed face consisting of fill, organics and glacial till with only 10-15 ft. (3-4.6 m) of cover and under a busy traffic route.

A deep tunnel could avoid these dangers, but required pumping of dry weather as well as wet weather flows at the upper end of Fort Point Channel and created more difficult connections along the route.

The third alternative, by open cut in Atlantic Avenue using either a single 102 to 120 in. (2.6-3.0 m) or two 72 in. (2.1 m) diameter pipes was considered, but due to lack of space was not feasible. Atlantic Avenue is a major access road between the expressway and downtown Boston, and as a result of having been relocated and narrowed is underlain by a maze of utilities.

While BWSC was evaluating possible options for the NESI-NB, the federal government reduced funding for CSO projects, and as a result the MDC placed the Fort Point Channel treatment facility in the distant future. Without this facility, the proposed NESI-NB would consolidate CSO's at the upper end of the channel, worsening its water quality. Because of the above, BWSC with U.S. EPA and Massachusetts Division of Water Pollution Control (MDWPC) permission, chose to design the dry weather pipe only. Since wet weather flows would no longer be carried, open cut now appeared feasible and more economical than either the channel or tunnel options. In July 1982, design of the NESI-NB along the Atlantic Avenue route began (Figure 2 - "Current Route").

Moving inland solved many problems, but brought with it a host of new ones. For one, the interceptor would now pass through the MBTA's South Station and FRA's Northeast Corridor $44 million Projects, which included construction of expressway access ramps and installation of new Washington-to-Boston commuter rail tracks and signal systems. The MBTA's work was already underway, and designs for the FRA's track and MTA's ramps nearing completion. After laying out several possible alignments, each of which met objections from one of these agencies, a compromise was worked out whereby the effected portion of the North Branch would be incorporated into the ramp contract. A push began to complete plans for this portion of the NESI-NB in time for MBTA's advertisement in June 1983.

Having finally found a solution for the South Station area, the BWSC was still left with the problem of traffic and utility interferences, including a subway station crossing, along Atlantic Avenue. The Boston Traffic and Parking Department (BTPD) required that two traffic lanes be maintained around South Station and for several blocks north and that construction, where two lanes could not be maintained, be limited to off peak traffic hours. Cross streets which provided the only convenient access to northern South Boston from downtown would also need to be kept open. Although the City had previously developed traffic plans for BWSC work, they could not provide this service for such a complex project. Therefore an amendment to CDM's design contract was necessary for development of a traffic management plan. This plan would have to meet BTPD's constraints yet still provide the contractor with sufficient room for installing 72 in. (1.8 m) diameter pipe, 23 feet (7 m) below the surface and a triple 42 in. (1.1 m) diameter barrel inverted siphon, 60 feet (18.5 m) deep.

Except for the area where the inverted siphon would be used to cross the subway station, pipe installation would be within a tightly sheeted trench to keep excavation as narrow as possible. Despite this, interference with utilities could not be avoided. Impact on electric company facilities would be especially great. Construction of the interceptor would conflict with their cables and manholes at 56 locations and would also require the relocation of 1,200 linear feet of 16 inch steam lines, 700 linear feet of 6 inch condensate lines, 1,200 linear feet of fuel lines, and 1,400 linear feet of distribution conduits. Not included are other lines which might be supported. The electric company concluded that it would take millions of dollars for them to complete the work required and, because of their budget, many years.

Reimbursement for such relocation costs now became a major issue and point of discussion. Private utilities normally relocated for BWSC as well as the City at their own expense. However, because of the magnitude of relocation required for this project, some were now considering legal or legislative action requiring the BWSC to bear the costs.

Major relocations would also be required of the telephone company, gas company, MBTA, and Western Union. Because of lack of space, relocation plans and schedules would need to be developed and coordinated for all these utilities. This again was beyond the original scope of the design contract.

Construction of the North Branch could significantly interfere with maintenance of utility services. Again the electric company was especially effected. Pipe excavation would pass by a major distribution station feeding the northeastern end of the city, including the financial district. Twenty four 15,000 volt cables leave that station, all crossing the proposed sewer route.

This was the status of design as of last spring. The design along the Atlantic Avenue route has been completed. It is still likely that another change may be needed before the final section of the North Branch goes to construction.

Construction Phase

To adapt to the availability of EPA and MDWPC funds, the project has been divided into four basic contracts constructed in phases. The table below includes the construction costs and pipe lengths for each contract.

The first contract for the Mount Vernon Street Sewer was awarded in April 1982 to be completed in August 1983. It involved open cut installation of 3,700 linear feet (1140 m) of 36 and 18 inch (.91 and .46 m) diameter pipe at depths from 15 to 25 feet (4.6 to 7.6 m) below the surface. Pre-stressed concrete cylinder PCC pipe was specified for 36 inch (.91 m) pipe to reduce infiltration and to withstand soil pressures.

Construction began at the MDC Diversion Structure. A 30 inch (.76 m) cast iron gas main passed over the connection point requiring

TABLE 2

CONSTRUCTION COSTS

Project Segment	Bid Date	Construction Costs (millions)	Total Length of Pipe (lin.ft./meters)
Mt. Vernon Street Sewer, Contract #1	12/81	$ 2.4	4,700/1,000
New Boston Main Interceptor, Contract #2	9/82	$ 6.5	2,700/830
New Boston Main Interceptor, Contract #3	Est.* 5/83	$37	9,800/3,000
New East Side Interceptor-South Branch, Contract #3B	Est.* 8/83	$13	2,600/800
New East Side Interceptor-North Branch (Railyard), Contract #4A	Est.* 7/83	$ 6	1,400/400
New East Side Interceptor-North Branch (Atlantic Avenue), Contract #4B	Est.* 8/84	$34	3,600/1,100

* Present estimates for future work including contingency and engineering during construction.

support and careful hand excavation for 20 feet (6 m) below the main. At the structure, three feet of concrete wall had to be drilled through and the time of connection coordinated with the MDC to coincide with low flow conditions.

Once beyond Station 0+00, construction went smoothly. The contractor successfully used a trench box and steel plates in conjunction with a well point system, both of which made it possible to complete pipe installation within seven months.

The same contractor bid on and was awarded Contract No. 2 in November 1982. This contract included the first 2,700 linear feet (823 m) of the NBMI west of the MDC Tunnel Headwork. The contract called for installation of the 102 in. (2.6 m) diameter reinforced concrete pipe by open cut, and either jacking or tunneling under major traffic intersections. In addition, work within the park was restricted to the winter months so as not to interfere with the local baseball season.

Construction began in January 1983. In order to handle the 16 ton (17.6 T) sections of pipe, the contractor used a six and one-half yard bucket Northwest backhoe. Progress was slow at first. The trench box originally used did not withstand lateral soil pressures for the depths, requiring that an enlarged and stronger box be designed and assembled on site. A mild, rainy winter robbed several weeks from the contractor's progress. Two of these days were spent unburying the backhoe after it sank into the soft soils. Further hampering progress in the park were picketing by the local labor unions, a trailer fire, and equipment vandalism. To compensate for lost time, the contractor worked two shifts and temporarily moved his key engineering staff to the field in order to meet the required restoration date of late April.

In addition to the time constraints and soil conditions, the section in the park required crossing under a 72 in. (1.8 m) brick combined sewer overflow originally bearing on wood piles. The contractor strapped the pipe and supported it in place while excavating beneath. Concrete encasement was used to ensure good bedding for the CSO after backfilling.

Along the western end, construction was within a 55 ft. (17 m) wide residential street with average house setbacks of 7 feet (2 m) from the road. The pipe was 15 feet (4.6 m) from the northern curbline. Much consideration was given to methods for limiting inconveniences to residents and to the few businesses along this street. However, the 11 ft. (3.4 m) width and 31 ft. (9.4 m) depth of excavation made disruption of normal activities inevitable.

At the traffic rotary between this street and the park, pipe was installed by soft ground tunneling using an open face shield and 11.5 ft. (3.5 m) diameter liner plates. Dewatering for the tunnel was achieved by a well point system. To avoid settlement to the roadway above the tunnel or to neighboring buildings, the level in observation wells was monitored. Tunneling progress averaged 10 ft. (3.6 m) per day.

Two live connections were necessary as part of this contract: one from a 36 x 54 in. (.9 x 1.4 m) brick South Boston Interceptor carrying an average flow of 6 mgd. and a second to the 9 x 7.25 ft. (2.8 x 2.2 m) reinforced concrete connector carrying BMDS flows to the MDC Tunnel. To make these connections, special junction structures were constructed. To avoid damaging the heavy MDC connector, the structure base slab was installed in sections. When these structures were completed, the timing for breaking into the existing pipes had to be coordinated with MDC pumping operations at Deer Island Wastewater Treatment Plant.

At the present time work still progresses on this contract, and completion is scheduled for October 1984.

A contract for construction of the remainder of the NBMI from the end of Contract No. 2 to the NESI-NB was awarded this summer. It will consist of 3016 linear feet (918 m) of 102 in. (2.6 m) pipe, 3720 feet (1133 m) of 84 in. (2.1 m) pipe and 813 feet (248 m) of 78 in. (2 m) pipe. Pile supports will be required for 1700 linear feet (518 m). The contract will begin with a triple barrel siphon under a subway tunnel consisting of three 54 in. (1.4 m) diameter P.C.C. pipes at the depth of 46 feet (14 m) below ground level.

Main concerns during construction of this contract will include maintenance of commuter rail service while construction crosses railroad property, traffic planning around the siphon jacking pits, construction of a junction structure within a major utility intersection, prolonged groundwater control and pile driving without damage to adjacent structures and utilities, and avoidance of dust around the Food Market meat distribution center. Construction of the special structures within this contract will also be difficult. One of these structures will be built around a 142 x 124 in. (3.6 x 3.2 m) CSO conduit which is pile supported. It will include installation of two 54 in. x 96 in. (1.4 x 2.4 m) flap valve tidegates. A second major structure will serve as a transition under the 204 x 120 in. (5.2 x 3.0 m) CSO conduit. A third major structure will be at the connection to the existing 36 x 60 in. (0.9 x 1.5 m) ESI between active rail lines.

Conclusions

Most of you probably do not consider the design of wastewater interceptors to be complex or expensive. However, our experience has shown that design for an urban area, especially an old city like Boston, can be extremely difficult and construction unusual.

An important factor complicating the design has been the preponderance of government agencies proposing projects without a central, more powerful organization for overall planning. Boston has no master plan, and, therefore, BWSC has borne the expense and frustration of having to redesign to accommodate various state and federal projects.

The interceptor design project has required extreme flexibility on the part of the BWSC staff. Because the Commission is self-supporting, our Finance Department has been able to adopt to schedule changes, and fund availability has not been a problem up to now. Having an experienced engineering staff has helped tremendously to speed up decisions when design changes were necessary as well as in the negotiations and acquisition of easements. This staff has also been able to successfully coordinate with the numerous state and federal agencies involved with the project.

Status as a Commission rather than a City agency has not always been beneficial. As a separate body, it has less authority in public lands than the former Sewer Division had within the Public Works Department. Therefore negotiations with utilities for relocation has been more difficult. Obtaining easements, rights of way and city permits has also been complicated by BWSC's independent status.

Despite all these obstacles, BWSC has managed to complete design within fifteen percent of the original 1980 estimate. Construction has gone well on the contracts awarded to date and has proceeded ahead of schedule. The most difficult parts of construction are still ahead of us, but the goal is to complete this project by 1988.

* * * * *

Appendix I
REFERENCES

1. Camp Dresser & McKee, Inc., Report on Improvements to the Boston Main Drainage System, September 1967.

2. Camp Dresser & McKee, Inc., Supplement to 1967 Report on Improvements to the Main Drainage System, November 1975, revised March 1976.

3. Clark, Eliot C., Main Drainage Works of the City of Boston, Rockwell and Churchill, Boston, 1885.

4. Haley & Aldrich, Inc., Reports on Geotechnical Engineering Studies prepared for CDM Contract Nos. 1, 2, 3, and 4, 1980-1983.

5. Metcalf & Eddy, Inc., Wastewater Engineering and Management Plan for Boston Harbor - Eastern Massachusetts Metropolitan Area - EMMA Study, November 1975.

Sewer Force Mains for A Congested Suburban Area

C. Peter Thomas, Member ASCE*

Abstract

This paper describes the major features as well as the problems addressed in the design of new sewer force mains to be installed on the East Bank of the Mississippi River in Jefferson Parish, Louisiana. Two influent force mains are for carrying sewage from present collection and treatment points to a new centralized treatment facility to be constructed in the future. An effluent force main is to carry treated sewage for discharge into the Mississippi River. 44,000 feet of pipeline ranging in sizes from 36 inches to 78 inches in diameter is to be installed under this project for the Jefferson Parish Department of Public Utilities (JPDPU).

Project Planning

Planning efforts detailed in the facility plan, prepared by Albert Switzer and Associates and subsequently reviewed by other consultants, document a comprehensive study of the present and future wastewater treatment needs for the unincorporated area on the East Bank of Jefferson Parish, Louisiana. At the present time wastewater undergoes primary treatment only at three locations before being discharged to adjacent canals which lead into Lake Pontchatrain. The new water quality standards for the canals and Lake Pontchatrain require that significant improvements to treatment processes be made.

Approaches considered in the facility plan include upgrading and expanding existing treatment processes and the construction of new facilities. A total of 174 alternatives for treatment of wastewater were considered initially using five different sites for locating treatment facilities. Seventy Five (75) alternatives were based on using the Mississippi River as the receiving stream, 75 alternatives, with advanced waste treatment (AWT) technology, using the existing canals as the receiving water body and the remaining 24 considered a combination of treatment methods. In order to evaluate these alternatives, feasible collection system routes and effluent discharge outlets were conceived. Five feasible routes were developed by considering many limiting factors related to availability of right-of-way, displacement of residents and underground utilities, costs related to the need for special and expensive construction, government regulations and environmental factors. Each of the five feasible routes were studied to determine utility conflicts, pavement removal and replacement, boring and jacking

*Project Engineer, N-Y Associates, 2700 Lake Villa Drive, Metairie, Louisiana 70001.

requirements, right-of-way acquisition and the need for special crossings. At least three routing alternatives for collection of influent and effluent discharge were examined for each treatment alternative. Present worth cost estimates of routes combined with present worth cost of associated pump stations were compared to determined the most cost effective route for each treatment plant alternative.

The final selected route was based on total costs of routing and associated pumping, environmental concerns, limiting legal problems associated with easement acquisition along private right-of-ways, and minimizing construction problems. The selected alternative consist of two major influent force mains which link three existing primary treatment facilities to one centralized treatment complex. Secondarily treated effluent from this complex is to be discharged into the Mississippi River through the effluent force main.

Selected Alternative

The pipeline route shown in Figure 1 ties in the three existing treatment facilities and three proposed main collection points to one centralized location. Two separate influent force mains are being used to accomplish this. This routing enables continued use of existing facilities thereby reducing the overall expansion of treatment facilities that would be required otherwise. Primary treatment of wastewater will be undertaken at the existing locations-Helois, West Napoleon and Camp Plauche before wastewater is pumped through the new influent force mains for secondary treatment at the new facility. Discharge of the effluent into the Mississippi River instead of drainage canals avoids the need for AWT treatment facilities to meet water quality standards.

It is anticipated that the new facility will not be built in the very near future, therefore existing treatment plants will be upgraded as best as possible to accomplish secondary treatment. The upgraded plants will be able to treat design flows to the secondary level and peak flows to the primary level. With the new force mains and pumping stations interconnecting discharges from Helois, West Napoleon and Camp Plauche, all effluent will be diverted from existing drainage canals and Lake Pontchatrain to the Mississippi River.

The sizes of the force mains range from 36 inches to 78 inches in diameter as shown in Figure 1. The velocities in the force mains are expected to range from 2 feet/sec to 8 feet/sec. The new pumping stations associated with the force mains are designed to pump constantly by use of variable speed pumps. Due to limited available space at existing treatment sites, there will be no storage of wastewater by these pumping stations or at the existing treatment plant sites. The major portion of the 24,000 feet of influent force mains to be installed are 42, 48, 66 and 72 inches in diameter. One influent force main is approximately 1800 feet in length while the other is 22,200 feet. The remaining 20,000 feet from the total length of 44,000 feet is a 78 inch diameter pipe for discharge of

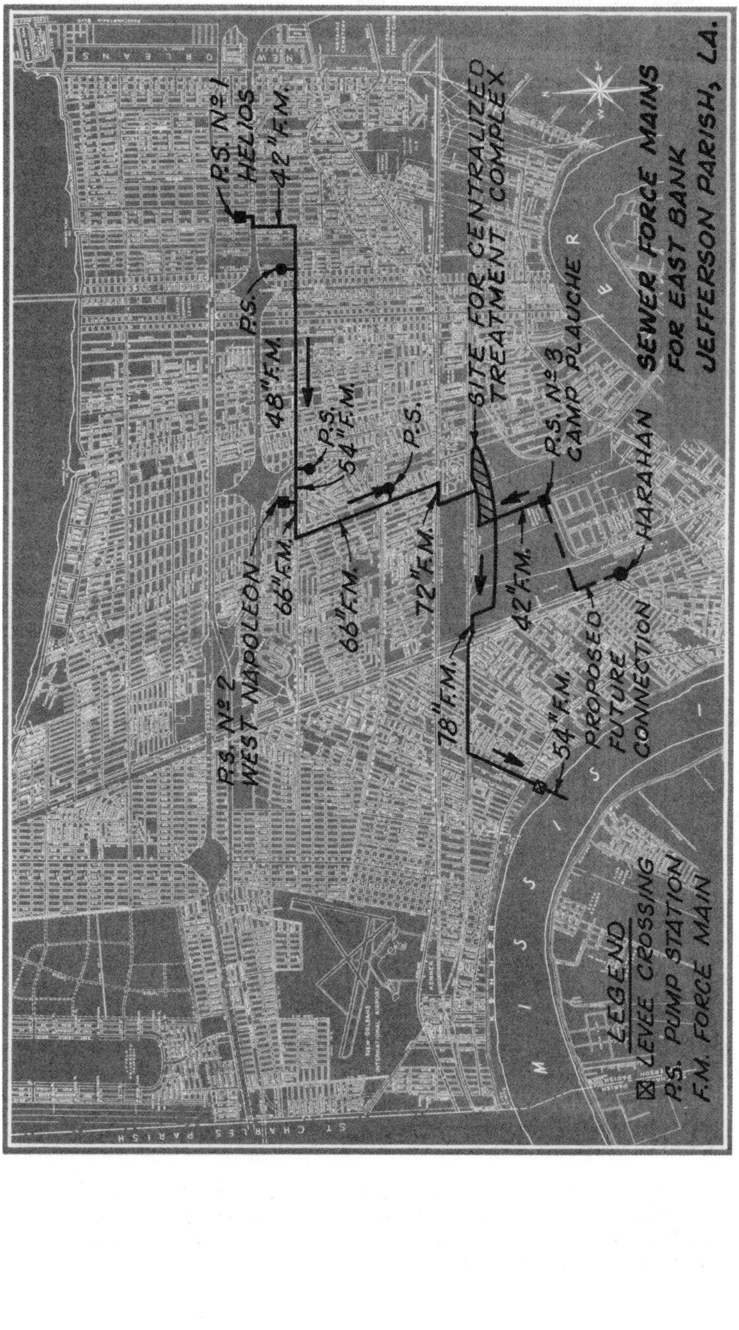
SEWER FORCE MAINS
FOR EAST BANK
JEFFERSON PARISH, LA.
LEGEND
LEVEE CROSSING
P.S. PUMP STATION
F.M. FORCE MAIN
P.S. No 1
HELIOS
42" F.M.
P.S.
48" F.M.
P.S.
54" F.M.
P.S.
P.S. No 2
WEST NAPOLEON
66" F.M.
66" F.M.
72" F.M.
78" F.M.
SITE FOR CENTRALIZED
TREATMENT COMPLEX
P.S. No 3
CAMP PLAUCHE
42" F.M.
HARAHAN
PROPOSED
FUTURE
CONNECTION
54" F.M.
ORLEANS
NEW
ST CHARLES PARISH

the effluent to the Mississippi River. The 36 inch and smaller pipes are to be used in special situations where only a relatively small quantity of the pipe is required for special crossings or connection to valves or pump discharges.

The design of the force main is based on the following general criteria:

a) Minimum cover of 4 feet for all pipelines except under special conditions.
b) Air-release valves at all high points.
c) Minimum clearance of 10 feet with all parallel water lines and 1-1/2 feet clearance with all intersecting water lines.
d) Test pressure of 75 psi and design working pressure of 45 psi for all lines.
e) Pipe materials can either be prestressed concrete cylinder pipe or steel pipe.
d) The use of thrust blocks will not be allowed for resisting thrust forces.

The general criteria outlined were followed for the design, however there were a number of special features of the design relating to pipe trenching, aerial and bore and jack crossings, a siphon system, oxygen injection, anchoring of the effluent force main in the river, existing utility conflicts and easement problems which had to be addressed.

Pipe Trenching

Soil conditions along the pipeline route can generally be described as poor to very poor, consisting of a layer of organic material, varying in thickness, which overlays a soft grey clay. At a few locations there are layers of sand varying in thickness from six inches to 2 feet, or there are mixes of sand and clay interspersed within the general organic material.

The groundwater table varies in elevation from several inches below ground level to the water level in the drainage canals. The groundwater level along the pipe route fluctuates considerably depending on precipitation levels and proximity to canals. Generally, the closer the location of the pipelines to canals and the higher the precipitation levels, the greater the fluctuation in groundwater levels. Fluctuating groundwater levels cause slight swelling and significant shrinkage of soil as the groundwater table rises and falls. Soil settlement of over two feet is not uncommon, with most of the settlement due to decomposition of subsurface organic materials. These adverse soil and groundwater conditions indicate that special trenches may be needed for installation of portions of the force mains.

Three types of trenches are designed in anticipation of the foundation material to be encountered during excavation. The three types of pipeline foundation conditions being designed for are based on the following expected trench conditions:

a) firm, dry sub-base upon which to place the pipe.
b) firm, but wet and mucky sub-base upon which to place pipe.
c) sinking foundation where no support can be derived from the sub-base material.

Because of the unstable soil conditions, proximity to other utilities, structures and canal embankments, sheeting and bracing of the trench will be needed for installation of the pipe. However, the design of sheeting and bracing is to be done by the contractor. Also, dewatering is the responsibility of the contractor and no specific recommendations for maintaining a dry trench are being made in the plans and specifications.

A sand shell mixture, 35% locally available river sand and 65% clam shell, will be used to embed the pipe, except where the pipe is located under pavement. For trench condition (b), described above, timber will be placed on the firm sub-base before the embedment material and pipe is placed. For trench condition (c), described above, sheet piling will have to be driven to sufficient depths and braces installed to support foundation timber, pipe, embedment and backfill material. At these locations the contractor will be paid for leaving the sheet piling and bracing in place. At other locations, no such payment will be made if the contractor elects to leave sheeting and bracing in place. The soils consultants indicated that at least 10 percent of the route will not have an adequate foundation and therefore require sheet piling, bracing and foundation timber so a trench can be built to support the pipe.

Utility and Other Structural Conflicts

Almost all of the influent force mains are located in parish or other public rights-of-way where existing water, sewer, drainage, electric and gas utilities are located. Approximately 75 percent of the effluent force main is located in right-of-way belonging to Illinois Central Gulf Railroad (ICGRR) and in some locations fairly close to existing railway facilities. Due to the potential for conflict with existing utilities and facilities, the force mains had to be carefully located.

Gas Utilities

There are numerous instances where the force mains are required to cross existing gas mains. In two places 18 inch high pressure gas lines belonging to United Gas Pipeline have to be crossed and in another place a rectifier bed is within 50 feet of the pipeline. In addition there are numerous locations where gas lines 12" and smaller belonging to Louisiana Gas Company are crossed. In one location the force main passes within 10 feet of a rectifier bed. All gas lines are cathodically protected by compressed current systems.

Due to the age of the gas lines, United Gas requires weldovers to be placed on the existing gas mains if constructed 18" or less from the force main. The weldovers are for support of the line in the event that the pipe becomes unsupported during construction. Although the design calls for the force main to be 2 feet away from

United Gas facilities, in one instance the United Gas pipeline is crossed by a lengthy bore and jack. Maintaining the alignment of this bore and jack, by the time it reaches the gas pipeline approximately 400 feet from where it is to begin, could be a real challenge for the contractor.

The other gas mains will be relocated by the local gas company where necessary. The use of sheeting and bracing will keep trench walls from collapsing, thereby the contractor should be able to avoid costly and dangerous ruptures of high pressure gas mains which from time to time will be adjacent to and parallel to the trench. Of more concern in the design phase was the question of what to do about electrical currents, in this area of low resistivity soil, which are present mainly as a result of the existing cathodic protection system of the gas mains.

Although for different reasons neither the gas companies nor the JPDPU desires a cathodic protection system to protect the force mains from corrosion. The gas company does not want additional cathodic protection systems which would only interfere with their existing cathodic protection and the JPDPU has not in the past and does not want to develop cathodic protection capability within its organization. However, considerable stray current is expected in the soil in the vicinity of the rectifier beds. Also there are other locations where the force main will be close to gas mains within the same rectifier zone. This means that there is some potential for stray current to corrode the pipeline should it be able to get onto and off the pipeline within the same rectifier zone. The local gas utilities use a reinforced coating to protect their gas mains and they indicate that the potential likelihood of corrosion is very small provided the force main, at least the metal therein, is adequately insulated. After considering the use of a polyethylene wrap around the pipe and relocation of either the force main or gas facilities, it was decided to pay careful attention to the pipe material and the exterior coating of the pipeline as the primary means of protection against corrosion. This is discussed elsewhere in the paper.

Water and Sewer Conflicts

Avoidance was the main approach to potential conflicts with water and sewer lines. Sewer lines generally do not present a problem. Gravity sewers are usually deeper than the proposed depth of the new force main. Conflicting force mains from small sewer pumping stations are easily relocated to cross over the new force main.

Water line conflicts include household connections, major transmission mains and all sizes in between. The most significant situation in the design phase involving a water line is where the centerline of the force main is within 10 feet of the centerline of a 60" water transmission main for approximately 1,000 feet. This distance between the two mains violates Ten State Standards, therefore it was decided to line the trench bottom and sides of the force main with a continuous sheet of 20 mil thick, reinforced PVC liner. Trench liner will be placed directly on the foundation before pipe, embedment material or backfill is placed in the trench. The thickness and type of liner is based on its purpose which is to prevent leakage of wastewater. PVC will be satisfactory as it will not be exposed to ultraviolet rays and therefore subject to deterioration and its thickness and reinforcement should be adequate to withstand tearing during installation of pipe.

Avoidance of water and sewer mains rather than removal and relocation is likely to cause less problems for the JPDPU due to reduced potential problems and associated costs caused by interruptions in service. Also removal and relocation can be expensive as most water mains are vinyl coated ductile or cast iron. Coating pipe with vinyl is an expensive means to guard against corrosion in this soil environment. In general careful attention was paid to locating, during the initial field surveys, all water lines in proximity to the force main so that the force main could be located in order to minimize removal and relocation of existing water mains.

Drainage Conflicts

Avoidance was also the primary approach used to address conflicts with drainage facilities. However, there were many more conflicts with drainage lines in certain parts of the project than with water or sewer main conflicts. The underground drainage system with in the more urban areas of the force main route is extensive. Also the general terrain of the area is flat and because all drainage leads to the various canals or associated box culverts one cannot relocate drainage pipe as easily as water or sewer mains and still have an effective drainage system. The flat terrian and elevations of the canal bottoms are constraining factors in the design of the drainage network and changes in elevation of drainage pipe at one point in the system can contribute to local flooding problems in this flat, high rainfall area.

A lengthy portion, approximately 10,000 feet, of this force main is located parallel to a drainage canal. There are numerous outfalls into this canal which drain large adjacent residential and commercial areas. The outfalls are located at various elevations and range in diameter from 15 inches to 60 inches with the most popular sizes being 24 and 30 inches. The elevations of these outfalls cannot be changed. The force main will be located under and perpendicular to most of the outfalls. Due to the size of most of these lines and the small clearence between them and the force main, it will be easier for the contractor to remove these outfalls and replace them after installing the force main. The contractor will be paid by unit prices for each size drainage line removed and replaced so as to avoid objections to this requirement in the field. All outfall drainage pipe replaced will be corrugated steel pipe regardless of what type of pipe is removed. Avoiding drainage outfall pipe along canal embankments resulted in more vertical directional alignment changes along the pipeline route than would otherwise be required. At each of these vertical directional alignment changes air release valves are required, more than actually needed to exhaust air from the line if the alignment were straighter.

For approximately 800 feet space is so cramped between within the public right-of-way that the influent force main could not be placed in the canal embankment without installation on private property. The alternatives are to place the pipe under the canal or extend a box culvert to close in the canal and place the force main in the backfill adjacent to the box culvert. The latter, and more expensive, but technically more desirable alternative was selected. Placing the force main under the canal would have provided future problems of access to the pipe should the utility decide to extend the box culvert as often happens when open canals become problems to maintain. Crossing canals and box culverts are for the most part being done with aerial pipe. Other special underground crossings of existing facilities will be bored and jacked.

Boring and Jacking

Pipe or pipe casings to be bored and jacked range in length from 40 feet under a drainage culvert to 400 feet under a railway yard belonging to Illinois Central Gulf Railroad. These are numerous major roadways to be crossed by the influent force main and many railway crossings for the effluent force main, most of which is located on railroad property. All pipe located under most roadways and under all railroad tracks shall be installed in bored and jacked steel casings ranging in diameter from 57 inches to 108 inches and having thicknesses of 3/8 to 1-5/16 inches.

Generally, boring in this soft grey clay is not expected to cause problems, because, the stiffest material encountered is likely to be a mixture of sand and stiff clay found in the railroad embankments. The longest bore will be a parallel bore of 60 inch diameter casing with the centerline of the parallel casings 15 feet from each other. This bore will cross sixteen (16) tracks, an 18 inch diameter gas main and 60 inch diameter water main. The force main is to be installed in the casing after it has been bored and jacked into place. The interior and exterior of the steel casings are to be coated with coal tar.

After pipe installation, the remaining voids between casing and pipe are to be filled, as much as possible, with pumped river sand, to guard against flotation, and the ends are to be sealed with masonry bricks. Casings associated with roadway crossings are also to be vented in accordance with the requirements of the local and state authorities.

Some of the bore and jack operations will have to be completed from jacking pits that will be somewhat smaller than usual due to confining structures close to the bore location. The Contractor will be subject to railway company regulations for jack and bores on railway property that require a halt in the boring operation every time a train passes on adjacent tracks. Finally, bores shall be conducted to completion or abandonment once started regardless of length.

Aerial Crossings

Much of the public servitude owned by parish is associated with the drainage system and a significant portion of the drainage system is open canals. Therefore since the force main is being installed on public servitude numerous canal crossings are unavoidable. All canal crossings are being done with aerial pipe. Except for the elevation of the effluent force main at the levee crossing, the elevations of the aerial pipe are the highest along the alignment. Elevations are determined based on the expected water surface elevations in the canals due to a 1 in 100 year storm. These elevations are basically estimated because runoff in canals is function of rainfall as well as the degree of urbanization of the drainage watershed which is constantly changing. However, the elevations are high enough that if the water reached up to the force mains at canal crossings, people in the East Bank of Jefferson Parish would have long lost their homes to flooding.

The width of the canal crossings range from approximately 70 feet to 170 feet. All aerial pipe needs to be pile supported by either four piles or six piles depending on the length of the crossing. Pilings in this situation are required to withstand tension forces due to the upward force of water on the bends of the pipe at the crossings. The pile tip elevations are determined primarily based on the need to withstand tension forces, therefore piling will generally be driven deep in order to develop enough friction. All piles will be vertical piles except at one location where batter piles are also required since the force main is making large horizontal charge in direction just prior to a canal crossing. All piles associated with canal crossings will be prestressed concrete piles. The pipe will be bolted with 1/2 inch thick steel straps attached to 1 inch diameter bolts embeded at least 12 inches into the concrete pile caps. The spacing of the pilings is done such that deflection of all the pipeline materials, likely to bid the job, will not exceed 1/360th of the span length.

Due to the elevation of pipe and the sharp change in vertical alignment associated with the canal crossings the interior of the pipe at these locations will be most susceptible to attack by sewer gases. This could especially occur if pumping is stopped periodically. A temporary halt in the pumping operation would result in the water tending to drain away from the high elevation at aerial crossings leaving the space to be filled by air and sewer gases. Therefore the interior of the pipe will have a thicker lining or special lining depending on the pipe material. For prestressed concrete cylinder pipe material, the T-Loc PVC liner is specified and for the steel pipe material the epoxy coating and concrete lining will be made thicker than used in pipe elsewhere.

Other Utilities

There are telephone and electric power lines to be avoided or relocated as the force mains are installed. These utilities do not pose any special problems. Except for a couple of telephone switching boxes along the pipeline routes, major telephone conduits are being avoided and the amount of relocation of existing electric poles is minimized. In order to avoid a major telephone conduit, the force main is to be installed under the pavement of a major local roadway instead of under the nuetral ground which separates the lanes of the road.

There are numerous overhead crossings of electric lines and a few of these lines are for primary feeders carrying 40,000 volts or more of power. The Contractor will simply have to exercise caution in his operations as the power company is unwilling to de-energize these primary feeders. These lines are much higher than the secondary lines which carry 8,000 volts or less. Due to the proximity of some of these secondary feeders to the pipeline route there will be some temporary de-energization required.

Decisions on how to handle conflicts, if any occurs, are to be made by the Contractor and local utilities.

Oxygen Injection

The influent force main from the Helois pumping station to the centralized treatment complex is approximately 22,000 feet in length. Due to this length combined with the possibility of interrupted pumping, sewage may become septic causing a significant build-up of potentially explosive gases in the pipeline. In order to guard against this potential for build up of gases, an oxygen injection system is being installed approximately halfway along the length of the influent force main. The point of injection is to be located in the vicinity of the Napoleon pumping station indicated in Figure 1. Liquid oxygen will be located at the pump station and is to be regularly injected through a 4 inch diameter stainless steel pipe into the force main. There are two injection points which are located in the bottom half of the pipe located in a valve box.

Valves

There are four valves located along the longer of the two influent force mains which are for isolation of a portion of the force main without taking the entire pipeline out of service. In addition, each pump can be isolated without affecting the remainder of the force main. There are six such valves to be installed on branch lines which connect to pump stations as shown in Figure 1.

All valves will be manually actuated ball valves able to stop flow in either direction. All valves are to be placed in pile supported reinforced concrete valve boxes, except at one location where pilings are not needed.

Levee Crossing

The highest point on the effluent force main will be where it crosses the Mississippi River levee. The Corps of Engineers regulations do not allow penetration of the levee with any structure or pipeline construction, therefore, the pipe will be installed over the top of the levee. This requirement increases the static level requirements of the pumps at the effluent pumping station considerably given that the rest of the terrain is generally quite flat. In order to save $500,000 over the next twenty years in pumping costs, a system for creating and maintaining a siphon at the levee crossing is to be constructed.

The main problem to be solved in the design of the levee crossing is the vacuum pump priming system that would exhaust air from the line during initial start-up and when pumping at the effluent station restarts after have been temporarily shutdown. Two vacuum pumps will be placed at the top of the levee to exhaust air from the parallel 54" discharge lines crossing the levee. The effluent pumps located at the new treatment plant site will pump water to the crest of the levee.

There is about 900 feet of force main located between the top of the levee and the end of the outfall in the river. Instead of a vacuum pumping tank normally used in conjunction with the vacuum pumps to effect a joining of the water column, it is thought to be cheaper and just as effective to install the last 700 feet of the force main below the average low water level in the river. Water will therefore be located within 200 feet of the vacuum pumps on the river side of the levee.

Waterhammer is also a problem to be addressed due to the height of the levee crossing and the steep gradient associated with the crossing. On the landward side at the bottom of the levee 48 inch diameter butterfly valves are to be installed on each pipe crossing the levee. The operation of these butterfly valves will be synchronized with the operation of the pumps at the effluent pumping station. The important feature of these valves will be the rapid closure of its disc when the pumps shutoff unexpectedly. The closing action of the valve disc will serve to damp any waterhammer waves that are bound to develop as the water column in the pipe separates at the top of the levee. The valves will close 90 percent of the way in less than 5 seconds after the pump shuts down and continuing closing at a much slower rate thereafter. Utilizing similiar type valves on the pump discharge, it is expected that waterhammer will be minimized.

Without a system to control waterhammer the design of the pipeline to withstand waterhammer pressures would have been very expensive. Waterhammer pressures at bends in the pipeline would be so high that they could not be resisted by only restraining the pipe. Thrust blocks to resist waterhammer forces would have to be pile supported by vertical and batter pilings in order to be effective in the poor soil environment. Finally, in addition to being expensive location of the effluent force main along a 5000 foot section of its route located between parallel railway tracks which are in some spots less than 20 feet from the pipeline would most likely not be permitted by the railway company.

In addition to the vacuum pumps and butterfly valves an air release vacuum breaker valve will also be installed on the force main at the top of the levee. This air release-vacuum break valve will exhaust air from the line during normal pumping operations. It is to be located on a 45 degree bend located at the top of the levee just before the pipe descends on the river side of the levee. It is in this bend that air is most expected to accumalate during normal pumping. The vacuum breaker component of this valve is also very important, because it is the means by which air will enter the pipeline in the event of a sudden shutdown of the pumps. A sudden shutoff of the pumps at the effluent pumping station followed by the separation of the water column at the top of the levee will leave a vacuum in the pipe, which unless immediately filled with air, could result in the collapse of the pipe walls. Operation of the butterfly valves on the landward side of the levee wall help to reduce the air requirements of the vacuum breaker valve from what they would otherwise be if the rush of water from the crown of the levee is not dampened. On the river side, some air will probably be released from the effluent in the pipe but how much depends on the water level in the force main which is a function of river stage. The higher the water level in the pipe the more air will be available to fill the vacuum. Although sewer gases are not normally expected in the effluent force main, the pipe of the levee crossing will be lined, as with the aerial crossing.

Effluent Discharge Outfall

At the point of discharge of the effluent into the Mississippi River the Corps of Engineer requirements include erosion protection for the bank of the river and a set of piers with barriers and navigation lights to guard against damage due to accidents with ships on the river.

Erosion protection will be accomplished by installing an outfall structure. The outfall structure is a reinforced concrete box, pile supported with 80 foot long piles, into which the effluent will discharge before being dissipated to the river. The main thrust of the water at the point of discharge will be absorbed by the reinforced concrete box structure.

The outfall structure and pipe in the river will be protected by a set of concrete dolphins pile caps in the river located in 180 degree radius around the end of the pipe 45 feet from the outfall structure. Each pier will consist of a pile cap wihch is supported by two battered piles and one vertical pile. A navigation light will be placed on each pile cap. The pile caps are designed to withstand impacts from wave action as well as ships in the event of an accident. Each of the piles are connected to other piles supporting adjacent pile caps by horizontal and diagonal steel pipe used as cross braces. The purpose of these braces is to transfer a portion of the impact force at any one pile cap to its two adjacent pile caps. This way the impact forces at any one pile cap is withstood in party by the whole interconnected system of pilings which support the dolphin pile caps.

Pipeline Material

Acceptable pipe materials included prestressed concrete

cylinder pipe or steel pipe. Transite and plastic pipe were not considered as viable alternatives for this job because of the sizes involved. Ductile iron was also dropped as viable alternative due to its uncompetitiveness compared with the other pipe materials. Without a cathodic protection system, ductile iron pipe will be uncompetitive because of the need for expensive exterior coating to protect the pipe against corrosion. All pipe will require cement mortar lining or a concrete core for the pipe interior. Prestressed concrete cylinder pipe and steel pipe will both have a cement mortar coating on the pipe exterior, but ductile iron will require bituminous coating together with encasement in a polyethylene tube or taping system which is significantly more expensive than a cement mortar coating.

Prestressed concrete cylinder pipe will include normally lined cylinder pipe in sizes up to and including 48 inches in diameter and embedded cylinder pipe is to be furnished for all larger diameters. All joints will be bell and spigot joints made with a snap ring. Thrust forces are to be controlled by using restrained joints only. Extra deep pipe bells will be allowed to accommodate slight changes in direction. All exterior mortar coating is to be made corrosion resistant using type V portland cement which is low in tri-alumina silicate. A PVC interior lining will be required at all aerial crossings as previously explained.

The shell thickness of steel pipe will not be less than 1/4 inch or have a diameter to thickness ratio that exceeds 165. The joints are to be lap joints welded on the inside of the pipe with a seal weld on the outside. Joints are to be tested as they are welded. In order to resist thrust forces the steel shell is to be made thicher and joints to withstand thrust forces are to be butt welded. The cement mortar lining of steel pipe is to be at least a 1/4 inch thick except at bends where cement lining is to be reinforced with wire mesh. In addition an epoxy coating is to be applied to the cement mortar. The thickness of the epoxy coating will be 30 mills except at aerial crossings where the thickness will be doubled. A 2 inch thick exterior cement mortar coating, made with type V cement, will be required to guard against corrosion. This is being done in accordance with the wishes of the JPDPU who desires to avoid a special system of cathodic protection. Finally at certain points along the route the steel pipe may have to be anchored to guard against flotation. An empty steel pipe is a lot lighter than an empty prestressed concrete cylinder pipe and it may tend to float in areas of high groundwater. Alternatives for anchoring the steel pipe include encasement in concrete or a pile supported structure to which the pipe is strapped or otherwise fastened.

Pipe is to be designed to resist dead and live loads. Live loads vary depending on location and at certain locations can consist of a combination of live leads. At certain locations the impact of passing trains and live loads due to construction equipment could be simultaneously in effect on the pipe. In addition, pipe will be fabricated with 4" diameter outlets for connection of air release valves and 24" diameter openings for access manholes.

Specifying Offshore Pipe Installation: A Case Study

Leonard C. Dawson* and Bruce A. Bennett**, M. ASCE

Abstract

The authors present recent experience in specifying a supply and installation contract for large diameter pipe for an offshore seawater installation in the Middle East. The discussion covers a range of civil engineering considerations including: design loads, pipe materials, codes and standards, design and fabrication, burial depth, and backfill requirements for pipelines. Site data include the review of environmental factors required for stability analyses. Common causes of submarine pipeline failures and engineering considerations for reducing the risks of failure are stated. The authors close with recommendations for inspection and certification of the completed installation.

Introduction

Specifying large diameter pipe for offshore installation is a specialized engineering task, which requires the combined knowledge and resources of the construction industry and multiple civil engineering disciplines including: hydraulic, structural, mechanical geotechnical, and environmental engineering. This paper discusses the principal specification features used for the supply and installation of twin 3660-mm (144 in) diameter pipelines extending more than one-half mile offshore located in an adverse offshore environment.

Large diameter pipes are being installed offshore throughout the world to provide access to deepwater sources of cooling water and for the disposal of heated water or other effluents. Where subsurface conditions are suitable, large pipe installation with minimal cover depth is the most economical means of providing this service.

---

*Senior Structural Engineer, Stone & Webster Engineering Corporation, Boston, Massachusetts.

**Project Engineer, National Hydro Corporation, Boston, Massachusetts.

Because of the indeterminate nature of the many variables encountered offshore, this type of work is best bid and awarded as a lump-sum contract, which places the responsibilities for detailed site investigations entirely on the Contractor. Weather conditions impose irregular and seasonal work schedules. The coordination of dredging and pipe laying frequently requires different work shifts. Seabed variations, tidal currents and tides, wind, and weather are among the major variables that the Engineer can never adequately research to litigate against claims of misrepresentation.

To maintain single responsibility consistent with the lowest cost installation, it is recommended that the Contractor be required to propose both the pipe and method of pipe installation. This permits the Contractor to match available installation equipment with the type and design of available pipe. This also provides the Contractor with maximum flexibility in the planning and execution of the work.

Independent engineering inspections and certifications are recommended during the progress of work to ensure conformance with the approved procedure and specifications. The authors emphasize that a competent inspection will guard against costly and difficult repairs or corrections after the offshore installation is completed.

Engineering Considerations

Engineering considerations complete with a conceptual design and a clearly defined scope of work showing the project boundaries must be defined in the specification.

The conceptual design forms the basis for the work. The extent of the work is supplemented with design criteria, standards, and specifications that define pipe materials, design and installation methods, and materials acceptable to the engineers.

Conceptual Design. The conceptual design is shown on engineering sketches or drawings as part of the bid specification. Figure 1A shows the proposed pipeline arrangement in plan and profile for the case study. The arrangement consists of two 3660-mm (144 in) diameter intake pipes. Pertinent information on pipeline orientation to the shoreline, intake location, points of interface with onshore structures, coordinates, elevation, and lengths is provided. Figure 1B shows proposed details on trenching, pipe burial, and protective backfill (6).

Twin pipelines were selected during the design development because of the large volume of cooling water (128,000 m^3/hr) required by the plant (7). In addition, the twin pipes permit backwash of one inlet structure without total interruption of flow to the plant. The two pipelines are shown buried for a minimum depth of 2 m (6 ft) throughout their entire length. A protective rock cover is also shown.

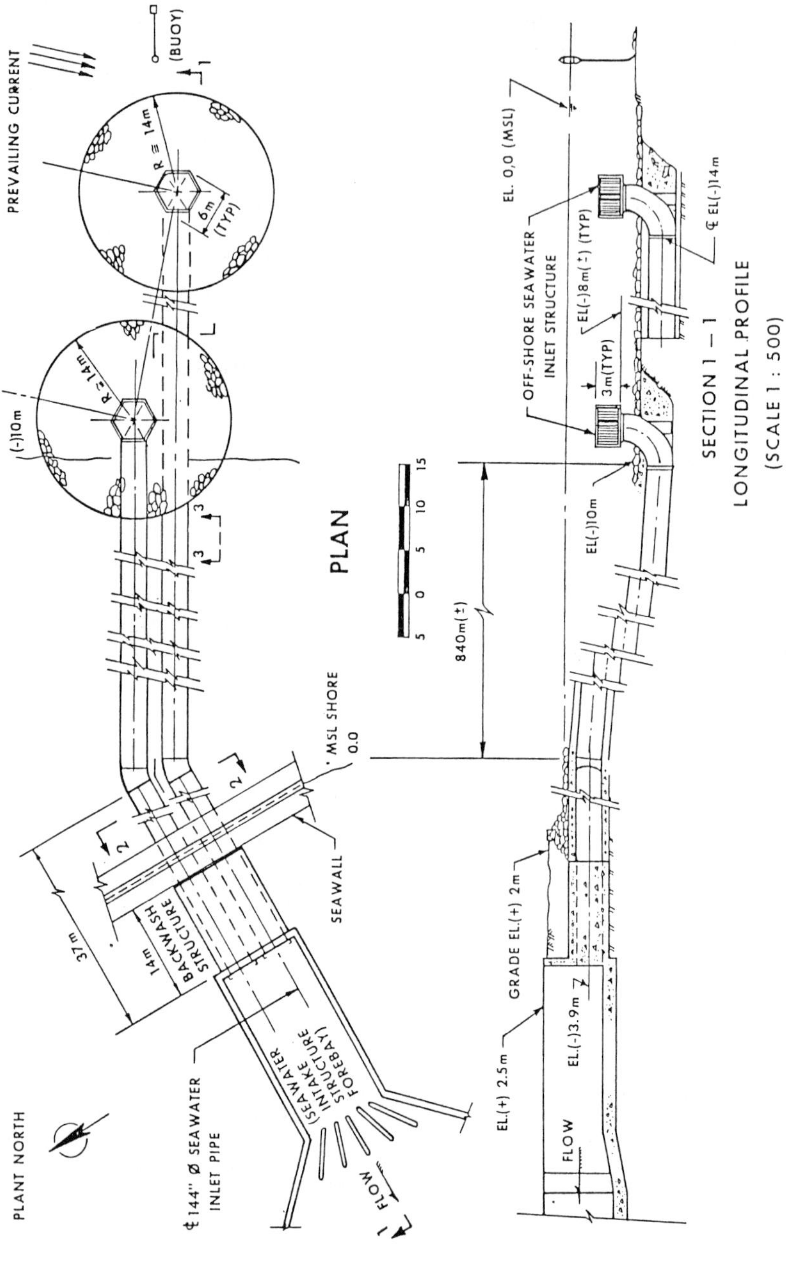

FIG. 1A SEAWATER INTAKE SYSTEM PLAN AND PROFILE

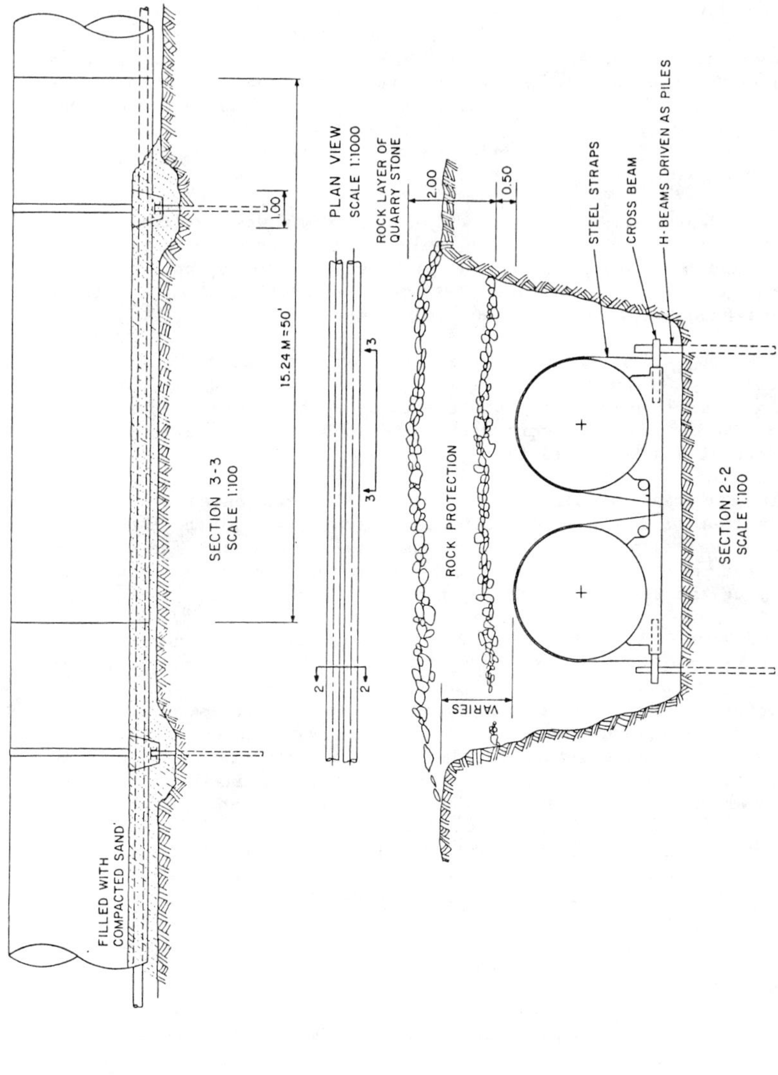

FIG. 1B SEAWATER INTAKE SYSTEM-SECTIONS

The offshore inlet structures are designed as velocity caps and are located beyond the surf zone, in 10 m (33 ft) of water. Seawater enters radially and a low intake velocity (about 0.6 m/s) reduces entrainment of seaweeds. A 14-m (50 ft) radius of rip-rap prevents strong currents from scouring the fill surrounding the intake elbows. The two inlets are oriented perpendicular to the prevailing current and are spaced far enough apart to prevent reimpingement of debris during backwash.

The conceptual design shown in Figure 1A was presented as a reference design for information only. The Contractors could adopt this design or develop their own, provided acceptable specifications and design calculations that meet the approval of the Consulting Engineer adequately describe and support the design.

The objective of the specification, drawings, and design criteria is to provide a contract document complete with sufficient detail to permit bidders to submit a lump-sum bid for the entire work and to provide the Engineers with a common basis for evaluating both a commercial and technical bid.

In the case study, several Contractors presented alternative designs that varied from concrete box culverts (in place of circular intake pipes) to pipes of different materials and size. The selected Contractor adopted the conceptual design presented in Figure 1A, confirming the results of the Engineers' earlier studies.

Scope of Work. The specification for this type of contract must clearly define the scope of work, the Contractor's responsibilities, and the responsibilities of others. In the case study, the Contractor was required to supply all materials, tools, transportation, equipment, and supplies necessary for designing, furnishing, and installing the offshore pipeline system. Appurtenances, such as chlorination piping, valves, fittings, and other fixtures required for a complete system, were also included in the Contractor's scope.

The Contractor's work included all excavation, dredging, backfilling, supplying, and placing concrete, stone, sand, and rock fill. The work also included but was not limited to the following tasks:

- Designing, supplying, and installing the offshore piping.
- Supplying and installing all equipment including chlorination system piping, cathodic protection system, all valves, valve pits, and fixtures required for operating the system.
- Furnishing, installing, maintaining, and removing all temporary construction facilities installed by the Contractor for this work.

- Furnishing all supervision, labor, materials, transportation, and equipment to perform all dredging, backfilling, removal and disposal of excess material, necessary for installing all offshore piping shown on the approved procedures, and drawings in the specifications.

- Performing all onshore beach earthwork including sheet-piling, excavating, dewatering, backfilling, temporary roads, and jetties as required for the installation of the onshore portion of the pipeline.

- Furnishing, driving, and cutting or removing any coffer-dams, bulkheads, or necessary construction sheet piling, including bracing or temporary retaining walls.

- Protecting the pipe during installation by anchorage and/or backfilling to maintain pipe protection, alignment, and grades.

- Performing all other work shown in the drawings or indicated in the specification.

The above tasks were listed in detail in the specification to reduce the risks of oversight and error by the Contractor.

Site Conditions. Site conditions listed in the specification included a description of the site location, its weather and sea conditions, and observations of all known seasonal working conditions in the area.

The description of the site location included its distance from major cities and ports, identified access roads, noted the existence of harbor facilities, and described onshore preparations that were to be made by others prior to the Contractor's work. The selected Contractors visited the sites during the Contractor's prequalification stage.

Weather and sea conditions were taken from historical records and collected from onsite investigations. Projects located in the Middle East or Africa usually do not have an abundance of data available to the Consulting Engineer. Some of the little known sources of data used in the case study included the U.S. Army Defense Mapping Service for topographic maps of coastal lands in the project area. Obtaining these maps required a letter of approval from the government of the host country. British Admiralty maps were a source of data for coastal bathymetry. These maps exist for most countries that are British Colonies or under British influence. Finally, the North Atlantic Marine Atlas provided wind data for the site located on the southern coast of the Mediterranean Sea. This wind data were also used to calculate wave heights and coastal surge.

The type of adverse environment in which the project was constructed is shown in the following list of pertinent data:

- Air temperatures ranging from 0 to 46°C with a maximum sun (black bulb) temperature of 80°C.
- Strong winds frequently rising to 45 knots with gust speeds up to 70 knots.
- Severe winter storms with waves up to 5 m (16 ft) height in the surf zone.
- Surf zone extending from shoreline to a distance of about 900 m (3000 ft) out to 9 m (30 ft) depth of water.
- Current strengths of up to 3 knots.
- Extremely large deposition of seaweeds near the beach.

Observations of seasonal working conditions were some of the most important data on site conditions presented in the specification. These observations provided direction and assistance to the Contractor in selecting equipment and in planning and scheduling construction activities. The only sources of this data were previous records and work experience in the area. The data were included in the specification as conditional information to be confirmed by the Contractors.

Typically, coastal sites have certain seasons that are favorable for working offshore. In the case study, autumn was the best season for working on the water (6). Sea breezes are slight during this season and no major storms occur until late November. The major storms occur during the winter season. These storms can last up to one week during which all offshore construction activities are interrupted. Periods of good weather are highly unpredictable during the winter season. Afternoon breezes with wave heights of 1 to 2 m (3 to 6 ft) occur during the spring and summer seasons. These waves will not cause significant downtime for large derrick barges or backup platforms, but would be a problem for small equipment and activities carried out in the surf zone. One unusual storm event known as Sirocco or Gibleh can occur during the summer season. This storm brings a hot, oppressive, dust-laden wind from the African desert that blows onto the Mediterranean coast.

Design Criteria. Design criteria were presented in the specification to highlight special requirements and conditions applicable to the offshore installation shown in Figure 1A.

The design criteria included the hydraulic requirements associated with pipe flow, the structural criteria requirements for pipe strength, soil properties and external loading conditions including surcharges expected from water or beach vehicles, and pipe protection requirements.

Hydraulic design criteria were stated in terms of the minimum and maximum required performance and operating parameters of the pipeline system. The criteria included maximum allowable velocities, maximum allowable head loss, acceptable head loss formula, and absolute roughness factors. Meeting these criteria was particularly important because of their relevance to the hydraulic design of the onshore intake structure and equipment settings (i.e., pumps and screens).

Pipe design pressures were referenced to the horizontal centerline of the pipe for this large diameter, low-pressure pipeline system. Because the system involved subaqueous pipelines, these design pressures represented the differential pressures between the interior and exterior of the pipe. Design pressures were stated for the normal operating pressure, maximum transient pressure, and maximum operating vacuum pressure. The latter could result from the offshore inlet becoming clogged and the onshore pumps subsequently pumping out the intake and offshore pipelines.

Offshore surcharge pressures from backfill material above the pipe were stated as maximum and minimum depths of fill, depths of water above pipe, and additional pressures resulting from wave action. In the beach area, AASHO H-20 wheel loading replaced offshore water surcharge pressures.

Soil properties and soil boring data were given, as available. The modulus of soil reaction was estimated to be equivalent to 400 psi for the subaqueous pipe and 700 psi for onshore pipe placed in a dry trench. A deflection lag factor equivalent to 1.5 was specified for the design.

Seawater chemistry, water temperatures, biological data, air temperatures, and other pertinent environmental data were listed as available.

Seismic design criteria were stated with reference to the Uniform Building Code Earthquake Requirements with the seismic zone.

The site data provided in the specification were based on the best available information. The Contractor was required to verify all site criteria and site-specific data.

The Contractors were required to include methods of preventing pipe flotation, during and after placement, as a result of liquefaction of the surrounding backfill under storm-wave or seismic activity (3) for their proposed design. Specific details were provided in their proposals, particularly in the surf zone.

The Engineers' accepted a minimum cover consisting of 1200 mm (4 ft) of suitable stone fill, topped with armor stone, to prevent pipe flotation or exposure by seabed scour or other wave or current induced change of seabed level. A 1800-mm (6 ft) cover was preferred to accommodate construction tolerances. The stone cover would be designed as a graded filter to prevent stone loss within the underlying sand or silt. The nature and depth of pipe burial became controversial in the case study. The Client chose to release the Contractor from the specified requirements during the bidding stage, as a cost saving measure. However, most of the Contractors, including the low Bidder, recognized the need for pipe burial and shore protection at the project site and adopted these requirements in their design.

Data Furnished by Bidder. The bid specification included a list of technical questions that summarized the specification's requirements and the required specific responses by the Bidder. Required submittals or drawings, details, procedures, and other documents were included in this list.

These questions served as a checklist of significant decisions made by the Bidders in completing their proposals. The list was divided into information required for the proposal, and a more detailed questionnaire to be completed after award of the contract.

Pipe Material Requirements. Pipe material standards and specifications were stated to provide acceptable standards for pipe design, fabrication, and construction materials. Standards and specifications were stipulated to provide a minimum qualitative basis of acceptability for all products and materials. Other standards and specifications were acceptable, if it could be shown that they were comparable to those specified. All materials proposed for construction had to be shown suitable for the proposed service and capable of meeting the requirements for the minimum service life without excessive operating or maintenance costs.

The Contractors could propose the use of any suitable material or design for the offshore pipe that would provide the required service and operating life. The Contractor's proposal had to describe the materials and proposed installation providing this service life.

It was assumed that the Contractor would select from prefabricated (8) steel, concrete or fiberglass pipe that would conform to one or more of the following pipe standards and specifications listed in the specification.

Pipe Standards and Specifications. Approved pipe materials and design standards were those of the American Water Works Association (AWWA). Other national standards and specifications were from the American Society for Testing and Materials (ASTM), American Concrete Institute (ACI), American Welding Society (AWS), and the American National Standards Institute (ANSI). An abbreviated version of the language used to specify the approved pipe materials (5) is discussed below.

Steel Pipe. Steel pipe will meet the requirements of AWWA-C200 Standard for Steel Water Pipe 152.4 mm (6 in) and larger. The pipe will be wrapped and coated externally and lined or coated internally in conformance with AWWA Standards. Cathodic protection will be provided, if required, by the materials or the environment.

The Contractor will identify the grade of steel plate, plate thickness, normal inside pipe diameter, type of field joints, minimum unit pipe lengths and all accessories.

Stiffeners will be provided to maintain the pipe cross-section during handling, storage, and placing.

Concrete Pipe. Concrete pipe will meet the requirements of AWWA-C301 Standard for Prestressed Concrete Pressure Pipe, Steel Cylinder Type for Water and Other Liquids, or AWWA-C302 Standard for Reinforced - Concrete Water Pipe - Noncylinder type, not prestressed. The two types will not be intermixed.

Slip type self-sealing subaqueous joints of bell and spigot design with neoprene gaskets are the preferred joints. The method of protecting integral steel bell and spigot material and other steel hardware from corrosion will be stated.

Fiberglass Pipe. Fiberglass pipe will conform to the requirements of AWWA-C950 Standard for Glass Fiber Reinforced Thermosetting Resin Pressure Pipe (RTRP pipe). To provide adequate beam strength where control, placement, and consolidation of backfill are limited, filament wound pipe is specified.

RTRP Grade 2 polyester design is specified for use in a seawater environment. A 'C' type reinforced, thermosetting, polyester resin liner is required to protect the glass-fiber filament windings; and a resin rich, exterior protective layer is required to minimize mechanical and corrosion damage. Stiffener ribs usually are used on large diameter pipes to minimize pipe wall thickness.

Joints will be either single or double neoprene gasketed, bell and spigot type.

Based on the above material, the selected Contractor chose fiberglass pipe in his design. Because the Contractor had not used this type material previously, the pipe manufacturer had a greater role in the project. The successful use of fiber glass pipe in this project resulted in proposals for the use of this material in other work bid by the Contractor.

Construction Considerations

The selected Contractor, a West German firm, had entire responsibility for the construction. The Contractor selected the type of pipe used and the installation equipment and submitted a detailed pipe installation procedure and installation schedule for the Engineers' approval.

The Contractor was required to demonstrate knowledge of the properties of the type of pipe proposed and the associated shipping, handling, and installation requirements including pipe placing and backfill. The installation procedure had to identify the common hazards the selected pipe would be subjected to during and after installation and also had to detail the safeguards provided to prevent damage or loss of the pipe.

Schedule and Construction Proposal. The specification included a construction schedule showing the proposed sequence of major events from the award of the contract to project completion. Because the work required coordination with other site contractors and suppliers, it was equally important that the dates of all major events were met as well as the final completion date.

The Contractor had to submit a detailed Critical Path Method (CPM) schedule or bar chart to show that his proposal conformed to the required schedule and that the schedule would meet the required completion date. The Contractor had to include a written description of the proposed method for executing the work, showing clearly the method of excavation, placing pipe, and backfilling. All proposed subcontractors and subcontracted equipment had to be identified for the Engineers' approval.

The Contractor was also required to submit a detailed listing of all major equipment proposed for the work and the source and type of all construction materials, including pipe, backfill, and armor stone.

Pipe Handling and Transportation. Because the pipe supplier, material, and place of manufacture was not known at the time the specification was issued for bids, it was difficult to specify all special delivery and handling requirements to cover each type of pipe material and means of transportation. To overcome this difficulty, the specification required the Contractor to submit handling and transportation procedures with the proposal. The pipe manufacturer and carrier were party to this document, as advised by the AWWA Manual No. 11, Steel Pipe Design and Installation (1).

For large diameter pipe, barge or ship transportation will result in the lowest delivery cost. The selected pipe manufacturer was an American firm whose plant was located near an East Coast port. This fact coupled with the site being located on the Mediterranean Sea resulted in ship transportation as the only means of conveying the pipe to the site.

Placing and Backfilling. Although construction methods available for dredging the trench are similar for all offshore pipes, placing and backfilling requirements for reinforced concrete, steel, or RTRP pipe materials are different because of the large differences in pipe materials, lengths, weights, and jointing methods. Table 1 presents data for 12-ft diameter commercial pipe as taken from the Contractor's proposals (4). Concrete pipe is designed as a rigid pipe and depends less on adjacent backfill for lateral support than the flexible conduit design of steel or RTRP pipe. Concrete and steel pipe are less likely to sustain damage from large stone placed next to the pipe wall than RTRP pipe, because large stone can pierce the pipe. Steel pipe coatings are particularly vulnerable to damage from improper handling, placement, and backfill.

An effective joint design is essential when placing large diameter pipes under water. Large pipe is too rigid lengthwise to permit draping the pipe from a barge deck and welding the joint in the dry. Joints have to be made under water; a slip type self-sealing gasket joint (Figure 2), which only requires pulling the joint together, is the most practical joint available.

TABLE 1. - Data for 12-Ft Diameter Commercial Pipe

Pipe Material	Pipe Wall Thickness (in.)	Unit Length, (ft)	Unit Weight (tons)	Weight Ratio	Weight of Pipe (lb/ft)
(1)	(2)	(3)	(4)	(5)	(6)
Concrete	10.500	12	38	20	6300
Steel	0.625	40	21	3	1061
RTRP	0.700	60	10	1	315

NOTE: 1 in = 25.4 mm 1 ft = 304.8 mm 1 lb = 2.205 kg

The number of joints for underwater pipe should be minimal and consistent with maximum pipe handling length. Of the three pipe materials, the use of RTRP pipe in the case study resulted in the least number of joints. The RTRP pipe was manufactured in 60-ft lengths, which is a reasonable value for pipe of this size. If the pipe section was much longer, more lifting equipment would have been required and the section would have been more unwieldly, particularly if exposed to underwater currents.

To protect large pipes from exposure or displacement as a result of flotation, they must remain buried during their entire working life. A minimum depth of 1800 mm (6 ft) was adopted in the case study for the shallow water, which are subjected to surf or heavy wave action. Backfill material and a heavy armor stone covering were placed as soon as possible. Less backfill material and covering were used in deep water where static underwater conditions exist.

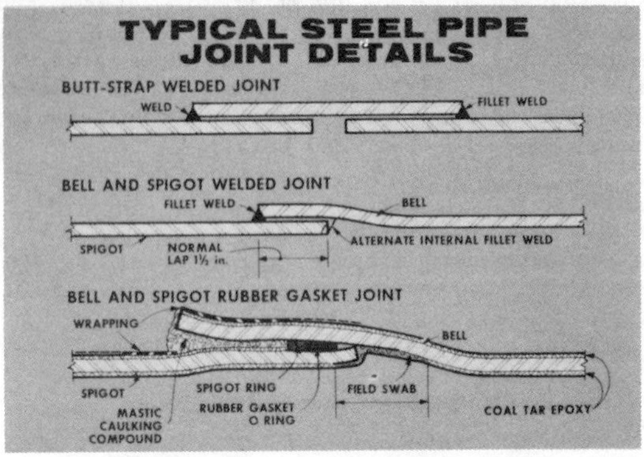

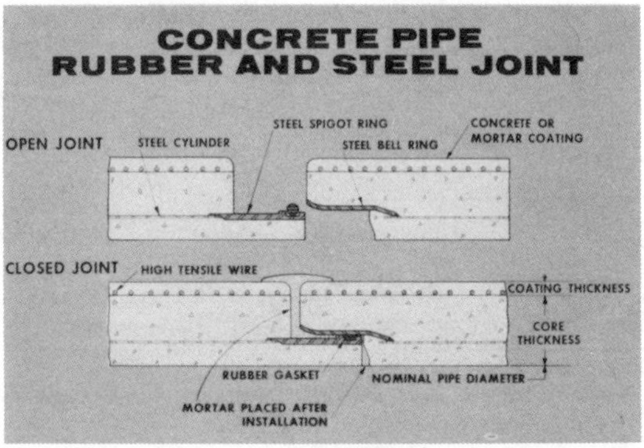

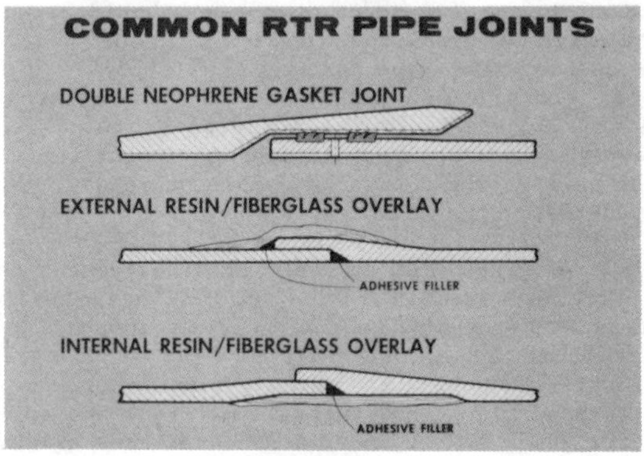

c. RTR PIPE

FIG. 2 PIPE JOINT DETAIL

In the immediate beach zone, the pipes were installed from the beach to a mean water depth of 1800 to 2400 mm (6 to 8 ft) where the jack-up platform shown in Figures 3A and 3B could complete the installation. The dredge was an integral part of the platform. The discharge from the dredge was used as fill over the completed pipe. A floating discharge line with a "Y" type surface discharge dropped the material on both sides of the pipe simultaneously.

When available, a jack-up and walking platform is superior to other methods. These stable platforms can operate continuously in water depths up to 18 m (60 ft), depending on the wave height. In the case study, this permitted worked to continue during rough sea conditions and helped to shorten the construction period.

The use of barges or other floating equipment to place pipe depends largely on sea conditions. Sea conditions in excess of a 600-mm (2 ft) swell can cause delays; likewise, sudden storms or prolonged adverse weather can result in damage to barges, equipment, and piping. Temporary trestles or movable platforms supported on the seabed can minimize such delays.

Pipes have been placed from floating barges using mobile equipment. Figure 4 shows a mechanical leveling and packing frame holding a pipe section ready to be lowered to the seabed.

Dredging Equipment. - The rate of pipe laying is normally determined by the rate of dredging. Dredging and pipe laying require close coordination because the dredged trench will rapidly fill if it remains open too long before pipe laying. In the case study, the dredged material was acceptable and backfill, dredging, and pipe laying were performed in alternate shifts, which eliminated interference and provided clear water conditions for pipe laying.

Because effective dredging equipment is critical to the performance of the work, the Contractors were required to identify, in their proposals, all dredging equipment by type, capacity, and availability.

Construction Hazards.- The pipelines in the case study were installed in an area where sea conditions were severe and unpredictable. The authors thoroughly reviewed the Contractor's qualifications, installation procedures, and previous experience in similar locations. A rigid installation procedure with a high level of mechanical equipment was avoided, because it would not provide adequate flexibility for improvements or modifications that are normally required to meet actual conditions.

Fiberglass pipe is not as abrasive resistant as steel pipe and care must be taken to prevent pipe movement in situations where abrasion can occur, particularly where concrete saddles are used to support this pipe. Placing of backfill was done carefully to prevent impact or puncture by large stone. The pipe was handled carefully to prevent damage to the bell, spigot ends, and exterior coating. Both steel and fiberglass have thin walls compared with concrete pipe and may fail by buckling or as a result of other damage if armor stone is dumped carelessly onto the pipe or backfilled trench.

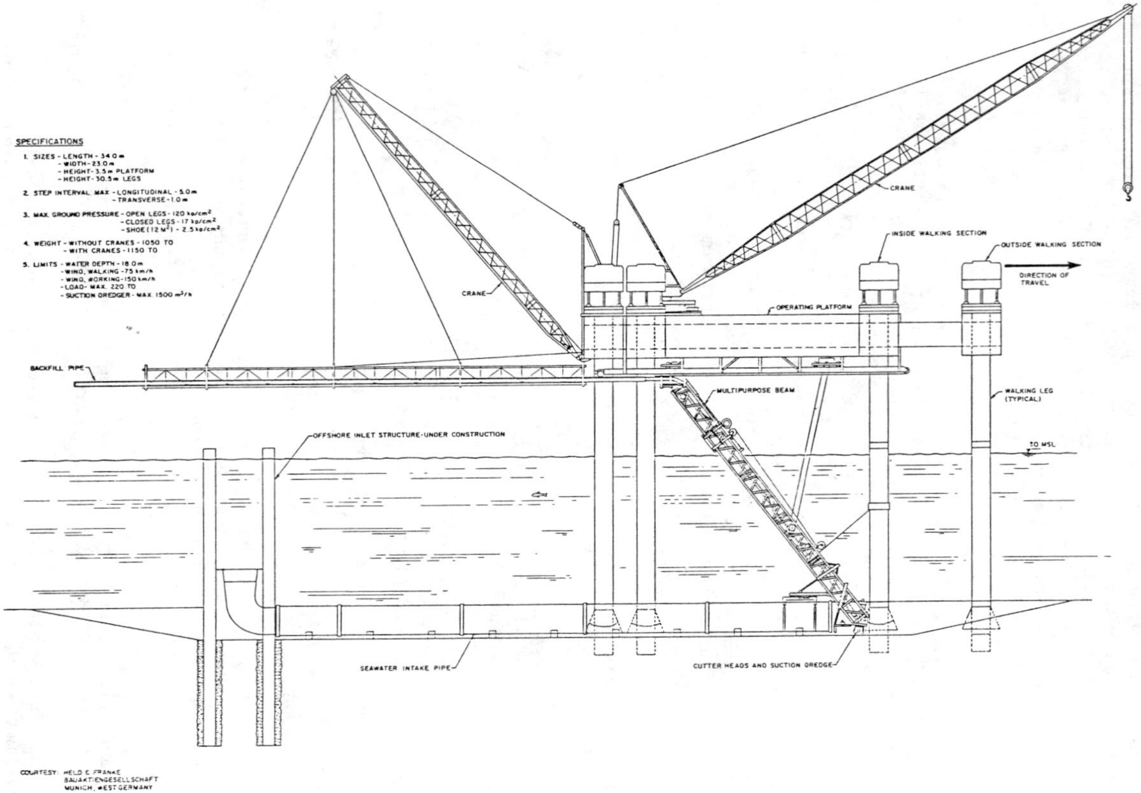

FIG. 3A JACK-UP AND WALKING PLATFORM-PROFILE

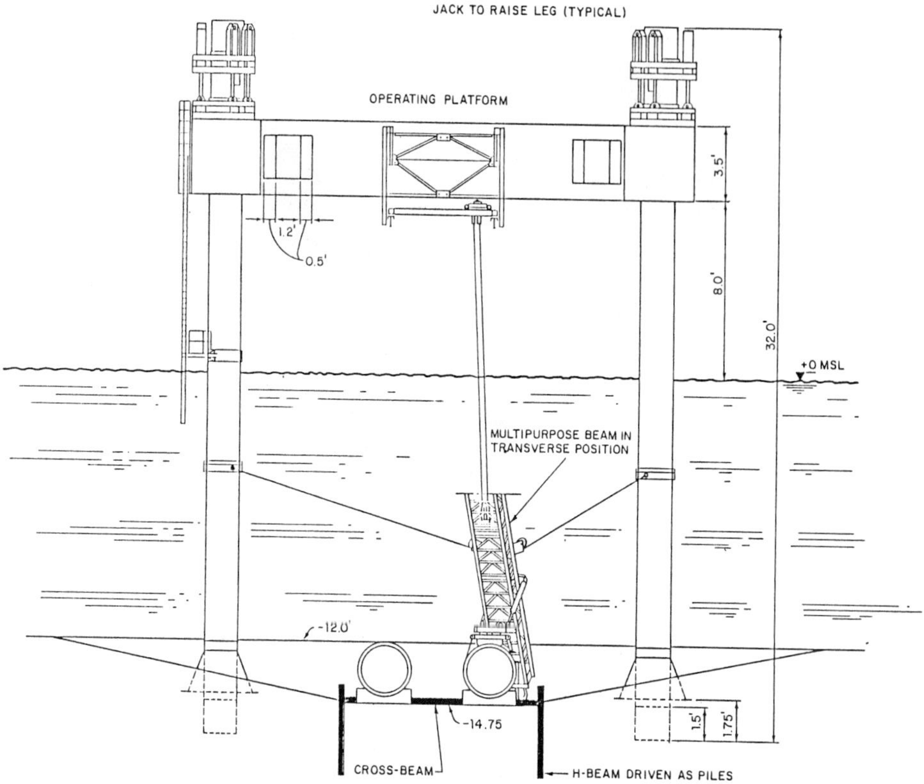

FIG. 3B JACK-UP AND WALKING PLATFORM-SECTION

Storm conditions have repeatedly destroyed large sections of pipe during pipe installation prior to completion of backfill.

Inspection and Certification

General. - Inspection and certification included shop and field testing of pipe and all other materials used in the work. Inspection of work in progress and verification of conformance with the contract is important and difficult when performed offshore, but it was essential in detecting deviations and errors before they were covered up. The specification identified and listed all tasks that required action by the inspection team.

FIG. 4 UNDERWATER LEVELING AND JACKING FRAME

Shop Testing. - The pipe fabricator was required to perform shop tests and provide test reports for all materials and products indicated by the referenced standards and as amplified in the specification. The Contractor was required to submit an inspection report to the Engineers, listing all tests and test results, prior to pipe delivery.

The RTRP pipe was tested in accordance with AWWA C-950 Standard for RTRP Pipe, Section 3. Hydrostatic leak tests, stiffness tests, hoop tensile tests, and axial strength tests were required, with an affidavit that all materials comply with the specification.

Dimensional checks were performed on each section of RTRP pipe including the neoprene gaskets and the steel fittings.

Material Tests. - Material testing of construction materials is desirable. These include tests of rock, stone, and other backfill materials. The test frequencies and acceptance criteria should be stated in the Contractor's proposal.

The Contractor should construct onshore test panels with equipment similar to that proposed for offshore use, to demonstrate the equipment's ability to control and place the backfill, cover stone, armor stone, and so forth to the specified dimensions.

The Engineers can randomly select materials for testing from the Contractor's stock piles or barges.

Inspection of Work in Progress. - The Contractor and his subcontractors were required to maintain an inspection, testing, and documentation program to ensure that all the materials meet specification requirements. Engineers or engineering representatives were allowed access to the engineering offices, shops, and work areas of the Contractor and subcontractors, and had access to information necessary to demonstrate that engineering, production, and inspection were proceeding in accordance with the schedule. Engineers also inspected material and equipment, to ensure that the Contractor's production and inspection procedures were being followed.

The Engineers' field inspector had similar duties to those mentioned above with access to the Contractors' and subcontractors' field construction offices, shops, and work areas. The inspector's authority was clearly defined in the specification and understood by the Contractor. Experienced Contractors will agree that thorough inspections protect them as well as the owner.

Verification of Work in Progress. - To verify the completed subaqueous work, it is desirable that the field inspection services be supplemented by a specialized underwater inspection agency qualified to inspect and to verify all underwater work. The limitations of these services and the procedures and methods developed for the project site should be clearly understood. The methods used in the case study included the use of underwater video tape cameras, and coordination of inspection and pipe laying activities.

The field inspection team was assigned the following additional tasks, which identified in the specifications:

- Perform a check of the line and grade of pipe before and after the completion of backfill.
- Witness procedures used in dredging, excavating, pipe laying, and backfilling to ensure that all work conforms to the specification and Engineer-approved procedures.
- Check pipe interior of completed pipe for alignment, joint closure, cleanliness, and circularity.
- Perform a check of bottom contours after dredging and at completion of the work.
- Perform checks on the level and line of pipe trench; check the depth of backfill layers placed around and over the pipe.

Conclusions

Currently, the trend is toward more comprehensive and carefully worded specifications. A construction review--the preferred term for inspection and certification--is particularly warranted for specifications covering offshore installation of large diameter pipe. Work progress inspection is vital in verifying that the work conforms to the specification requirements. Owners must realize that inspections identify and correct nonconformances before they develop into problems or lead to full-scale failure. The cost for these services are minimal when compared with the total construction cost of the offshore installation.

The authors concur with the views expressed at the 1981 ASCE Construction Division specialty conference (2). Briefly summarized, the views state that risk and liability are reduced through more detailed specifications and inspections. We have stated pertinent features that we believe formed good specification requirements and whose application resulted in a satisfactory offshore installation. Such a specification benefited the Engineers, the Owner, and also protected the Contractor.

Appendix I. - References

1. American Water Works Association, "Steel Pipe Design and Installation," Manual No. 11, American Water Works Association, Denver, Colorado, 1964.

2. ASCE News, Civil Engineering Magazine, May 1981, p 78.

3. Christian, J.T.; Taylor, P.K.; Yen, J.K.C.; and Erali, D.R. "Large Diameter Underwater Pipeline for Nuclear Power Plant Designed Against Soil Liquefaction Offshore Technology Conference, Houston, May 1974.

4. Dawson, L. C., "What Type Pipe for Your Cooling Water System," presented at the American Power Conference, Chicago, Illinois, April 1982.

5. Dawson, L.C. and Bennett, B.A., "Specifying Large Diameter Pipe for Offshore Installation," presented at the ASCE National Convention, New Orleans, Louisiana, October 1982.

6. Dravo Van Houten, Inc., "Report on Proposed Pipeline Installation at Ras Lanuf, Libya," Dravo Van Houten, Inc., New York, N.Y., January 1980.

7. Stone & Webster Engineering Corporation, "Final Report, Conceptual Design of Seawater Supply Systems, Ras Lanuf Complex - Ethylene Project," Stone & Webster Engineering Corporation, Boston, Massachusetts, February 1979.

8. Svenson, G., "What Contractors Think About Prefabricated Piping," Civil Engineering Magazine, January 1978, pp 71-74.

EBDA Outfall Meets Challenges of San Francisco Bay

H. B. Tennant,[1] M.ASCE, D. G. Gray,[2] M.ASCE,
and O. Fenton,[3] M.ASCE

Abstract

The $140 million East Bay Dischargers Authority Project (EBDA) is one of California's largest subregional wastewater treatment and disposal programs. It is also an important part of the effort to improve water quality in San Francisco Bay. The final segment of EBDA's 31-mile (50 km) pipeline conveyance system is a 7-mile-long (11.3 km) reinforced concrete outfall, eight feet in diameter, constructed in waters up to 25 feet (7.6 m) deep.

Throughout planning, design, and construction of the outfall, various geotechnical, environmental, and administrative problems were posed. This paper describes both technical issues (settlement, liquefaction, diffuser design) and institutional issues (presence of U.S. Navy anchorage system, environmental restrictions during construction). Resolutions to these problems are discussed. Key features include 1) extensive geotechnical investigations, 2) unique outfall hydraulics modeling using the Corps of Engineers' Bay-Delta Model, 3) development of alignment and diffuser criteria, 4) coordination of a large construction barge fleet, 5) use of laser technology for construction, and 6) development of unique equipment for underwater outfall inspection.

The paper also discusses how the outfall has performed since the EBDA project began operation in 1981 and how the outfall performed during record storms occurring in the winters of 1982 and 1983.

---

1. Chief Civil Engineer, Kennedy/Jenks Engineers, 657 Howard Street, San Francisco, California 94105
2. Principal Engineer, Harding Lawson Associates, 7655 Redwood Boulevard, Novato, California 94948
3. Manager, East Bay Dischargers Authority, 2651 Grant Avenue, San Lorenzo, California 94580

Introduction

The East Bay Dischargers Authority (EBDA) Project is one of the largest subregional wastewater management programs in the United States. A joint powers agency comprising the Castro Valley, Oro Loma, and Union Sanitary Districts, and the Cities of San Leandro and Hayward, California, initiated the project in 1971. Placed in operation ten years later, the system has 1) eliminated shallow water discharges to tidal flats and sloughs, 2) upgraded sewage treatment, and 3) consolidated the effluent of EBDA's five member agencies into one deep-water outfall in San Francisco Bay.

Twelve alternatives ranging from continued independent treatment and disposal to total consolidation were examined by EBDA before the scope of the project was defined. The implemented plan involved improving three existing treatment plants, contructing one new plant, and converting two plants to raw sewage pump stations. Six existing shallow water outfalls were replaced by a 30-mile-long (483 km) conveyance system culminating in a $20 million, 96-inch-diameter (2.4 m) concrete outfall and discharge structure (diffuser) that extends seven miles (11.3 km) into the Bay as shown in Figure 1.

This paper examines the geotechnical and environmental challenges involved in the planning, design, and construction of the EBDA outfall. These challenges include 1) soft soils, 2) zero underwater visibility, 3) seasonal fog and winds, 4) localized tidal currents, 5) seismic hazards, 6) regulatory constraints, 7) interference with existing improvements, and 8) a highly corrosive environment.

The paper also examines the practical engineering solutions determined by the two prime consultants, Kennedy Engineers and Jenks and Harrison [now Kennedy/Jenks Engineers(K/J)], as well as those developed by the geotechnical subconsultant, Harding Lawson Associates (HLA).

Planning

A detailed study was undertaken to identify feasible outfall alignment alternatives. Seven combinations of routes were evaluated for hydraulic efficiency, construction cost, time of construction, and ease of obtaining rights-of way. The study considered three diffuser locations approximately two miles (3.2 km) apart. With the water depth at all three locations constrained by designated ship anchorages in depths over 20 feet (6.1 m), diffuser performance was recognized as the controlling criterion.

To supplement current studies and computer modeling data, hydraulic modeling was performed by Waterfront Design Associates, using the U.S. Army Corps of Engineers Bay-Delta Model. A cost-effective alignment was selected that would connect to a dechlorination facility near the San Leandro Marina (Figure 2). A portion of the adopted alignment paralleled the marina's boat channel, thereby saving on dredging costs for construction access.

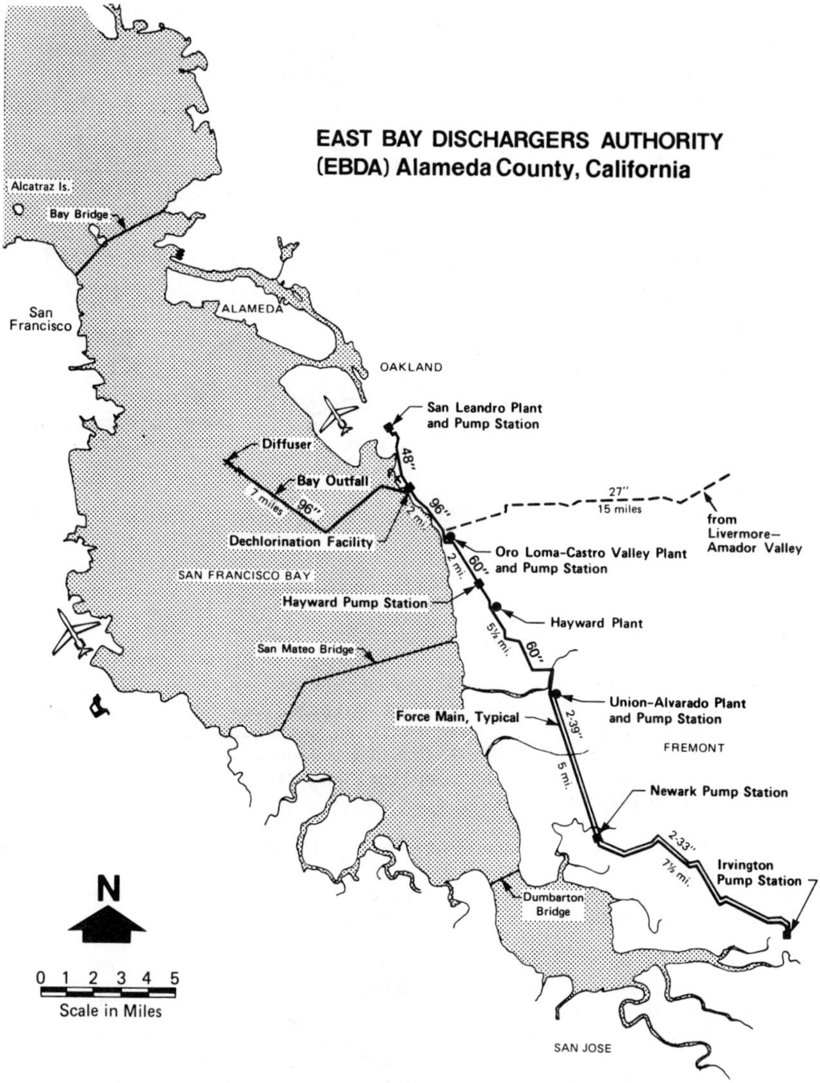

Figure1
Vicinity Map

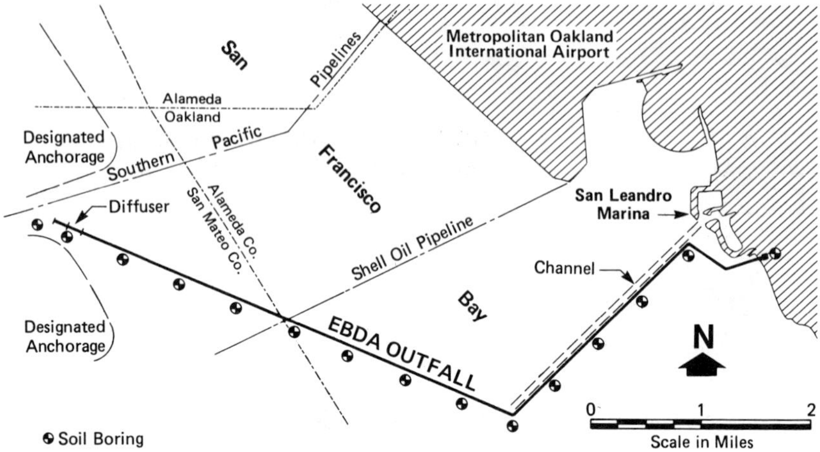

Figure 2
Site Plan

Geotechnical Investigations

A comprehensive geotechnical investigation was undertaken to develop data for project planning and detailed design. The investigation consisted of 1) reviewing existing geological data, 2) performing site-specific investigations including geophysical surveys and conventional drilling and sampling, and 3) evaluating faulting and seismic-related problems.

After reviewing published data, HLA performed a marine geophysical investigation to determine water depths, bottom conditions (i.e., obstructions and uniformity), and subbottom characteristics, including the thickness of soft bay sediments.

Fathometer data showed that the bay floor slopes very gently toward the center of the Bay with water depths ranging up to 25 feet (7.6 m) at the diffuser. Side-scan sonar data indicted that the zone at least 300 feet wide (91.5 m) along the alignment was free of any foreign obstructions or significant topographic undulations. Seismic profiles showed several strong reflecting horizons up to depths of 100 feet (30.5 m). These horizons normally indicate changes in material, consolidation, or accoustical characteristics of the sediments. One of the horizons was continuous and, due to its irregularity, appeared to be an old erosional surface. The depth to this horizon generally agreed with previous data regarding the depth of soft bay sediments (bay mud) overlying more consolidated sands and older bay clays. The profiles also indicated a general thickening of the bay mud to about 25 feet (7.6 m) near the end of the marina channel. The mud thicknesses decreased toward the diffuser end of the outfall.

HLA conducted a site-specific soil boring program to confirm the geophysical interpretations and to collect samples for measuring and evaluating physical soil properties. One on-shore test boring and 14 off-shore borings were drilled approximately 1/2 mile (.8 km) apart to provide general coverage of the alignment and to gather data at specific areas of interest as indicated by the seismic profiling. The test borings detected up to 24 feet (7.3 m) of bay mud near the end of the marina channel (see Figure 2). These data agreed with those from the geophysical subbottom profiling.

Laboratory testing confirmed the relatively weak and compressible nature of the bay mud. Along the outfall alignment, the mud consists of a clayey silt and silty clay that is normally consolidated. Its shear strength ranges from less than 150 pounds per square foot (psf) (7.2 kPa) near the surface to about 500 psf (24 kPa) at a depth of about 20 feet (6.1 m). The older bay clays under the mud are overconsolidated by a few thousand psf and generally have shear strengths in excess of 2000 psf (96 kPa). The sandy soils are slightly silty or clayey and medium dense to dense.

Because the San Francisco Bay area is seismically active, HLA evaluated regional faulting and the seismic behavior of the site soils. There are two major fault zones near the outfall. The San Andreas fault lies approximately 15 miles (24.2 km) west of the site and the Hayward fault approximately 4 miles (6.4 km) to the east. Although neither fault crosses the outfall alignment, both are a potential source of strong earthquake shaking that could result in liquefaction of the soils surrounding the pipeline. Liquefaction could cause lateral and longitudinal displacement of the pipe. However, because the sands encountered in the borings were medium dense to dense or had an appreciable clay content, HLA concluded that the chances of significant liquefaction were small.

Design Criteria

The planning studies and site-specific investigation data were used by K/J and HLA to develop design criteria and details to accommodate the adverse site conditions and anticipated construction difficulties.

On-site excavated materials (clay, bay mud and sand) were not suitable for bedding. The silts and clay would be soft if placed below water; the sands would be loose after placement and therefore be subject to liquefaction in the event of an earthquake. HLA recommended that the pipe be supported on imported bedding material placed below and along the sides of the pipeline up to the spring line. Bedding material consisting of coarse (1-1/2-inch-diameter $\pm$ [25 mm]) gravels was selected 1) to distribute the weight of the pipe uniformly over the natural soils, 2) to provide lateral stability in the event of an earthquake, and 3) to provide a relatively smooth grade on which to lay the pipe (dredging operations typically result in very irregular bottom contours).

Relatively large lateral thrust forces (up to 150 kips [668 kN]) occur where flowing effluent changes direction at bends in a pipeline. Such forces can be resisted either by constructing thrust blocks or by designing the alignment with large radii so thrust forces can be distributed along the periphery of the pipe. Because bay mud yields under sustained thrust loads, the use of thrust blocks would require excavating the mud or supporting the blocks on piles. It was decided to construct the alignment with large radii curves to avoid the cost of thrust blocks and the potential for differential settlement between the pipe and thrust blocks. With the wide curves, the thrust forces on any individual pipe section would be relatively small and could be resisted by friction between the pipe and bedding.

Soil samples were taken for corrosivity evaluation during the geotechnical drilling program. In addition to the obvious corrosivity of the sea water environment, high sulfate and chloride values revealed by laboratory analyses of the bay mud indicated severe corrosion potential from that source. Since the pipeline would convey secondary treated effluent, which had the characteristics of fresh water, interior corrosion was not considered to be a major problem.

Design Details

Piping materials considered for use in the outfall included steel, reinforced concrete, fiberglass, and reinforced plastic mortar (RPM). Steel was eliminated because of the cost of cathodic protection for such a large, long pipeline. Although of excellent corrosion resistance, flexible fiberglass and RPM pipes were also eliminated, as they would require extensive special bedding designs to provide adequate lateral support in bay mud.

Reinforced concrete bell-and-spigot pipe, with its rigid design and favorable corrosion resistance, was selected. The pipe conforms to ASTM Standard C361, with the exception that minimum exterior cover over the reinforcing was specified for corrosion protection to be two inches (50 mm) instead of one inch (25 mm). For seismic flexibility and the possibility of long-term differential subsidence, neoprene rubber gasketed joints were specified. Figure 3 portrays the typical pipe joint. Two gaskets were used at each joint, permitting individual joint testing during pipe laying. For use in long-radius (400 feet [120 m]) curves, beveled joints up to 2-1/2 degrees were specified.

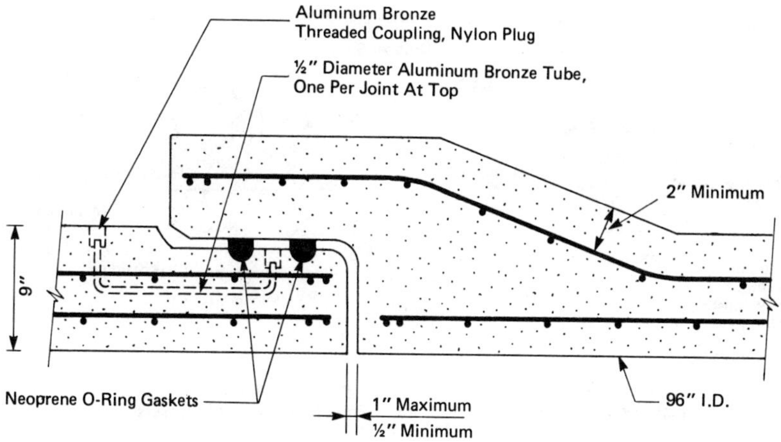

Figure 3
Typical Pipe Joint

Based on the design criteria, K/J and HLA designed trench sections to provide the necessary support for the pipe while minimizing the volume of excavation. Where special features were not required, the typical trench section specified (see Figure 4) provided for a minimum of 2 feet (.6 m) of gravel [1-1/2-inch (38 mm) maximum size] beneath the pipe, extending to slightly above the spring line for lateral support. No further backfill was required.

At the diffuser section, that is, the final 2,000 feet (610 m) of the outfall, armor rock (12-inch [.3 m] maximum) was added over the bedding to four feet (1.2 m) above the pipe to support the diffuser risers and to protect the concrete pipe from errant anchor chains. (See Figure 5.)

Special provisions were made in other locations as well. Along that portion of pipeline parallel to the marina boat channel, bedding material was brought up two feet (.6 m) over the pipe to minimize hull damage to small boats accidently leaving the channel. Where the outfall crosses under the channel, 12-inch (.3 m) armor rock was required over the pipeline to protect it against possible damage during periodic channel maintenance dredging.

For corrosion resistance and ease of fabrication, K/J designed diffuser risers (6-inch [150 mm]) and ports (3-inch [75 mm]) using PVC pipe. (See Figure 5.) This diffuser incorporated flexible rubber spools and breakaway joints for added flexibility, localization of damage in case of breakoff, and ease of repair.

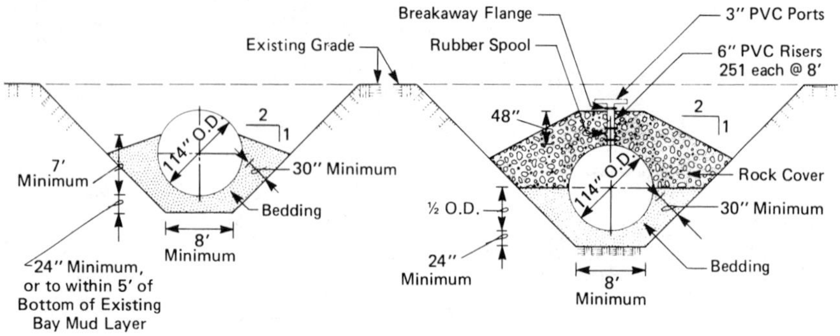

Figure 4
Typical Trench Section

Figure 5
Trench Section at Diffusers

To permit diver access to the interior of the pipeline, five manholes were incorporated into the design: one at the shore, one near the mouth of the marina channel, and three at the diffuser section. All of the submerged manhole risers have a 48-inch-inside-diameter (1.2 m) and are made of reinforced concrete cast integrally with the pipe. Three of the manholes have a fixed concrete cover, secured with eight aluminum bronze bolts, nuts, and washers. The manhole located at the extreme end of the outfall was designed to perform a pressure relief function as well as an access point. Constructed of reinforced concrete and weighing 4-1/2 tons (4080 kg), the tapered cover extends down into the riser much like a champagne cork. Since there are no bolts or guides, the underwater work of replacing this cover is simplifed.

Construction

The $16 million outfall construction contract, awarded to the joint venture of Willamette-Western Corporation and Santa Fe-Pomeroy, was performed in two separate but coordinated operations: 1) trench excavation and spoils disposal, performed by Willamette-Western Corporation and 2) pipe and bedding placement, performed by Santa Fe-Pomeroy. The contractor excavated the trench with a barge-mounted Model 4500 Manitowoc crane, equiped with a 5-cubic-yard (3.8 m^3) clam shell bucket. Work proceeded in two phases: excavation of a 100-foot-wide (30 m) access trench to allow work operations to proceed in shallow water and excavation of the pipe trench itself. The side walls of the trench were excavated on approximately a 1:1 stepped slope. The contractor loaded the excavated material onto bottom-dump barges and disposed of it in a designated area near Alcatraz Island 15 miles (24 km) away; side casting of excavated soil was prohibited by regulatory agencies. The trench alignment was maintained by sighting between control posts installed by survey and relocated in sequence with the work. The contractor was able to maintain proper depth to a surprising degree of accuracy by relating paint marks on the bucket chain to tidal variations. Daily progress was monitored by means of a sextant, verified by intermittent survey. A deck hand took continuous soundings during the trenching operations to check the proper depth of the trench. Divers verified the excavation intermittently.

Outfall pipe joints were designed to safely accomodate the amount of rotation that would occur with a differential thickness of up to 5 feet (1.5 m) of soft material underlying any one section of pipe. Based on the boring and geophysical studies, it was anticipated that excessive mud would exist in two areas. Overexcavation and replacement of the mud was specified in these areas, as shown on Figure 6. However, three additional areas underlain by excess soft soil were detected during construction. An accurate profile of the soft material was determined in each area by probing from the deck of a work boat.

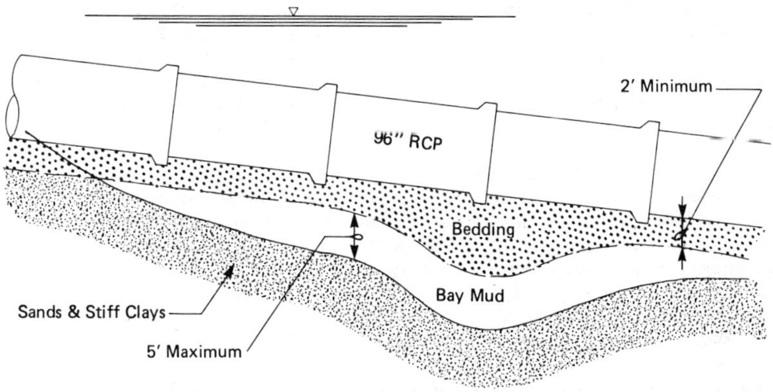

Figure 6
Profile at Deep Mud Zones

Once the profile was developed, a change order was issued to overexcavate the trench leaving 5 feet (1.5 m) or less of bay mud below the new trench bottom. The overexcavated material was replaced by bedding rock.

Following the trench dredging, the main pipelaying operations were performed from the 150 ton (136,000 kg) derrick barge "Seminole". These operations included trench cleaning, setting the

24-foot-long (7.3 m) pipe sections, and placing the bedding beneath and alongside the pipe.

A separate 1-1/2 cubic yard (1.2 m^3) clamshell crane at the bow of the "Seminole" accomplished any spot trench excavation that remained. Just ahead of pipelaying, an air-lift pump removed the slurry-like "slop" in the bottom of the trench.

Pipe manufactured at Santa Fe-Pomeroy's yard in Petaluma, California, was delivered 40 miles (64 km) by barge, 12 lengths per load. The "Seminole's" main crane transferred each 40-ton (36,300 kg) piece from the pipe barge to its place in the trench. A laser system that featured pile-supported instrument platforms and a crosshair target on the pipelaying sling assembly closely controlled vertical and horizontal alignment. With each spigot aligned into the preceding bell, the joint was quickly pulled home by the use of a bulkhead and a small pump which created a reduced internal pressure. A movable conveyor delivered the bedding material from a gravel barge into the trench, while a diver jetted it around the pipe.

Using the tube connection shown in Figure 3, the contractor hydrostatically tested each joint between the gaskets to 20 psi (138 kPa) shortly after laying. Thus, it was determined that no displaced gasket, cracked bell, or other defect existed at each joint as the work progressed. Each test tube was closed with a nylon plug, to allow retesting at any point in the pipe's lifetime.

As part of the full-time resident engineering services, Ames and Associates of San Rafael, California, provided a diving crew to inspect the underwater work. Divers checked and recorded trench conditions, bedding continuity, and joint gaps. Since visibility in the turbid bay waters was virtually zero, the firm developed innovative techniques to observe and measure the in-place work. These innovations included the use of special feeler gauges, a lighted lucite viewing prism, and a pneumatic fathometer.

Construction of the offshore portion of the outfall began in April 1978. The contractor completed the outfall in 29 months, 1 month ahead of schedule. Rainy weather did not cause serious delays, but high winds and dense fog occasionally stopped the work. In addition, bay currents, although vital in meeting dispersion criteria at the diffuser, periodically disrupted the construction and inspection diving, especially during flood tides. Despite the hazards inherent in the project, no serious injuries were incurred.

Project Performance

In April 1981, after all elements of the EBDA project had been completed, the entire system was placed in operation. During the record storms of the winters of 1981-82 and 1982-83, the outfall handled peak flows of over 200 million gallons (758 million liters) per day, exceeding the design flows forecast for 1997.

During construction, a moderate (magnitude 5.9) earthquake occurred about 12 miles (19 km) east of the outfall. In spite of noticeable ground shaking, no pipe displacement, slope failure, or joint leaks were detected in the project.

In October 1982, boaters reported a mysterious "boiling" in the Bay, near the San Leandro Marina channel. EBDA recognized it as a leak in the pipeline. Initial speculation was that it was caused by a broken pipe from a collision of a barge or boat, corrosion failure of manhole fasteners, or vertical or lateral movement of the soft soils in the area and the resultant separation of a pipe joint.

When a diver investigated, he discovered that a concrete manhole cover had lifted up over the anchor bolt nuts, apparently because washers had not been installed. Using air bags, the diver lifted the 1.2 ton (1100 kg) cover for inspection and reinstalled it using plate washers of aluminum bronze. Except for abrasion, the bolts and nuts were found in excellent condition. The diver checked the other bolted manholes and found that all washers were in place and that the aluminum bronze nuts and bolts were in excellent condition.

While the manhole cover was off, the diver inspected interior joint gaps. They were found to be uniform all around, an indication that no differential movement had occurred.

Conclusions

Thorough planning studies established an outfall alignment that would reduce costs and take advantage of strong currents for effluent dilution. Site-specific investigations using geophysics were an effective method of obtaining continuous subsurface data in an environment where conventional drilling alone would have been cost-prohibitive. The data from the investigations confirmed the presence of corrosive soft soils and seismic hazards. Design features to deal with these conditions included extra thick concrete pipe; flexible pipe joints; long radius curves; accessible diffuser ports for simplifed underwater maintenance; overexcavation of deep, soft soils; and use of imported bedding material.

Post-construction inspections and the outfall's exposure to seismic shaking are testimony that a reinforced concrete pipeline can be constructed in shallow bay mud if flexible joints are provided and a proper foundation maintained. In addition, the suitability of aluminum bronze, in lieu of more expensive metals such as monel, has been verified for marine environments.

Offshore Geotechnics: Trans-Panama Pipeline System

H. R. Al-Alusi[1], M. ASCE, and G.R. Thiers[2], M. ASCE

ABSTRACT

The Trans-Panama Pipeline crosses the Panama Isthmus near the Costa Rican border. The Atlantic terminus is near the village of Chiriqui Grande.

The loading system at the Chiriqui Grande Terminal consists of two single-point mooring buoys located about 2,000 meters from the shoreline, with an offshore pipeline laying on the seafloor. This paper describes geotechnical conditions, design concepts and installation methods for the offshore pipeline, end manifold and mooring buoys.

Phase 1 studies -- seafloor geophysical surveying and geotechnical investigations -- were carried out to study the practicality of the offshore system proposed. Seven test holes were drilled in this phase, ranging in depth from 12 meters to 42 meters (39.4 ft to 138 ft) into seabed.

Phase 2 studies -- an extensive field investigation and a laboratory testing program -- developed specific engineering properties of the subsoil needed to enable designing against liquefaction of the seabed and settlement of the pipeline, end manifold and buoy anchors. In this phase, the field program consisted of drilling 13 test holes with a maximum depth of 42 meters (138 ft.). Standard Penetration Tests were carried out in all test holes. Vane shear tests were performed in the field and in the laboratory. Disturbed and undisturbed samples were collected and tested under static and dynamic conditions.

---

1 Principal Geotechnical Engineer, International Engineering Company, Inc., San Francisco, California.

2. Chief, Soils and Foundations Department, International Engineering Company, Inc., San Francisco, California.

In the analysis portion of the Phase 2 studies, the use of the measured soil properties in the design of the pipeline is discussed, including studies of soil liquefaction, marine slope stability, and short and long-term differential settlements between the pipeline and the mooring buoy units. Geotechnical recommendations regarding offshore pipeline design are presented.

INTRODUCTION

The Trans-Panama Pipeline crosses the Panama Isthmus near the Costa Rican border. The Atlantic terminus, near the village of Chiriqui Grande, overlooks the Chiriqui Lagoon which was selected as a suitable location for sheltering loading vessels. Figure 1 shows the approximate location of the pipeline across the Panama Isthmus, and the location of the Atlantic terminus. This paper presents soil properties and geotechnical aspects related to the design of the offshore facilities at the Atlantic Terminus.

From the shore side manifold two marine pipelines branch out to two pipeline end manifolds (PLEM) located about 2,000 meters (6,560 ft) from the shoreline (See Figure 2). Each PLEM unit rests on the seabed and is connected to a single-point mooring buoy (SPM) as shown in Figure 3. The marine pipes are 91 cm (36 in.) in diameter with a 7.3-cm (2.88-in.) thick concrete jacket. The pipelines recieve oil by gravity through the shoreside manifold from three crude tanks located on a nearby hilltop.

The pipeline was designed to be laid on seabed without trenching. For installation purposes a 3-meter (9.8 ft) deep by 21-meter (68.9 ft) wide canal was dredged inland. A push-pad was constructed at the land end of this canal for assembling and pushing the temporarily buoyed pipeline seaward.

HYDROGRAPHIC AND GEOPHYSICAL SURVEY

To determine water depths, seabottom composition and the location of submerged obstructions along the sea approach to the mooring area and in the mooring area itself a hydrographic and geophysical survey was carried out. This survey covered the area immediately offshore of the village of Chiriqui Grande, as shown in Figure 2. Detailed sounding and sidescan sonar data were acquired in a 3-km (North-South) by 4-km (East-West) area of the lagoon where the proposed mooring and oil loading facilities would be located. Water depth ranged from about 2 m (6.5 ft) near the beach line to about 25 m (82 ft) at the PLEM locations. Sampling of the seafloor and shallow sub-bottom materials was carried out along the proposed pipeline alignments and at the location of the PLEM units. Samples were acquired in three ways.

1. Bottom sampling employing a Shipek Model 860 sediment sampler.

2. Gravity coring employing a Benthos Model 2172 piston corer equipped with a 1.5-m (6-ft) long core barrel.

3. Grab sampling by divers.

Shipek and grab samples and gravity cores were acquired at each PLEM site and at several locations in the mooring area and along the pipeline routes. In addition, during dives made at each core location, 1.5-m (6-ft) grounding rods were used to probe the bottom. The Shipek and grab samples recovered were stored in sample bags. The gravity cores recovered were stored in capped plastic core barrel liners. Locations of various sample points are shown on Figure 2.

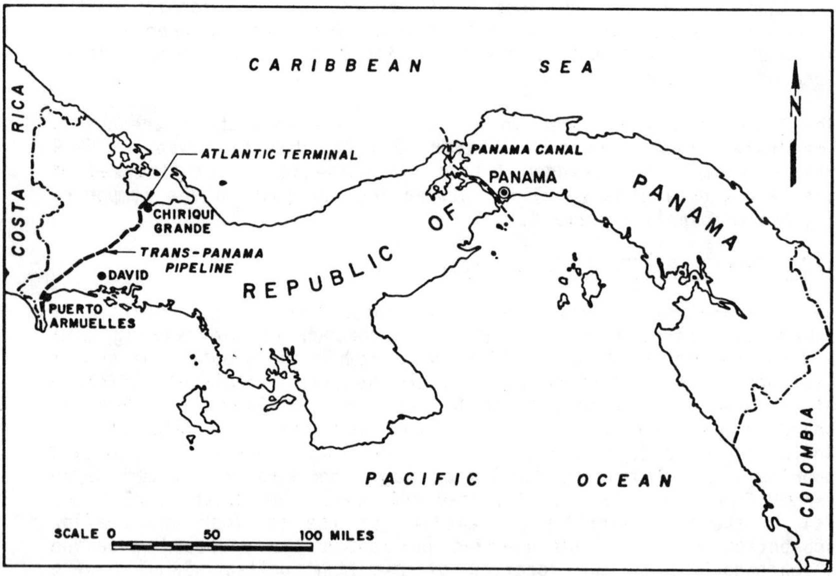

Figure 1 - GENERAL MAP OF PANAMA

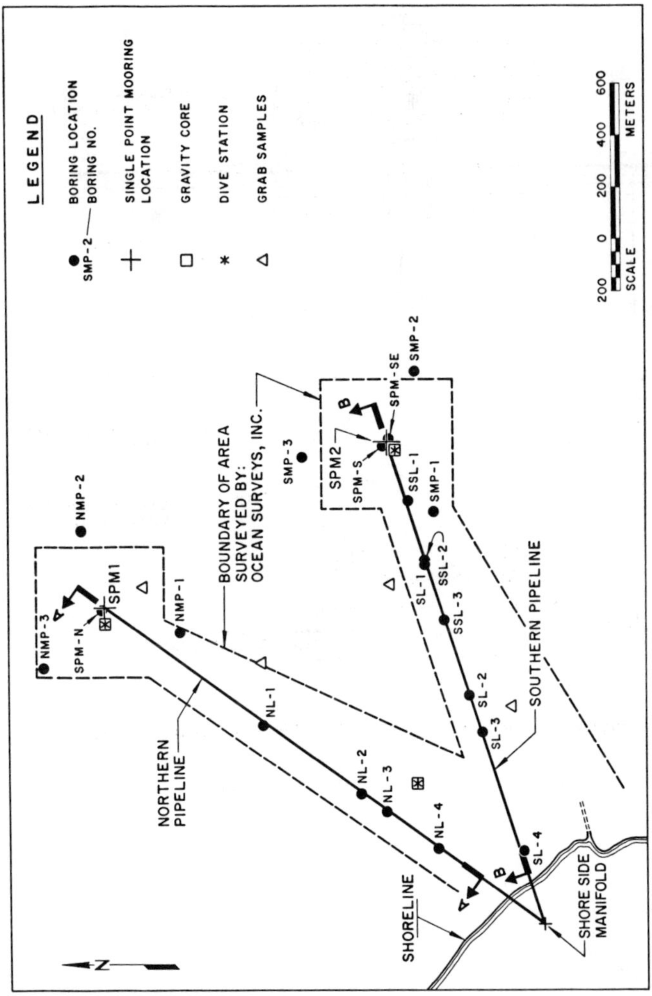

Figure 2 - LOCATION OF TEST HOLES AND SEABED SAMPLES

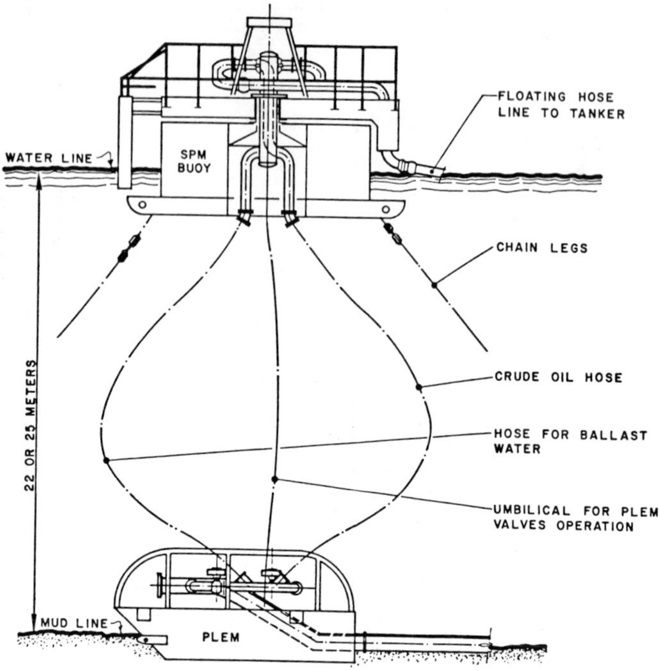

Figure 3 - ARRANGEMENT OF PLEM UNIT AND SPM BUOY

All samples were visually described upon recovery before transporting them to the laboratory. Field data and laboratory classification test results obtained from this program were used with other information from test boring and field and laboratory testing in developing the soil profiles presented in Figure 4.

GEOLOGY AND SEISMOLOGY OF THE CHIRIQUI GRANDE AREA

The rocks underlying the area in the vicinity of the Chiriqui Lagoon are comprised by Tertiary volcanics, intrusives, and sedimentary sequences. The soils are recent unconsolidated sediments and alluvium. The volcanics include agglomerates, andesites, and basalts. The intrusives are primarily granodiorites and diorites. The sedimentary rocks include limestones, sandstones, conglomerates and shales of the Miocene Gatun-Uscari formation. The Gatun-Uscari formation is exposed in a Northwesterly trending fold belt northwest of Chiriqui Lagoon and on the islands north of the lagoon, and may underlie the thick sediments of the lagoon itself.

Although the Trans-Panama pipeline crosses areas of well-known high seismic activity at the Pacific terminal and along the Costa Rican border (see Figure 1), the Atlantic Terminal at Chiriqui Grande lagoon is of low to medium seismic activity and there are no recorded historical events within 17 km (10.6 miles) of the lagoon area (12). The local design earthquake is of magnitude 7.0 at a distance of 20 km (12.5 miles), giving peak acceleration of 0.25g and predominant period of 0.31 seconds at the site. The regional design earthquake is of magnitude 8.3 at a distance of 100 km (62.5 miles), giving peak acceleration of 0.15g and predominant period of 0.65 seconds at the site. For pseudo-static analyses a seismic coefficient of 0.12g was used.

DRILLING, FIELD TESTING AND SAMPLING

The geotechnical field work included test borings, soundings, sampling and in-situ testing. In Phase I of the geotechnical investigations seven test holes were drilled to a maximum depth of 42 meters (138 ft), along the pipeline routes and at the PLEM locations. The primary purpose of these holes was to explore the practicality of the proposed offshore system. These test holes revealed information which was used for the preliminary design of the offshore pipeline and to plan the field work for Phase II. In Phase II, 13 test holes were drilled to a maximum depth of 42 meters (138 ft). The locations of all 20 test holes are shown on Figure 2. The test holes were drilled using pontoon-mounted rigs. The holes were advanced by wash-boring using a fish tail bit through cohesive soil materials and a tricone rotary bit for all other soil conditions. Standard Penetration Tests (SPT) were carried out following the procedures of ASTM method D-1586. Shelby tube samples were obtained from cohesive soil strata. A pocket torvane device was used in the field to obtain shear strength values at the time of Shelby tube recovery. One measurement by divers was made to obtain an in-situ

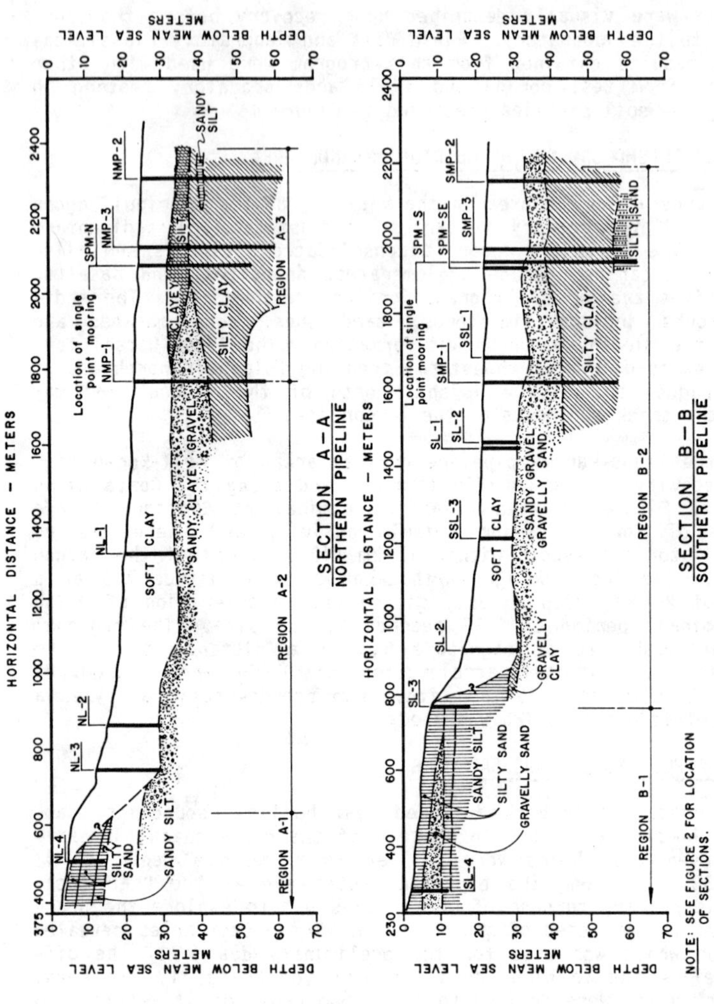

Figure 4 - SOIL PROFILES ALONG THE PIPELINES

determination of shear strength using a manual vane shear device. (The results of all shear strength determinations are presented in Figure 5.) All recovered soil samples were visually classified in the field. Thirty-five samples were sent to the United States for laboratory testing. Laboratory testing was performed by Ertec Western Inc., Long Beach, California and by N. H. Williams and W. N. Houston of the University of California at Berkeley, California (8).

SOIL CONDITIONS

The field investigation program revealed that the seafloor is blanketed by a layer of very soft to soft clay along the routes of both pipelines except the 500 meters (1,640 ft) near shore along the Southern pipeline, which is covered by a layer of very loose silty sand. These two layers and the various underlying layers are shown in Figure 4. Based on the general soil characteristics, the profile of the Northern line was divided into three regions: A-1, A-2 and A-3, and the profile of the Southern line was divided into two regions: B-1 and B-2, as presented in Figure 4.

NORTHERN LINE

The seabed in Region A-1 slopes very gently at 1.0 to 1.5 percent seaward. The soil near the shore consists of one meter of very soft clay, which increases in thickness to many meters at the seaside of Region A-1 underlain by interbedded layers of soft to stiff sandy silt and loose to medium dense silty sand with the stiffer and denser materials occuring below Elevation -12.5 meters (-41 ft).

Region A-2 presents a steeper seabed profile in the initial 200 meters (656 ft) where water depth increases from six meters to about 13 meters (19.7 to 43 ft). Thereafter, the seabed slopes more gradually (0.8 percent). The thickness of the surficial clay layer ranges from one meter to 17 meters (3.3 to 55.8 ft). The Penetration Resistance "N" value was consistently less than 1 throughout this layer. A very dense sandy-clayey gravel layer with "N" values greater than 100 was encountered below the clay. Its thickness was determined to be 4 meters (13.1 ft) in Boring NMP-1 and 10 meters (32.8 ft) in Boring SMP-1.

Region A-3 is similar to Region A-2 except as follows:

1) Water depth increases gradually seaward, reaching 25 meters (82 ft) at the PLEM location.

2) Thickness of the very soft surficial clay ranges between 7 meters and 9 meters (23 ft. and 29.5 ft).

3) The material below the surficial clay is medium stiff to stiff clayey silt, reaching 5.5 meters (18 ft) in thickness in Boring SPM-N.

SOUTHERN LINE

The seafloor slope and water depths in Region B-1 are very similar to that in Region A-1 (1 to 1.5 percent and 0 to 7 meters (23 ft) respectively) and except for the absence of the surficial one meter layer of very soft clay, the region has a generally similar soil profile, as presented in Figure 4

Region B-2 is essentially the same as Region A-2 with the following exception: At Boring SL-2 the very soft surficial clay layer is underlain by stiff gravelly clay, extending at least 0.6 meters (2 ft) below the bottom of the very soft clay.

SURFICIAL DISCONTINUITY

During the geophysical survey program it was revealed that a discontinuity, a shelf-like feature which "parallels" the shoreline, exists. This 2-km long (1.25 mile) discontinuity, which is at about 500 meters (1,640 ft) and about 800 meters (2,624 ft) from the shoreline along the Southern line and Northern line respectively, with a maximum elevation difference from top to bottom of the discontinuity of 24 meters (78.7 ft), required special attention, including determining if the discontinuity was caused by a fault.

In Phase II, test holes were drilled at the top and the toe of the shelf-like feature to examine the continuity of the subsoil layers. It was found that below a certain depth, approximately 30 meters (98.4 ft), the subsoil layers are continuous on both sides of the shelf; i.e., the discontinuity only involves the soil layers at depths of less than 30 meters (98.4 ft). The cause of the abrupt elevation drop and discontinuity of the upper soil layers was determined not to be caused by a fault based on the following reasoning:

1. To produce a fault scarp of a height of 24 meters (78.7 ft), an earthquake much greater than magnitude 8 would have been required; the 1964 Alaska earthquake of magnitude 8.5 gave an uplift of 15 meters (49.2 ft). A high fault scarp in "loose" under water sediments is much more difficult to preserve than one in base rock. If this elevation drop is a result of earthquake uplift, it must have been produced by a great event of very recent time. No recent historical seismic activity of great magnitude has been recorded at this location. The largest historical event of magnitude 8.3 in this region occurred at least 100 km (62.5 miles) away.

2. If this discontinuity is the result of an earthquake rupture, it should continue deep into the base rock underneath, because the focal depth of an earthquake of magnitude 8 or greater is not likely to be shallower than 10 km (6.25 miles). However, the cross sections of Figure 4 do not show such a deep continuation of this discontinuity.

3. A fault that is able to generate a shock of magnitude 6 or greater should be at least 10 km (6.25 miles) long or longer. The subject discontinuity is about 2 km (1.25 miles) long.

The stability of the marine slope (caused by this discontinuity) was studied and determined to be of no significant concern to the construction or the safety of the pipelines as discussed later.

SOIL PROPERTIES

Classification test results for the major soil layers are presented in Table 1. For the design of the pipelines and the PLEM units, the surficial very soft clay layer, which blankets the profiles of both pipelines except region B-1, and the surficial very loose silty sand layer, which blankets region B-1, are the most important layers of the soil profiles. For the very soft clay, laboratory testing program was carried out to assess the strength under static conditions. The static testing program revealed that the very soft to soft clay has an undrained shear strength of 12 to 36 kN/m^2 (0.02 to 0.75 ksf), and an effective angle of internal friction of 31 degrees with zero apparent effective cohesion. Sensitivity of this clay was 1.8. Clay content ($< 2\mu$) ranged between 35 to 73 percent. The natural moisture content was between 85 and 114 percent while the Liquid Limit and the Plasticity Index were between 78 to 127 percent and 37 to 62 percent, respectively. Total unit weight was 1.4 to 1.5 g/cm^3 (87 to 93 pcf).

For comparison purposes Table 2 presents laboratory test results for five soft marine clays including Chiriqui Grande Clay. Although the parameters measured for Chirique Grande clay generally fall within the ranges covered by the total group of other marine clays listed, no one particular clay matches it property for property.

The Modulus of Elasticity for the Chiriqui Grande clay ranged between 475 and 8500 kN/m^2 (10 and 1809 ksf) which is on the lower side of the range for typical values reported for very soft to soft clays (4).

For the silty sand layer, laboratory testing showed that the soil of this layer has a total unit weight that ranges between 1.7 to 1.8 g/cm^3 (105 to 115 pcf). The relative density was estimated to be between 40 and 80 percent based on a standard relationship between "N", relative density, and overburden pressure (6).

To evaluate the pipeline behavior, it was necessary to determine the loss of strength normally associated with soil liquefaction resulting from earthquake-induced or wave-induced cyclic loading in the surficial layers.

For the dynamic soil properties of the surficial clay layer, three samples were first subjected to cyclic loading, and then tested for static strength. Each sample was initially loaded for 60 cycles,

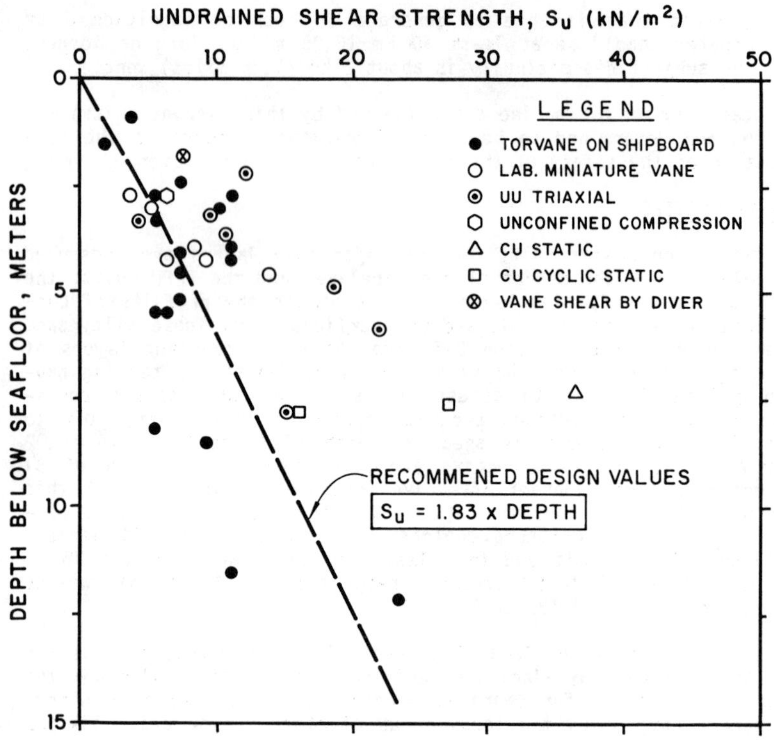

Figure 5 - UNDRAINED SHEAR STRENGTH FOR CHIRIQUI GRANDE CLAY

with a pulsating deviator stress selected so that the cyclic stress ratio would be approximately 0.35. Then the ratio was increased to approximately 0.5 and cyclic loading continued until the maximum total axial strain was approximately 10 percent. Finally the static strength remaining after cyclic loading was determined. Table 3 presents the results of the static and cyclic triaxial tests. Figure 6 presents the relationship between total strain and number of cycles.

For the silty sand layer, four cyclic simple shear tests were performed on two samples from Shelby tube and two remolded samples to evaluate liquefaction potential. A sinusoidal shear stress with a frequency of 1.0 hz was applied until the sample liquified. The relationship between stress ratio and number of cycles required to induce initial liquefaction is shown in Figure 7.

LIQUEFACTION POTENTAIL AND LOSS OF STRENGTH

Surficial Clay Layer

The results of the cyclic-static tests (see Table 3) indicate that no catastrophic loss of strength will result from earthquakeor wave-induced cyclic loading. The results of (S_u/P') for the soil after the cyclic loading described above, to (S_u/P')static for undistrubed soil was 0.88 for the sample consolidated to 49 kN/m^2 (1.05 ksf) and 0.74 for the sample consolidated to 98 kN/m^2 (2.1 ksf). These results indicate that approximately 75 to 90 percent of the undrained static strength will be available immediately after the expected earthquake loading. Relatively large deformations resulted from moderate cyclic loads as shown on Figure 6. Also, a significant buildup of pore pressure due to cyclic loading was reported (8).

SURFICIAL SILTY SAND LAYER

The susceptibility of this layer to liquefaction was studied using three simple evaluation methods based on relative density and overburden stresses. These methods were used for comparison of results and confirmation. The procedure outlined by Seed and Idriss (9), the procedure outlined by Seed (10), and the laboratory results from cyclic simple shear tests were used for this evaluation. For an earthquake of magnitude 8.3 at a distance of 100 km (62.5 miles), some of the silty sand was found to be liquifiable. The pipeline design proceeded on the assumption that the upper layer of sand is liquifiable under the effect of the design earthquake.

DESIGN AND CONSTRUCTION CONSIDERATIONS

Based on the geotechnical, hydrographical and geophysical investigations, the pipeline was designed to lay on the seabed without trenching or cover. The PLEM units were designed to be assembled in the factory and connected to the pipeline after ballasting and

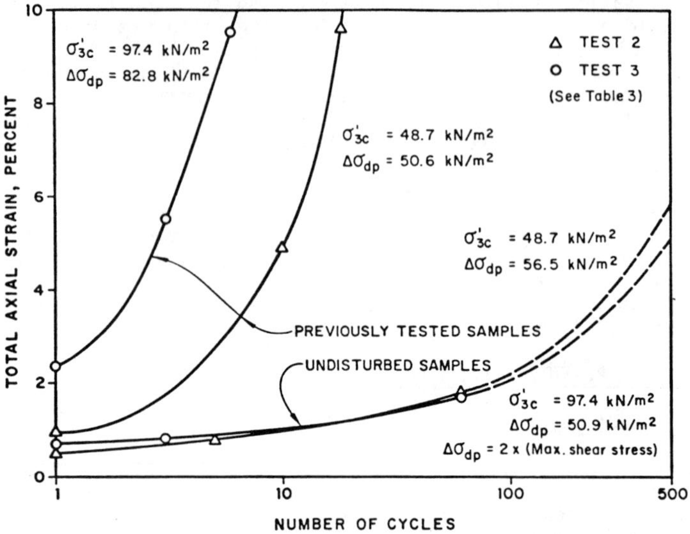

Figure 6 - RELATIONSHIP BETWEEN TOTAL AXIAL STRAIN AND NUMBER OF CYCLES

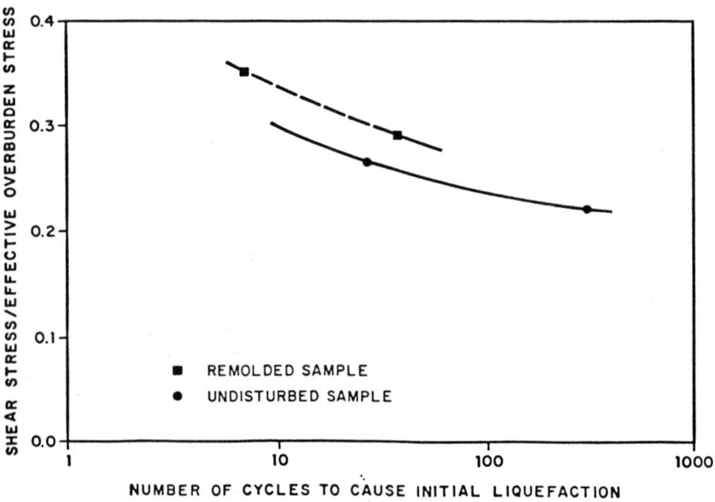

Figure 7 - RELATIONSHIP BETWEEN STRESS RATIO AND NUMBER OF CYCLES

resting on the seabed. The units were designed to be so ballasted as to result in differential settlements which are tolerable for the joint connecting the pipeline to the unit. Each SPM buoy was designed to be held in position by six anchors distributed equidistantly around the unit. The weight of each anchor is 15 tons. The depth, strength, and unit weight of the surficial clay layer were determined to be suitable for the placement and function of these anchors, and therefore, dimensioning of anchors proceeded, based on these findings.

The concrete-coated pipelines were assembled in the inland canal. The concrete coating was designed to allow floatation of the empty pipeline in sea water. After ballasting, each pipeline was floated to the required location to be connected to its PLEM unit. Calculations indicated that the pipelines would penetrate the surficial clay layer a distance of approximately 30 to 45 cm (12 to 18 in.). During construcion, the ballasting proceeded slowly enough to eliminate inertial forces which might have resulted in unfavorable penetration levels.

No special requirement was needed for the pipelines in crossing the sea floor discontinuity. For other reasons the pipelines cross the disconuity at a favorable angle of between 30 and 60 degrees, a range which is also the range recommended for pipelines crossing faults (7).

SUMMARY AND CONCLUSIONS

Field survey for the offshore loading-unloading facilities at the Atlantic terminus of the Trans-Panama Pipeline started at an early stage of the project schedule to acquire survey data for the preliminary design of the facilities, and to plan for the coming field work. The hydrographical, geophysical, geological, and seismological studies were performed and their findings were used in planning the Phase I geotechnical field work and laboratory testing. The planning of Phase II geotechnical investigation was based on the results of Phase I study and on the specific design requirements for the facilities. Based on these studies the following conclusions are warranted:

1. The undrained shear strength of the surficial very soft to soft clay is sufficient for the use of the proposed hook anchors.

2. The total and differential settlements of the PLEM units and the pipelines can be maintained within acceptable limits.

3. The loss of strength caused by liquefaction due to dynamic loading is not more than 25 percent for the surficial clay, and the pipeline will behave within acceptable limits upon liquefaction of the surficial silty sand layer which underlies the quarter of the Southern pipeline nearest to the shore.

4. The seabed elevation discontinuity was not a fault-induced feature. Considering the angle at which both pipelines cross this discontinuity, no special design was needed.

The construction of the facilities proceeded as designed. The facilities have been used continuously since September 1982.

TABLE 1

BASIC PROPERTIES OF MAJOR SOIL LAYERS

Soil Type and Consistency	Natural Moisture Content %	Unit Weight g/cm³ (pcf)	Gradation <76 μ %	Gradation <2 μ %	Liquid Limit %	Plasticity Index
Surficial clay, very soft to soft	85-114	1.4-1.5 (87-93)	80-98	35-73	78-127	37-62
Surficial Silty Sand, Loose to Medium Dense	29-50	1.7-1.8 (105-115)	0-70	0-20	-	-
Sandy Gravel/ Gravelly Sand Medium Dense to Dense	-	-	35	0	-	-
Silty Clay Medium Stiff to Stiff	108-111	1.7-1.8 (110-115)	94-97	12-22	126-148	83-101

TABLE 2
TYPICAL PROPERTIES OF SOFT MARINE CLAYS

Property	Chiriqui Grande(8) Panama	Singapore Clay (1) Singapore	James Bay Clay (2) Canada	Leda Clay (3) Canada	Norwegian Clay(2 & 3) Norway	San Francisco Bay Mud U.S.A. (11)
Natural moisture content %	85-114	50- 83	22- 38	28- 50	27- 40	85- 96
Liquid Limit %	78-127	50- 90	26- 38	20- 45	25- 36	88
Plasticity Index %	37- 62	30- 50	5- 18	5- 20	6- 20	45
Unit Weight, g/cm^3	1.4-1.5	1.3-1.9	1.8-2.0	1.7-2.0	1.6	1.5
(pcf)	(87- 93)	(81-119)	(112-125)	(106-125)	(100)	(94)
Gradation:						
%, < 76μ	80- 98					
%, < 2μ	35- 73	60- 70	-	-	-	-
Undrained shear strength, kN/m^2	1- 36	14- 72	19- 77	38- 96	7- 28	19
(ksf)	(0.02-0.75)	(0.3-1.5)	(0.4-1.6)	(0.8-2.0)	(0.15-0.6)	(0.4)
Sensitivity	1.8	1- 6	2- 12	7- 32	42	8
Effective Cohesion kN/m^2	0	0	0	0	0	0
Effective angle of internal friction, degrees	31	12- 16	30- 32	25- 30	26- 30	32*

* Calculated based on data given by Duncan and Buchignani (5) assuming effective cohesion equals zero.

TABLE 3
RESULTS OF CONSOLIDATED-UNDRAINED TRIAXIAL TESTS
ON CHIRIQUI GRANDE CLAY

Test No.	Type* of Test	σ'_{3c} kN/m^2	σ'_{3f} kN/m^2	$\frac{S_u}{\sigma'_3}$	$\left(\frac{\sigma'_1}{\sigma'_3}\right)_f$	$\frac{(S_u/P')}{(S_u/P')_{static}}$	Mode of Failure
1	Static	97.4	29.3	0.381	3.53	1.00	Shear Plane
2	Cyclic-Static	47.5	12.3	0.336	3.67	0.88	Shear Plane
3	Cyclic-Static	97.4	24.2	0.282	3.27	0.74	Bulge

* The cyclic loading was accomplished by initially loading each sample with 60 pulsating deviator stress cycles selected so that the cyclic stress ratio would be approximately 0.35. Then the ratio was increased to approximately 0.5 and maintained until the maximum axial strain was approximately 10 percent. Finally the static strength remaining was determined.

σ'_{3c} = Effective consolidation stress

σ'_{1f} & σ'_{3f} = Effection major and minor principal stress at failure respectively

S_u = Unconsolidated - Undrained shear strength.

P' = Average of effective major and minor principal stresses

Appendix 1 - References

1. Ahmad, S. A., and Peaker, K. R., "Geotechnical Properties of Soft Marine Singapore Clay," International Symposium on Soft Clay, Bangkok, Thailand, July 5-6, 1977.

2. Bjerrum, L., "Geotechnical Properties of Norwegian Marine Clays," Geotechnique, Vol. 4, 1954, pp. 49-69.

3. Bjerrum, L., "Stability of Natural Slopes in Quick Clay," Publication No. 10, Norwegian Geotechnical Institute, 1955.

4. Bowles, J. E., Foundation Analysis and Design, Third Edition, McGraw Hill Publishing Company, 1982, p. 67.

5. Duncan, J. M. and Buchignani, A. L., "Failure of Underwater Slope in San Fransisco Bay," Journal of the Soil Mechanics and Foundations Division, September, 1973.

6. Gibbs, H. J., and Holtz, W. G., Research on Determining the Density of Sands by Spoon Penetration Testing, 4th ICSMFE, Vol. 1, 1957, pp. 35-39.

7. Kennedy, R. P. and Short, S. A, "Seismic Design of Oil Pipeline System," Journal of the Technical Councils of ASCE, April 1979, pp.119-134.

8. Offshore Pipelines and Buoy Areas, Trans-Panama Pipeline, Geotechnical Report, International Engineering Company, Inc., San Francisco, California, October, 1981.

9. Seed, H. B., and Idriss, I. M., "Simplified Procedure for Evaluating Soil Liquefaction Potential," Journal of the Geotechnical Division, ASCE, Vol. 97, No. SM 9, (1971), pp. 1249-1279.

10. Seed, H. B. "Soil Liquefaction and Cyclic Mobility Evaluation for Level Ground during Earthquakes," Journal of the Geotechnical Division, ASCE, February, 1979, pp. 201-255.

11. Thiers, G. R. , and Seed, H. B., "Strength and Stress-Strain Characteristics of Clays Subjected to Seismic Loading Conditions," ASTM Special Technical Publication - 450,1968.

12. Trans-Panama Pipeline System, Seismicity Study for Petroterminal de Panama, International Engineering Company, Inc., San Francisco, California, August, 1981.

TRANS-PANAMA PIPELINE SYSTEM GEOTECHNICAL STUDIES

R. W. Heneks[1], M. ASCE. and G. R. Thiers[2] M. ASCE.

ABSTRACT: Rapid acquisition and analysis of geotechnical data were essential to the fast-track design and construction of the trans-Panama pipeline system, and geotechnical studies continued throughout construction of the project. The pipeline, 130 km (80 miles) in length, was designed and contructed during a 20 month period, ending in October 1982. The pipeline crosses the isthmus in a northeasterly direction from Puerto Armuelles on the Pacific Ocean to Chiriqui Grande on the Carribean Sea.

The pipeline, two pump stations and the Atlantic terminal were constructed on varied terrain at elevations from sea level to 1219 m (4,000 ft.). Thick, lateritic soils are present on most of the mountainous part of the project, and the Caribbean coastal area is primarily swamp. Geotechnical investigations were made on land and in Chiriqui Lagoon to design on-shore and off-shore facilities, and to facilitate construction.

I. INTRODUCTION

The Trans-Panama Pipeline system, owned by Northville Industries, Corp., Republic of Panama, and CBI Industries, Inc., is located in western Panama, near the Costa Rican border. Completed in October, 1982 by Consorcio Oleoducto Panama, a consortium of Morrison-Knudsen International Company, Boise, Idaho and Constructora Urbana, S.A. of Panama, the line extends 130 km (80 miles) from the Pacific Ocean at Puerto Armuelles northeast to Chiriqui Grande on the Caribbean Sea (Fig. 1).

Originating at an existing terminal at Puerto Armuelles, the pipeline rises from near sea level to an elevation of nearly 1,219 m (4,000 ft) where it crosses the continental divide in the Serrania de Tabasara mountains and then descends again to sea level at Chiriqui Grande, 13 km (eight miles) to the northeast.

---

1 Principal Geological Engineer, International Engineering Company, Inc., San Francisco, California.

2. Chief, Soils and Foundation Department, International Engineering Company, Inc., San Francisco, California.

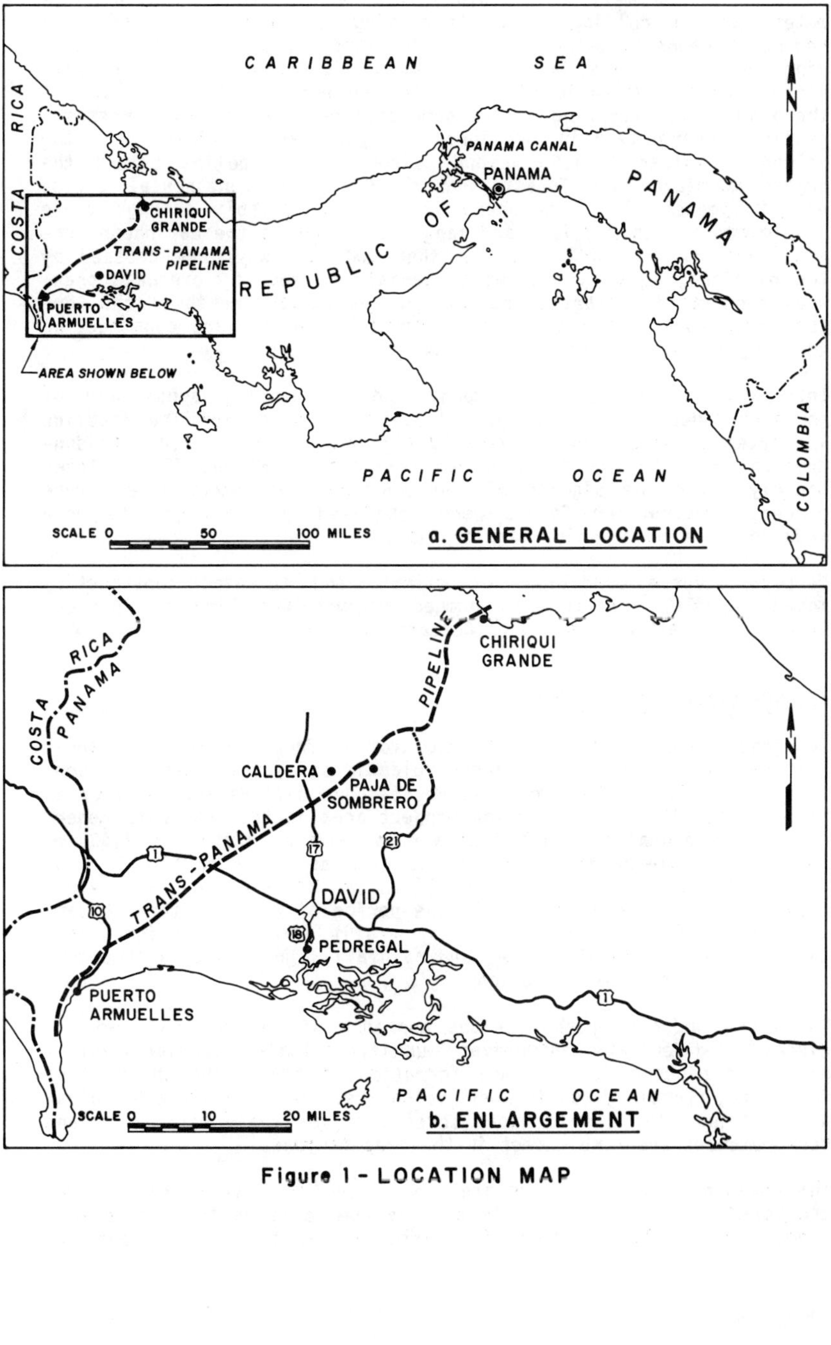

Figure 1 - LOCATION MAP

Terrain is extremely varied along the route. The first ten kilometers are in rolling hills, traversing uplifted marine sediments and beach deposits. The next 20 kilometers cross a low, gently sloping region of rice farms and banana plantations on the outwash plain from the volcanic highlands to the north. The next part of the alignment, extending to the pump station near Caldera, crosses a gently sloping terrain developed on pyroclastic flows from the volcano, "Volcan Baru." Beyond Caldera, the pipeline enters the rugged mountain rain forests and rises to the continental divide near kilometer 110, from where it descends steeply through dense jungle and across mudflats and mangrove swamps to the Caribbean terminal at Chiriqui Grande. More than 100 waterways are crossed by the pipeline, many by underwater burial and some by overhead span. When construction started, no road access existed north of the Fortuna area (just south of the continental divide) where construction for a hydroelectric project is in progress.

International Engineering Company, an engineering subsidiary of Morrison-Knudsen Co., started a feasibility study for the pipeline in September, 1980, and provided design engineering after Morrison-Knudsen received the construction contract in March, 1981. Immediately after the signing of the contract, equipment orders were placed, construction forces were mobilized and arrangements were made to start geotechnical exploration.

Despite adverse conditions of climate and terrain, construction moved rapidly, and oil was pumped through the line to Chiriqui Grande into waiting tankers on October 1, 1982.

II. GEOGRAPHY AND GEOLOGY

Topography along the pipeline route varies from nearly level agricultural land to the sharp-ridged volcanic spine that marks the continental divide. Slopes on some mountainous divides are as steep as 50 percent. The climate in the project area is tropical with generally heavy annual rainfall that varies from as little as 1,400 mm (55 in.) on the Pacific coast to as much as 6,000 mm (236 in.) in the mountains. In general, rainfall on the Pacific side is heaviest from May through November, while the period of December through January is relatively dry. Rain is generally more uniform throughout the year on the Atlantic side, and rainfall is nearly a daily occurrence in the Cordilleran region.

The rocks underlying the various portions of the Project Area include: Mid-Tertiary intrusives, undifferentiated Tertiary volcanics, the Miocene Gatun-Uscari formation, the Pliocene Charco Azul formation, Tertiary-Quarternary volcanics, and Pleistocene volcanics. Generally, a thick soil cover overlies these rock units and rock outcrops are rare except in the deep canyons.

The dominant feature of the area is the volcanic mountainous area, the Cordillera Central. It in turn is dominated by the active volcano, "Volcan Baru," which is surrounded on the lower slopes by large deposits of unconsolidated volcanic ejecta composed of pyroclastic bombs, blocks, tuffaceous sands, and other material. The main part of the volcanic chain is a part of the La Fortuna anticline with an axis trending east-west. Granitic rock types are exposed in several places near the axis of the structure, where erosion has cut though the thick andesitic volcanics.

Major fault systems are present near the crest of the anticline and thick gouge zones may be seen along the construction road to Fortuna Dam. Regional fault systems are also marked by the ranges of low hills near David and Puerto Armuelles.

III. PROJECT DESCRIPTION

Major features of the system include:

- The 130-kilometer long, welded steel 36-inch and 40-inch diameter pipeline.
- Oil storage and tanker-loading facilites at Chiriqui Grande.
- Pumping stations at Puerto Armuelles and Caldera.

Forty-inch diameter pipe makes up the first 86 km (53 mi) of line from Puerto Armuelles pump station to Caldera pump station. At Caldera, the line changes to a 36-inch diameter pipe for the climb over the mountains and the descent down the north side to the crude oil storage facilities at Chiriqui Grande. Except for water crossing spans, the entire pipeline is buried. Pipe used in buried river crossings was concrete coated in the field. Offshore lines were factory-coated with concrete. Pipe laid in a dredged channel from the base of the hill below the crude storage tanks to the shoreline manifold was not concrete coated, but is held in place on the channel bottom with screw anchors.

Pump stations were constructed at Puerto Armuelles and at Caldera. Each station is equipped with four pumps driven by 5,000- horsepower motors. At Puerto Armuelles, electric power is provided by eight diesel engine generators. Power for the Caldera pump station is supplied by the Panama National Power Company.

Both on-shore and off-shore facilities were constructed at the Atlantic terminal. Structures were required for the following facilities:

- Crude oil storage
- Ballast water receiving, storage and treatment
- Shoreline distribution manifold
- Unloading dock and causeway

Other structures constructed at Chiriqui Grande include a 450-m (1,476-ft) rock causeway, control and office building, recovered oil tanks, warehouse, hose-handling building, air strip and camp. A pile-supported cargo dock at the end of the causeway provides loading facilities in water with a depth of 6 m (20 ft).

IV. GEOTECHNICAL INVESTIGATIONS

The fast-track design and construction of the pipeline made it necessary to acquire and analyze geotechnical data rapidly to provide design parameters in advance of construction. Geotechnical data acquisition was an on-going part of the pipeline project and IECO's engineers and geologists were present on site from start of exploration to completion of construction. The pipeline facilities were constructed in areas of varied terrain with varied soils conditions. Geotechnical investigations were made to aid final site selection and structures design and to select construction methods.

A. Onshore Geotechnical Studies

Geotechnical studies for on-shore facilties included a review of published data, air and surface reconnaissance, air photo interpretation, seismicity study, soils tests and detailed field investigations. Field investigations included geologic mapping, electrical resistivity, and sampling from test pits and boreholes.

After a preliminary pipeline alignment had been laid out on topograpic maps, a brief reconnaissance was made in January 1981. This reconnaissance consisted of aerial flights over the planned route and on-ground visits to locations readily accessible by established roads. The report of this preliminary reconnaissance was the basis for planning further investigations and it described conditions anticipated to be met during construction.

1. Crude Oil Storage Tank Site

a. Field exploration

In order to start construction of crude oil storage tanks by a date that would permit their completion in time to receive oil from the pipeline, it was necessary to rapidly select a suitable site and to start excavation as soon as possible. Exploration for a tank site started the latter part of April, 1981. Sites were evaluated based upon the following considerations:

- o Proximity to Chiriqui Lagoon
- o Elevation high enough to provide gravity flow to buoys without excessive pressure.
- o Sufficient area without excessive excavation.
- o Suitable soil conditions.

- Miscellaneous buildings for offices, control, hose-handling and warehousing.

- Oil spill impound basin and fire protection reservoir

- Offshore loading

In the crude oil storage area, three steel tanks with floating roofs were erected to hold a total of 2,500,000 barrels of oil. Each tank is 92 m (302 ft) in diameter and 20 m (65 ft) in height. Oil from the 36-inch pipeline enters a receiving manifold from where it flows by gravity through the distribution manifold to the selected tank. One of the tanks is at an elevation of 247 m (810 ft.) and two are at 235 m (771 ft.). Concrete walls and shotcrete-surfaced earth dikes surround each tank to collect storm water and oil spillage. The enclosed area around each tank provides storage for a minimum of 10 percent of the tank capacity. Power for lighting, pumps and controls is provided by diesel engine generators.

A lined earth reservoir to store water for industrial and fire-fighting use was constructed on a nearby hill at an elevation of 436 m (1430 ft). Water is pumped to the reservoir from the Margarita River at the base of crude tank hill. A radio tower was erected on a higher hill near the reservoir as part of a communications network linking all parts of the pipeline system.

An earth-fill dike, constructed at the base of the hill below the crude oil tanks, provides a basin to impound spilled oil in excess of that contained within the individual tank dikes in event of a catastrophic spill.

Ballast water from tankers is treated to remove oil and sludge and cleansed water is then discharged into the ocean. Raw ballast received from tankers via 30-inch diameter submarine lines is stored in three 120,000-barrel tanks erected on a shoreline site at an elevation of 23 m (75 feet). Ballast treatment is provided by a dissolved air floation unit, a chemical mixing building and a holding pond. The treating facilities have a capacity of 12,000 barrels per hour.

Crude oil flows by gravity from the crude storage tanks through two 36-inch lines 3.5 kilometers (2.2 miles) to the shore line distribution manifold. At the shore line manifold, oil flow is controlled and the crude is delivered by two 36-inch submarine lines to two mooring buoys located about 2,000 m (6,560 ft) from shore.

Each mooring buoy has a design loading capacity of 60,000 barrels per hour. Each buoy is anchored in about 22 m (72 ft) of water with six anchors. The submarine crude oil and ballast lines enter a pipeline end manifold (PLEM) on the sea bottom under each buoy and 508 mm (20-in) diameter hoses connect the PLEM to the buoy.

The area chosen for a prospective crude tank site was on the north slope of a hill about 3.5 km (12 mi) west of Chiriqui Grande village. The hill has a maximum elevation of about 500 m (1,640 ft). Although the selected site was a little higher than desired, it was the only nearby location that provided sufficient area with topography permitting excavation to finish grade by the established deadline.

Preliminary exploration of the site, consisting of five boreholes, from 15 m (49 f.) to 29 m (95 ft) in depth and two hand-dug test pits, each 5 m (16 ft) in depth, showed soil conditions apparently suitable for the proposed facilites. Tentative locations were selected for three tanks, and 17 additional boreholes were drilled on these locations to determine engineering properties for foundation design and analysis of slope stability and settlement. IECO geotechnical engineers supervised the drilling, testing, and sampling and logged the boreholes and testpits. Standard penetration tests were made in all boreholes, soil samples were obtained using split spoon samplers and Shelby tubes, and rock was cored with double-tube core barrels. Soil samples and cores were visually examined and soil samples were tested in laboratories in Panama and the United States.

Many of the borings did not encounter ground water, and measurement of the water levels that developed indicated that the water table could normally be expected to be well below the proposed tank foundation levels. In the final configuration, two tanks were placed at a final grade of 235 m (771 ft) and the third at 247 m (810 ft), requiring cuts as great as 25 m (82 ft).

The tank farm site is typical of tropical areas where thick residual soils (laterites) have developed by weathering of igneous rocks. The thick soil, from 12 m (39 ft) to 34 m (112 ft) in thickness is underlain by granodiorite. Because of the deep weathering, no outcrops of in-place bedrock were seen near the site, but huge boulders of granodiorite have accumulated in the gullies and stream courses, and dioritic boulders of various sizes are scattered randomly over the ground surface. Many boulders were encountered during drilling, reducing productivity and, sometimes, making it necessary to abandon boreholes and to drill new holes alongside.

The suburface profile is uniform across the site, generally following site topography. Six soil and rock zones were recognized from the subsurface exploration -- three upper soil zones, separated by a transition zone from weathered and unweathered rock zones below. The upper stratum, occurring to a depth as great as 3 m (10 ft), is a clayey silt (MH), with organic material, yellow to brown, soft to medium stiff, and damp to moist. The second horizon, from one to four meters in depth is a clayey-sandy silt (MH), mottled, dark brown to red, soft to medium stiff, moist to wet, with traces of cemented silt. The third zone is a sandy silt (ML) with silty sand pockets, white to gray, medium dense to dense, moist to wet, with scattered weathered diorite boulders (corestones). Consi- dered to be a saprolite, this zone contains relict rock structure, and is marked with black micaceous zones. The fourth zone is a transition from saprolite to weathered bedrock. It is a sandy silt (ML), light brown to white and gray with micaceous zones, dense to very dense, moist to wet. Diorite boulders are more common in this zone than in the overlying zone. This fourth zone contains numerous voids and soft rock or silt zones and was often recognized by drilling-water losses when encountered in a borehole. Below this zone of severely weathered rock there is a gradual transition downward through a zone of partly weathered rock to unweathered bedrock. The partly weathered rock zone contains joints along which the rock is altered, and there is discoloration of the rock in recovered cores. The unweathered rock shows no alteration of mineral constituents and there is only some slight discoloration along joints.

b. Investigation Results

Representative samples were tested in laboratories in the United States for classification, engineering properties and chemical content.

Based upon field observations and results of laboratory tests, the following generalizations can be made regarding soils conditions and properties at the crude tank site:

- Natural moisture content, which varied from 36% to 50%, decreased with depth.
- Total density increased downward, varying from 101 pounds per cubic foot in the upper stratum to as much as 110 pounds per cubic foot in the fourth stratum.
- The plasticity index decreased with depth.
- Sand content increased with depth.
- Shear strength increased with depth.

The calculated ultimate bearing capacity for the tank sites produced factors of safety greater than required for the three cases considered -- end of construction with full tank load, long term condition with full tank load, and long term condition with full tank load plus seismic load. Calculated settlement values were small, less than 8 cm (3 in.), and maximum differential settlement along the ring wall was computed to be 8 cm (3 in.) in 75 m (246 ft). Actual settlements recorded when the tanks were hydrotested were much smaller than those predicted, possibly due to the presence of better foundation conditions than those assumed for design.

Stability of cut and fill slopes was analyzed by standard methods, assuming 2 horizontal:1 vertical cuts and 1.5 horizontal to 1 vertical fills. Calculated factors of safety were satisfactory for all conditions investigated. Because of the high erodibility of the silty soils, excavated slopes were protected during construction with plastic ground sheets, and finish slopes were covered with wire-reinforced shotcrete. The tank sites were overexcavated, plastic sheets were placed over the cut surface and river-run gravel fill was placed to bring the surface to finish grade.

2. Ballast Treatment Area

An early decision was made to locate the ballast treatment facilities on a small, isolated hill on the shore of Chiriqui Lagoon at the north edge of Chiriqui Grande village. Bedrock exposed in the hillside at shoreline showed the site would probably provide an excellent foundation for the three 120,000 barrel cone-roof ballast tanks and appurtenant structures, and excavated rock would provide fill material for roads and structural fill. Excavation of soil and shooting of the rock was started as soon as equipment was available, and the shot-rock was used as fill to construct a road inland through the swamp. The site elevation was reduced to a finish grade of 23 m (75 ft) from its original elevation of 53 m (175 ft). Material excavated from the hill was hard igneous rock, including basalt, rhyolite, andesite, and agglomerate. The blocky, shot-rock was an excellent fill material, and was used to construct foundation fills up to 11 m (36 ft) in depth. In addition to its use as road fill and at the ballast treating area, the shot-rock was also used to construct the 400-m (1312-ft) long causeway, and as fill at the shore manifold, warehouse and control building sites.

Because the finished grade for the ballast tanks was primarily rock, geotechnical exploration in this area was limited to drilling a series of shallow holes on the circumference of the two southerly tanks, where bedrock was not exposed on the excavated surface, to determine the depth to bedrock and the soil conditions. The strength of fine grained soils present was estimated using a pocket penetrometer. Ground surface beneath the ballast tanks was brought to grade with shot-rock and river-run gravel and a final course of sand compacted with vibratory compacting equipment. After the tanks were erected, a stabilizing layer of shotcrete was placed on the gravel surface around the tank circumference for a distance of 1 m (3 ft) from the tank wall.

3. Shore Manifold Area

Like most of the coastal mud flat north of Chiriqui Grande, the site selected for the shore manifold was a swamp with dense underbrush and mangrove trees. The soil profile consists of an upper zone, with a thickness of about 1.5 m (5 ft), of saturated soft, clayey silt with organic material underlain by a gray silt that grades with depth from soft in the upper part to stiff at a depth of about 3 m (10 ft). Zones of sandy silt and shell remains are present.

A road to the site was constructed using shot-rock from the ballast tank hill excavation as fill. After trees were felled and removed, backhoes removed the upper soft soils and vegetation, and the site was brought to finished grade with shot-rock and river-run gravel fill.

4. Crude Oil Spill Impound Basin

In event of a catastrophic occurrence it is necessary to have a means of containing spilled oil for economic and environmental reasons. At Chiriqui Grande terminal, topography dictated that the most economical and least hazardous location would be at the base of the hill below the crude tank area.

An earthen dike, 800 m. (2624 ft) in length and 3 m (10 ft) in height was constructed with selected river-run gravel to provide storage capacity in excess of 1.25 million barrels. The dike was constructed with side slopes of 2 horizontal to 1 vertical and a crest width of 3 m (10 ft). In-situ permeability tests, performed at the impound basin site, showed the natural ground to be nearly impermeable to crude oil. Normally saturated, the surface soil, a clayey silt, acts as a natural barrier to transmission of crude oil. Permeability tests performed in the laboratory on the river-run silty gravels indicated that seepage to be expected through the dike in event of crude oil collection would be within acceptable limits. Experience with the river-run gravel at crude tank hill, showed the permeability of the material reduced dramatically with compaction and subsequent exposure.

Storm water from the diked area around each crude tank flows through valved 24-inch storm drains into a canyon on the north side of the tank farm. Storm water held within the diked areas can be released into the canyon, or it can be picked up at the drain outlets for recovery of spilled oil. In case of a large oil spill that cannot be contained within the diked areas, the oil will flow over weirs in the dike walls to the canyon.

By excavating a small divide and using the excavated material to build a small levee, drainage from the crude tank area was redirected into an adjacent canyon that drains to the impound basin. Two 24-inch diameter valved drains were placed through the impound basin dike to permit draining of the basin, maintaining a reserve storage capacity at all times.

Some dike maintenance will be required to repair eroded areas of the dike caused by heavy rains.

5. Fire-water Reservoir and Microwave Tower

Other facilities constructed near crude tank hill include a water reservoir and a microwave tower. The plastic-lined earth reservoir provides about 7,000,000 gallons of water for industrial and fire-fighting use. It was constructed on a hill near the crude tank storage area, excavating in sandy silts to a bottom elevation of 436 m (1430 ft). The tower was erected on an adjacent hill at an elevation of about 500 m. (1640 ft), providing line of sight to a repeater on Volcan Baru.

Soils conditions at these sites were similar to those at crude tank hill, but excavation depths were much less and fewer boulders were encountered. Pocket pentrometer tests provided confirmation of soils conditions assumed for foundation design.

6. Pump Stations

a. Puerto Armuelles

Geotechnical investigation of the Puerto Armuelles pump station site was conducted during the period of 11 August to 13 August, 1981. Exploration consisted of three borings, ranging in depth from 5.1 m (16.7 ft) to 6.3 m (20.7 ft) and a test pit, excavated to a depth of 3.6 m (12 ft). The site is located a short distance from the Pacific Ocean beach about seven km (4.4 mi.) south of Puerto Armuelles. The site is underlain by up-lifted marine sediments and beach deposits. Bedded sediments exposed in the vicinty are primarily claystones and mudstones. The materials encountered in the boreholes and test pits consisted of medium stiff to stiff silty clays and clayey silts with some medium dense to dense sand and gravel. Site soils conditions were suitable to found the proposed structures.

b. Caldera

The Caldera pump station site is located approximately three km (1.9 mi.) northeast of the village of Caldera. The site is on a hill at an elevation of about 420 m (1378 ft) above mean sea level. The soils at the Caldera pump station site are residual soils developed from the undifferentiated volcanic rocks underlying this region.

The site was investigated with a test pit dug to a depth of 4.5 m (15 ft) and with two boreholes, one 14 m (46 ft) in depth and the other 13 m (43 ft) in depth. In general, descending from the surface, the soils consist of an upper stiff, sandy, clayey silt with some organic material; an underlying very stiff, sandy, clayey silt; and a very stiff to hard, clayey to sandy silt. Lenses of silty sand with rock fragments are present locally. Field and laboratory tests showed the soils conditions to be suitable for the structures proposed.

7. Pipeline

a. Routing

The pipeline was routed through a pass in the mountains rather than following a straight line from the Pacific to the Atlantic. Although the selected route is about 22 km (14 mi.) longer than the straight-line route, the construction cost was estimated to be less because of reduced mountain construction. The maximum elevation on the chosen route, 1200 m (3936 ft), is about 700 m (2236 ft) less than the direct route over the mountains, and annual power costs were estimated to be approximately $6,200,000 lower.

Following the aerial reconnaissance of the proposed route, the route through the mountains was studied using aerial photographs, and the access road, already under construction east of the Fortuna Dam site, was inspected to assess the performance of cut slopes.

The interpretation of aerial photographs located many existing landslides and identified numerous faults along the pipeline route. The alignment of the pipeline between approximate station 86 and station 112 was subsequently changed, moving northwesterly to a higher elevation. The revised route avoided the steep slopes along Rio Chiriqui and missed most of the identified landslides. The only landslides crossed by the pipeline are ancient, stable, reforested slides that should not present a hazard to the pipeline. The faults identified are discussed below.

Because of lack of access and the concealment of the ground by dense vegetation and heavy tree cover, it was necessary to make ground investigations of the route on a continuing basis as construction proceeded.

Although the pipeline crosses varied terrain underlain by a variety of sedimentary and igneous rocks, right-of-way excavation was primarily in thick, residual soils, and little trenching in rock was required.

b. Slope stability

From Puerto Armuelles to Caldera pump station the pipeline is constructed on gently sloping ground and there are no steep cuts to produce slope stability problems. From Caldera pump station to Chiriqui Grande, the pipeline crosses mountainous terrain with great relief that required steep cut-slopes at many locations. Along most of this mountainous portion of the route, the pipeline is in deeply weathered Tertiary volcanics, comprised chiefly by ash, tuff, agglomerate and basalts. Locally, there are Tertiary andesites, and intrusive granodiorites and quartz diorites. In general, the surficial soil is a brown to red, medium stiff to very stiff, sandy silt with an overlying layer of medium stiff to very stiff brown clay containing organic material. The total thickness of surficial materials is as much as 25 m (82 ft) or greater. Most pipeline cuts did not penetrate below highly weathered rock into the underlying less weathered, hard bedrock.

Minor landslides were a problem during the construction of the pipeline through the mountainous region, but the quantities of material involved were not great. Some bedding plane discontinuities were present, but most slides were not deep-seated and were confined to the surficial soils. Benches were used to improve stability of higher cutslopes and to reduce erosion.

Numerous faults were crossed by the pipeline in the mountainous region, but none are of regional magnitude. No indications of recent activity were found on any of these faults and no special treatment was taken for fault crossings.

c. Seismic considerations

An historical seismicity study showed peak accelerations for design earthquakes for the pipeline system to vary from 0.15 g at Chiriqui Grande to 0.55 g at Puerto Armuelles. Design magnitudes varied from 7.0 for local earthquakes to 8.3 for regional earthquakes. Foundation designs were based on these parameters.

At most faults, the pipeline crossed at an angle that would not give maximum compressive stresses. The limited burial depth (generally, 1 m cover) and the use of a moderately dense, essentially cohesionless backfill material provide capacity for pipe movement.

B. Off-Shore Geotechnical Studies

Chiriqui Lagoon is shallow with a gently sloping bottom that receives sediment from numerous silt-laden streams that empty into it. The maximum depth of the lagoon is about 21 fathoms.

Ocean Surveys, Inc. conducted a detailed marine survey of the area of interest immediately offshore of the village of Chiriqui Grande between April 28, 1981 and June 6, 1981. A drilling program, supervised by IECO engineers, was performed during the period of May to September, 1981. Results of these investigations provided the basis for engineering and construction of the offshore pipelines and loading facilities.

Twenty-three borings were drilled to depths ranging from 12 m (39 ft) to 62 m (203 ft). Two borings were drilled to determine soil conditions at the site for the causeway and the pile-supported loading dock. The rest of the boreholes were located to investigate the two mooring points and the submarine pipe routes to each mooring point. All drilling was directed by IECO engineers, who logged the boreholes and supervised sampling and testing. Representative samples were tested in soils laboratories in the United States to evaluate engineering properties. Dynamic testing showed no catastrophic loss of strength, normally associated with soil liquefaction, will result from expected earthquake-induced cyclic loading.

In general, the soil profiles along the lines to the two mooring points are similar, and along each line the sea floor slopes gently seaward for a distance of about 400 m (1312 ft) from shoreline, at which point the seaward gradient steepens. This change in slope probably marks a change in sedimentation, with a build-up of coarser materials shoreward. Toward the shore, the soils are comprised in descending order, by an upper, soft, silty clay, 1 m (3 ft) in thickness; a gray-green silty sand, 2.4 m (8 ft) in thickness; a gray-green sandy silt, 4.6 m (15 ft) in thickness; and a silty sand, with some clay. Seaward from the abrupt slope, the soils generally consist of an upper, dark-gray, very soft silty clay, 11 m (37 ft) in thickness, underlain by a dark-gray sandy gravel, 3.4 (11 ft) in thickness, and a silty clay with layers of fine gravel. Suspended silt forms a very soft layer above the defined lagoon floor. This layer becomes gradually denser with depth, grading into the soft bottom muds. The vertical position of the PLEMs in the soft bottom was controlled at each mooring by ballasting.

V. SUMMARY AND CONCLUSIONS

The trans-Panama pipeline, a fast track design and construction project, was completed on schedule under adverse conditions in a tropical environment. Geotechnical studies for facility site selection, structures design, selection of construction methods and construction monitoring were performed on a continuing basis until completion of the project. Geotechnical exploration of varied soil conditions at remote locations provided a continuous flow of information to revise designs and adjust construction methods as work progressed. Because of climate and soil conditions, it will be necessary to monitor ground water levels, stability of slopes and erosion of cuts and fills to ensure continued satisfactory performance.

Slurry Pipeline Operations in Cold Regions - Some Aspects

Anders Sellgren*

Abstract

International experience has demonstrated that iron ore slurries can be shut down in a pipeline and restarted again without difficulties after several days. However, in cold regions shutdown and restart activities are critical factors because of the risk of freezing.

In this study the temperature of the flowing slurry along the pipeline and the cooling process of a shutdown period were simulated by mathematical models.

Background operating data were provided by preliminary considerations for a pipeline in northern Sweden and viscometric tests with iron ore slurries at different temperatures. The experimental results showed that great temperature-dependent changes in the rheological properties can be expected during restart procedures.

Introduction

About ten years of international experience has shown excellent technical and economic benefits of long-distance pipeline transportation of ores, in situations where the ore has been ground to its final dressing stage with a maximum particle size of 0.1-0.2 mm. In connection with a possible opening of a new iron ore mine north of the arctic circle in Sweden the LKAB mining company analyzed the feasibility of pipeline transportation of 2.5 Mtonnes of ore to the coast (250 km). Figure 1.

No long-distance slurry pipelines operate in locations where the climate causes considerable risks of freezing. In Sweden, the ground mainly consists of moraine and the maximum frost line in ground covered with snow is between 0.5 and 2.0 m. In ground without snow the frost line will be considerably deeper. Preliminary studies indicate that the pipe laying level of a 250 km iron ore slurry pipeline with a diameter of about 0.25 m would be 1.0-1.3 m (Jansson 1977).

*Ass. Professor, Department of Water Resources Engineering (WREL), Luleå University of Technology, S-951 87 Luleå, Sweden.

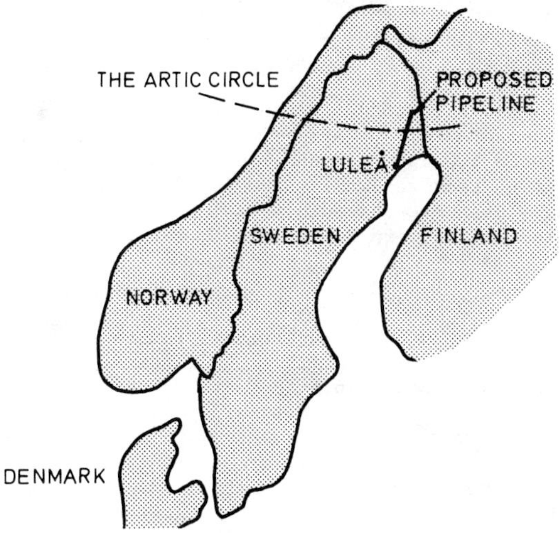

Monthly temperatures in 1981/82

Jul	15.8 °C
Aug	14.2
Sep	8.5
Oct	2.8
Nov	-3.8
Dec	-15.3
Jan	-14.8
Feb	-6.0
Mar	-2.4
Apr	1.3
May	6.2
Jun	10.1

Figure 1. Location of a proposed iron ore slurry pipeline in northern Sweden and climate data for Luleå.

International experience has shown that slurries can be shut down in a pipeline and started again after several days without difficulties. However, in cold regions, shutdown and restart activities are critical factors because of the risk of freezing.

The new mine was never opened because of the decreased demand for iron products on the world market. However, the studies performed indicated that the risk of freezing must be carefully considered in the case of a detailed design. For example, the following factors should be evaluated

- use of insulation at certain locations
- increase of the slurry temperature in the concentrator during winter
- installation of heating cables along the pipeline.

Objectives and assumptions

The overall objective was to evaluate the risk of freezing at shutdown in a slurry pipeline of a length of 250 km and with an inner diameter of 0.25 m. The aim was to simulate the thermal conditions along the pipeline and the cooling at a typical shutdown period. The thermal processes were modelled mathematically and hydraulic data were obtained experimentally with a representative iron ore slurry.

The restart process of a long-distance pipeline full of slurry involves complex transient conditions with great local changes of the slurry properties. Therefore, the aim was also to evaluate the rheological behavior of a high-concentrated slurry at temperatures close to freezing.

The simulation was based on the following simplified assumptions. The pipe laying level was 1.25 m in moraine with an average thermal conductivity of 1.8 W/mK and with a water content of 8 %.

The assumed climatic conditions corresponded to late fall and beginning of winter. During the late fall the air temperature is about $0^{o}C$ and close to the temperature in the upper ground layers. Furthermore, it was assumed that the average temperature in December was $-15^{o}C$ with no snow cover on the ground. These are reasonable assumptions when risks of freezing are studied in northern Sweden.

Hydraulic design considerations

In most industrial slurries economic considerations dictate the use of high concentrations and particle size distributions weighted significantly towards the fine particle end to provide a slurry with good pumping characteristics and high solids transport rate. Most systems are operated in such a way that the particles are nearly uniformly distributed, i.e. the settling rate of the solid component is small under typical flow conditions. The flow can generally be described as consisting of a homogeneous fluid medium formed by the water and the finer particles. The coarser particle sizes then act heterogeneously in this medium.

The operating velocity must be high enough to ensure turbulent flow conditions, because even fine-grained industrial minerals contain traces of coarse particles capable of settling at laminar flow conditions. The range of velocities where the flow changes from laminar to turbulent conditions is very sensitive to the system rheology, which means that the mixture may behave in a non-Newtonian way. To ensure turbulent conditions in the system, the transition velocity must be established. The rheological properties are normally identified in laboratory viscometric studies.

Viscous non-Newtonian effects exert considerable influence on the deposition conditions and the pipeline resistance to the flow. Generally, it is not sufficient to characterize the slurry rheologically, or the fine particle portion of it, in order to determine a suitable operating velocity. The deposition velocity may be significantly greater than the transition velocity, especially in larger pipe sizes. In order to predict the deposition velocity it is normally necessary to perform pipe loop tests.

Long-distance pipeline transportation systems for ore concentrates are today operating at velocities in the range of 1.5-1.8 m/s in pipes with diameters of 0.2-0.4 m. The solids concentrations by weight are in the range of 60-64 % which corresponds to concentrations by volume of 23-27 %.

The cooling rate at shutdown

Heat is built up along the pipeline after the operation has started. Fluctuations from a steady-state situation are influenced by the climatic variations or by changes in the operating conditions.

The slurry temperature from the ore concentrator at the beginning of the pipeline is normally about 20°C. The temperature distribution was now simulated in sections where the temperature of the flowing slurry had dropped to 10°C and 5°C. The cooling was modelled for a shutdown period after 24 days of continuous operation with a ground surface temperature of -15°C.

The non-stationary cooling of the warm soil around a buried pipeline with a ground temperature below 0°C means that the frostline is moving downward. The phase change from water to ice in the soil also changes the thermal properties in the ground. The different regions, the phase change interface, and boundary conditions are schematically shown in Figure 2.

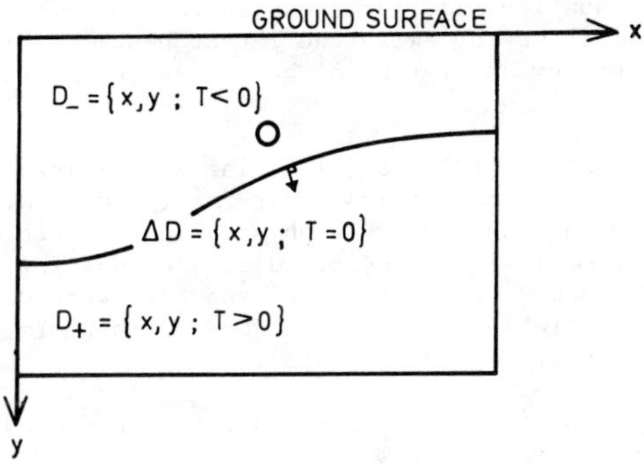

Figure 2. Schematic representation of the movement of the frostline.

The heat transfer in an x-y plane in a homogeneous medium is defined by the following relationship, known as the heat conduction equation

$$\frac{\partial}{\partial x}\left(\lambda \frac{\partial T}{\partial x}\right) + \frac{\partial}{\partial y}\left(\lambda \frac{\partial T}{\partial y}\right) - \rho c \frac{\partial T}{\partial t} + S = 0 \qquad (1)$$

where λ, ρ, and c are thermal conductivity, soil density, and heat capacity, respectively. T is temperature, t is time, and S is heat generated per unit time and volume.

Equation (1) was applied to the domains shown in Figure 2

$$\frac{\partial}{\partial x}\left(\kappa \frac{\partial T}{\partial x}\right) + \frac{\partial}{\partial y}\left(\kappa \frac{\partial T}{\partial y}\right) + q = \frac{\partial T}{\partial t} \qquad (2)$$

where the thermal diffusivity is defined by

$$\kappa = \frac{\lambda}{c\rho}$$

$$\kappa = \begin{cases} \kappa_- \text{ on } \{D_- = x,y;\ T<0\} \\ \kappa_+ \text{ on } \{D_+ = x,y;\ T>0\} \end{cases}$$

The diffusivities κ_- and κ_+ correspond to frozen and unfrozen soil, respectively, and q is a source term. The jump in thermal properties is accommodated by an energy balance at the phase change interphase.

$$\Delta D = \{x,y: T=0\}, \quad L \cdot U_n = \lambda_- \left(\frac{\partial T}{\partial n}\right)_- - \lambda_+ \left(\frac{\partial T}{\partial n}\right)_+ \tag{3}$$

where U_n is the speed of propagation of the frost line in the direction of its normal n and where λ_- and λ_+ are related to frozen and unfrozen conditions, respectively.

The left side of equation (3) expresses liberation/uptake of heat due to phase change in time dt with the latent heat L. The right side terms express outflow and inflow of heat, respectively, at the same time.

Mathematically, equations (2) and (3) define a non-linear heat conduction problem with moving boundaries. Therefore, numerical procedures must be adapted in order to treat physically acceptable problems. Different numerical methods are discussed by, for example, Lunardino (1981). The numerical procedures used here were developed and firstly adapted to problems of airflow cooling in storm water pipeline systems in winter (Sellgren, 1983).

The upper boundary in Figure 2 corresponded to the ground surface with a constant temperature of -15°C. The under boundary was chosen to a depth of 8 m where the temperature was constant, approximately 4°C. The initial temperatures in the ground were assumed to vary linearly from 0°C at the ground surface. The model was adopted for a situation where a shutdown occurred 24 days after the operation started. The cooling in sections along the pipeline where the temperature of the flowing slurry is 10°C and 5°C is shown in Figure 3.

It follows from the results shown in Figure 3 that temperatures below 0°C occurred in the pipe after some days up to a week, dependent on the initial temperature of the still slurry. The results showed that there is a risk of freezing with a shutdown for several days, if the temperature of the flowing slurry is allowed to drop to about 5°C along the pipeline.

Slurry temperature along a buried pipeline

A constant heat transfer of q W/m from a pipe at a depth H below a horizontal ground surface in a half-infinite homogeneous medium represents a classical steady-state solution to equation (1) (without phase change)

$$T - T_R = q\,R \tag{4}$$

where T is the temperature at the outer radius of the pipe and T_R is the temperature at the ground surface. R is the thermal

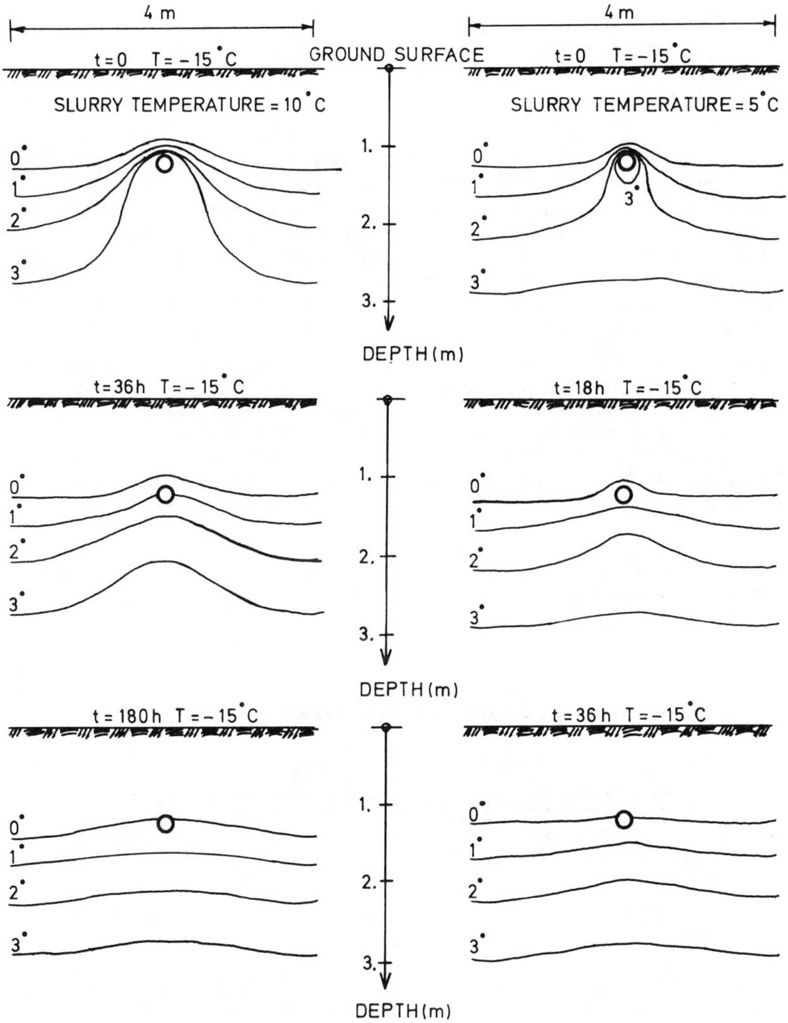

Figure 3. The cooling process after shutdown of the pipeline. Temperature profiles at sections where the temperature of the flowing slurry was 10°C (left) and 5°C (right). Moraine with a water content of 8 % and thermal diffusivities in frozen and unfrozen soil of $0.82 \cdot 10^{-6}$ m^2/s and $0.70 \cdot 10^{-6}$ m^2/s, respectively.

resistance in the ground:

$$R = \frac{\ln \frac{4H}{D}}{2\pi\lambda} \tag{5}$$

where D is the pipe diameter and λ is the average thermal conductivity of the ground. Equation (4) yields strictly only if D<<H, however, for the problem considered here it is a reasonable approximation. Furthermore, the thermal resistance in the steel pipe and between the slurry and the inner of the pipe wall turned out to be quite small. In this way, equation (4) provides a reasonable relationship for the steady-state temperature difference between the ground surface and the flowing medium in the pipe.

However, there is a serious physical limitation in the relationship because the influence of the natural thermal gradient in the ground is not included. This was accounted for by Jansson (1977), based on heat flow superposition principles and field data. He suggested that the soil temperature without pipe on the burial depth should represent T_R in equation (4). For the problem considered here, with the temperatures shown in Figure 3 after 24 days, a reasonable value of T_R is 0^oC.

The temperature distribution along the pipeline will now be determined based on equation (4). T is the temperature in the slurry at a distance z from the inlet (Figure 4).

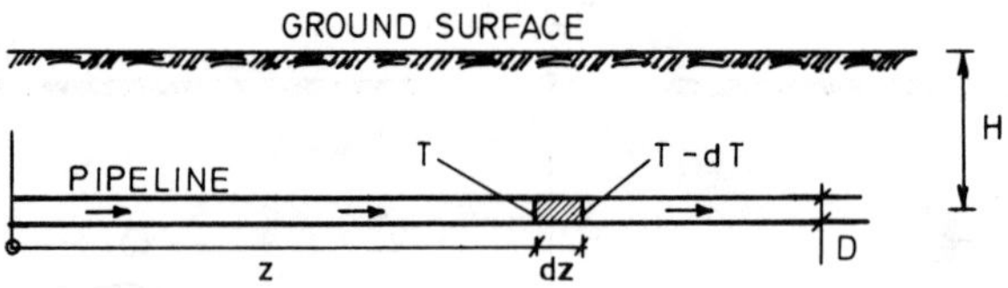

Figure 4. Steady-state heat transfer with a varying temperature along the pipe.

The necessary pressure gradient, p , to provide the transportation in the horizontal pipeline is given by the following equation

$$p = \rho g i \tag{6}$$

where ρ is the density of slurry, g is the acceleration due to gravity, and i is the hydraulic gradient in metres of mixture per metre of pipe. The density of slurry is given by

$$\rho = \rho_o(1 + C(s_s - 1)) \tag{7}$$

where C is the solids concentration by volume and

$$s_s = \frac{\rho_s}{\rho_o}$$

ρ_s = density of solids

ρ_o = density of water

The outflow of heat in the element dz in Figure 4 is qdz minus the friction production of heat, P (W/m),

$$P = Q p dz \tag{8}$$

where Q is the flow rate of slurry. The outflow is balanced by a decrease in the temperature of the slurry dT. The corresponding decrease in heat is

$$Q \rho c dT \tag{9}$$

where c is the heat capacity of the slurry (J/kgK):

$$c = c_s C_w + c_w (1-C_w) \tag{10}$$

where C_w is the concentration by weight

$$C_w = \frac{C \cdot \rho_s}{\rho}$$

and c_s and c_w are the heat capacity of solids and water, respectively. The balance is expressed by equations (4), (8), and (9) in the following way

$$R^{-1}(T-T_R)dz - Qpdz + Q\rho cdT = 0 \tag{11}$$

Rearrangement, separation of variables, and integration of equation (11) result in the following expression of the temperature profile

$$T = T_R + F + [(T_o - T_R) - F] \exp(-Az) \tag{12}$$

where $F = RQp$ and $A = (RQ\rho c)^{-1}$ are parameters.

The capacity of the pipeline or the mass flow rate of ore, M_s, is related to the volumetric flow rate and the concentration by the following equation

$$Q = \frac{M_s}{\rho_s C}$$

Thus, with a given throughput of solids, Q, ρ, and c are functions of the solids concentration. The flow rate and the concentration determine the hydraulic gradient if the pipe diameter is given.

In order to determine the hydraulic gradient experimental tests were carried out with a representative iron ore product. The rheological properties were identified in viscometric tests.

Non-Newtonian properties are less severe under turbulent pipe flow conditions and Newtonian methods may apply. A method of determining the turbulent friction loss is to use the conventional Moody-diagram, where Reynolds number is calculated with a limiting viscosity obtained as an asymptotic value, μ_R. Most industrial slurries behave so that μ_R adequately describes the turbulent flow behavior.

The experimentally determined values of a representative viscosity for different temperatures are shown in Figure 5 together with the particle size distribution.

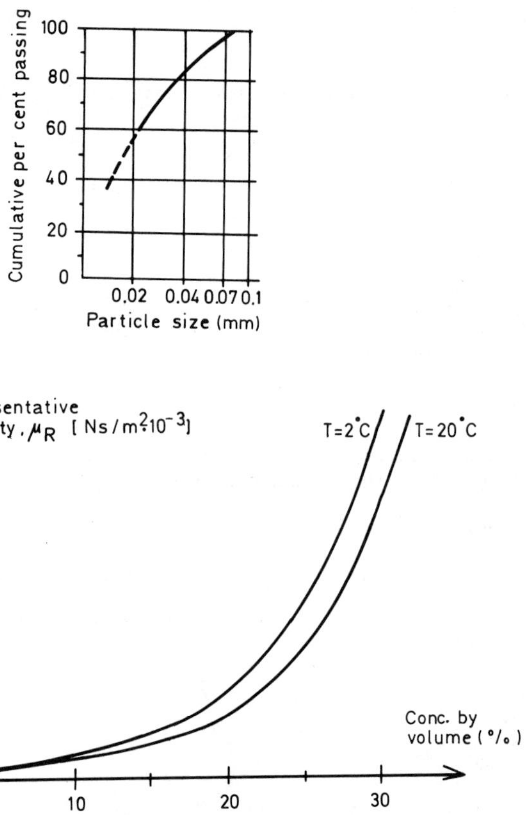

Figure 5. Measured data of a representative viscosity versus the solids concentration at different temperatures. The particle size distribution is also shown. The density of the iron ore product, $\rho_s = 5000\ kg/m^3$.

The results in Figure 5 showed that the increase in μ_R with a decrease in temperature from 20°C to about 2°C could be directly related to the change in water viscosity from $1 \cdot 10^{-3}$ to $1.75 \cdot 10^{-3}$ Ns/m^2. With a solids concentration of 25 % and an operating velocity of 1.5 m/s then the hydraulic gradient was determined with Colebrook's formula (or Moody's diagram) with the viscosities from Figure 5. It followed that the flow was in the turbulent zone and that the increase in viscosity of 60 %, due to temperature changes, corresponded to an increase in the hydraulic gradient of less than 10 %.

The additional operating data were determined and p = 196 Pa/m, Q = 0.0736 m^3/s. Furthermore, the heat capacity was calculated from equation (10) with c_s = 600 and c_w = 4180 J/kgK. Finally, the temperature profile was determined by equation (12) with T_R = 0°C. The result is shown in Figure 6.

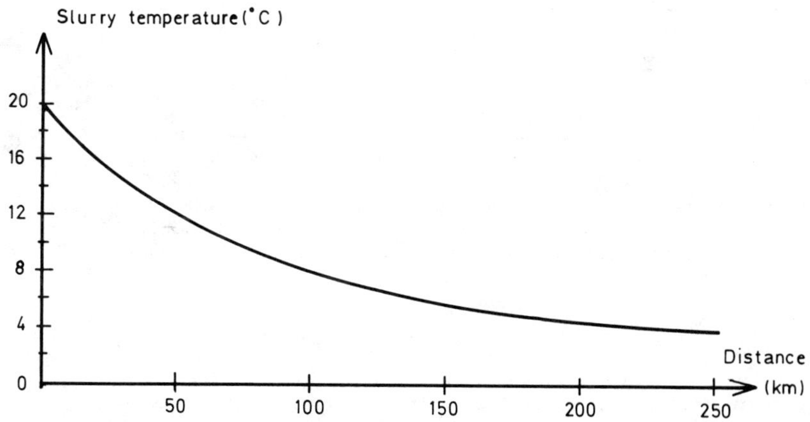

Figure 6. Temperatures of the flowing slurry along the buried pipeline.

It follows from Figure 6 that the temperature drops to about 10°C after 75 km and to values of 4-5°C at the end of the pipeline. In this interval heat transfer to the ground is balanced by friction heat produced by the flowing slurry. Therefore, with respect to the results shown in Figure 3, a shutdown of some days means that the slurry temperature will be close to 0°C.

Slurry properties close to 0°C

The startup procedure means that very high solids concentrations may occur locally. With high concentrations, the rheological properties dominate the hydraulic behavior. In order to establish the rheological properties viscometric tests were carried out. The experimental results are represented in a rheogram in Figure 7.

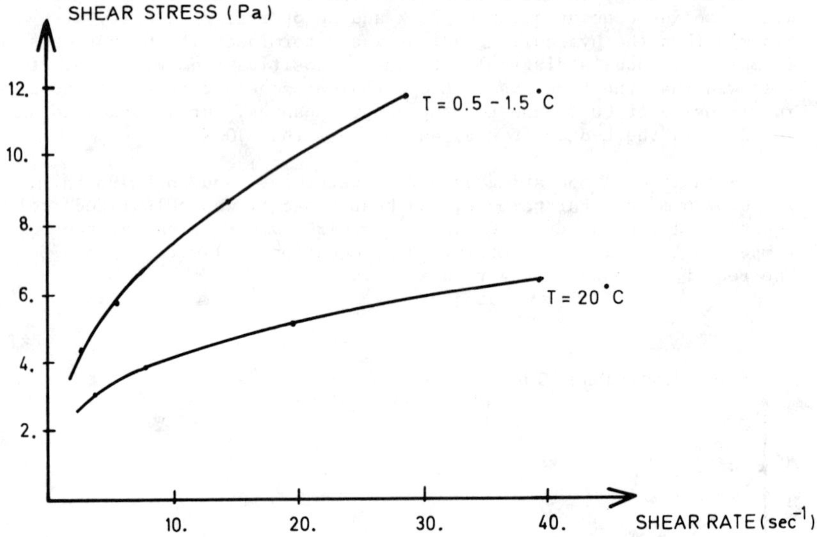

Figure 7. Rheogram for an iron ore slurry at a concentration by volume of 35 % and temperatures of 20°C and 0.5-1.5°C. The measurements were carried out with a rotational viscosimeter (Brookfield).

Analysis of the experimental results in Figure 7 (details of procedures omitted here, see for example Sellgren 1982) showed that transportation of the slurry at a concentration by volume of 35 % (C_w = 73 %) at 1.5 m/s is in the laminar/turbulent transitional region at temperatures at 0.5-1.5°C. Flow in the transitional region is not well-defined, not even for Newtonian fluids. Thus, the restart process is further complicated if the temperature of the slurry is near 0°C.

Conclusions

Viscosimetric tests showed that the increase in viscosity of the pumped slurry caused by a drop in temperature from 20°C to 2°C was directly related to the increase in the water viscosity. The corresponding increase in friction resistance at typical operating conditions in the pipeline was less than 10 %.

The temperature of the flowing slurry along the pipeline at continuous operation and the cooling process at a shutdown period were simulated with mathematical models. In the simulations, the temperature was assumed to be -15°C at a ground surface practically free of snow cover over a period of about one month.

With a normal slurry temperature of 20°C from the ore concentrator, the temperature dropped to about 4°C at the end of the pipeline (250 km). The friction heat produced by the flowing slurry balanced the heat transfer to the ground at this temperature.

The modelling of the cooling after a shutdown at a section where the temperature of the flowing slurry was 5°C, showed that the temperature at the pipe was close to 0°C within some days. Therefore, a shutdown with the pipeline full of slurry involves certain risks of freezing.

Startup of a pipeline after a shutdown in cold climate means that the temperature of the pumped slurry can be close to 0°C. Experimental results with a highly concentrated slurry showed that great temperature-dependent changes of the rheological properties of the slurry can be expected at restart procedures.

Freezing problems must be carefully considered in the design of a slurry pipeline in cold areas. Hereby, mathematical modelling of the temperature processes in the ground at the pipeline provides an effective tool for systematic assessments of risks and technical arrangements.

References

Jansson, L-E. (1977): Friction Loss as Frost Protection in Buried Pipelines, Report, VBB, Stockholm (in Swedish).

Lunardini, V.J. (1981): Heat Transfer in Cold Climates, Van Nostrand Reinhold, N.Y.

Sellgren, A. (1982): Rheological Analysis of Industrial Slurries. An Investigation of the Pipeline Transportation Characteristics of Two Limestone Products. Report, Series A no 105, Dept. of Water Resources Eng., Luleå University of Technology, Luleå.

Sellgren, A. (1983): Temperature Processes in Buried Storm Water Pipelines. An Investigation of Airflow Cooling. Report, Series A no 115, Dept. of Water Resources Eng., Luleå University of Technology, Luleå.

WATER SUPPLY LINES IN THE ARCTIC AND SUBARCTIC

Robert Alan Haviland*
Associate Member, ASCE

ABSTRACT

Water line design for arctic and subarctic construction is in a state of continuing development. Depending on design criteria and site conditions, various materials and techniques are feasible for water pipeline construction.

To meet the project requirements for four remote Air Force sites in Alaska, an insulated water supply line was designed using a drain-back feature or dry pipe design. The final design specified a 2-inch diameter galvanized steel carrier pipe, insulated with 3 3/4-inches of polyurethane foam protected by a 10-inch diameter corrugated metal pipe. Preheating is provided by an electrical heat tape inside a metal conduit attached to the steel carrier pipe. A second heat tape and conduit was supplied as a backup. The 20-foot long pipe sections are supported at joints by a 12-inch vertically adjustable 3-inch diameter steel pipe pile. Carrier pipes were joined with grooved pipe-type couplings surrounded by a urethane foam ring and metal coupling band.

The design was based on a low temperature of -65°F and winds in excess of 100 m.p.h. Each site required supplying a new 100,000 gallon insulated water storage tank from an existing well or water infiltration gallery. Pipeline distances varied from 900 feet to 1500 feet. Of the 100,000 gallons, 80,000 gallons were reserved for fire protection and 20,000 gallons for domestic supply.

INTRODUCTION

In 1979, the Alaska Air Command (AAC), U.S. Air Force, prepared a project book on the Minimally Attended Radar (MAR) project. This project was the first phase in converting the existing remote Air Force radar stations to an automated concept. This automation was to reduce the station manning from over a hundred personnel to 12 permanent personnel and up to 8 transients. The project

* Civil Engineer, Alaska District, Corps of Engineers, Pouch 898, Anchorage, AK 99506

locations for Phase I were Indian Mountain A.F.S., Sparrevohn A.F.S., Tatalina A.F.S., and Cape Romanzof A.F.S.

Phase I of the MAR project required new water lines from existing water sources to the water storage tanks at the new support facilities. The design process resulted in an insulated, above-ground, heat-taped water line, with a drain-back or dry-pipe feature. The MAR I project was designed by the Alaska District, Corps of Engineers as design agent for AAC. The contract was awarded in 1982, with completion scheduled for autumn 1984, see Figures 1 and 2.

CRITERIA AND CONSTRAINTS

User requirements called for an above-ground, insulated water line with accessible internal heat cables. An above-ground, heat-traced line was preferred in order to provide a high degree of accessibility. While an above-ground line allowed for easier routine and emergency maintenance, the cost was increased exposure to mechanical damage and thermal heat loss. Standardization was facilitated between the four sites, by the use of the worst case climatic data as the design parameters. The Indian Mountain site had the lowest recorded temperature of -65°F, and a maximum freezing index of 7434°F days.[1] The maximum wind velocity occurred at Sparrevohn and was recorded at 103 m.p.h.

Soils in the areas vary from good gravel fill to silts. Most of the areas consist of frost-susceptible material.[2] Scattered permafrost[3] is present at the Indian Mountain site. To traverse the varying terrain, a standard pipe pile was designed to support the above-ground line. The early preliminary (30%) design called for a 6-inch x 6-inch wood pile. Due to the deep embedment required, easier workability, and remote site

---

[1]The freezing index is a design parameter equal to the number of degrees below freezing multiplied by the number of days the temperature is below freezing. In comparison to the design value of 7434°F days, the average freezing indicies for Anchorage and Juneau are 2200°F days and 200°F days respectively.

[2]Frost-susceptible material is defined by the Corps of Engineers as soil having over 3% material passing the number 200 sieve. Frost susceptibility is significant as this type of soil is liable to wick water toward a freezing front which results in a frost heave.

[3]The standard definition of permafrost is any ground material remaining in a frozen state for at least two years.

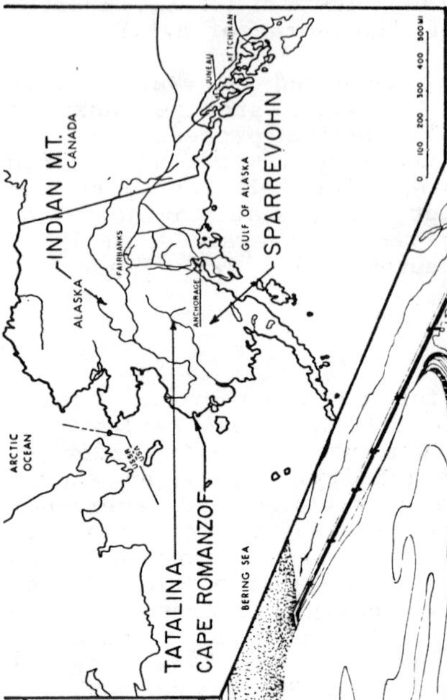

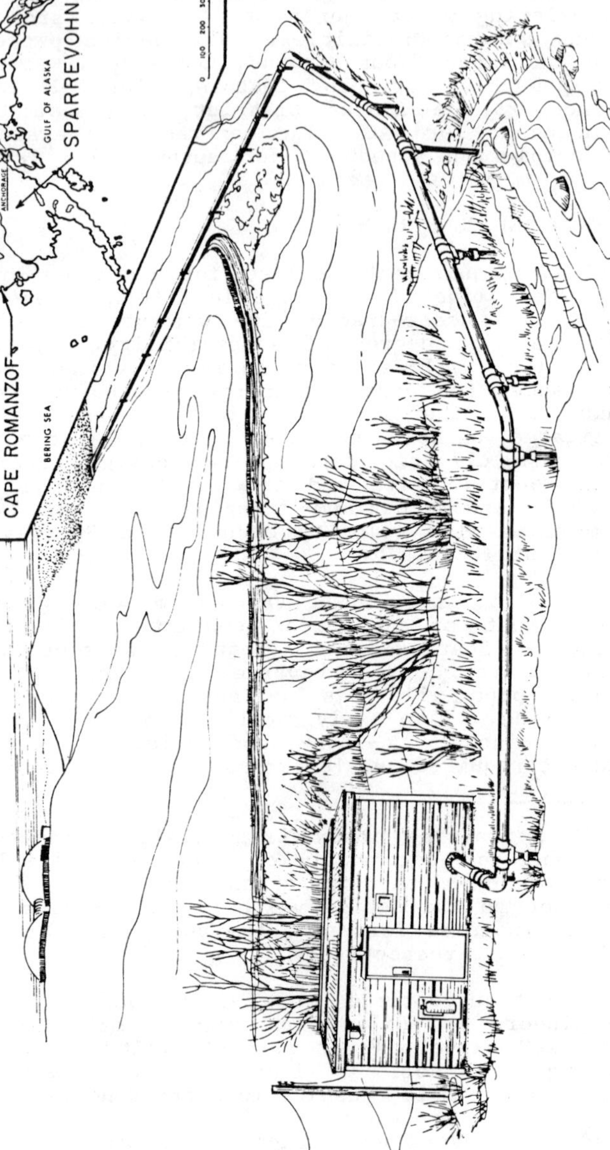

Figure 1, Location Map. Figure 2, Artist Conception.

logistics, a 3-inch diameter steel pipe pile replaced the earlier design. The steel pipe pile was also designed for limited vertical adjustment to maintain slopes for the drain-back feature.

As water storage tanks were to be used, the production rates of the existing well and infiltration galleries[4] were used to size the submersible pumps and carrier pipe. A 2-inch diameter pipe was designed for all four sites.

SEQUENCE OF OPERATION

When there is adequate water in the storage tank, the water line is cold or dormant. The water supply system is designed to operate automatically and begins operation when a signal from a sensor in the water storage tank indicates a 10,000 gallon drop. This is approximately half of the domestic reserve. Upon activation, if temperature sensors on the water line indicate a carrier pipe temperature below 33°F, the primary electric heat tape is switched on (the heat tape shuts off at 37°F and cycles back on at 33°F), the solenoid drain-back valve in the gallery pumphouse closes, and the submersible pump begins to fill the water line and storage tank. Once the storage tank is filled, the pump shuts off, the solenoid valve opens, and the line drains back into the gallery. Fifteen minutes after drain-back is complete, the heat tape is shut off by a timer. Between fillings, there is no energy demand by the system. The intent of the dry pipe design is to limit the energy demand on the station's generators. Depending on water usage, the system will operate every week or two weeks for approximately four hours.

FINAL DESIGN

Pipeline (see Figure 3)

<u>Thermal Design</u>.

In the early design stages, a circulating glycol system was considered to heat the water line for winter operation. A steam trace was also considered, however, electric heat tapes were selected for final design to

---

[4]Water infiltration galleries are prevalent throughout the north and are used to collect surface water, water shallower than thirty feet, for consumption. The material most often used is perforated, corrugated-metal pipe (CMP) from two feet to four feet in diameter. Vertically and/or horizontally installed CMP is backfilled with granular material adjacent to streams or other water bearing locations.

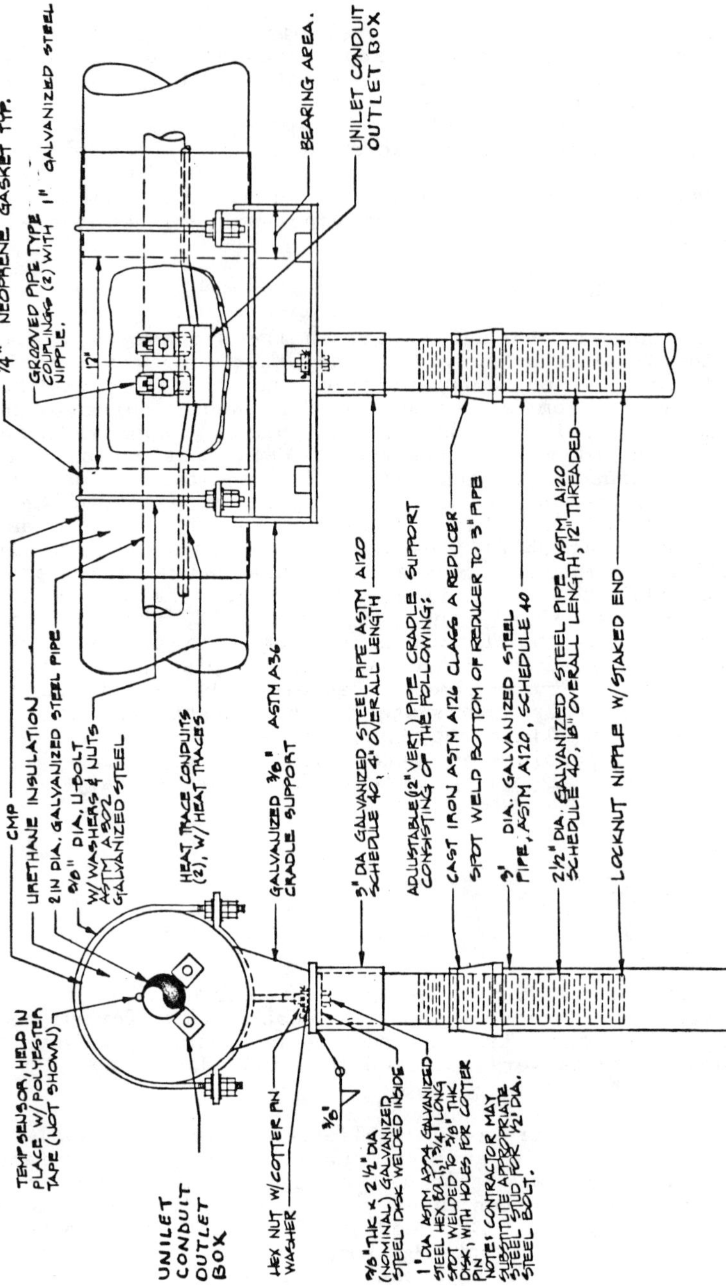

Figure 3, Water Line

avoid the requirements for circulating pumps and heaters or a steam boiler. The heat tapes were operationally simple and required less maintenance. As these stations are both remote and minimally manned, the selection of electric heat tapes was to avoid the necessity of inspection and maintenance of the pumps, heaters, or boilers required for the other systems. The heat tape system was designed to operate automatically with no moving, mechanical parts for possible failure. Two heat tapes were installed to provide system redundancy and the required degree of reliability. During winter operation, a major heat tape failure may be remedied if the system is switched to the second heat tape. Inspection and repair of the failed heat tape can then wait for less severe weather conditions.

To utilize dimensional materials, a standard 10-inch corrugated metal pipe was selected as the exterior jacket with approximately 3 3/4-inches of polyurethane foam insulation, the coldest site required a 3.6 watts per foot heat tape which includes a safety factor of two.

To insure that extreme high temperatures would not occur within the pipe system if a temperature sensor malfunctioned, a test program was conducted with 5-foot lengths of insulated water line.

Three set-ups were tested.[5] These were: (1) empty 2-inch insulated water pipe with Econotrace FP-13 standard parallel resistance cable, rating at 4 watts per foot and producing 4.5 watts per foot; (2) water-filled 2-inch insulated water pipe with Econotrace FP-13 standard resistance cable; and (3) water-filled 2-inch insulated water pipe with Econotrace FP-10 standard parallel resistance cable, rated 10 watts per foot and producing 11 watts per foot. The procedure for all three set-ups was similar and can be summarized as follows: A completed 5-foot length of insulated water line was placed in a chest freezer set at -20°F and the thermistors monitored. After a length of time for stabilization, the power to the heat tapes was switched on. The thermistors were monitored as the temperatures rose. After what was considered stabilization or burn out, the experiment ceased and inspection of the heat tape and insulation made.

Results are summarized in Figures 4 and 5. The 11 watts per foot, FP-10 heat tape did burn out upon reaching a temperature of 455°F inside the EMT. The FP-13 heat tape had no such problem with either the water-filled or empty 2-inch pipe. The concern with the electric heat tapes preheating a dry pipe, was possible burn out due to

---

[5]The use of any brand name product does not imply any type of endorsement.

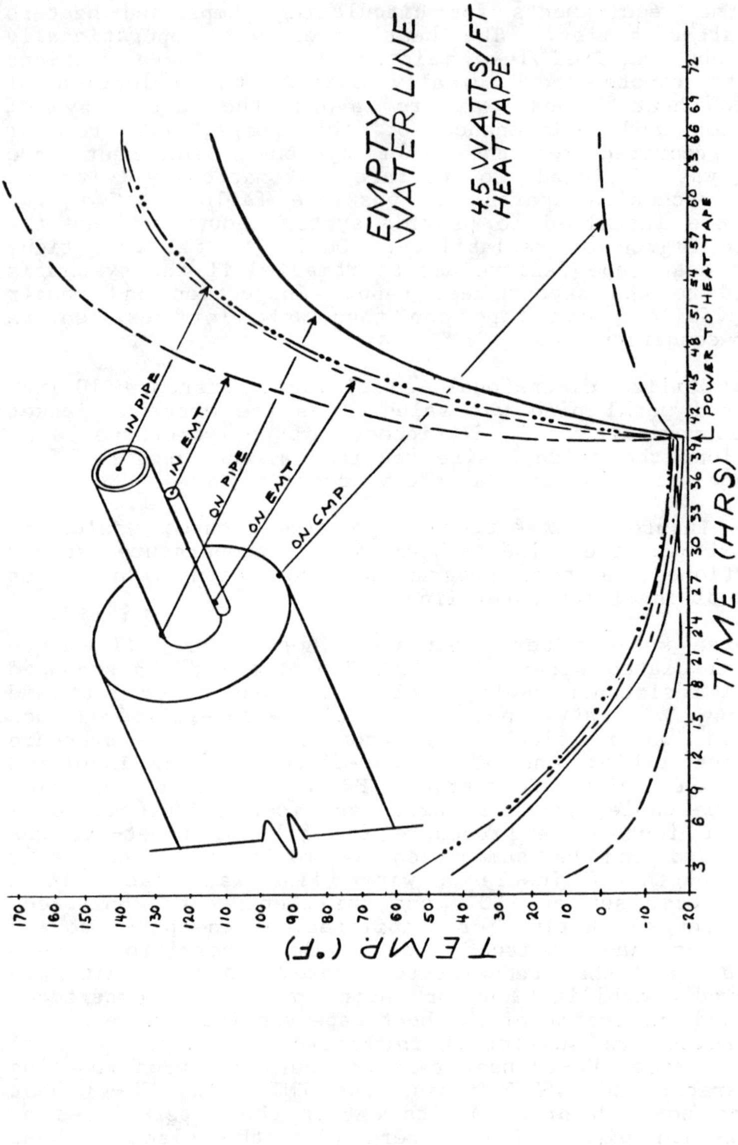

Figure 4, Insulated Water Line Test

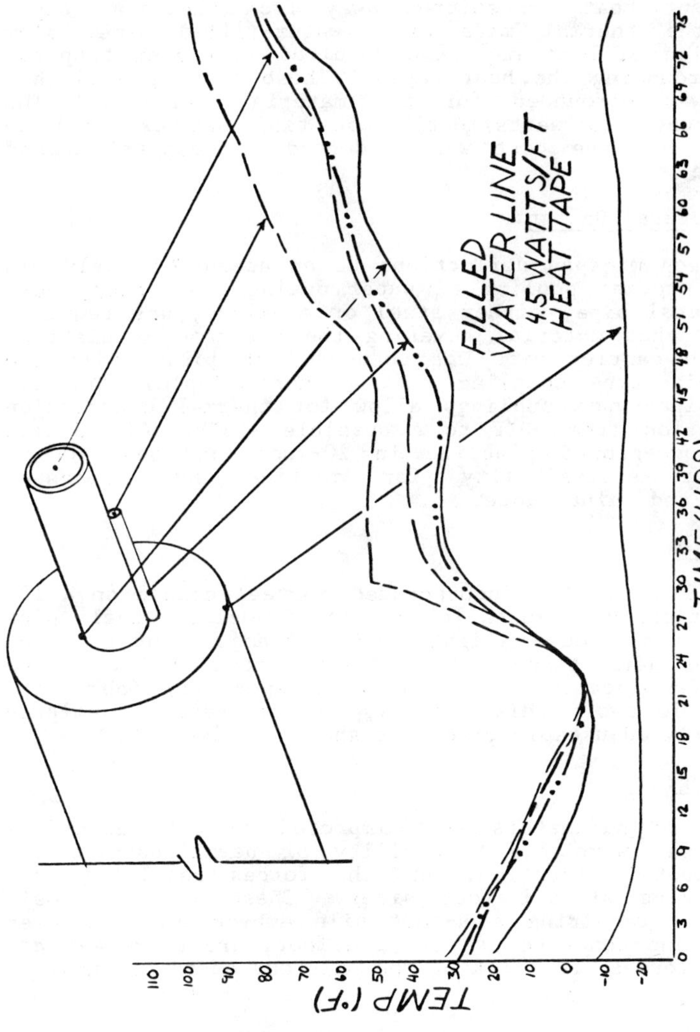

Figure 5, Insulated Water Line Test

insufficient heat transmitted away from the heat tape. Without the thermal mass of a water-filled pipe, slow transmission of heat may have resulted in higher temperatures surrounding the heat tapes. These tests proved this concern was unfounded for the material selected. The higher rated (10 watts/foot) heat tape was expected to burn out, and the test was conducted for comparison and future use.

Structural Design.

To keep midspan deflections to an acceptable value in order to prevent pooling of water during drain-back, corrugated metal pipe, either steel or aluminum, was required for the jacket material covering the urethane insulation. The 2-inch carrier pipe was connected at joints with two grooved pipe-type couplings and a short nipple. The two grooved pipe-type couplings allow for thermal contraction and expansion from 40°F to a possible -65°F. The design of an above-ground pipeline in 20-foot sections was to allow for accessibility for routine and emergency operation and maintenance.

Cradle.

The final design incorporated a steel cradle on a 12-inch vertically adjustable 3-inch diameter steel pipe pile. This allowed a slight amount of adjustment to maintain water line slopes. The cradle provided for angled water line placement with shims or specially fabricated cradles. To avoid this process, the contractor submitted an angularly adjustable cradle as shown in Figure 6.

Piling.

The two parameters that impacted on the subsurface pile design, were the possibility of permafrost at the Indian Mountain location, and the forces resulting from frost jacking at all four sites. These were the main reasons for requiring a 9-foot pile embedment. The deep embedment and concrete at the pile foot, are to resist any adfreeze[6] forces from jacking the pile out of the ground.

---

[6]Adfreeze forces result from the adherence of frozen soil to the pile. Jacking results from the progression of the seasonal frost level into the ground. As the soil around the pile freezes, it adheres to the pile. With the freezing, water in the soil expands. The direction of least resistance is upward, so the freezing soils expand toward the surface carrying the pile with it. The amount of annual jacking depends on freezing indices, soil type, and pile material. Pole or frost jacking has been designed around by using non-adhering coatings on the pile, additional weight to counter-balance the uplifting

COST ESTIMATE

Although actual construction costs are not available, the water line construction cost estimate total for the four sites was $454,500 for material and labor only. Freight for the 146,000 pounds of material was not included in the above figure. The total water line length for the four sites of 5,000 feet averages to $91.00 per lineal foot. The accepted bid for the entire project was approximately 14 percent lower than the estimated contract amount.

FUTURE CONSIDERATIONS

Further efforts to simplify the design may include the use of a high density polyethylene carrier pipe with a CMP jacket for rigidity, a shallow buried system with a polyethylene jacket in lieu of the corrugated metal pipe jacket, and an electric ball valve in place of the solenoid valve. There was some initial concern over using a polyethylene carrier pipe with a metal jacket, due to the differences in thermal expansion between the two materials, and the large temperature difference that may be expected. There have been some projects that have used high density polyethylene carrier pipe, urethane insulation, and a metal jacket. Reports on their performance indicate no thermal expansion or contraction problems. To our knowledge, there have been no controlled tests conducted to determine if differential axial thermal expansion and contraction would be a problem with this type of insulated pipe, or to define any design guidelines for this type of composite pipe.

CONCLUSION

The design of a water supply system must consider the type of service being provided, climatic constraints, topographical constraints, system demands, soils information, accessibility, and the degree of reliability required. The available construction materials, system designs, and buried or above-ground configurations should be evaluated to provide an adequate system and still remain economically feasible.

REFERENCE

Jaeger, Clare L. "Test Program for Insulated Water Pipe and Heat Tracing," Alaska District, Corps of Engineers, unpublished report, 1982.

(Footnote 6 cont'd) force, or the addition of a bearing area (such as cast concrete at the bottom of the pile) to resist the uplifting force with the weight of the soil above the bearing area.

VERTICAL VS. HORIZONTAL CYLINDERS IN WAVES

John H. Nath*, F ASCE

Abstract

Pipelines are exposed to various wave and current conditions in the ocean. The spatial orientation of the pipeline with respect to the flow conditions is important in determining the wake condition produced from the flow around the pipe. In addition, marine biofouling can increase the hydrodynamic forces considerably. This paper presents results of research at Oregon State University (OSU) on laboratory tests at moderately high Reynolds numbers (up to $3x10^5$) and Keulegan-Carpenter numbers (up to 25) for vertical and horizontal cylinders in waves. The force differences are displayed with the force transfer coefficients from the Morison equation and with maximum force coefficients. The tests were performed on smooth, sand roughened and artificially marine roughened cylinders. The results for the vertical cylinder show significant increases in both C_d and C_m and in the maximum forces.

BACKGROUND SUMMARY

Pipelines and structural members of offshore platforms utilize the *circular* cylinder on which this paper concentrates. They can have infinite orientations with respect to the ambient waves and currents. Our attention is constrained to waves acting on two common orientations of the cylinder -- vertical and horizontal with the axis parallel to the crest of the waves. The characteristics of ocean waves are not the same as the idealized long crested laboratory waves, nor are all cylinders just vertical or horizontal. For these and other reasons, the force transfer coefficients presented here should be considered with caution.

Of the many other aspects of wave forces on cylinders, one is the influence of marine growths, or biofouling (11, 12, 13). Here we consider smooth, sand roughened (with k/D = .02) and an artificially marine roughened cylinder, where k is the average sand grain height and D is the smooth cylinder diameter.

Pipelines at sea can be horizontal or vertical or at other orientations. Usually they are on, or close to , the ocean bottom and they are vertical within the confines of an ocean platform. Horizontal pipelines may exist above the ocean bottom in the form of somewhat flexible conduit hoses. This paper addresses the question of the differences in wave forces for single, vertical vs. horizontal

*Professor, Dept. of Civil Engineering, Oregon State University, Corvallis, Oregon 97331.

pipelines utilizing the time dependent Morison equation and a time *independent* maximum force equation.

Planar oscillatory flow can be generated in a U-tube as in (16) or at the node of a standing wave as in (7), and possibly in other flow conditions. It is often assumed that the flow around vertical cylinders in waves is planar oscillatory, although there is a vertical component to the water particle motion. This component may have a small effect on the horizontal forces on smooth vertical cylinders; however, it may have an important effect on roughened cylinders. Possibly the vertical flow over a rough cylinder will generate disturbances that will modify the wake characteristics in some presently unknown way so as to reduce the general momentum in the fluid and increase the horizontal force on the cylinder.

Sarpkaya (19) oscillated a cylinder normal to the direction of planar oscillatory flow to simulate a horizontal cylinder in wavy flow. The normal oscillations were characterized with the maximum velocities, w_m, and the maximum in-line fluid velocities were u_m. Thus, quasi-orbital motion was characterized with $\Omega = w_m/u_m$. The drag coefficient was larger for $\Omega = 0$ than for $.3 < \Omega < .6$ by about 100% at $K = 6$, 50% for $10 \lessapprox k \lessapprox 20$ and approaching 0% for $K > 100$. The inertia coefficient was smaller and β was 784 , where $\beta = R/K$, R is the Reynolds number and K is the Keulegan-Carpenter number, defined as

$$R = \frac{U_m D}{\nu} \tag{1}$$

and

$$K = \frac{U_m T}{D} \tag{2}$$

where this paper defines U_m as the maximum horizontal velocity, ν is the kinematic viscosity, and T is the fundamental wave period.

Ramberg and Niedzwecki (15) performed experiments in waves with horizontal cylinders and formulated force transfer coefficients that smoothed the observed wave-to-wave variations with

$$C_F = \frac{(F_x)_{rms}}{1/2\ \rho\ DL(u_{rms})^2} \tag{3}$$

for the force component in-line with the waves (x-direction) and

$$C_L = \frac{(F_z)_{rms}}{1/2\ \rho\ DL(w_{rms})^2} \tag{4}$$

for the vertical component (z-direction). Their results were approximately 20-25% smaller than predictions that used coefficients from planar oscillatory flow.

Holmes and Chaplin (6) used an experimental setup where a smooth vertical cylinder was rotated in a circular orbit at the end of an arm. The cylinder also rotated about its axis so that the north-south line of the cylinder remained in the north-south direction. Thus, by changing the length of the rotating arm and the speed of rotation they hoped to simulate a stationary cylinder in waves with the axis of the cylinder parallel to the wave crest. (The effects from the surface wave that forms where the cylinder penetrates the water surface were not examined.) Their drag coefficients were about 20% less than those from Sarpkaya's planar oscillatory flow results. The values for C_m were from about 40% to 200% lower than those for oscillatory flow. They also determined that there could have been an influence from the cylinder encountering its own wake.

Heideman, et al (5) found that drag coefficients from the Exxon ocean test structure were comparable with steady state flow values from other investigators. They were also concerned with the scatter that is apparent in experiments utilizing ocean measurements. They formulated a *random wake encounter* model for accounting for some of the scatter, wherein the wave was washed back either onto the cylinder or onto the current meters. Because of the three-dimensional aspects of ocean waves, the wake could create a greater or smaller flow acting on the cylinder than that measured. Garrison (4) professed that the wake effect in planar oscillatory flow could create drag coefficients that are around 20-60% greater than those for steady flow conditions. He also predicted that inertia coefficients in planar oscillatory flow will be considerably reduced from that for the constant acceleration case. He developed the Navier Stokes equations, with respect to coordinate axes moving with the cylinder, for a cylinder in oscillatory motion. His work showed they are identical with the case of a fixed cylinder in planar oscillatory flow, and that the kinematic conditions were therefore the same.

First consider a cylinder which is moving in otherwise still fluid, as shown in Fig. 1. The idealized wake considered is of such a character that if the speed, V, were constant, the wake would be steady with respect to a coordinate axis moving with the cylinder. (Actually, the vortex generation and shedding, of a scale equal to the cylinder diameter, produces an unsteady flow added to the idealized wake.) The cylinder imparts momentum to the fluid resulting in the in-line wake velocities, v(x,z,t). The cylinder motion creates a relatively high shear stress with the fluid within the boundary layer. After separation, the shear stress diffuses into the ambient fluid through the distributed vortex shear region, imparting momentum to the ambient still fluid. The resulting velocity profile is approximated by Fig. 1. The momentum in the wake is larger than the momentum in the ambient still fluid. For any fixed coordinate, x, the magnitude of v_{max} will decay with time and δ will increase in dimension through turbulent diffusion. If the cylinder reverses its course, as shown by the dashed circle, it will encounter its own wake on the return trip. Thus, the relative velocity acting on the cylinder will be greater than the cylinder velocity. If the cylinder oscillation amplitude is large, the return flow effect may be small because as the cylinder reaches its maximum velocity, there may have been sufficient time for the wake to decay

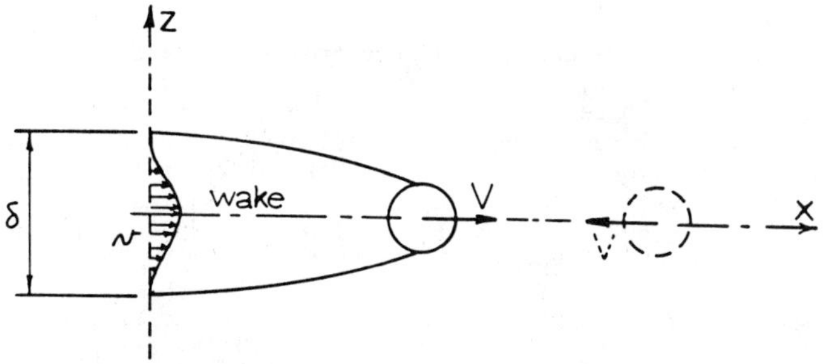

Fig. 1 - Cylinder oscillating in otherwise still fluid

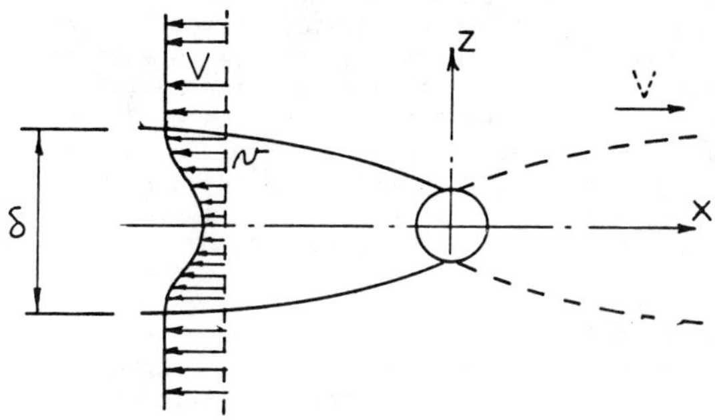

Fig. 2 - Two-dimensional planar oscillatory flow past a stationary cylinder

so that v_{max}/V is small. The $(V_{max}+v_{max})^2$ term will not be much larger than the V_{max}^2 term. On the other hand, there may be certain values of K that would make v_{max}/V an appreciable correction to the V^2 term. It is important to note that v and V are not in phase.

Figure 2 represents the condition when the fluid is accelerating past the cylinder for confined flow as in a U-tube experiment. Again, the wake has been grossly idealized, excluding considerations for vortex generation and shedding of a scale equal to the cylinder diameter (the Karman vortex street). Momentum is diffused *into* the wake from the ambient flow. Thus, if V is sinusoidal, or at least periodic, then v is periodic and dependent on V for magnitude as per Schlichting (20) and for momentum as per Batchelor (1).

Large scale vortices *will* be generated and shed from the cylinders in either case and will retain angular momentum to be swept back and forth upon the cylinders. These vortices are the real source of the problem of variations in the force transfer coefficients in the Morison equation, as presented by Sarpkaya (18). Some insight as to the manner in which the vortices are shed and then washed back onto the cylinder can be formed from the graphic portrayals of Stansby (21), Chaplin (3), Bearman, et al (2), and Sarpkaya and Isaacson (17).

A number of investigators have carefully developed repeatable waves for experimentation. The water surface profiles and the particle velocities were measured and found to be reasonably repeatable from wave to wave in a given sequence. Even so, the *forces* on the cylinder specimens varied (sometimes greatly) from wave to wave in a given sequence. The same phenomenon was experienced at OSU. This may be due to a change in the number (or the phase) of vortices that are shed from the cylinder in sequential waves. In addition, the manner in which the vortices are swept back upon the cylinder is not completely regular (and is sometimes quite random) as evidenced by the flow visualization and *transverse force measurements* studies of many investigators; and since the vortex shedding is so critical to the determination of the forces acting on the cylinder, it should be *expected* that there will be a variation in the total force acting on the cylinder in sequential waves.

This was true for this study and the phenomenon is sensitive to the conditioning of the data toward either the drag or the inertia component. When K is small, the inertia coefficients have small scatter from one wave to another but the drag coefficients can have wide scatter. When K is large, the drag coefficients have small scatter from wave to wave but the inertia coefficient has considerable scatter.

The coefficients for vertical vs. horizontal cylinders obtained for this study have a slightly different basis from each other. The computational details of analysis were the same and for both cases four sequential waves were measured, from which three were selected for analysis. However, for the horizontal cylinder only a single wave was analyzed -- the one with the lowest rms error between the measured water surface profile and the stream function theory. For the vertical cylinder, all three waves were analyzed and the coefficients were averaged. If one of the three was widely different from the other two, it was discarded, as were negative coefficients. Therefore, the horizontal cylinder data show more scatter than for the vertical cylinder.

Regarding cylinder roughness, this paper is restricted to the smooth aluminum cylinder (SMC), a sand roughened cylinder (SRC, wherein k/D = .02) and an artificially marine roughened cylinder (AMRC). These roughnesses are shown in Fig. 3. Based upon circumferential measurements, the approximate k/D for the AMRC was .09.

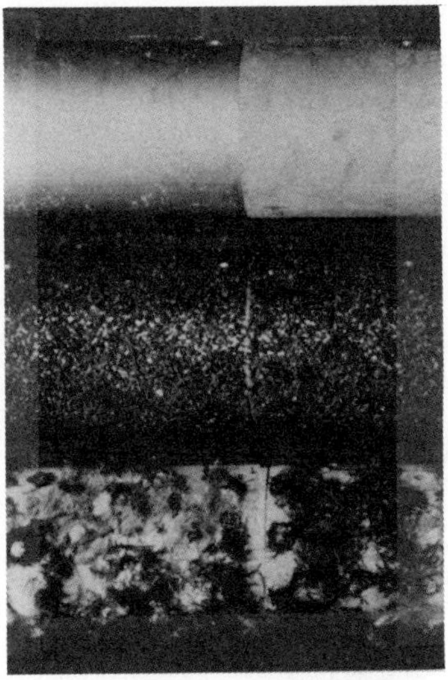

Fig. 3 - Smooth, sand roughened and artificially marine roughened surfaces

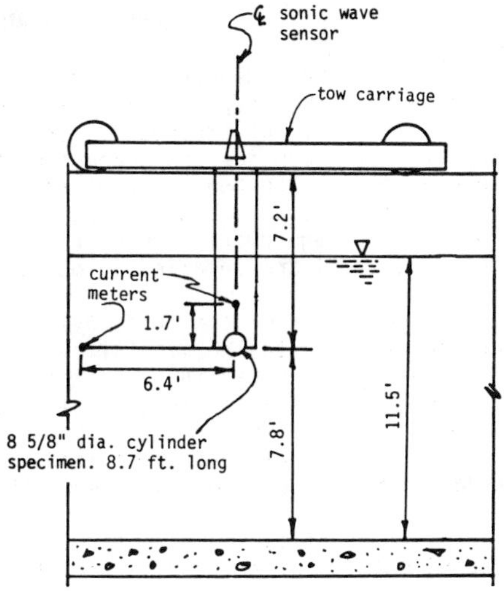

Fig. 4 - Setup for the horizontal cylinder testing

Fig. 5 - Calibration in air to the East for the artificially marine roughened cylinder. Nominal diameter 8 5/8". Final calibrations for data analysis were made with the strain gages submerged.

EXPERIMENTS AND ANALYSIS

The OSU Wave Research Laboratory is a flume 340 ft. long, 12 ft. wide, and 15 ft. deep. Ordinarily, the water depth is 11.5 ft. so there is 3.5 ft. of freeboard. The test length in the flume is about 125 ft. after considering space for the wave generator and its evanescent modes, the beach for dissipating wave energy, and space for other activities. The cylinders tested were within this 125 ft. test length. The equipment (tow carriage, suspension system, calibration methods, etc.) are detailed in other publications (9, 10, 11, 12, 13).

Figure 3 shows the three cylinder roughnesses tested. The general arrangement for the horizontal cylinder is shown in Fig. 4. The vertical cylinder had the identical surface roughness as the horizontal cylinder and the 2-ft. long local force transducer (12) was centered at exactly the same elevation as the center of the horizontal cylinder. Figure 5 shows the vertical cylinder (AMRC) during a calibration. The calibrations were linear and repeatable (10, 14).

In (12) it is shown that for the range of waves produced in the experiment, the rms error between the measured velocity vector and that predicted by the stream function theory varied from about 9 to a high of 18% (with an isolated value of 22%). Although these percentages seem a bit high, it should be kept in mind that a very stringent criterion is established for this calculation dependent upon the magnitude *and direction* of the total velocity vector. Most of the error was due to the vertical velocities. The horizontal velocity rms error ranged from about 4% to 12%. The *mean* errors were much smaller. The kinematics measurements showed that a systematic mean flow existed at the level of the cylinder test sections toward the wave board. The magnitude ranged from .03 up to .4 fps depending upon the wave characteristics. This reverse flow was taken into account in the reduction of the data in conjunction with the stream function theory. For the vertical cylinder, it turned out there was less error in using the stream function theory than the measured velocities (which were made opposite the center of the local force transducer, parallel to the wave crests). The value of Ω ranged from about .4 to .96.

The experiments were in waves such that $2 < K < 25$ and $.3\times10^5 < R < 1.8\times10^5$. In this paper, only the results for $15 < K < 25$ for C_d and C_m will be reported in order to compare with the results of $K = 20$ from other publications. The complete results for the full range can be obtained in (10 and 14). For the horizontal cylinder tests, the cylinder was sufficiently distant from the bottom of the wave flume to nullify proximity effects and was submerged sufficiently below the trough of the waves to nullify or minimize the free surface effects, according to (9).

The analysis for the horizontal cylinders followed the vector approach to the Morison equation,

$$\vec{F}(t) = C_d\, D\, \frac{\rho}{2}\, \vec{q}|q| + C_m \frac{\pi D^2}{4} \rho \vec{q}' \qquad (5)$$

where $\vec{q} = \vec{u} + \vec{w}$ (6)

with $\vec{u}$ and $\vec{w}$ the horizontal and vertical instantaneous velocity vectors, C_m is the inertia coefficient $(1 + C_a)$, and

$$\vec{q}\,' = \frac{\partial \vec{q}}{\partial t} + u \frac{\partial \vec{q}}{\partial x} + w \frac{\partial \vec{q}}{\partial z} \quad . \tag{7}$$

The Morison equation used for the vertical cylinder was

$$F(t) = C_d\, D\, \frac{\rho}{2}\, u|u| \;+\; C_m \frac{\pi D^2}{4}\, \rho \dot{u} \, . \tag{8}$$

In both cases the force transfer coefficients were determined by minimizing the error with the usual least squares technique.

It is interesting to note that in (15) the vector approach for the horizontal cylinder, wherein $\Omega = 1.0$, reduces to a linear form. Thus, in deep water waves an interesting linearity can be achieved for utilization in random wave studies, etc. However, for this work $.4 < \Omega < 1.0$ and nonlinearities can be expected. In the analysis for the vertical cylinder, it was assumed that there was no attenuation of velocities and accelerations with respect to depth *over the length of the two-foot long test section*. The error thus introduced was studied and found to be usually much less than 1% and up to 2% for a maximum case. Therefore, the same stream function kinematics and dynamics were used in the analysis for the vertical cylinder as for the horizontal one.

It is also instructive to compare total *forces* as well as C_d and C_m. A maximum horizontal force coefficient was therefore defined for both the horizontal and vertical cylinders as

$$C_x = \frac{2\, F_{x\ max.}}{L\, D\, \rho\, u_{max}^2} \tag{9}$$

RESULTS

The horizontal cylinders were towed on several occasions in steady state motion up to $R \simeq 7 \times 10^5$. The results are given in (10) through (14). There was modest scatter, but results were consistent. The following plots for C_d include the steady tow results for a best fit curve by eye. The drag and inertia coefficients for the test specimens are shown in Figs. 6 through 11. Samples of output appears in (10, 11, 12 and 14). Figure 12 is a composite of information from Heideman (5) and from Miller (8) which show the results for post critical drag coefficients vs. relative roughness. Onto this the aluminum smooth cylinder results have been superimposed at a $k/D = 4 \times 10^{-4}$ by means of fitting the steady state tow results into those by Miller (8).

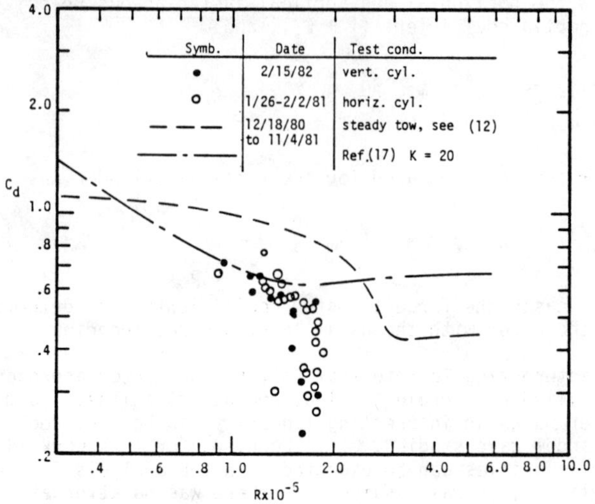

Fig. 6 - Smooth aluminum pipe cylinder drag coefficients, 15 < K < 25.

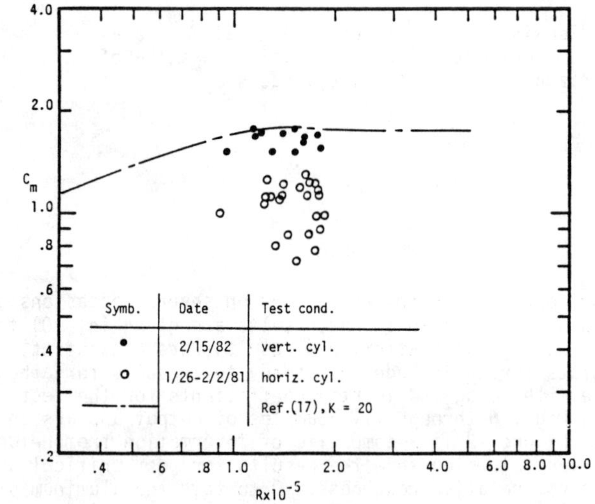

Fig. 7 - Smooth aluminum pipe cylinder inertia coefficients, 15 < K < 25.

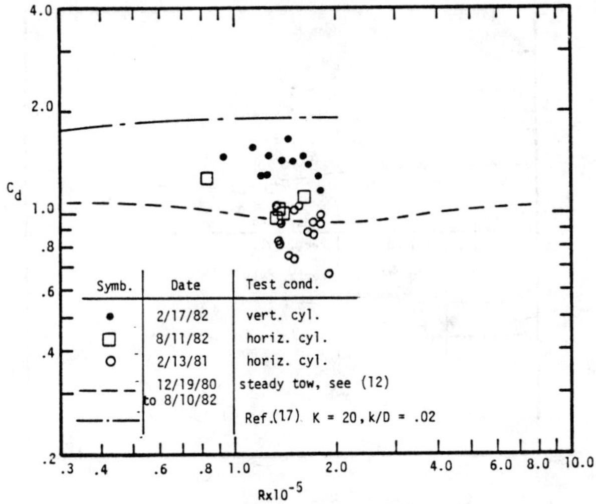

Fig. 8 - Sand roughened cylinder drag coefficients for 15 < K < 25, k/D = .02.

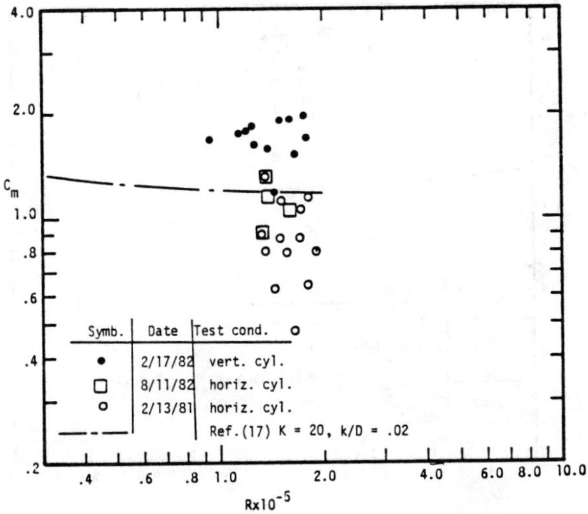

Fig. 9 - Sand roughened cylinder inertia coefficients for 15 < K < 25, k/D = .02.

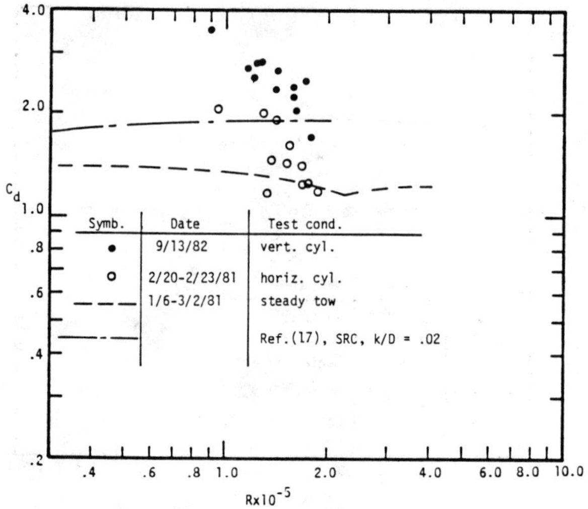

Fig. 10 - Artificially marine roughened cylinder drag coefficients for 15 < K < 25.

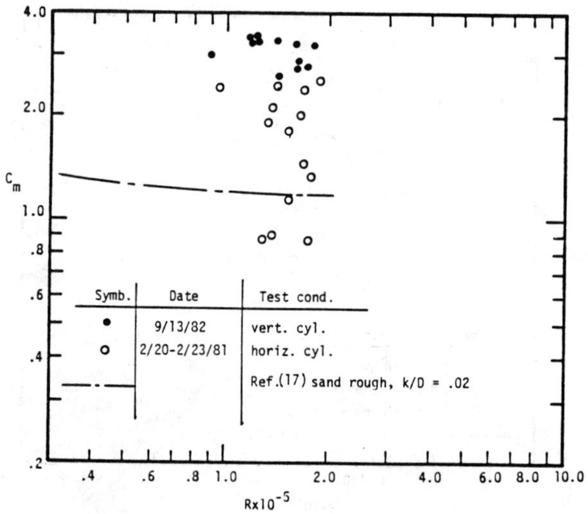

Fig. 11 - Artificially marine roughened cylinder inertia coefficients for 15 < K < 25.

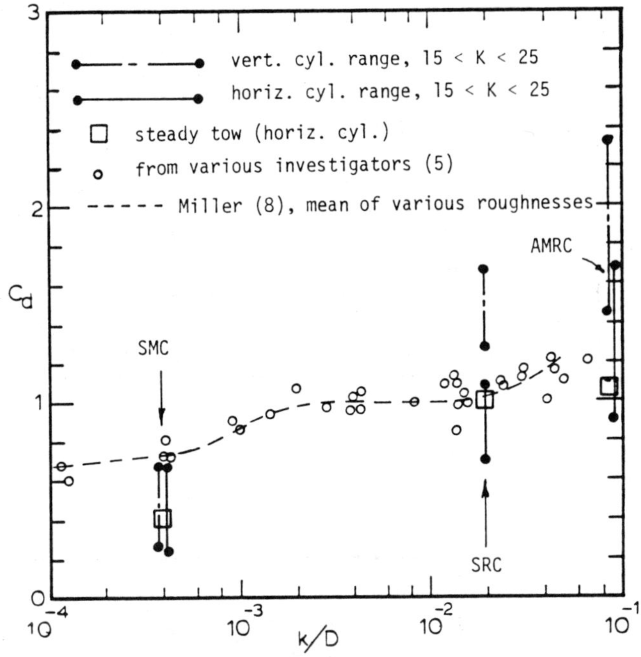

Fig. 12 - Drag coefficients vs. relative roughness for post-critical flow from Ref. (5) with present data superimposed (steady tow and wavy flow). Data reduced to values for the equivalent diameter.

Figure 6 shows no appreciable difference between vertical vs. horizontal cylinder results for the SMC. The scatter at the higher R values is accentuated by the log-log plot. Agreement is only fair compared to results from ref. (17). The vertical C_m values are decidely higher than the horizontal C_m's in Fig. 7, and they evidently influence the total force significantly as seen later. The SRC and AMRC show increases in both C_d and C_m for the vertical cylinder. Figure 12 shows a considerable increase in C_d values for the vertical vs. horizontal cylinder. A summary of the average values of these coefficients and for the full range of K tested is presented as Table 1.

The maximum force coefficients are shown in Figs. 13 through 17. For K > 20 the "mean" curves indicate an increase in the maximum force coefficient of approximately 60, 70 and 100% for the SMC, SRC and AMRC, respectively. Figures 16 and 17 show that relatively mild roughnesses increase the forces considerably for higher K values, as would be expected, and the more severe roughnesses increase the maximum forces throughout the range of K. The curves indicate that C_x will probably become independent of K for some value of K > 20.

Table 1. Summary of force coefficient averaged over the entire data set.

		C_d		C_m	
		Vert.	Horiz.	Vert.	Horiz.
SMC	2 < K < 25	0.50	0.60	1.67	1.06
	15 < K < 25	0.51	0.52	1.64	1.09
SRC	2 < K < 25	1.37	0.95	1.71	0.90
	15 < K < 25	1.41	0.94	1.70	0.86
AMRC	2 < K < 25	2.77	1.86	2.91	1.66
	15 < K < 25	2.49	1.58	3.04	1.78

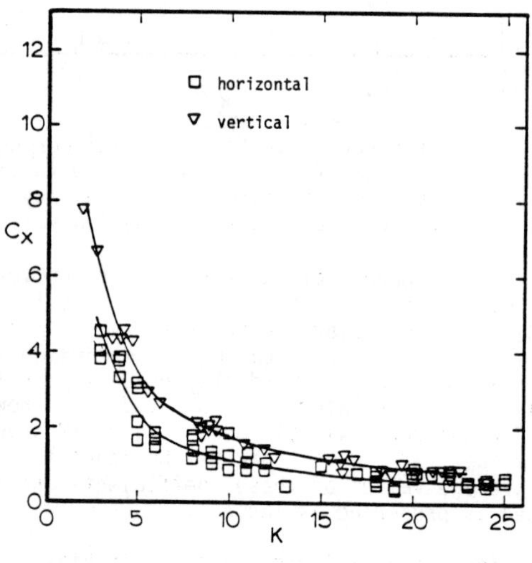

Fig. 13 - Smooth cylinder maximum force coefficients

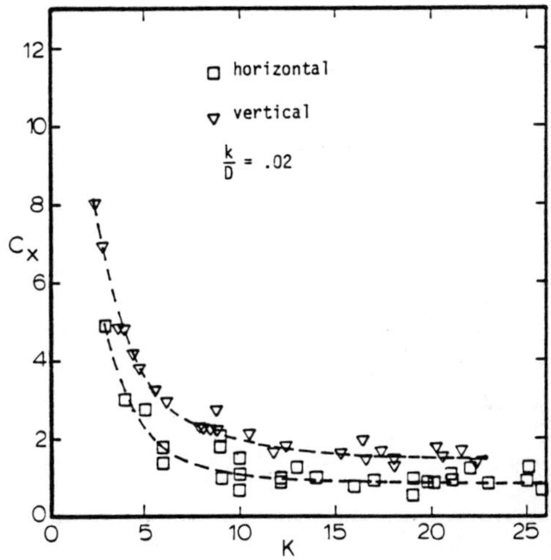

Fig. 14 - Sand roughened cylinder maximum force coefficients

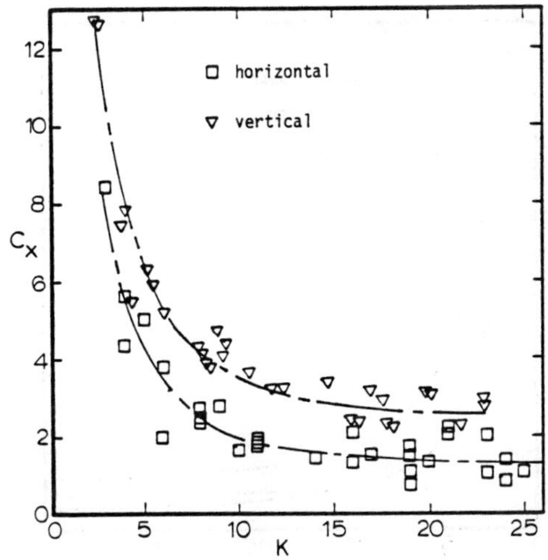

Fig. 15 - Artificially roughened cylinder maximum force coefficients

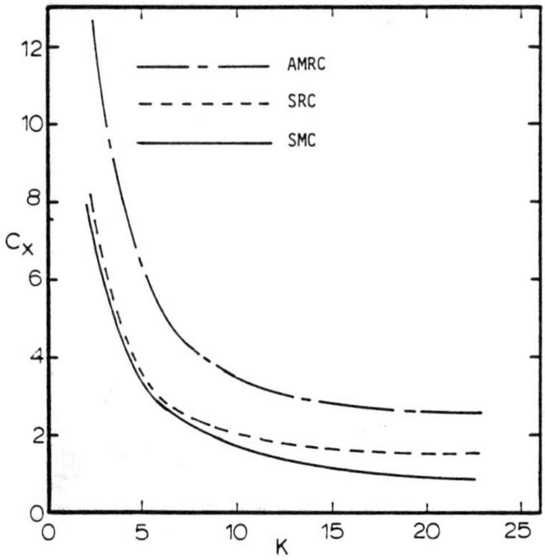

Fig. 16 - Vertical cylinder maximum force coefficients

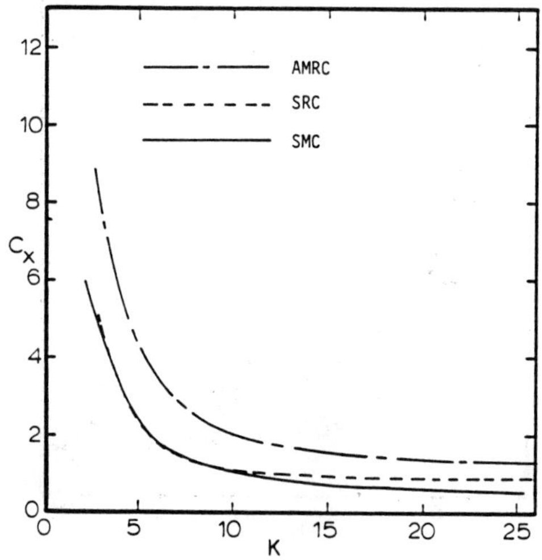

Fig. 17 - Horizontal cylinder maximum force coefficients

CONCLUSIONS

Vortex shedding is somewhat unpredictable and there may not be an integer number of vortices for a wave. Therefore, when there are several waves in a sequence of periodic waves, each one may not have the same number of vortex sheddings and the phase of these vortices may be different from the phase of the waves. Therefore, it is to be expected that there will be scatter in the results of the force transfer coefficients from wave to wave in a sequence of periodic waves.

The force coefficients for vertical cylinders are larger than for a horizontal cylinder based upon the form of the Morison equation assumed for both cylinders. Generally, this difference seems to be larger for the rougher cylinders than for the smoother cylinders. More importantly, the measured *maximum forces* are higher for vertical cylinders than for horizontal ones, in periodic waves.

ACKNOWLEDGEMENTS

Portions of this work were supported by the American Petroleum Institute, the National Science Foundation and the OSU Sea Grant College Program. The author's views presented here do not necessarily represent those of any of the funding agencies. The author is grateful for the able assistance of C. Baumgartner, M. Hsu and C. Teng. Equipment was constructed by T. Noland and laboratory work was supervised and done by L. Crawford. D. Standley developed the major part of the software for data processing. Thanks are also due Dr. R. T. Hudspeth for stimulating discussions about the wake encounter problem.

REFERENCES

1. Batchelor, G. K., "An Introduction to Fluid Dynamics," Cambridge University Press, 1970, Ch. 5.

2. Bearman, P. W., et al, "The Role of Vortices in Oscillatory Flow about Bluff Cylinders," Proc. of International Symposium on Hydrodynamics in Ocean Engineering, The Norwegian Institute of Technology, Trondheim, August 24-28, 1981.

3. Chaplin, J. R., "Boundary Layer Separation from a Cylinder in Waves," Proc. of International Symposium on Hydrodynamics in Ocean Engineering, The Norwegian Institute of Technology, Trondheim, August 24-28, 1981.

4. Garrison, C. J., "A Review of Drag and Inertia Forces on Circular Cylinders," Proc. of the Offshore Technology Conference, Houston, Texas, May 5-8, 1980.

5. Heideman, J. C., Olsen, O. A. and Johansson, P. I., "Local Wave Force Coefficients," Proc. of the Specialty Conference, Civil Engineering in the Oceans IV, Vol. 2, September 1979.

6. Holmes, P. and Chaplin, J. R., "Wave Loads on Horizontal Cylinders," Proc. of 16th International Conference on Coastal Engineering, Hamburg, Germany, 1978.

7. Keulegan, G. H. and Carpenter, L. H., "Forces on Cylinders and Plates in an Oscillating Fluid," Journal of Research of the National Bureau of Standards, Vol. 60, No. 5, May 1958.

8. Miller, B. L., "The Hydrodynamic Drag of Roughened Circular Cylinders," Royal Institution of Naval Architects, Spring Meetings, April 1976.

9. Nath, J. H. and Yamamoto, T., "Forces from Fluid Flow Around Objects," Proc. of the 14th Coastal Engineering Conference, Copenhagen, Denmark, June 1974.

10. Nath, J. H., "Hydrodynamic Coefficients for Cylinders with Pronounced Marine Growths," API PRAC Project 80-31 Final Report to the American Petroleum Institute, Oregon State University, Dept. of Civil Engineering, October 1981.

11. Nath, J. H., "Hydrodynamic Coefficients for Macro-Roughnesses," Proc. of the Offshore Technology Conference, Houston, TX, 1981.

12. Nath, J. H., "Heavily Roughened Horizontal Cylinders in Waves," Proc. of the Conference on Behavior of Offshore Structures, MIT, August 1982.

13. Nath, J. H. and Wankmuller, R. N., "Wave Forces on Kelp Covered Horizontal Cylinders," Proc. of the Ocean Structural Dynamics Symposium, Oregon State University, September 8-10, 1982.

14. Nath, J. H., "Hydrodynamic Coefficients for Cylinders with Pronounced Marine Growths," API PRAC Project 80-31B-Vertical Cylinder, Final Report to the American Petroleum Institute, Oregon State University, Dept. of Civil Engineering, January 1983.

15. Ramberg, S. E. and Niedzwecki, J. M., "Horizontal and Vertical Cylinders in Waves," Ocean Engineering, Vol. 9, No. 1, pp 1-15, 1982.

16. Sarpkaya, T., "The Hydrodynamic Resistance of Roughened Cylinders in Harmonic Flow," The Royal Institute of Naval Architects, April 1977.

17. Sarpkaya, T. and Isaacson, M., Mechanics of Wave Forces in Offshore Structures, Van Nostrand and Reinhold Co., 1981.

18. Sarpkaya, T., "Morison's Equation and the Wave Forces on Offshore Structures," Report CR 82.008, Naval Civil Engineering Laboratory, Port Hueneme, California, December 1981.

19. Sarpkaya, T., Personal Correspondence, 1982.

20. Schlichting, H., "Boundary Layer Theory," McGraw-Hill, 1960, Chapter 23.

21. Stansby, P. K., "The Force of a Cylinder in Sinusoidal Flow," The American Society of Mechanical Engineers, Winter Annual Meeting, November 15-20, 1981, Washington, DC, Reprint No. 81-WA/FE-27.

NONLINEAR TRANSIENT RESPONSE OF DEEP-OCEAN PIPELINE

By M. Zimmerman[1], J. Leonard[2], J. Tedesco[3], R. Hudspeth[4] and L. Borgman[5]

ABSTRACT

A suite of computer programs has been developed for the prediction of dynamic stresses and motions of bottom-laid deep-ocean pipelines in a random wave environment. As part of that suite, a finite element program has been developed for the prediction of large displacements of pipelines subjected to episodic hydrodynamic loads and sliding on irregularly contoured marine sediments.

Nonlinear elastic-plastic soil springs simulate transverse and axial bottom resistance from both cohesionless and cohesive sediments. Inclusion of a pressure differential across the pipe wall modifies tensile stresses and influences the flexural stiffness of the finite element model through a geometric stiffness matrix. Hydrodynamic added mass and nonlinear viscous drag forces are included using a relative motion form of the Morison equation. The formulation results in a three-dimensional finite element model of the pipeline constrained to follow the irregularly contoured ocean bottom.

The equations of motion during each episode of extreme wave activity are integrated using the Newmark method of implicit integration. This method allows selection of larger time step sizes without concern for the numerical stability of the computations. Within each time step a Newton-Raphson iteration scheme is used to achieve equilibrium while accurately tracking the nonlinear path-dependent response. An updated Lagrangian formulation shifts the pipeline reference configuration and separates rigid-body movements from pipeline deformations. This provides a large deflection, small strain, transient analysis of the pipeline motion which is capable of computing significant pipeline motions that have been accumulated over the design lifetime of the pipeline.

[1]Grad. Research Asst., Dept. of Civil Engrg. and Ocean Engrg. Program, Oregon State Univ., Corvallis, OR 97331.
[2]Prof. Dept. of Civil Engrg. and Ocean Engrg. Program, Oregon State Univ., Corvallis, OR 97331.
[3]Asst. Prof., Dept of Civil Engrg. and Ocean Engrg. Program, Oregon State Univ., Corvallis, OR 97331
[4]Assoc, Prof., Dept. of Civil Engrg. and Ocean Engrg. Program, Oregon State Univ., Corvallis, OR 97331.
[5]Prof., Dept. of Statistics, Univ. of Wyoming, Laramie, WY 82070.

INTRODUCTION

Traditional design practices for offshore pipelines attempt to insure stability along the entire pipeline system. In shallow water, the standard practice requires the burial of pipeline systems, resulting in increased construction costs from trenching operations. Uncertainties regarding the effectiveness of trenching operations [Ells (9)], coupled with a desire to minimize construction costs, has generated an increased interest in bottom supported pipeline systems for deeper waters [Gravesen and Nielson (10); Karal and Halvorsen (14); and Lambrakos (15)]. The design forces used to evaluate the stability of bottom supported pipelines include; hydrodynamic fluid pressures, resistive soil forces, and the resistive stiffness of the pipeline. An analysis based on these criteria often results in a pipeline design which requires a weight coating to develop additional resistive forces to maintain stability. Pipelines designed in this manner will experience significant motion when subjected to extreme wave events. The forces which determine the dynamic response of the pipeline system are 1) the hydroelastic fluid forces, 2) the resistive soil forces and 3) the inertial and resistive forces of the pipeline.

A method is presented for predicting the dynamic stresses and displacements which develop in a pipeline subject to a peak wave event. For the purposes of this investigation, the return period of the extreme wave episodes is considered to be long with respect to the duration of the episode. Stipulation of a long return period allows sequential solution for the dynamic response of the pipeline to a series of peak wave events. The sequential solutions are coupled by the residual stress distribution within the pipeline. However, because the time between successive events is considered to be long, the resistive soil forces are assumed to be independent of stress history.

The primary features of the methodology are 1) the ability to predict the large deflection response of the pipeline, 2) the ability to model nonlinear hydroelastic fluid forces, 3) the ability to independently model cohesive and cohesionless sediments, and 4) the ability of the model to adapt to irregular bottom contours. The methodology does not address possible time-dependent effects on resistive soil forces [Karal (12) and Karal (13)] nor does it attempt to define the actual stress distributions which arise as result of pipe laying operations. In an actual design these factors would warrant additional consideration.

FORMULATION FOR DYNAMIC EQUILIBRIUM

A continuous ocean pipeline is depicted in Figure 1 along with both the global (X,Y,Z) and local (τ,η,β) coordinate systems. The local tangential direction (τ), shown in Figure 1, coincides with the longitudinal axis of the pipe, and the binormal direction (β) is perpendicular to the ocean floor. The normal direction (η) completes the right handed triad for the intrinsic pipeline system.

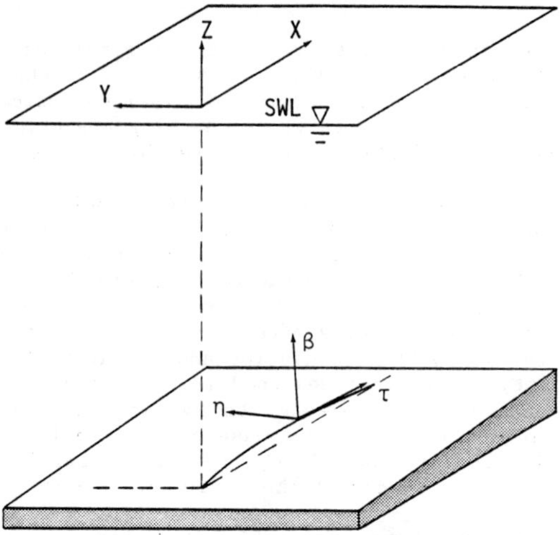

Figure 1- Definition Sketch of Global (X,Y,Z) and Local (τ,η,β) Coordinate Axes.

For such a system, the partial differential equations for the prediction of small deflection response of the pipeline [Chajes (6) and Clough and Penzien (7)] are as follows:

$$\frac{\partial^2}{\partial x^2}\left(EI\,\frac{\partial^2 u_i}{\partial x^2}\right) + p\,\frac{\partial^2 u_i}{\partial x^2} + m\,\frac{\partial^2 u_i}{\partial t^2} = r_i \qquad i = 2,3 \qquad (1a)$$

$$m\,\frac{\partial^2 u_i}{\partial t^2} - \frac{\partial}{\partial x}\left(EA\,\frac{\partial u_i}{\partial x}\right) = r_i \qquad i = 1 \qquad (1b)$$

and

$$GJ\,\frac{\partial^2 u_i}{\partial x^2} = r_i \qquad i = 4 \qquad (1c)$$

in which m = total mass per unit length; x = position along the pipe segment; t = time; E = elastic modulus; I = cross-sectional moment of inertia; A = cross-sectional area; G = shear modulus; J = torsional inertia; p = tensile axial force on the steel pipe (e.g. from pressure differential across the pipe section); u_i = i^{th} component of displacement, where u_i is defined in Figure 2; and r_i = i^{th} component of applied load per unit length, i.e. fluid and sediment load. Equations (1) describe biaxial bending, axial deformation, and torsional response, respectively. The effect of structural damping, shear deformation, and rotational inertia are neglected in Eqs. (1). Simplification of the general load representations and the imposition of constraints on pipeline motions, to be introduced later, will reduce the significance of Eq. (1c). However, torsional rigidity will be retained at this time in order to obtain a full three-dimensional formulation for the pipeline response.

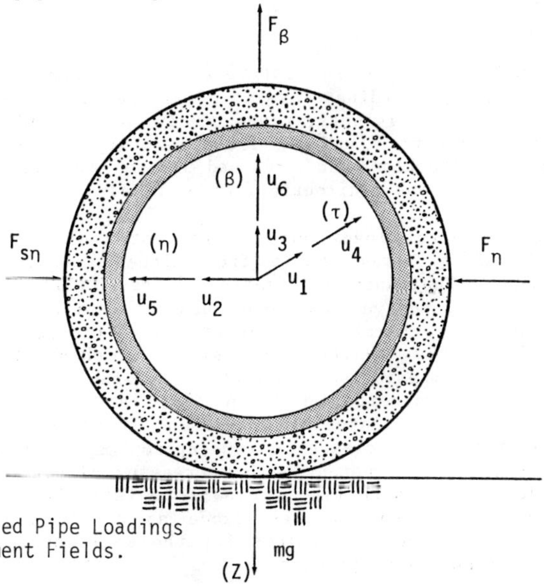

Fig. 2 Applied Pipe Loadings and Displacement Fields.

In the current solution algorithm, extension of the controlling equations to include finite deflections is performed by adopting a convected coordinate system which tracks system displacements and separates rigid body motions from pipeline deformations. Even for small deflection analysis, however, Eqs. (1a) and (1b) are nonlinear since the applied environmental loads are nonlinear functions of the pipeline displacement or its derivatives. These environmental loadings, r_i, consist of wave-induced hydroelastic pressures and resistive sediment forces.

Focusing first on the wave-induced components of the environmental loads, a relative-motion Morison equation is employed to define fluid loadings in terms of a hydrodynamic lift, drag and inertial forces. Using the relative-motion Morison equation, the dynamic wave-induced pressure force per unit length are given by

$$F_\beta = .5 C_L \rho_o D(v - \dot{u}_2)^2 \qquad \text{(lift)} \qquad (2a)$$

$$F_\eta = .5C_D\rho_o D(v - \dot{u}_2)|v - \dot{u}_2| + .25\rho_o \pi D^2 C_M[\dot{v} - (1 - \frac{1}{C_M})\ddot{u}_2]$$

(drag and inertia) (2b)

$$F_\tau = 0 \qquad (2c)$$

in which F_τ, F_η, F_β = the distributed fluid loadings corresponding to the τ, η, and β coördinate system of Figure 1; C_D, C_L, C_M = drag, lift and inertial coefficients, respectively; ρ_o = fluid mass density; D = pipeline diameter; v and $\dot{v}$ = fluid velocity and acceleration in the intrinsic η direction; and, $\dot{u}_2$ and $\ddot{u}_2$ are the pipeline velocity and acceleration in the η direction, respectively.

Environmental loads due to sediment resistance along the pipeline, $F_{s\tau}$ and $F_{s\eta}$, may result from either cohesive or cohesionless materials. Published information [Karal (12)] indicates that the behavior of cohesive and cohesionless materials are significantly different; and, therefore, sediment force formulations should reflect these differences. Although cohesive and cohesionless sediments display different resistive properties, available literature [Anand and Agarwal (1); Audibert and Nyman (2); Karal (12); Lyons (15); and Wantland (18)] indicates that sediment resistance for both types of material tends to increase to some approximate maximum value and then remain relatively constant with increasing displacement. In view of the tendency for both types of soil to fail plastically, it was decided to model sediment resistance as a linear-elastic, perfectly plastic force. Actual choices for the soil models are discussed as part of the solution approach.

SOLUTION PROCEDURE

Analytical solutions to Eqs. (1) are generally not possible for the complicated environmental loadings or pipeline geometries associated with most practical ocean engineering problems. It is, therefore, necessary to evaluate the response of the pipeline numerically. For numerical analysis, it is convenient to note that a solution to the equilibrium Eqs. (1) is tantamount to satisfying Hamilton's minimization principle [Hildebrand (11)]. The statement of this principle is given by [Clough and Penzien (7)]

$$\int_{t_o}^{t_1} \delta(T - V)\, dt + \int_{t_o}^{t_1} \delta W_{\eta c} = 0 \qquad (3)$$

in which t_o and t_1 = initial time and an arbitrary time, respectively; δ = variation during the interval t_o to t_1; T = total kinetic energy; V = total potential energy of the system and any conservative load; and W_{nc} = work of nonconservative forces.

The numerical analysis technique employed in this investigation was the finite element method. More specifically, the continuous pipeline was discretized as a series of three-dimensional beam elements to which Hamilton's principle was applied to yield the equilibrium expression for each element. In matrix notation the expression for linear structural response at time t + Δt is

$$[\mathbf{m}]\ {}^{t+\Delta t}\{\ddot{\mathbf{d}}\} + ([\mathbf{k}_E] + {}^{t+\Delta t}[\mathbf{k}_G])\ \{\mathbf{d}\} = {}^{t+\Delta t}\{\mathbf{r}\} \tag{4}$$

in which $[\mathbf{m}]$ = the mass representation for an element; $[\mathbf{k}_E]$ and $[\mathbf{k}_G]$ = the elastic and geometric stiffnesses, respectively of the element; $\{\mathbf{d}\}$ = a vector of generalized coordinates associated with the element displacement fields; $\{\mathbf{r}\}$ = a vector of generalized forces; and in which $(\ddot{\ })$ = differentiation twice with respect to time. Extension of Eq. (4) to evaluate nonlinear structural response and nonlinear loadings is performed by adopting a convected coordinate system for each element [Belytsehko and Hsieh (5) and Cook (8)] and iteratively evaluating the equilibrium expression obtained by expanding Eq. (4) in a linear Taylor series [Bathe (4)]. An iterative solution in this manner separates rigid body movement from element deformation, and allows evaluation of nonlinear terms such as geometric stiffness and applied loads.

Coupling of the dynamic response of an individual element to the complete finite element model is achieved by transforming the equilibrium expressions for individual elements into mutual coordinate systems for simultaneous solution of the generalized coordinates which produce equilibrium. Anticipating that the pipeline will slide only along a locally flat ocean bottom makes it advantageous to describe the generalized coordinates with respect to this ocean bottom. If it is assumed that the ocean bottom is a slowly varying surface over the domain of the pipeline system, it then becomes possible to describe the shared generalized coordinates of any two adjacent elements in terms of a nodal coordinate system at each node. Orienting the nodal coordinate systems at each node such that two of the coordinate axes are in the plane of the ocean bottom makes it possible to restrict the nonzero generalized coordinates so that all of the nodes remain in the plane of the ocean bottom. Imposition of this constraint reduces the number of generalized coordinates at each node from six to three and requires that the pipeline response at each node be limited to sliding in a locally flat facet of the ocean bottom.

Temporal integration of Eq. (3) is performed numerically by application of the Newmark-Beta method. This method is an extension of the linear acceleration method (Wilson - Θ). The Newmark method is based (4) on the following assumptions:

$${}^{t+\Delta t}\dot{\mathbf{d}} = {}^{t}\dot{\mathbf{d}} + [(1 - \lambda)\ {}^{t}\ddot{\mathbf{d}} + \lambda\ {}^{t+\Delta t}\ddot{\mathbf{d}}]\Delta t \tag{5a}$$

$${}^{t+\Delta t}\mathbf{d} = {}^{t}\mathbf{d} + {}^{t}\dot{\mathbf{d}}\Delta t + [(1/2 - \alpha)\ {}^{t}\ddot{\mathbf{d}} + \alpha\ {}^{t+\Delta t}\ddot{\mathbf{d}}]\Delta t^2 \tag{5b}$$

in which λ and α are integration parameters; superscripts $t + \Delta t$, and t denote temporal increment number; and Δt = time step. For the commonly adopted values of $\lambda = 1/2$ and $\alpha = 1/4$; this method integrates the kinematics over a time step in which the acceleration is assumed to have a constant average value. The integration method is an implicit scheme which eliminates some of the constraints on the selection of a simulation time step.

The finite element formulation used in the solution algorithms employs a lumped mass representation for the structural mass; and utilizes lumped hydroelastic forces and linear-elastic perfectly plastic soil springs to model applied environmental loads. The generalized wave-induced hydrodynamic fluid forces applied at both ends of an element are represented by the relative motion Morison equation. Using a lumped-force formulation, these hydrodynamic pressures, which are assumed to be constant over one-half the element length, are summed over the half-element lengths tributary to a particular node and applied as concentrated forces at the element ends. Because lumped forces are uncoupled, the element load vector, $\{\mathbf{r}\}$, contains no moment components.

Modeling of sediment resistance along a pipeline using discrete nonlinear springs has been discussed by Audibert, Lai and Bea (2), Audibert and Nyman (3), and Nyman (17). This type of soil model was selected for computational efficiency and adaptability to both cohesive and cohesionless sediments. Spring characteristics, which are determined independently for cohesive and cohesionless materials, reflect the capacity and yield displacement for a given sediment. As stated previously, this simulation is directed toward evaluating the pipeline response during peak wave episodes which have a long return period. Therefore, hysteresis is relaxed between episodes. Furthermore, because only the gross dynamic behavior of the pipeline during these peak wave events is of primary interest, it is possible to use order of magnitude approximations for the yield characteristics of the sediment springs without greatly modifying the gross dynamics of the pipeline.

The spring capacity for cohesionless sediments is defined by the Coulomb friction capacity of the sediment [Anand and Agarwal (1), and Lyons (16)]. Using a Coulomb friction model, the capacity of the soil spring at each end of an element is given by

$$SC_i = \mu_i \left(W_s - F_\beta\right) \frac{L}{2} \qquad i = 1,2 \tag{6}$$

in SC_i = sediment spring capacity; (i = 1: τ component of capacity; i = 2: η component of capacity); μ_i = coefficient of friction for τ or η response; W_s = submerged weight of pipe per unit length; F_β = hydrodynamic lift force; and L = element length. The corresponding yield points for the cohesionless soil model were chosen as 0.2% and 2.0% of the pipeline diameter, respectively, for axial and transverse deflections [Anand and Agarwal (1), and Lyons (16)].

The cohesive sediment capacity for the spring model are evaluated using a formulation developed by Swanson and Jones (18). The settlement of the pipeline section is determined by balancing the maximum stress under the pipeline, f_d, with the bearing capacity of the cohesive sediment, q_d. These stresses are represented by

$$f_d = N_\sigma \frac{W_s}{b} \tag{7a}$$

and

$$q_d = c\left(1 + \frac{.35s}{b}\right) N_c + \gamma_e s \tag{7b}$$

in which N_σ = stress coefficient (= 2.5); W_s = submerged pipe weight; b = width of bearing below the pipe section; N_c = bearing capacity factor (= 5.14); c = average cohesion; s = depth of settlement; and γ_e = effective unit weight of the sediment. Equating Eqs. (7a) and (7b) gives an expression which may be evaluated iteratively to determine the pipe settlement, s [Swanson and Jones (18)].

Having obtained the pipeline settlements, passive sediment pressures determine the lateral capacity, SC_2, while skin friction determines the axial capacity, SC_1. Cohesive capacities established in this manner are given by Swanson and Jones (18) as

$$SC_1 = ca \tag{8a}$$

$$SC_2 = \left(3 + \sqrt{\frac{D}{s} - 1}\right) + \frac{3}{2}\gamma_e s^2 \qquad \text{for } \frac{s}{D} \leqslant .5 \tag{8b}$$

$$SC_2 = cs\left(3 + \frac{D}{2s}\right) + \frac{3}{2}\gamma_e s^2 \qquad \text{for } \frac{s}{D} > .5 \tag{8c}$$

in which SC_1 = axial sediment capacity; SC_2 = lateral sediment capacity; a = contact area; and D = pipe diameter. The corresponding yield points for the cohesionless soil model were estimated as 2.0% of the pipe settlement for both axial and lateral sediment resistance [Nyman (17)].

NUMERICAL EXAMPLES

To allow greater flexibility in simulating a variety of possible problems, the solution algorithms developed utilize a grid system to define the bathymetry, soil characteristics, and fluid kinematics to which the pipeline is exposed. Soil characteristics and fluid kinematics which define the environmental loadings on the pipeline are assigned along the pipeline using a best linear unbiased estimator to

Table 1. Pipeline, Sediment, and Wave Data

Overall pipe diameter (D)		24.0 in.
Steel pipe section:	diameter	20.0 in.
	wall thickness	0.5 in.
	Youngs Modulus (E)	30.0×10^6 psi.
Specific gravity		1.2
Internal pressure		1440.0 psi.
Pipeline length between terminal constraints		2500.0 ft.
Drag, lift, and inertial coefficients (C_D, C_L, C_M)		1.0, 1.0, 2.0
Sediment cohesion (c)		40 psf.
Sediment effective weight (γ_e)		21 pcf.
Water depth (h)		200.0 ft
Wave height (H)		30.0 ft.
Wave period $(2\pi/\omega)$		12.0 sec.
Phase shift (Φ)		90°
Wave angle from X axis (Θ)		90°, 60°, 30°

interpolate within the grid network. Fluid kinematics are obtained from an FFT simulation of directional seas. Fluid kinematics are updated at time increments that are determined by the FFT simulation of the directional seas. If these wave simulation time increments are longer than the FEM numerical integration time steps, fluid kinematics are treated as constant values during these shorter time increments.

Three representative example problems are presented. The input parameters for three example problems are given in Table 1. The grid system for each problem was a horizontal plane, and sediment characteristics within the grid were uniform. Fluid kinematics were calculated from linear wave theory, assuming a monochromatic wave traveling at an angle Θ to the global X axis, i.e.

$$N_w\ (X,Y,t) = \frac{H}{2} \cos \kappa X \cos \Theta + \kappa Y \sin \Theta - \omega t + \Phi) \qquad (9)$$

in which N_w = water surface profile; H = wave height; K = wave number (rads/length); w = wave frequency (rads/time); and Φ = phase shift. Fluid kinematics and structural response were calculated at a 0.2 sec. time step.

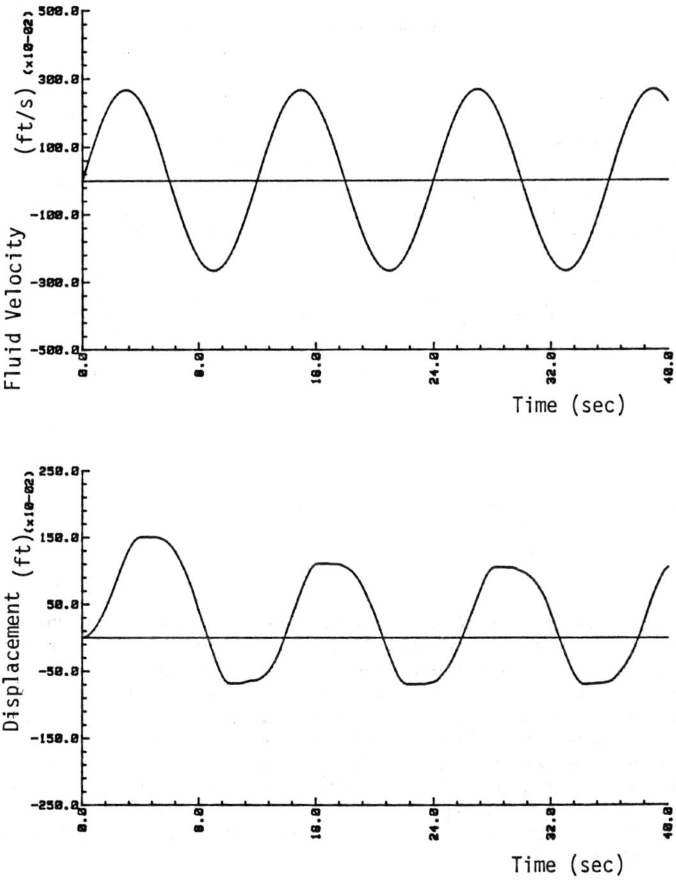

Figure 3. Effect of water particle velocity on pipeline response: x=1250.0 ft. (θ=90 deg.).

The finite element model of the pipeline system in each example problem consisted of 50 elements. The boundary conditions at the pipe terminals were specified as fixed (i.e. $u_2 = u''_2 = 0$ at x = 0, and at x = 2500 ft). The longitudinal axis of the pipeline in the pre-loading stage was selected to coincide with global X axis, therefore, the wave angle Θ corresponds to the angle of wave attack relative to original configuration of the pipeline.

The calculated deflections at pipeline midpoint (x = 1250 ft) and the horizontal water particle relocity are shown in Figure 3 for a normally incident wave (Θ = 90°). The maximum deflections and longitudinal stresses and their location along the pipeline are given in Table 2 for both normally incident and obliquely incident (Θ = 60°, 30°) waves.

Table 2. Computed Results for a Monochromatic Wave

Wave Angle	Max values for $0 < x < 2500$ ft.				Max values for $x = 1250$ ft.	
	Lateral displ.		Longitudinal stress[1]			
θ	$\lvert u_{2\,max} \rvert$	$x\ (u_{2\,max})$	σ_{max}	$x\ (\sigma_{max})$	$\lvert u_{2\,max} \rvert$	σ_{max}
degrees	ft	ft	ksi	ft	ft	ksi
90	1.77 (t = 39.8 sec)	570.0	19.6 (t = 39.8 sec)	0.0	1.51 (t = 4.0 sec)	14.1 (t = 27.2 sec)
60	.74 (t = 33.6 sec)	2220.0	18.6 (t = 37.0 sec)	2500.0	.36 (t = 14.0 sec)	14.5 (t = 33.2 sec)
30	0.006 (t = 8.2 sec)	853.3	13.7 (t = 7.2 sec)	853.3	-	-

1. Initial stress = 13.5 ksi

An investigation of the results from the example problems under consideration reveals that the directionality of a wave field greatly affects both the magnitude and location of the maximum pipeline response. The initial transient motion appears to be an important factor in determining the net lateral displacement of the pipeline; however, these transients effects are not as significant when evaluating maximum stress levels or maximum displacements resulting from a series of peak waves. Furthermore, these examples illustrate that even for a normally incident wave, the maximum pipeline deflections do not necessarily occur at the midpoint of the pipeline [Karal and Halvorsen (14)].

SUMMARY

A methodology has been presented for predicting the nonlinear transient response of bottom layed pipeline systems. The applicability of the solution algorithms to a variety of general environmental conditions has been outlined, and example problems have been presented for a simplified set of environmental parameters. The effects of environmental conditions, such as an irregular ocean bottom or a random wave environment for directional seas, on pipeline response are currently under investigation.

Results from the examples considered herein, and results from other investigators [Karal and Halvorsen (14) and Lambrakos (15)] indicate that for certain pipeline systems a potential reduction in pipeline fabrication costs may be realized by decreasing the weight coating to less than that required to insure stability during the entire life of the pipeline. The analytical methods presented here are intended to aid in evaluating a potential pipeline design, however, further investigation of the coefficients and parameters used in the problem formulation is warranted before these results can be applied to actual ocean engineering problems.

ACKNOWLEDGMENTS

The authors wish to express their gratitude to the American Gas Association for its funding of the current research and to D.C. Crowell, D.A. Degenhardt, W.T. Jones, and K. Lambrakos of the AGA Random Seas Project Committee for their guidance and assistance.

APPENDIX I. - REFERENCES

1. Anand, S. and Agarwal, S., "Field and Laboratory Studies for Evaluating Submarine Pipeline Frictional Resistance," Proceedings of the Twelfth Offshore Technology Conference, OTC, No. 3781, Vol. II, 1980, pp. 371-382.
2. Audibert, J., Lai, N., and Bea, R., "Design of Pipelines - Sea Bottom Loads and Restraints," Pipeline in Adverse Environments, Vol. 1, Proceedings of the ASCE Pipeline Division Specialty Conference, New Orleans, LA, January 15-17, 1978, pp. 184-203.
3. Audibert, J. and Nyman, K., "Soil Restraint Against Horizontal Motion of Pipes," Journal of the Geotechnical Engineering Division, ASCE, Vol. 103, No. GT10, October 1977, pp. 1119-1142.
4. Bathe, K., "Solution of Equilibrium Equations in Static Analysis," and "Solution of Equilibrium Equations in Dynamic Analysis," Finite Element Procedures in Engineering Analysis, Prentice-Hall, Inc., Englewood Cliff, NJ, pp. 490-491 and pp. 511-551.
5. Belytsehko, T. and Hsieh, B., "Nonlinear Transient Finite Element Analysis with Convected Coordinates," International Journal for Numerical Methods in Engineering, Vol. 7, No. 3, 1973, pp. 255-271.
6. Chajes, A., "Torsional Buckling," Principles of Structural Stability Theory, Prentice-Hall, Inc., Englewood Cliffs, NJ, 1974, pp. 196-201.
7. Clough, R. and Penzien, J., "Overview of Structural Dynamics," and "Partial Differential Equations of Motion," Dynamics of Structures, McGraw-Hill, Inc., New York, NY, 1975, pp. 10-11 and pp. 293-306.
8. Cook, R. "Introduction to Nonlinear Problems," Concepts and Applications of Finite Element Analysis, 2nd Edition, John Wiley and Sons, Inc., New York, NY, 1981, pp. 351-359.
9. Ells, J., "Scours and Spanning Threaten Sea Lines," Oil and Gas Journal, Vol. 73, No. 27, July 7, 1975, pp. 67-71.

10. Gravesen, H. and Nielson, R., "Some Environmental Design Aspects of Marine Pipelines for Danish North Sea Service," Offshore Oil and Gas Pipeline Technology, European Seminar, London, England, February 1981.
11. Hildebrand, F., "Calculus of Variations and Applications," Methods of Applied Mathematics, Second Edition, Prentice-Hall, Inc., Englewood Cliffs, NJ, 1965, pp. 119-144.
12. Karal, K., "Lateral Stability of Submarine Pipelines," Preprints, Ninth Annual Offshore Technology Conference, OTC 2967, May 1977.
13. Karal, K., "Time Effects on Lateral Resistance to Pipelines Supported by Cohesive Soils," Proceedings of the International Offshore Mechanics/Artic Engineering Symposium, Energy-Sources Technology Conference and Exhibition, Houston, TX, ASME, 1983.
14. Karal, K. and Halvorsen, S., "Dynamics of the Water-Pipeline-Soil Interaction," Transaction ASME, Journal of Energy Resources Technology, Vol. 104, 1982, pp. 307-312.
15. Lambrakos, K., "Marine Pipeline Dynamic Response to Waves from Directional Wave Spectra," Ocean Engineering, Vol. 9, No. 4, 1982, pp. 385-405.
16. Lyons, C., "Soil Resistance to Lateral Sliding of Marine Pipelines," Fifth Annual Offshore Technology Conference, Houston, TX, OTC 1876, May 1973.
17. Nyman, K., "Soil Response Against the Horizontal-Vertical Motion of Pipes," ASCE Preprint 82-534, ASME Convention, New Orleans, LA, October 25-29, 1982.
18. Swanson, R. and Jones, W., "Mudslide Effects on Offshore Pipelines," Transportation Engineering Journal, ASCE, Vol. 108, No. TE6, November 1982, pp. 585-605.
19. Wantland, G., O'Neill, M., Reese, L. and Kalajian, E., "Lateral Stability of Pipelines in Clay," Eleventh Annual Offshore Technology Conference, Houston, TX, OTC 3477, pp. 1025-1034.

APPENDIX II. - NOTATION

The following symbols are used in this paper:

a = contact area between pipeline and soil per unit length of pipe;

A = cross-sectional area of steel pipeline;

b = width of bearing below the pipe cross section;

c = cohesion of sediment;

C_D, C_L, C_M = drag, lift and inertia coefficients, respectively;

$\mathbf{d}$ = component of the generalized coordinate vector;

$\{\mathbf{d}\}$ = vector of generalized coordinates for an element;

D = outer pipe diameter;

E = modulus of elasticity of steel pipe section;

f_d = stress under pipe section;

F_τ, F_η, F_β, = tangential, normal, and binormal hydrodynamic loading per unit length, respectively;

$F_{s\tau}$, $F_{s\eta}$ = axial and lateral sediment resistance per unit length of pipe, respectively;

G = shear modulus of steel pipe;

h = water depth;

H = wave height;

I = moment of inertia of steel pipe section;

J = torsional inertia of steel pipe section;

$[\mathbf{k}_E]$ = elastic stiffness of an element;

$[\mathbf{k}_G]$ = geometric stiffness of an element;

L = length of a pipe element;

m = mass per unit length of pipe;

$[\mathbf{m}]$ = mass representation for an element;

N_c = bearing capacity factor (= 5.14);

N_σ = stress coefficient (= 2.5);

p = axial load in pipeline;

q_d = bearing capacity of a cohesive sediment;

r_i = i^{th} component of the continuous environmental load;

$\{\mathbf{r}\}$ = vector of generalized element loads;

s = pipeline settlement in cohesive sediment;

SC_i = i^{th} component of sediment capacity;

t = time;

t_o = initial time;

t_1 = arbitrary time;

Δt = time step for numerical integration;

T = total kinetic energy;

u_i = i^{th} component of displacement or rotation along pipeline;

v = component of fluid velocity normal to pipeline;

V = potential energy of th system and any conservative load;

W_{nc} = work of nonconservative forces;

W_s = submerged pipeline weight per unit length;

x = position along pipeline;

X, Y, Z = global coordinate axes;

α = Newmark integration parameter;

β = component of intrinsic coordinate system which is perpendicular to ocean bottom;

δ = variation of;

η = component of intrinsic coordinate system normal to pipeline and parallel to ocean bottom;

N_w = water surface profile;

γ_e = effective unit weight of soil;

λ = Newmark integration parameter;

κ = wave number (2π/length)

μ_i = coefficient of friction for the ith component of cohesionless sediment capacity;

τ = component of intrinsic coordinate system along pipeline axis;

ω = wave frequency (rads/time);

Φ = phase shift and

Θ = angle of wave advance relative to X axis

σ = longitudinal pipeline stress

Superscripts

t = previous time step;

$t+\Delta t$ = current time step;

$\dot{}$ = differentiation with respect to time; and

$'$ = differentiation with respect to longitudinal position along the pipeline.

SUBJECT INDEX

Page number refers to first page of paper

AUTHOR INDEX

Page number refers to first page of paper